Applied Optimal
Control & Estimation

Prentice Hall and Texas Instruments
Digital Signal Processing Series

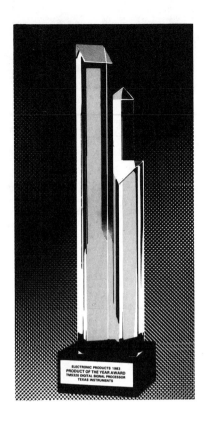

DOTE, *Servo Motor and Motion Control Using Digital Signal Processors* (1990)

HUTCHINS/PARKS, *A Digital Signal Processing Laboratory Using the TMS320C25* (1990)

JONES/PARKS, *A Digital Signal Processing Laboratory Using the TMS32010* (1988)

KUN-SHAN LIN, EDITOR, *Digital Signal Processing Applications with the TMS320 Family, Volume 1* (1987)

KUN-SHAN LIN, EDITOR, *Digital Signal Processing Applications with the TMS320 Family, Volume 2* (future)

LEWIS, *Applied Optimal Control and Estimation* (1992)

PAPAMICHALIS, *Practical Approaches to Speech Coding* (1987)

TI SDP ENGINEERING STAFF, *First-Generation TMS320 User's Guide* (1988)

TI SDP ENGINEERING STAFF, *Second-Generation TMS320 User's Guide* (1988)

Applied Optimal Control & Estimation
Digital Design & Implementation

Frank L. Lewis

The University of Texas at Arlington

Prentice Hall, Englewood Cliffs, New Jersey 07632

Library of Congress Cataloging-in-Publication Data

LEWIS, FRANK L.
    Applied optimal control & estimation : digital design & implementation / Frank L. Lewis.
        p. cm.
    Includes bibliographical references and index.
    ISBN 0-13-040361-X
    1. Automatic control. 2. Control theory. 3. Estimation theory.
I. Title: Applied optimal control and estimation.
TJ213.L43   1992   91-29202
629.8'9--dc20   CIP

Editorial/production supervisor
  and interior designer: **Karen Bernhaut**
Cover designer: **Lundgren Graphics**
Prepress buyer: **Mary McCartney**
Manufacturing buyer: **Susan Brunke**
Acquisitions editor: **Karen Gettman**

© 1992 by Texas Instruments Incorporated

Published by Prentice-Hall, Inc.
A Simon & Schuster Company
Englewood Cliffs, New Jersey 07632

The publisher offers discounts on this book when ordered in bulk quantities.
For more information, write:

   Special Sales/Professional Marketing
   Prentice-Hall, Inc.
   Professional & Technical Reference Division
   Englewood Cliffs, New Jersey 07632

All rights reserved. No part of this book may be
reproduced, in any form or by any means,
without permission in writing from the publisher.

Printed in the United States of America

10 9 8 7 6 5 4 3 2 1

ISBN 0-13-040361-X

Prentice-Hall International (UK) Limited, *London*
Prentice-Hall of Australia Pty. Limited, *Sydney*
Prentice-Hall Canada Inc., *Toronto*
Prentice-Hall Hispanoamericana, S.A., *Mexico*
Prentice-Hall of India Private Limited, *New Delhi*
Prentice-Hall of Japan, Inc., *Tokyo*
Simon & Schuster Asia Pte. Ltd., *Singapore*
Editora Prentice-Hall do Brasil, Ltda., *Rio de Janeiro*

To Theresa
without whom there would be no book
and little of anything else

Contents

List of Tables — xv
List of Examples — xvii
Preface and Acknowledgments — xxi

PART I INTRODUCTION

1. Introduction to Modern Control Theory — 3

  1.1 A Brief History of Automatic Control — 3
      Water Clocks of the Greeks and Arabs, 5
      The Industrial Revolution, 6
      The Birth of Mathematical Control Theory, 9
      Mass Communication and the Bell Telephone System, 10
      The World Wars and Classical Control, 11
      The Space/Computer Age and Modern Control, 12
      Computers in Controls Design and Implementation, 15
      The Union of Modern and Classical Control, 16

  1.2 The Philosophy of Classical Control — 17
  1.3 The Philosophy of Modern Control — 18
      References for Chapter 1 — 21

2. Review of Linear State-Variable Systems — 25

2.1 Continuous-Time Systems — 26
    Frequency-Domain Solution, 32
    Time-Domain Solution, 35
    Natural Modes and Stability, 36
    Computer Simulation of Continuous Systems, 43

2.2 Discrete-Time Systems — 50
    Solution of Discrete State Equation, 50
    Discrete Natural Modes and Stability, 52

2.3 System Properties — 53
    Minimum-Phase Systems, 53
    Reachability, 59
    Observability, 63

2.4 Realization and Canonical Forms — 66
    State-Space Transformations, 67
    Jordan Normal Form, 72
    Reachable Canonical Form, 82
    Observable Canonical Form, 85
    Minimality and Gilbert's Realization Method, 87

2.5 Feedback Control — 91
    State-Variable Feedback, 92
    Stabilizability and Detectability, 105
    Output Feedback, 106

Problems for Chapter 2 — 100

PART II CONTINUOUS-TIME CONTROL

3. Optimal Control of Continuous-Time Systems — 119

3.1 The General Continuous-Time Optimal Control Problem — 120
    Solution of the Nonlinear Optimal Control
        Problem, 120
    Two-Point Boundary-Value Problems, 127

3.2 Continuous-Time Linear Quadratic Regulator — 136
    Open-Loop Control, 136
    Closed-Loop Control, 143

3.3 Steady-State and Suboptimal Control — 157
    Steady-State Control, 157
    Suboptimal Control, 161
    Eigenstructure LQR Design, 167

3.4 Minimum-Time and Constrained-Input Design   172
    Nonlinear Minimum-Time Problems, 173
    Linear Quadratic Minimum-Time Design, 173
    Constrained-Input Design, 175

Problems for Chapter 3   187

4. Output-Feedback Design   191

4.1 Linear Quadratic Regulator with Output Feedback   192
    Output Feedback LQR Design Equations, 193
    Closed-Loop Stability, 196
    Deriving the LQR Using Full State Feedback, 198
    Solution of the Design Equations, 198

4.2 Tracking a Reference Input   204
    Compensators of Known Structure, 204
    Linear Quadratic Servo Design Using Output Feedback, 206

4.3 Tracking by Regulator Redesign   229
    Deviation System, 230
    Regulator Redesign Adding Feedforward Terms, 230
    Tracking a Unit Step, 231

4.4 Command-Generator Tracker   233
    Tracking, 234
    Tracking with Disturbance Rejection, 239

4.5 Explicit Model-Following Design   240
    Regulator with Model Following, 240
    Tracker with Model Following, 242

Problems for Chapter 4   245

PART III   DIGITAL CONTROL

5. Digital Control By Continuous Controller Redesign   251

5.1 Simulation of Digital Controllers   252

5.2 Discretion of Continuous Controllers   254
    Bilinear Transformation, 255
    Matched Pole-Zero, 258
    Digital PID Controller, 266
    Digital Control of Processes with Time Delays, 270

5.3 Sampling, Hold Devices, and Computation Delays — 278
    Sampling and Aliasing, 279
    Hold Devices, 282
    Computation Delay, 287
    Modified Design of Continuous Controllers for Discretization, 288

5.4 Minimum-Time Control — 295

Problems for Chapter 5 — 296

6. Implementation of Digital Controllers — 298

6.1 Actuator Saturation and Windup — 299

6.2 Quantization and Roundoff — 306
    Deterministic Analysis of Quantization Effects, 306
    Stochastic Analysis of Quantization Effects, 313

6.3 Overflow and Scaling — 313
    Overflow Handling, 314
    Signal Scaling, 316
    Filter Coefficient Scaling, 318
    Measurement Noise, 319

6.4 Controller Realization Structures — 326
    Parallel and Cascade Implementations, 326
    Second-Order Modules, 328
    Delta Form, 333

6.5 Digital Signal Processor Subroutines — 336
    FIR Filter, 337
    First- and Second-Order IIR Filters, 339
    PID Controller, 342

6.6 Digital Signal Processor Control Implementation Example — 342

Problems for Chapter 6 — 353

7. Digital Control by Direct Discrete-Time Design — 355

7.1 Discretization of Continuous Systems — 356
    Sampling the Plant, 356
    Sampling Systems with Time Delays, 363
    Sampling the Transfer Function, 365

7.2 Discretization of the Performance Index — 369
    Sampling the PI Weighting Matrices, 370
    Discretization of Time-Domain Performance Specifications, 371

7.3 Practical Considerations in Sampling 373
    Nonminimum-Phase Zeros, 373
    Loss of Observability, 375
    Selecting the Sample Period, 377

7.4 Discrete Design Techniques 377
    Discrete Classical Design, 378
    Tracking by Regulator Redesign, 390
    Linear Quadratic Tracker with Output Feedback, 396
    Linear Quadratic Regulator with Output Feedback, 406
    Linear Quadratic Regulator with State Feedback, 411

Problems for Chapter 7 413

PART IV FREQUENCY-DOMAIN TECHNIQUES

8. Robust Design 419

8.1 Multivariable Loop Gain and Sensitivity 421
    A Typical Feedback System, 421
    Closed-Loop Transfer Relations, 423
    Sensitivity, Cosensitivity, and Loop Gain, 424

8.2 Multivariable Bode Plot 424
    Singular Values, 425
    Bode Magnitude Plot, 426

8.3 Frequency-Domain Performance Specifications 429
    Bandwidth, 430
    L_2 Operator Gain, 430
    Low-Frequency Specifications, 432
    High-Frequency Specifications, 437
    Model Reduction and Stability Robustness, 440
    Robustness Bounds for Plant Parameter Variations, 446

8.4 Robust Output-Feedback Design 447

Problems for Chapter 8 453

PART V OBSERVERS, FILTERS, AND DYNAMIC REGULATORS

9. State Estimators 459

9.1 Output-Injection Observer Design 460
    Digital Observer, 460
    Continuous Observer, 470

9.2 Reduced-Order Observers — 474
Digital Observer, 474
Continuous Observer, 478

9.3 Discrete Kalman Filter — 481
Linear Discrete Stochastic Systems, 482
Kalman Filter Derivation, 484
Kalman Filter Formulations, 489
Implementation of the Kalman Filter, 491
Suboptimal Steady-State Kalman Filter, 492

9.4 Digital Filtering of Continuous-Time Systems — 493
Discretization of Continuous Stochastic Systems, 493
Nonwhite Noise and Shaping Filters, 496
Design and Implementation Example, 497

9.5 Continuous Kalman Filter — 512
Kalman Filter Derivation, 513
Filter Properties and Implementation, 515
Steady-State Kalman Filter, 516

Problems for Chapter 9 — 523

10. Multivariable Dynamic Compensator Design — 526

10.1 Linear-Quadratic-Gaussian Design — 527
LQG Formulation, 528
The Separation Principle, 530
Transfer Function Description of the LQG Regulator, 531
Digital LQG Regulator, 532

10.2 LQG/Loop-Transfer Recovery Robust Design — 548
Guaranteed Robustness of the Linear-Quadratic Regulator, 548
Recovery of Robust Loop Gain at the Input, 553
Nonminimum-Phase Plants and Parameter Variations, 558
Recovery of Robust Loop Gain at the Output, 558

Problems for Chapter 10 — 572

APPENDICES

A. Computer Software — 574

A.1 Time Response and Controls Simulation — 574
A.2 Discretization of Continuous-Time Systems — 578

	A.3 Ackermann's Formula for Pole Placement	581
	A.4 Output-Feedback Design	581
B.	**Review of Matrix Algebra**	**585**
	B.1 Basic Facts	585
	B.2 Partitioned Matrices	586
	B.3 Quadratic Forms and Definiteness	587
	B.4 Singular Value Decomposition	588
	B.5 Matrix Calculus	590
C.	**Review of Probability Theory**	**593**
	C.1 Mean and Variance	593
	C.2 Two Random Variables	594
	C.3 Random Processes	595
	C.4 Spectral Density and Linear Systems	596
D.	**The Texas Instruments TMS320C25 Digital Signal Processor**	**598**
	D.1 Architectural Overview	598
	D.2 Addressing Modes	600
	D.3 Instruction Set	603
	D.4 Using the TMS320C25 in Digital Control	604
	D.5 Multiply-and-Accumulate	605
	D.6 Delaying and Storing Signals	608
	D.7 Coefficient and Variable Scaling	609
	D.8 Overflow Management	611
	References	**613**
	Index	**619**

List of Tables

2.1-1 Solution of the State Equation
2.4-1 State-Space Transformation and Canonical Forms
3.1-1 Continuous Nonlinear Optimal Controller
3.2-1 Open-Loop Linear Quadratic Controller
3.2-2 Continuous Linear Quadratic Regulator
4.1-1 LQR with Output Feedback
4.1-2 Optimal Output-Feedback Solution Algorithm
4.2-1 LQ Tracker with Output Feedback
5.2-1 PID Parameters Using Transient-Response Method
5.2-2 PID Parameters Using Stability-Limit Method
5.3-1 Padé Approximants to $e^{-s\Delta}$
5.3-2 Approximants to $(1 - e^{-sT})/sT$
6.4-1 Elements of Second-Order Modules
6.4-2 Difference Equation Implementation of Second-Order Modules
7.1-1 Table of Z-Transforms
7.3-1 Numerators of $H^s(z)$ For Small T vs. Relative Degree
7.4-1 Discrete LQ Tracker with Output Feedback
7.4-2 Discrete LQR with Output Feedback
7.4-3 Discrete LQR with State Feedback
9.3-1 Discrete-Time Kalman Filter: Time Update/Measurement Update Formulation

- **9.3-2** Discrete-Time Kalman Filter: Alternative Measurement Update Equations
- **9.3-3** Discrete-Time Kalman Filter: Recursive A Priori Formulation
- **9.5-1** Continuous-Time Kalman Filter
- **10.1-1** Continuous-Time LQG Dynamic Regulator
- **10.1-2** Digital LQG Dynamic Regulator
- **10.2-1** Continuous-Time LQG/LTR Design Algorithms

List of Examples

Examples for Chapter 2

- **2.1-1:** State Model for a Pendulum
- **2.1-2:** State Model for a Circuit
- **2.1-3:** DC Motor with Compliant Coupling
- **2.1-4:** Systems Obeying Newton's Law
- **2.1-5:** Damped Harmonic Oscillator
- **2.1-6:** Simulation of van der Pol Oscillator
- **2.1-7:** Simulation of Damped Harmonic Oscillator
- **2.1-8:** Simulation of DC Motor with Compliant Coupling
- **2.3-1:** Cayley-Hamilton Theorem
- **2.4-1:** State-Space Transformation in the 2-Body Problem
- **2.4-2:** Pole/Zero Cancellation and JNF Realization
- **2.4-3:** Real Jordan Form in Controller Implementation
- **2.4-4:** Analysis and Simulation Using the RCF
- **2.4-5:** Realization and Minimality for SISO Systems
- **2.4-6:** Gilbert's Method for Multivariable Minimal Realizations
- **2.5-1:** Control of Inverted Pendulum Using Ackermann's Formula
- **2.5-2:** Multivariable Decoupling Using Eigenstructure Assignment

Examples for Chapter 3

- **3.1-1:** Optimal Control of a Nonlinear Steering System
- **3.1-2:** Shortest Distance Between Two Points
- **3.1-3:** Numerical Solution of Hamiltonian System for Armature-Controlled DC Motor
- **3.1-4:** Unit Solution Method for Scalar System
- **3.2-1:** Open-Loop Control of a DC Motor
- **3.2-2:** Open-Loop Control of Systems Obeying Newton's Laws
- **3.2-3:** LQR for DC Motor Using Scalar Model
- **3.2-4:** LQR for Armature-Controlled DC Motor
- **3.3-1:** Steady-State LQ Design for Systems Obeying Newton's Laws
- **3.3-2:** Receding Horizon Control—an Easy Way to Stabilize a System
- **3.3-3:** Suboptimal Control of a DC Motor Using Scalar Model
- **3.3-4:** Suboptimal Control for Systems Obeying Newton's Laws
- **3.3-5:** Eigenstructure LQR Design for Systems Obeying Newton's Laws
- **3.4-1:** Minimum-Time Control of Systems Obeying Newton's Laws

Examples for Chapter 4

- **4.1-1:** Multivariable Regulator Using Output Feedback
- **4.2-1:** Multivariable Regulator with Gain Element Constraints
- **4.2-2:** Proportional-Plus-Integral Compensator for Tracking
- **4.2-3:** Speed Control of an Armature-Controlled DC Motor
- **4.2-4:** Control of an Inverted Pendulum
- **4.3-1:** Tracking by Regulator Redesign
- **4.4-1:** Track Following for a Disk Drive Head-Positioning System

Examples for Chapter 5

- **5.2-1:** BLT and MPZ on a First-Order System
- **5.2-2:** Digital Inverted Pendulum Controller via BLT
- **5.2-3:** Digital Control of Process with Time Delay
- **5.3-1:** Inverted Pendulum Digital Control Using a FOH
- **5.3-2:** Digital Inverted Pendulum Controller via Modified Continuous Design

Examples for Chapter 6

- **6.1-1:** Digital PI Controller with Antiwindup Compensation
- **6.2-1:** Multiple Equilibria and Limit Cycles Under Quantization
- **6.2-2:** Quantization Effects in a Digital Inverted Pendulum Controller
- **6.3-1:** Overflow Oscillations
- **6.3-2:** Overflow Oscillations and Measurement Noise in a Digital Motor Speed Controller
- **6.4-1:** Pole Sensitivity to Coefficient Quantization of Second-Order Modules
- **6.4-2:** Direct Versus Parallel Implementation
- **6.6-1:** DSP Implementation of Digital Controller for an Inverted Pendulum

Examples for Chapter 7

- **7.1-1:** Discretization of Scalar System
- **7.1-2:** Discretization of Newton's System
- **7.1-3:** Discretization of Damped Harmonic Oscillator
- **7.1-4:** Sampling Using a First-Order Hold
- **7.1-5:** Discretization of Newton's System with Time Delay
- **7.1-6:** Discretization of a Transfer Function
- **7.2-1:** Digital Controls Design Using Ackermann's Formula
- **7.3-1:** Loss of Reachability and Observability Through Sampling
- **7.4-1:** Deadbeat Control and Intersample Behavior of a System Obeying Newton's Laws
- **7.4-2:** Digital Tracker by Regulator Redesign for Motor Speed Control
- **7.4-3:** Digital Inverted Pendulum Controller via Discrete LQ Design
- **7.4-4:** Multivariable Digital Controller Using Discrete LQR Design

Examples for Chapter 8

- **8.2-1:** MIMO Bode Magnitude Plots
- **8.3-1:** Precompensator for Balancing and Zero Steady-State Error
- **8.3-2:** Model Reduction and Stability Robustness
- **8.4-1:** Aircraft Pitch Rate Control System Robust to Wind Gusts and Unmodeled Flexible Mode

Examples for Chapter 9

- **9.1-1:** Digital Observer for Disk Drive Head-Positioning System
- **9.1-2:** Continuous Observer for Disk Drive Head-Positioning System
- **9.2-1:** Reduced-Order Digital Observer for Disk Drive Head-Positioning System
- **9.2-2:** Reduced-Order Continuous Observer for Disk Drive Head-Positioning System
- **9.4-1:** Digital Kalman Filter for Disk Drive Head-Positioning System
- **9.4-2:** Steady-State Kalman Filter
- **9.4-3:** DSP Implementation of the Kalman Filter
- **9.5-1:** Continuous Filtering for Damped Harmonic Oscillator

Examples for Chapter 10

- **10.1-1:** Digital LQG Regulator for Inverted Pendulum
- **10.1-2:** DSP Implementation of Digital LQG Regulator
- **10.2-1:** LQG/LTR Design of Aircraft Lateral Control System

Preface

Our interactions with nature are a peculiar affair. Taking our knowledge from a noisy and uncertain macrocosm, we must convert observations into useful information about the processes we are interested in. Having observed and understood, we are next faced with the task of controlling an essentially uncooperative world to our ends. These two aspects of interaction have been formalized in the philosophy of modern automatic controls as the dual notions of estimation theory and control theory.

The study of history is a curious endeavor unlike study in other fields; it is only as we move *further* from events that they can be placed into perspective and understood as part of an overall pattern. Thus, by now it is possible to understand the developments in automatic control theory at least in the first half of the century. With the advent of a "general theory of systems" in the early 1900s, control theory was provided with its context. With the advent of mass communications and the telephone long lines after about 1910, the theory of the frequency domain was developed. With the advent of gunnery, aircraft, and missile systems during the world wars, the Classical Theory of Control Systems was developed. With the advent of satellites and aerospace systems after 1957, the Modern Theory of Control Systems evolved.

During the past few years, we have seen the evolution of the personal computer (PC), sophisticated controls design and simulation software, and the digital signal processor (DSP). This has placed the design and implementation of control and estimation systems, even for complicated multi-input/multi-output processes, well within the reach of the individual engineer working alone in his or her office.

This book was motivated by a desire to bring to the individual researcher and engineer working at a PC all the power of modern control theory for design, simulation, and actual controls implementation. It was designed for a first-year graduate course, though the material has been presented as well in advanced undergraduate courses. The book has also been designed as a handbook for practicing engineers, about which more will be said later.

Prerequisites for the book involve a familiarity with classical control theory and ordinary differential equations. The reader is helped by having as well some knowledge of matrix algebra, state-variable systems, and, for the last part of the book, probability theory. However, the coverage of these topics in Appendix B, Chapter 2, and Appendix C, respectively, helps to prevent a lack of knowledge in these areas from becoming a major obstacle.

The multivariable matrix equations for modern control systems design are not simple. In practical situations they are multiple nonlinear coupled equations whose solution yields the control and estimation gains. This book tries to show that, in fact, these equations are straightforward to derive, and so to understand, as well as to solve using today's software design tools.

Theoretical derivation is the way to familiarity with the controls design equations, and examples provide the way to intuition and an understanding of controls design techniques. Therefore, in this book we have provided derivations where they are useful, referring the reader who is not interested in the derivations to the tables, where the design equations are collected. An attempt has been made to keep all derivations at the level of providing insight without delving to a level of mathematics that will generate distractions.

The intent of the tables of design equations is to make the book useful as a reference for engineers. The List of Tables makes it easy to find the equations needed for any application. The many examples are intended to show the use of the tables in control system design. Some examples use simple systems (Newton's law) to provide intuition. Some examples use actual systems (disk drive, DC motor, aircraft) to provide practical insight. The List of Examples makes it easy to find illustrations relevant to a given application. Simple software subroutines given in the examples show how to stimulate controllers and estimators to verify their performance before they are placed into service on actual processes. The longer computer programs are collected in Appendix A. The usefulness of modern design packages such as MATLAB, MATRIX$_x$, and so on is illustrated. In some cases, the reader is left to write the design software, with guidance being given on some of the details.

A major thrust in the book is to show how to implement controllers and estimators on actual systems. Several actual implementation examples are given. The Texas Instruments TMS320 C25 DSP has been selected to show that multivariable control systems are straightforward to implement on a PC using inexpensive fixed-point processors. The C25 is briefly described from a controls applications point of view in Appendix D.

The book is organized into parts. Part I provides by way of introduction a discussion on the history of automatic controls as well as an overview of state-variable systems. Part II contains the foundations of modern continuous-time control theory. Chapter 3 focuses on the feedback control of systems where all the internal states can be measured. Since in practical applications only some state information is directly provided in the form of the measured outputs, Chapter 4 covers output-feedback design, where the design equations become more complicated but the control systems become more practical. Standard PC software for solving the design equations makes output-feedback design direct and very applicable.

The main emphasis of the book is digital controls design, so that it was tempting to entitle the book "Applied Digital Control." In Part III we introduce the reader to this important field, which has revolutionized since about the 1960s the control of industrial processes and aerospace systems. Two design techniques are covered. In the "Continuous Redesign Approach" of Chapter 5, we show how an existing continuous-time controller can be sampled and converted into a digital controller. Since most classical controls engineers think in continuous terms, this technique has the advantage of simplicity. On the other hand, the "Direct Discrete Design Approach" of Chapter 7 offers the advantage of more accurate performance and greater insight. Chapter 6 presents practical implementation aspects of digital control systems.

Part IV consists of only one chapter, yet has great importance. As well as incorporating classical frequency-domain notions into modern controls design, it shows how to guarantee performance in the face of measurement noise, disturbances, and modeling inaccuracies. That is, this Part ties modern control theory to the exigencies of controlling processes in a real world.

The recovery of information from incomplete measurements in an uncertain world is the subject of Part V. Here, we discuss the design of estimators and observers that use only the available measurements to provide the information necessary for control. The Kalman filter is introduced, and its actual implementation on a DSP is illustrated. In Chapter 10 we show how to combine the notions in Part V with the control notions from the first portions of the book to design multivariable dynamic compensators using the linear-quadratic-Gaussian (LQG) approach.

ACKNOWLEDGMENTS

The implementation work in this book has been performed by Andy Lowe, whose insight and attention to detail often forced me to work long hours to redo the controls designs for use on actual systems. He could without requiring much justification be considered a co-author of the book. My other Ph.D. students have often gone out of their way to obstruct my progress toward a final version of the manuscript by raising questions and demonstrating a shrewdness that resulted in major rewritings of several sections of the book. Through their discontent with cursory explanations, they have made, often unbeknownst to them, major contributions to the final product.

Though they may not be aware of it, Demetrius Paris, Roger Webb, and William Sangster at the Georgia Institute of Technology have made contributions both in terms of moral support and hardware contributions. John Mills, Bob Mitchell, and John McElroy at The University of Texas at Arlington have been kind enough to continue this sustenance since September of 1990. Raynette Taylor at UTA has managed to keep affairs going on several fronts while I devoted the time necessary for the preparation of the manuscript. Ron Schafer and Tom Barnwell of Georgia Tech and Atlanta Signal Processors, Inc. (ASPI) assisted in the beginning stages by providing the DSP hardware for the implementation experiments. A great debt of thanks goes to Irfan Ahmed of Texas Instruments, who, over a glass of wine at the Automatic Control Conference in Atlanta, proposed in 1988 the concept of the book. Without him this project would never have been undertaken.

On a technical level I am indebted to Brian Stevens, whose mature engineering wisdom in control systems design and simulation has been a continuing source of knowledge and intuition.

Finally, my greatest appreciation goes to Theresa, whose sweet temperament and steady character have come to the rescue during the low points of the day-to-day work associated with writing a book. She has given me the motivation to bring the project to completion, while during the same period giving birth to a child, obtaining her Ph.D., and publishing several journal papers.

PART I

INTRODUCTION

In this first part of the book we provide an introduction to modern control theory. Chapter 1 outlines the history and philosophy of automatic control. A main objective is to compare and contrast classical control theory and modern control theory.

Chapter 2 provides a review of state-variable systems. Only the concepts needed in this book are mentioned. This, together with the reviews of matrix algebra and probability theory in the appendices, provides the background required for the rest of the book.

The meat of the book does not appear until Chapter 3, where we begin our discussion of control systems. Thus, Part I should be treated as background material that may be quickly skimmed by those comfortable with the basic notions of modern system theory.

1

Introduction to Modern Control Theory

SUMMARY

In this chapter we introduce modern control theory by two approaches. First, a short history of automatic control theory is provided. Then, we describe the philosophies of classical and modern control theory.

INTRODUCTION

Feedback control is the basic mechanism by which systems, whether mechanical, electrical, or biological, maintain their equilibrium or homeostasis. In the higher life forms, the conditions under which life can continue are quite narrow. A change in body temperature of half a degree is generally a sign of illness. The homeostasis of the body is maintained through the use of feedback control [Wiener 1948]. A primary contribution of C. R. Darwin during the last century was the theory that feedback over long time periods is a key factor in the evolution of species. In 1931 V. Volterra explained the balance between two populations of fish in a closed pond using the theory of feedback.

Feedback control may be defined as the use of difference signals, determined by comparing the actual values of system variables to their desired values, as a means of controlling a system. An everyday example of a feedback control system is an automobile speed control, which uses the difference between the actual and the desired speed to vary the fuel flow rate. Since the system output is used to regulate its input, such a device is said to be a *closed-loop control system*.

In this book we shall show how to use *modern control theory* to design feedback control systems. Thus, we are concerned not with natural control systems, such as those that occur in living organisms or in society, but with man-made control systems such as those used to control aircraft, automobiles, satellites, robots, and industrial processes.

Realizing that the best way to understand an area is to examine its evolution and the reasons for its existence, we shall first provide a short history of automatic control theory. Then, we give a brief discussion of the philosophies of classical and modern control theory.

The references for Chapter 1 are at the end of this chapter. The references for the remainder of the book appear at the end of the book.

1.1 A BRIEF HISTORY OF AUTOMATIC CONTROL

There have been many developments in automatic control theory during recent years. It is difficult to provide an impartial analysis of an area while it is still developing; however, looking back on the progress of feedback control theory it is by now possible to distinguish some main trends and point out some key advances.

Feedback control is an engineering discipline. As such, its progress is closely tied to the practical problems that needed to be solved during any phase of human history. The key developments in the history of mankind that affected the progress of feedback control were:

1. The preoccupation of the Greeks and Arabs with keeping accurate track of time. This represents a period from about 300 B.C. to about A.D. 1200.
2. The Industrial Revolution in Europe. The Industrial Revolution is generally agreed to have started in the third quarter of the eighteenth century; however, its roots can be traced back to the 1600s.
3. The beginning of mass communication and the First and Second World Wars. This represents a period from about 1910 to 1945.
4. The beginning of the space/computer age in 1957.

One may consider these as phases in the development of man, where he first became concerned with understanding his place in space and time, then with taming his environment and making his existence more comfortable, then with testing his place in a global community, and finally with his place in the cosmos.

At a point between the Industrial Revolution and the world wars, there was an extremely important development. Namely, control theory began to acquire its written language—the language of mathematics. J. C. Maxwell provided the first rigorous mathematical analysis of a feedback control system in 1868. Thus, relative to this written language, we could call the period before about 1868 the *prehistory* of automatic control.

Following Friedland [1986], we may call the period from 1868 to the early 1900s the *primitive period* of automatic control. It is standard to call the period from then until 1960 the *classical period*, and the period from 1960 through present times the *modern period*.

Let us now progress quickly through the history of automatic controls. A reference for the period 300 B.C. through the Industrial Revolution is provided by Mayr [1970],

1.1 A Brief History of Automatic Control

which we shall draw on and at times quote. See also Fuller [1976]. Other important references used in preparing this section included Bokharaie [1973] and personal discussions with J. D. Aplevich of the University of Waterloo, K. M. Przyłuski of the Polish Academy of Sciences, and W. Askew, a former Fellow at LTV Missiles and Space Corporation and vice-president of E-Systems.

Water Clocks of the Greeks and Arabs

The primary motivation for feedback control in times of antiquity was the need for the accurate determination of time. Thus, in about 270 B.C. the Greek Ktesibios invented a *float regulator* for a water clock. The function of this regulator was to keep the water level in a tank at a constant depth. This constant depth yielded a constant flow of water through a tube at the bottom of the tank which filled a second tank at a constant rate. The level of water in the second tank thus depended on time elapsed.

The regulator of Ktesibios used a float to control the inflow of water through a valve; as the level of water fell the valve opened and replenished the reservoir. This float regulator performed the same function as the ball and cock in a modern flush toilet. A float regulator was used by Philon of Byzantium in 250 B.C. to keep a constant level of oil in a lamp.

During the first century A.D., Heron of Alexandria developed float regulators for water clocks. The Greeks used the float regulator and similar devices for purposes such as the automatic dispensing of wine, the design of syphons for maintaining constant water level differences between two tanks, the opening of temple doors, and so on. These devices could be called "gadgets" since they were among the earliest examples of an idea looking for an application.

In 800 through 1200 various Arab engineers such as Al-Jazarī, the three brothers Musa, and Ibn al-Sā'ātī used float regulators for water clocks and other applications. During this period, the important feedback principle of "on/off" control was used, which comes up again in connection with minimum-time problems in the 1950s (see Section 3.4).

When Baghdad fell to the Mongols in 1258, all creative thought along these lines came to an end. Moreover, the invention of the mechanical clock in the fourteenth century made the water clock and its feedback control system obsolete. (The mechanical clock is not a feedback control system.) The float regulator does not appear again until its use in the Industrial Revolution.

Along with a concern for his place in time, early man had a concern for his place in space. It is worth mentioning that a pseudo-feedback control system was developed in China in the twelfth century for navigational purposes. The south-pointing chariot had a statue that was turned by a gearing mechanism attached to the wheels of the chariot so that it continuously pointed south. Using the directional information provided by the statue, the charioteer could steer a straight course. We call this a pseudo-feedback control system since it does not technically involve feedback unless the actions of the charioteer are considered as part of the system. Thus, it is not an automatic control system.

The Industrial Revolution

The Industrial Revolution in Europe followed the introduction of *prime movers*, or self-powered machines. It was marked by the invention of advanced grain mills, furnaces, boilers, and the steam engine. These devices could not be adequately regulated by hand, and so arose a new requirement for automatic control systems. A variety of control devices were invented, including float regulators, temperature regulators, pressure regulators, and speed control devices.

J. Watt invented his steam engine in 1769, and this date marks the accepted beginning of the Industrial Revolution. However, the roots of the Industrial Revolution can be traced back to the 1600s or earlier with the development of grain mills and the furnace.

One should be aware that others, primarily T. Newcomen in 1712, built the first steam engines. However, the early steam engines were inefficient and regulated by hand, making them less suited to industrial use. It is extremely important to realize that the Industrial Revolution did not start until the invention of improved engines and automatic control systems to regulate them. That is, the advent of such feedback control devices marks the true beginning of the Industrial Revolution.

The millwrights

The millwrights of Britain developed a variety of feedback control devices. The fantail, invented in 1745 by British blacksmith E. Lee, consisted of a small fan mounted at right angles to the main wheel of a windmill. Its function was to point the windmill continuously into the wind.

The mill-hopper was a device that regulated the flow of grain in a mill depending on the speed of rotation of the millstone. It was in use in a fairly refined form by about 1588.

To build a feedback controller, it is important to have adequate measuring devices. The millwrights developed several devices for sensing speed of rotation. Using these sensors, several speed regulation devices were invented, including self-regulating windmill sails. Much of the technology of the millwrights was later developed for use in the regulation of the steam engine.

Temperature regulators

Cornelis Drebbel of Holland spent some time in England and a brief period with the Holy Roman Emperor Rudolf II in Prague, together with his contemporary J. Kepler. Around 1624 he developed an automatic temperature control system for a furnace, motivated by his belief that base metals could be turned to gold by holding them at a precise constant temperature for long periods of time. He also used this temperature regulator in an incubator for hatching chickens.

Temperature regulators were studied by J. J. Becher in 1680, and used again in an incubator by the Prince de Conti and R.-A. F. de Réaumur in 1754. The "sentinel regis-

1.1 A Brief History of Automatic Control

ter" was developed in America by W. Henry around 1771, who suggested its use in chemical furnaces, in the manufacture of steel and porcelain, and in the temperature control of a hospital. It was not until 1777, however, that a temperature regulator suitable for industrial use was developed by Bonnemain, who used it for an incubator. His device was later installed on the furnace of a hot-water heating plant.

Float regulators

Regulation of the level of a liquid was needed in two main areas in the late 1700s: in the boiler of a steam engine and in domestic water distribution systems. Therefore, the float regulator received new interest, especially in Britain.

In his book of 1746, W. Salmon quoted prices for ball-and-cock float regulators used for maintaining the level of house water reservoirs. This regulator was used in the first patents for the flush toilet around 1775. The flush toilet was further refined by Thomas Crapper, a London plumber, who was knighted by Queen Victoria for his inventions.

The earliest known use of a float valve regulator in a steam boiler is described in a patent issued to J. Brindley in 1758. He used the regulator in a steam engine for pumping water. S. T. Wood used a float regulator for a steam engine in his brewery in 1784. In Russian Siberia, the coal miner I. I. Polzunov developed in 1765 a float regulator for a steam engine that drove fans for blast furnaces. By 1791, when it was adopted by the firm of Boulton and Watt, the float regulator was in common use in steam engines.

Pressure regulators

Another problem associated with the steam engine is that of steam-pressure regulation in the boiler, for the steam that drives the engine should be at a constant pressure. In 1681 D. Papin invented a safety valve for a pressure cooker which he used in 1707 as a regulating device on his steam engine. Thereafter, it was a standard feature on steam engines.

The pressure regulator was further refined in 1799 by R. Delap and also by M. Murray. In 1803 a pressure regulator was combined with a float regulator by Boulton and Watt for use in their steam engines.

Centrifugal governors

The first steam engines provided a reciprocating output motion that was regulated using a device known as a cataract, similar to a float valve. The cataract originated in the pumping engines of the Cornwall coal mines.

J. Watt's steam engine with a rotary output motion had reached maturity by 1783, when the first one was sold. The main incentive for its development was evidently the hope of introducing a prime mover into milling. Using the rotary output engine, the Albion steam mill began operation early in 1786.

A problem associated with the rotary steam engine is that of regulating its speed of revolution. Some of the speed regulation technology of the millwrights was developed and extended for this purpose. In 1788 Watt completed the design of the centrifugal flyball governor for regulating the speed of the rotary steam engine. This device employed two pivoted rotating flyballs which were flung outward by centrifugal force. As the speed of rotation increased, the flyweights swung further out and up, operating a steam flow throttling valve that slowed the engine down. Thus, a constant speed was achieved automatically.

The feedback devices mentioned previously either remained obscure or played an inconspicuous role as a part of the machinery they controlled. On the other hand, the operation of the flyball governor was clearly visible even to the untrained eye, and its principle had an exotic flavor that seemed to many to embody the nature of the new industrial age. Therefore, the governor reached the consciousness of the engineering world and became a sensation throughout Europe. This was the first use of feedback control of which there was popular awareness.

Around 1790 in France, the brothers Périer developed a float regulator to control the speed of a steam engine, but their technique was no match for the centrifugal governor, and was soon supplanted.

It is worth noting that the Greek word for governor is $\kappa\upsilon\beta\varepsilon\rho\nu\alpha\nu$. In 1947, Norbert Wiener at MIT was searching for a name for his new discipline of automata theory—control and communication in man and machine. In investigating the flyball governor of Watt, he also investigated the etymology of the word $\kappa\upsilon\beta\varepsilon\rho\nu\alpha\nu$ and came across the Greek word for steersman, $\kappa\upsilon\beta\varepsilon\rho\nu\eta\tau\eta\varsigma$. Thus, he selected the name *cybernetics* for his fledgling field.

The pendule sympathique

Having begun our history of automatic control with the water clocks of ancient Greece, we round out this portion of the story with a return to mankind's preoccupation with time.

The mechanical clock invented in the fourteenth century is not a closed-loop feedback control system, but a precision open-loop oscillatory device whose accuracy is ensured by protection against external disturbances. In 1793 the French-Swiss A.-L. Breguet, the foremost watchmaker of his day, invented a closed-loop feedback system to synchronize pocket watches.

The pendule sympathique of Breguet used a special case of speed regulation. It consisted of a large, accurate precision chronometer with a mount for a pocket watch. The pocket watch to be synchronized is placed into the mount slightly before 12 o'clock, at which time a pin emerges from the chronometer, inserts into the watch, and begins a process of automatically adjusting the regulating arm of the watch's balance spring. After a few placements of the watch in the pendule sympathique, the regulating arm is adjusted automatically. In a sense, this device was used to transmit the accuracy of the large chronometer to the small portable pocket watch.

1.1 A Brief History of Automatic Control

The Birth of Mathematical Control Theory

The design of feedback control systems up through the Industrial Revolution was by trial and error together with a great deal of engineering intuition. Thus, it was more of an art than a science. In the mid-1800s, mathematics was first used to analyze the stability of feedback control systems. Since mathematics is the formal language of automatic control theory, we could call the period before this time the prehistory of control theory.

Differential equations

In 1840, the British Astronomer Royal at Greenwich, G. B. Airy, developed a feedback device for pointing a telescope. His device was a speed control system that turned the telescope automatically to compensate for the earth's rotation, affording the ability to study a given star for an extended time.

Unfortunately, Airy discovered that by improper design of the feedback control loop, wild oscillations were introduced into the system. He was the first to discuss the *instability* of closed-loop systems, and the first to use *differential equations* in their analysis [Airy 1840]. The theory of differential equations was by then well developed, due to the discovery of the infinitesimal calculus by I. Newton (1642–1727) and G. W. Leibniz (1646–1716), and the work of the brothers Bernoulli (late 1600s and early 1700s), J. F. Riccati (1676–1754), and others. The use of differential equations in analyzing the motion of dynamical systems was established by J. L. Lagrange (1736–1813) and W. R Hamilton (1805–1865).

Stability theory

The early work in the mathematical analysis of control systems was in terms of differential equations. J. C. Maxwell analyzed the stability of Watt's flyball governor [Maxwell 1868]. His technique was to linearize the differential equations of motion to find the *characteristic equation* of the system. He studied the effect of the system parameters on stability and showed that the system is stable if the roots of the characteristic equation have *negative real parts*. With the work of Maxwell we can say that the theory of control systems was firmly established. E. J. Routh provided a *numerical technique* for determining when a characteristic equation has stable roots [Routh 1877].

The Russian I. I. Vyshnegradsky [1877] analyzed the stability of regulators using differential equations independently of Maxwell. In 1893, A. B. Stodola studied the regulation of a water turbine using the techniques of Vyshnegradsky. He modeled the actuator dynamics and included the delay of the actuating mechanism in his analysis. He was the first to mention the notion of the *system time constant*. Unaware of the work of Maxwell and Routh, he posed the problem of determining the stability of the characteristic equation to A. Hurwitz [1895], who solved it independently.

The work of A. M. Lyapunov was seminal in control theory. He studied the stability of nonlinear differential equations using a generalized notion of energy in 1892 [Lyapunov 1892]. Unfortunately, though his work was applied and continued in Russia, the time was not ripe in the West for his elegant theory, and it remained unknown there until approximately 1960, when its importance was finally realized.

The British engineer O. Heaviside invented operational calculus during the period 1892–1898. He studied the transient behavior of systems, introducing a notion equivalent to that of the *transfer function*.

System theory

It is within the study of *systems* that feedback control theory has its place in the organization of human knowledge. Thus, the concept of a system as a dynamical entity with definite "inputs" and "outputs" joining it to other systems and to the environment was a key prerequisite for the further development of automatic control theory. The history of system theory requires an entire study on its own, but a brief sketch follows.

During the eighteenth and nineteenth centuries, the work of A. Smith in economics [*The Wealth of Nations*, 1776], the discoveries of C. R. Darwin [*On the Origin of Species By Means of Natural Selection*, 1859], and other developments in politics, sociology, and elsewhere were having a great impact on the human consciousness. The study of Natural Philosophy was an outgrowth of the work of the Greek and Arab philosophers, and contributions were made by Nicholas of Cusa (1463), Leibniz, and others. The developments of the nineteenth century, flavored by the Industrial Revolution and an expanding sense of awareness in global geopolitics and in astronomy, had a profound influence on this Natural Philosophy, causing it to change its personality.

By the early 1900s, A. N. Whitehead [1925], with his philosophy of "organic mechanism", L. von Bertalanffy [1938], with his hierarchical principles of organization, and others had begun to speak of a "general system theory." In this context, the evolution of control theory could proceed.

Mass Communication and the Bell Telephone System

At the beginning of the twentieth century there were two important occurrences from the point of view of control theory: the development of the telephone and mass communications, and the world wars.

Frequency-domain analysis

The mathematical analysis of control systems had heretofore been carried out using differential equations in the *time domain*. At Bell Telephone Laboratories during the 1920s and 1930s, the *frequency-domain* approaches developed by P.-S. de Laplace (1749–1827), J. Fourier (1768–1830), A. L. Cauchy (1789–1857), and others were explored and used in communication systems.

A major problem with the development of a mass communication system extending over long distances is the need to periodically amplify the voice signal in long telephone lines. Unfortunately, unless care is exercised, not only the information but also the noise is amplified. Thus, the design of suitable repeater amplifiers is of prime importance.

To reduce distortion in repeater amplifiers, H. S. Black demonstrated the usefulness of *negative feedback* in 1927 [Black 1934]. The design problem was to introduce a phase shift at the correct frequencies in the system. Regeneration theory for the design of stable amplifiers was developed by H. Nyquist [1932]. He derived his *Nyquist stability criterion* based on the polar plot of a complex function. H. W. Bode in 1938 used the magnitude and phase *frequency-response plots* of a complex function [Bode 1940]. He investigated closed-loop stability using the notions of *gain and phase margin*.

The World Wars and Classical Control

As mass communications and faster modes of travel made the world smaller, there was much tension as men tested their place in a global society. The result was the world wars, during which the development of feedback control systems became a matter of survival.

Ship control

An important military problem during this period was the control and navigation of ships, which were becoming more advanced in their design. Among the first developments was the design of sensors for the purpose of closed-loop control. In 1910, E. A. Sperry invented the gyroscope, which he used in the stabilization and steering of ships, and later in aircraft control.

N. Minorsky [1922] introduced his three-term controller for the steering of ships, thereby becoming the first to use the *proportional-integral-derivative (PID)* controller. He considered nonlinear effects in the closed-loop system.

Weapons development and gun pointing

A main problem during the period of the world wars was that of the accurate pointing of guns aboard moving ships and aircraft. With the publication of "Theory of Servomechanisms" by H. L. Házen [1934], the use of mechanical control theory in such problems was initiated. In his paper, Házen coined the word *servomechanisms*, which implies a master–slave relationship in systems.

The Norden bombsight, developed during World War II, used synchro repeaters to relay information on aircraft altitude and velocity and wind disturbances to the bombsight, ensuring accurate weapons delivery.

MIT Radiation Laboratory

To study the control and information processing problems associated with the newly invented radar, the Radiation Laboratory was established at the Massachusetts Institute of Technology in 1940. Much of the work in control theory during the 1940s came out of this lab.

While working on an MIT/Sperry Corporation joint project in 1941, A. C. Hall recognized the deleterious effects of ignoring noise in control system design. He realized that the frequency-domain technology developed at Bell Labs could be employed to confront noise effects, and used this approach to design a control system for an airborne radar. This success demonstrated conclusively the importance of frequency-domain techniques in control system design [Hall 1946].

Using design approaches based on the transfer function, the block diagram, and frequency-domain methods, there was great success in controls design at the Radiation Lab. In 1947, N. B. Nichols developed his Nichols Chart for the design of feedback systems. With the MIT work, the theory of linear servomechanisms was firmly established. A summary of the MIT Radiation Lab work is provided in *Theory of Servomechanisms* [James, Nichols, and Phillips 1947].

Working at North American Aviation, W. R. Evans [1948] presented his *root locus* technique, which provided a direct way to determine the closed-loop pole locations in the s-plane. Subsequently, during the 1950s, much controls work was focused on the s-plane, and on obtaining desirable closed-loop step-response characteristics in terms of rise time, percent overshoot, and so on.

Stochastic analysis

During this period also, *stochastic techniques* were introduced into control and communication theory. At MIT in 1942, N. Wiener [1949] analyzed information processing systems using models of stochastic processes. Working in the frequency domain, he developed a statistically optimal filter for stationary continuous-time signals that improved the signal-to-noise ratio in a communication system. The Russian A. N. Kolmogorov [1941] provided a theory for discrete-time stationary stochastic processes.

The classical period of control theory

By now, automatic control theory using frequency-domain techniques had come of age, establishing itself as a paradigm (in the sense of Kuhn [1962]). On the one hand, a firm mathematical theory for servomechanisms had been established, and on the other, engineering design techniques were provided. The period after the Second World War can be called the classical period of control theory. It was characterized by the appearance of the first textbooks [MacColl 1945; Lauer, Lesnick, and Matdon 1947; Brown and Campbell 1948; Chestnut and Mayer 1951; Truxal 1955], and by straightforward design tools that provided great intuition and guaranteed solutions to design problems. These tools were applied using hand calculations, or at most slide rules, together with graphical techniques.

The Space/Computer Age and Modern Control

With the advent of the space age, controls design in the United States turned away from the frequency-domain techniques of classical control theory and back to the differential equation techniques of the late 1800s, which were couched in the *time domain*. The reasons for this development are as follows.

Time-domain design for nonlinear systems

The paradigm of classical control theory was very suitable for controls design problems during and immediately after the world wars. The frequency-domain approach was appropriate for *linear time-invariant* systems. It is at its best when dealing with *single-input/single-output* systems, for the graphical techniques were inconvenient to apply with multiple inputs and outputs.

Classical controls design had some successes with nonlinear systems. Using the noise-rejection properties of frequency-domain techniques, a control system can be designed that is *robust* to variations in the system parameters, and to measurement errors and external disturbances. Thus, classical techniques can be used on a linearized version of a nonlinear system, giving good results at an equilibrium point about which the system behavior is approximately linear.

Frequency-domain techniques can also be applied to systems with simple types of nonlinearities using the *describing function* approach, which relies on the Nyquist criterion. This technique was first used by the Pole J. Groszkowski in radio transmitter design before the Second World War and formalized in 1964 by J. Kudrewicz.

Unfortunately, it is not possible to design control systems for advanced nonlinear multivariable systems, such as those arising in aerospace applications, using the assumption of linearity and treating the single-input/single-output transmission pairs one at a time.

In the Soviet Union, there was a great deal of activity in nonlinear controls design. Following the lead of Lyapunov, attention was focused on time-domain techniques. In 1948, Ivachenko had investigated the principle of *relay control*, where the control signal is switched discontinuously between discrete values. Tsypkin used the phase plane for nonlinear controls design in 1955. V. M. Popov [1961] provided his *circle criterion* for nonlinear stability analysis.

Sputnik—1957

Given the history of control theory in the Soviet Union, it is only natural that the first satellite, Sputnik, was launched there in 1957. The first conference of the newly formed International Federation of Automatic Control (IFAC) was fittingly held in Moscow in 1960.

The launch of Sputnik engendered tremendous activity in the United States in automatic controls design. On the failure of any paradigm, a return to historical and natural first principles is required. Thus, it was clear that a return was needed to the time-domain techniques of the "primitive" period of control theory, which were based on differential equations. It should be realized that the work of Lagrange and Hamilton makes it straightforward to write nonlinear equations of motion for many dynamical systems. Thus, a control theory was needed that could deal with such nonlinear differential equations.

It is quite remarkable that in almost exactly 1960, major developments occurred independently on several fronts in the theory of communication and control.

Navigation

In 1960, C. S. Draper invented his inertial navigation system, which used gyroscopes to provide accurate information on the position of a body moving in space, such as a ship, aircraft, or spacecraft. Thus, the sensors appropriate for navigation and controls design were developed.

Optimality in natural systems

Johann Bernoulli first mentioned the *Principle of Optimality* in connection with the Brachistochrone Problem in 1696. This problem was solved by the brothers Bernoulli and by I. Newton, and it became clear that the quest for optimality is a fundamental property of motion in natural systems. Various optimality principles were investigated, including the minimum-time principle in optics of P. de Fermat (1600s), the work of L. Euler in 1744, and Hamilton's result that a system moves in such a way as to minimize the time integral of the difference between the kinetic and potential energies.

These optimality principles are all *minimum principles.* Interestingly enough, in the early 1900s, Einstein showed that, relative to the 4-D space–time coordinate system, the motion of systems occurs in such a way as to *maximize* the time.

Optimal control and estimation theory

Since naturally occurring systems exhibit optimality in their motion, it makes sense to design man-made control systems in an optimal fashion. A major advantage is that this design may be accomplished in the time domain. In the context of modern controls design, it is usual to minimize the time of transit, or a quadratic generalized energy functional or *performance index*, possibly with some constraints on the allowed controls.

R. Bellman [1957] applied *dynamic programming* to the optimal control of discrete-time systems, demonstrating that the natural direction for solving optimal control problems is backwards in time. His procedure resulted in closed-loop, generally nonlinear, feedback schemes.

By 1958, L. S. Pontryagin had developed his *maximum principle*, which solved optimal control problems relying on the calculus of variations developed by L. Euler (1707–1783). He solved the minimum-time problem, deriving an on/off relay control law as the optimal control [Pontryagin, Boltyansky, Gamkrelidze, and Mishchenko 1962]. In the United States during the 1950s, the calculus of variations was applied to general optimal control problems at the University of Chicago and elsewhere.

In 1960, three major papers were published by R. Kalman and co-workers, working in the United States. One of these [Kalman and Bertram 1960], publicized the vital work of Lyapunov in the time-domain control of nonlinear systems. The next [Kalman 1960a] discussed the optimal control of systems, providing the design equations for the *linear quadratic regulator (LQR)*. The third paper [Kalman 1960b] discussed optimal filtering and estimation theory, providing the design equations for the *discrete Kalman filter.* The *continuous Kalman filter* was developed by Kalman and Bucy [1961].

In the period of a year, the major limitations of classical control theory were overcome, important new theoretical tools were introduced, and a new era in control theory had begun; we call it the era of *modern control*.

The key points of Kalman's work are as follows. It is a time-domain approach, making it more applicable for time-varying linear systems as well as nonlinear systems. He introduced linear algebra and matrices, so that systems with multiple inputs and outputs could easily be treated. He employed the concept of the internal system state; thus, the approach is one that is concerned with the internal dynamics of a system and not only its input/output behavior.

In control theory, Kalman formalized the notion of optimality in control theory by minimizing a very general quadratic generalized energy function. In estimation theory, he introduced stochastic notions that applied to nonstationary time-varying systems, thus providing a recursive solution, the Kalman filter, for the least-squares approach first used by C. F. Gauss (1777–1855) in planetary orbit estimation. The Kalman filter is the natural extension of the Wiener filter to nonstationary stochastic systems.

Classical frequency-domain techniques provide formal tools for control systems design, yet the design phase itself remained very much an art and resulted in nonunique feedback systems. By contrast, the theory of Kalman provided optimal solutions that yielded control systems with guaranteed performance. These controls were directly found by solving formal matrix design equations which generally had unique solutions.

It is no accident that from this point the U.S. space program blossomed, with a Kalman filter providing navigational data for the first lunar landing.

Nonlinear control theory

During the 1960s in the United States, G. Zames [1966], I. W. Sandberg [1964], K. S. Narendra [Narendra and Goldwyn 1964], C. A. Desoer [1965], and others extended the work of Popov and Lyapunov in nonlinear stability. There was an extensive application of these results in the study of nonlinear distortion in bandlimited feedback loops, nonlinear process control, aircraft controls design, and eventually in robotics.

Computers in Controls Design and Implementation

Classical design techniques could be employed by hand using graphical approaches. On the other hand, modern controls design requires the solution of complicated nonlinear matrix equations. It is fortunate that in 1960 there were major developments in another area—digital computer technology. Without computers, modern control would have had limited applications.

The development of digital computers

In about 1830 C. Babbage introduced modern computer principles, including memory, program control, and branching capabilities. In 1948, J. von Neumann directed the construction of the IAS stored-program computer at Princeton. IBM built its SSEC stored-program machine. In 1950, Sperry Rand built the first commercial data processing machine, the UNIVAC I. Soon after, IBM marketed the 701 computer.

In 1960 a major advance occurred—the second generation of computers was introduced which used *solid-state technology*. By 1965, Digital Equipment Corporation was building the PDP-8, and the *minicomputer* industry began. Finally, in 1969 W. Hoff invented the *microprocessor*.

Digital control and filtering theory

Digital computers are needed for two purposes in modern controls. First, they are required to solve the matrix design equations that yield the control law. This is accomplished off-line during the design process. Second, since the optimal control laws and filters are generally time-varying, computers are needed to *implement* modern control and filtering schemes on actual systems.

With the advent of the microprocessor in 1969 a new area developed. Control systems that are implemented on digital computers must be formulated in discrete time. Therefore, the growth of *digital control theory* was natural at this time.

During the 1950s, the theory of *sampled data systems* was being developed at Columbia University by J. R. Ragazzini, G. Franklin, and L. A. Zadeh [Ragazzini and Zadeh 1952; Ragazzini and Franklin 1958], as well as by E. I. Jury [1960], B. C. Kuo [1963], and others. The idea of using digital computers for *industrial process control* emerged during this period [Åström and Wittenmark 1984]. Serious work started in 1956 with the collaborative project between TRW and Texaco, which resulted in a computer-controlled system being installed at the Port Arthur oil refinery in Texas in 1959.

The development of nuclear reactors during the 1950s was a major motivation for exploring industrial process control and instrumentation. This work has its roots in the control of chemical plants during the 1940s.

By 1970, with the work of K. Åström [1970] and others, the importance of digital controls in process applications was firmly established.

The work of C. E. Shannon in the 1950s at Bell Labs had revealed the importance of sampled data techniques in the processing of signals. The applications of *digital filtering theory* were investigated at the Analytic Sciences Corporation [Gelb 1974] and elsewhere.

The personal computer

With the introduction of the PC in 1983, the design of modern control systems became possible for the individual engineer at an individual work station. Thereafter, a plethora of software control systems design packages was developed, including ORACLS, Program CC, Control-C, PC-Matlab, MATRIX$_x$, Easy5, SIMNON, and others.

The Union of Modern and Classical Control

With the publication of the first textbooks in the 1960s, modern control theory established itself as a paradigm for automatic controls design in the United States. Intense activity in research and implementation ensued, with the IRE and the AIEE merging, largely through the efforts of P. Haggerty at Texas Instruments, to form the Institute of Electrical and Electronics Engineers (IEEE) in the early 1960s.

With all its power and advantages, modern control was lacking in some aspects. The guaranteed performance obtained by solving matrix design equations means that it is often possible to design a control system that works in theory without gaining any *engineering intuition* about the problem. On the other hand, the frequency-domain techniques of classical control theory impart a great deal of intuition.

Another problem is that a modern control system with any compensator dynamics can *fail to be robust* to disturbances, unmodeled dynamics, and measurement noise. On the other hand, robustness is built in with a frequency-domain approach using notions like the gain and phase margin.

Therefore, in the 1970s, especially in Great Britain, there was a great deal of activity by H. H. Rosenbrock [1974], A. G. J. MacFarlane and I. Postlethwaite [1977], and others to extend classical frequency-domain techniques and the root locus to multivariable systems. Successes were obtained using notions like the characteristic locus, diagonal dominance, and the inverse Nyquist array.

A major proponent of classical techniques for multivariable systems was I. Horowitz, whose *quantitative feedback theory* developed in the early 1970s accomplishes robust design using the Nichols chart.

In 1981 seminal papers appeared by J. Doyle and G. Stein [1981] and M. G. Safonov, A. J. Laub, and G. L. Hartmann [1981]. Extending the work of MacFarlane and Postlethwaite [1977], they showed the importance of the *singular value* plots versus frequency in robust multivariable design. Using these plots, many of the classical frequency-domain techniques can be incorporated into modern design. This work was pursued in aircraft and process control by M. Athans and others. The result is a *new control theory* that blends the best features of classical and modern techniques. A survey of this *robust modern control theory* is provided by P. Dorato [1987].

1.2 THE PHILOSOPHY OF CLASSICAL CONTROL

Having some understanding of the history of automatic control theory, we may now briefly discuss the philosophies of classical and modern control theory.

Developing as it did for feedback amplifier design, classical control theory was naturally couched in the *frequency domain and the s-plane*. Relying on transform methods, it is primarily applicable for *linear time-invariant systems*, though some extensions to nonlinear systems were made using, for instance, the describing function.

The system description needed for controls design using the methods of Nyquist and Bode is the magnitude and phase of the frequency response. This is advantageous since the frequency response can be experimentally measured. The transfer function can then be computed. For root locus design, the transfer function is needed. The block diagram is heavily used to determine transfer functions of composite systems. *An exact description of the internal system dynamics is not needed* for classical design; that is, only the input/output behavior of the system is of importance.

The design may be carried out by hand using graphical techniques. These methods impart a great deal of intuition and afford the controls designer with a range of design possibilities, so that the resulting control systems are not unique. The design process is an engineering art.

A real system has disturbances and measurement noise, and may not be described exactly by the mathematical model the engineer is using for design. Classical theory is natural for designing control systems that are robust to such disorders, yielding good closed-loop performance in spite of them. Robust design is carried out using notions like the gain and phase margin.

Simple compensators like proportional-integral-derivative (PID), lead-lag, or washout circuits are generally used in the control structure. The effects of such circuits on the Nyquist, Bode, and root locus plots are easy to understand, so that a suitable compensator structure can be selected. Once designed, the compensator can be easily tuned on line.

A fundamental concept in classical control is the ability to *describe closed-loop properties in terms of open-loop properties*, which are known or easy to measure. For instance, the Nyquist, Bode, and root locus plots are in terms of the open-loop transfer function. Again, the closed-loop disturbance rejection properties and steady-state error can be described in terms of the return difference and sensitivity.

Classical control theory is difficult to apply in multi-input/multi-output (MIMO), or multiloop systems. Due to the interaction of the control loops in a multivariable system, each single-input/single-output (SISO) transfer function can have acceptable properties in terms of step response and robustness, but the coordinated control motion of the system can fail to be acceptable.

Thus, classical MIMO or multiloop design requires painstaking effort using the approach of *closing one loop at a time* by graphical techniques. A root locus, for instance, should be plotted for each gain element, taking into account the gains previously selected. This is a trial-and-error procedure that may require multiple iterations, and it does not guarantee good results, or even closed-loop stability.

The multivariable frequency-domain approaches developed by the British school during the 1970s, as well as quantitative feedback theory, overcome many of these limitations, providing an effective approach for the design of many MIMO systems.

1.3 THE PHILOSOPHY OF MODERN CONTROL

Modern controls design is fundamentally a time-domain technique. A *state-space model* of the system to be controlled, or plant, is required. The linear version is a first-order vector differential equation of the form

$$\dot{x} = Ax + Bu \qquad (1.3.1)$$
$$y = Cx,$$

where $x(t)$ is a vector of internal variables or system states, $u(t)$ is a vector of control inputs, and $y(t)$ is a vector of measured outputs. It is possible to add noise terms to

1.3 The Philosophy of Modern Control

represent process and measurement noises. Note that the plant is described in the time domain.

The power of modern control has its roots in the fact that the state-space model can as well represent a MIMO system as a SISO system. That is, $u(t)$ and $y(t)$ are generally vectors whose entries are the individual scalar inputs and outputs. Thus, A, B, C are *matrices* whose elements describe the system dynamical interconnections.

Modern controls techniques were first firmly established for linear systems. Extensions to nonlinear systems can be made using the Lyapunov approach, which extends easily to MIMO systems, dynamic programming, and other techniques. Open-loop optimal controls designs can be determined for nonlinear systems by solving nonlinear two-point boundary-value problems.

Exactly as in the classical case, some fundamental questions on the performance of the closed-loop system can be attacked by investigating *open-loop properties*. For instance, the open-loop properties of controllability and observability of (1.3.1) (Chapter 2) give insight on what it is possible to achieve using feedback control. The difference is that, to deal with the state-space model, a good knowledge of matrices and linear algebra is required.

To achieve suitable closed-loop properties, a feedback control of the form

$$u = -Kx \qquad (1.3.2)$$

may be used. The feedback gain K is a matrix whose elements are the individual control gains in the system. Since all the states are used for feedback, this is called *state-variable feedback*. Note that multiple feedback gains and large systems are easily handled in this framework. Thus, if there are n state components (where n can be very large in an aerospace or power distribution system) and m scalar controls, so that $u(t)$ is an m-vector, then K is an $m \times n$ matrix with mn entries, corresponding to mn control loops.

In the standard linear quadratic regulator (LQR), the feedback gain K is chosen to minimize a quadratic time-domain performance index (PI) like

$$J = \int_0^\infty (x^\mathrm{T} Q x + u^\mathrm{T} R u) \, dt. \qquad (1.3.3)$$

The minimum is sought over all state trajectories. This is an extension to MIMO systems of the sorts of PIs (ITSE, ITAE, etc.) that were used in classical control. Q and R are weighting matrices that serve as *design parameters*. Their elements can be selected to provide suitable performance.

The key to LQR design is the fact that, if the feedback gain matrix K can be successfully chosen to make J finite, then the integral (1.3.3) involving the norms of $u(t)$ and $x(t)$ is bounded. If Q and R are correctly chosen, well-known mathematical principles then ensure that $x(t)$ and $u(t)$ go to zero with time. This *guarantees closed-loop stability* as well as bounded control signals in the closed-loop system.

It can be shown (see Chapter 3) that the value of K that minimizes the PI is given by

$$K = R^{-1} B^\mathrm{T} S \qquad (1.3.4)$$

where S is an $n \times n$ matrix satisfying the *Riccati equation*

$$0 = A^T S + SA - SBR^{-1}B^T S + Q. \tag{1.3.5}$$

Within this LQ framework, several points can be made. First, as long as the system (1.3.1) is controllable and Q and R are suitably chosen, the K given by these equations guarantees the stability of the closed-loop system

$$\dot{x} = (A - BK)x + Bu. \tag{1.3.6}$$

Second, this technique is easy to apply even for multiple-input plants, since $u(t)$ can be a vector having many components.

Third, the LQR solution relies on the solution of the *matrix design equation* (1.3.5), and so is unsuited to hand calculations. Fortunately, many design packages are by now available on digital computers for solving the Riccati design equation for S, and hence for obtaining K. Thus, computer-aided design is an essential feature of modern controls.

The LQR solution is a formal one that gives a unique answer to the feedback control problem once the design parameters Q and R have been selected. In fact, the engineering art in modern design lies in the selection of the PI weighting matrices Q and R. A body of theory on this selection process has evolved. Once Q and R are properly selected, the matrix design equation is formally solved for the unique K that guarantees stability.

Observe that K is computed in terms of the open-loop quantities A, B, Q, R, so that modern and classical design have this feature of determining closed-loop properties in terms of open-loop quantities in common. However, in modern control, all the entries of K are determined at the same time by using the matrix design equations. This corresponds to closing *all the feedback control loops simultaneously*, which is in complete contrast to the one-loop-at-a-time procedure of classical controls design.

Unfortunately, formal LQR design gives very little intuition on the nature or properties of the closed-loop system. In recent years, this deficiency has been addressed from a variety of standpoints.

Although LQR design using state feedback guarantees closed-loop stability, all the state components are seldom available for feedback purposes in a practical design problem. Therefore, *output feedback* of the form

$$u = -Ky \tag{1.3.7}$$

is more useful. LQR design equations for output feedback are more complicated than (1.3.5), but are easily derived (see Chapter 4).

Modern output-feedback design allows one to design controllers for complicated systems with multiple inputs and outputs and many control loops by formally solving matrix design equations on a digital computer.

Another important factor is the following. Although the state feedback (1.3.2) involves feedback from all states to all inputs, offering no structure in the control system, the output feedback control law (1.3.7) can be used to design a compensator with a desired dynamical structure, regaining much of the intuition of classical controls design.

Feedback laws like (1.3.2) and (1.3.7) are called *static*, since the control gains are constants, or at most time-varying. An alternative to static feedback is to use a dynamic compensator of the form

$$\dot{z} = Fz + Gy + Eu \qquad (1.3.8)$$
$$u = Hz + Dy.$$

The inputs of this compensator are the system inputs and outputs. This yields a closed loop and is called *dynamic output feedback*. The design problem is to select the matrices F, G, E, H, D for good closed-loop performance. An important result of modern control is that closed-loop stability can be guaranteed by selecting $F = A - LC$ for some matrix L which is computed using a Riccati design equation similar to (1.3.5). The other matrices in (1.3.8) are then easily determined. This design is based on the vital *separation principle* (Chapter 10).

A disadvantage with design using $F = A - LC$ is that then the dynamic compensator has the same number of internal states as the plant. In complicated modern aerospace and power plant applications, this dimension can be very large. Thus, various techniques for *controller reduction* and *reduced-order design* have been developed.

In standard modern control, the system is assumed to be exactly described by the mathematical model (1.3.1). In actuality, however, this model may be only an approximate description of the real plant. Moreover, in practice there can be disturbances acting on the plant, as well as measurement noise in determining $y(t)$.

The LQR using full state feedback has some important robustness properties to such disorders, such as an infinite gain margin, 60 degrees of phase margin, and robustness to some nonlinearities in the control loops (Chapter 10). On the other hand, the LQR using static or dynamic output-feedback design has no guaranteed robustness properties. With the work on robust modern control in the early 1980s, there is now a technique (LQG/LTR, Chapter 10) for designing robust multivariable control systems. LQG/LTR design incorporates rigorous treatments of the effects of modeling uncertainties on closed-loop stability, and of disturbance effects on closed-loop performance.

With the work on robust modern design, much of the intuition of classical controls techniques can now be incorporated in modern multivariable design.

With modern developments in *digital control theory* and *discrete-time systems*, modern control is very suitable for the design of control systems that can be implemented on microprocessors (Part III of the book). This allows the implementation of controller dynamics that are more complicated as well as more effective than the simple PID and lead-lag structures of classical control.

With recent work in *matrix-fraction descriptions* and *polynomial equation design*, a MIMO plant can be described not in state-space form, but in input/output form. This is a direct extension of the classical transfer function description and, for some applications, is more suitable than the internal description (1.3.1).

REFERENCES FOR CHAPTER 1

AIRY, G. B., "On the Regulator of the Clock-Work for Effecting Uniform Movement of Equatorials," *Memoirs of the Royal Astronomical Society*, vol. 11, pp. 249–267, 1840.

ÅSTRÖM, K. J., *Introduction to Stochastic Control Theory*, New York: Academic Press, 1970.

ÅSTRÖM, K. J., and B. WITTENMARK, *Computer-Controlled Systems: Theory and Design*, Englewood Cliffs, N.J.: Prentice Hall, 1984.

BELLMAN, R., *Dynamic Programming*, Princeton, N.J.: Princeton University Press, 1957.

BERTALANFFY, L. VON, "A Quantitative Theory of Organic Growth," *Human Biology*, vol. 10, pp. 181–213, 1938.

BLACK, H. S., "Stabilized Feedback Amplifiers," *Bell Syst. Tech. J.*, 1934.

BODE, H. W., "Feedback Amplifier Design," *Bell System Tech. J.*, vol. 19, p. 42, 1940.

BOKHARAIE, M., *A Summary of the History of Control Theory*, Internal Rept., School of Elect. Eng., Ga. Inst. of Technology, Atlanta, GA 30332, 1973.

BROWN, G. S., and D. P. CAMPBELL, *Principles of Servomechanisms*, New York: Wiley, 1948.

CHESTNUT, H., and R. W. MAYER, *Servomechanisms and Regulating System Design*, vol. 1, 1951; vol. 2, 1955, New York: Wiley.

DESOER, C. A., "A Generalization of the Popov Criterion," *IEEE Trans. Autom. Control*, vol. AC-10, no. 2, pp. 182–185, 1965.

DORATO, P., "A Historical Review of Robust Control," *IEEE Control Systems Magazine*, pp. 44–47, April 1987.

DOYLE, J. C., and G. STEIN, "Multivariable Feedback Design: Concepts for a Classical/Modern Synthesis," *IEEE Trans. Automat. Contr.*, vol. AC-26, pp. 4–16, Feb. 1981.

EVANS, W. R., "Graphical Analysis of Control Systems," *Trans. AIEE*, vol. 67, pp. 547–551, 1948.

FRIEDLAND, B., *Control System Design: An Introduction to State-Space Methods*, New York: McGraw-Hill, 1986.

FULLER, A. T., "The Early Development of Control Theory," *Trans. ASME (J. Dynamic Systems, Measurement, & Control)*, vol. 98G, no. 2, pp. 109–118, June 1976.

FULLER, A. T., "The Early Development of Control Theory II," *Trans. ASME (J. Dynamic Systems, Measurement & Control)*, vol. 98G, no. 3, pp. 224–235, September 1976.

GELB, A., ed., *Applied Optimal Estimation*, Cambridge: MIT Press, 1974.

HALL, A. C., "Application of Circuit Theory to the Design of Servomechanisms," *J. Franklin Inst.*, 1946.

HÁZEN, H. L., "Theory of Servo-mechanisms," *J. Franklin Inst.*, 1934.

HURWITZ, A., "On the Conditions Under Which an Equation Has Only Roots With Negative Real Parts," *Mathematische Annalen*, vol. 46, pp. 273–284, 1895.

JAMES, H. M., N. B. NICHOLS, and R. S. PHILLIPS, *Theory of Servomechanisms*, New York: McGraw-Hill, MIT Radiation Lab. Series, Vol. 25, 1947.

JURY, E. I., "Recent Advances in the Field of Sampled-Data and Digital Control Systems," *Proc. Conf. Int. Federation Automat. Control*, pp. 240–246, Moscow, 1960.

KALMAN, R. E., "Contributions to the Theory of Optimal Control," *Bol. Soc. Mat. Mexicana*, vol. 5, pp. 102–119, 1960a.

KALMAN, R. E., "A New Approach to Linear Filtering and Prediction Problems," *ASME J. Basic Eng.*, vol. 82, pp. 34–45, 1960b.

KALMAN, R. E., and R. S. BUCY, "New Results in Linear Filtering and Prediction Theory," *ASME J. Basic Eng.*, vol. 80, pp. 193–196, 1961.

KALMAN, R. E., and J. E. BERTRAM, "Control System Analysis and Design via the 'Second

Method' of Lyapunov. I. Continuous-time Systems," *Trans. ASME J. Basic Eng.*, pp. 371–393, June 1960.

KOLMOGOROV, A. N., "Interpolation and Extrapolation," *Bull. Acad. Sci. USSR*, Ser. Math. vol. 5, pp. 3–14, 1941.

KUHN, T. S., *The Structure of Scientific Revolutions*, Chicago: University of Chicago Press, 1962.

KUO, BENJAMIN C., *Analysis and Synthesis of Sampled-Data Control Systems*, Englewood Cliffs, N.J.: Prentice Hall, 1963.

LAUER, H., R. N. LESNICK, and L. E. MATDON, *Servomechanism Fundamentals*, New York: McGraw-Hill, 1947.

LYAPUNOV, M. A., "Problème Général de la Stabilité du Mouvement," *Ann. Fac. Sci. Toulouse*, vol. 9, pp. 203–474, 1907. (Translation of the original paper published in 1892 in *Comm. Soc. Math. Kharkow* and reprinted as Vol. 17 in *Ann. Math Studies*, Princeton University Press, Princeton, N.J., 1949.)

MACCOLL, L. A., *Fundamental Theory of Servomechanisms*, New York: Van Nostrand, 1945.

MACFARLANE, A. G. J., and I. POSTLETHWAITE, "The Generalized Nyquist Stability Criterion and Multivariable Root Loci," *Int. J. Contr.*, vol. 25, pp. 81–127, 1977.

MAXWELL, J. C., "On Governors," *Proc. Royal Soc. London*, vol. 16, pp. 270–283, 1868.

MAYR, O., *The Origins of Feedback Control*, Cambridge: MIT Press, 1970.

MINORSKY, N., "Directional Stability and Automatically Steered Bodies," *J. Am. Soc. Nav. Eng.*, vol. 34, p. 280, 1922.

NARENDRA, K. S., and R. M. GOLDWYN, "A Geometrical Criterion for the Stability of Certain Nonlinear Nonautonomous Systems," *IEEE Trans. Circuit Theory*, vol. CT-11, no. 3, pp. 406–407, 1964.

NYQUIST, H., "Regeneration Theory," *Bell Syst. Tech. J.*, 1932.

PONTRYAGIN, L. S., V. G. BOLTYANSKY, R. V. GAMKRELIDZE, and E. F. MISHCHENKO, *The Mathematical Theory of Optimal Processes*, New York: Wiley, 1962.

POPOV, V. M., "Absolute Stability of Nonlinear Systems of Automatic Control," *Automat. Remote Control*, vol. 22, no. 8, pp. 857–875, 1961.

RAGAZZINI, J. R., and G. F. FRANKLIN, *Sampled-Data Control Systems*, New York: McGraw-Hill, 1958.

RAGAZZINI, J. R., and L. A. ZADEH, "The Analysis of Sampled-Data Systems," *Trans. AIEE*, vol. 71, part II, pp. 225–234, 1952.

ROSENBROCK, H. H., *Computer-Aided Control System Design*, New York: Academic Press, 1974.

ROUTH, E. J., *A Treatise on the Stability of a Given State of Motion*, London: Macmillan & Co., 1877.

SAFONOV, M. G., A. J. LAUB, and G. L. HARTMANN, "Feedback Properties of Multivariable Systems: The Role and Use of the Return Difference Matrix," *IEEE Trans. Auto. Cont.*, vol. 26, no. 1, pp. 47–65, 1981.

SANDBERG, I. W., "A Frequency-Domain Condition for the Stability of Feedback Systems Containing a Single Time-Varying Nonlinear Element," *Bell Syst. Tech. J.*, vol. 43, no. 4, pp. 1601–1608, 1964.

TRUXAL, J. G., *Automatic Feedback Control System Synthesis*, New York: McGraw-Hill, 1955.

VYSHNEGRADSKY, I. A., "On Controllers of Direct Action," *Izv. SPB Tekhnolog. Inst.*, 1877.

WHITEHEAD, A. N., *Science and the Modern World*, Lowell Lectures (1925), New York: Macmillan, 1953.

WIENER, N., *The Extrapolation, Interpolation, and Smoothing of Stationary Time Series with Engineering Applications*, New York: Wiley, 1949.

WIENER, N., *Cybernetics: Or Control and Communication in the Animal and the Machine*, Cambridge: MIT Press, 1948.

ZAMES, G., "On the Input-Output Stability of Time-Varying Nonlinear Feedback Systems, Part I: Conditions Derived Using Concepts of Loop Gain, Conicity, and Positivity," *IEEE Trans. Automatic Control*, vol. AC-11, no. 2, pp. 228–238, 1966.

ZAMES, G., "On the Input-Output Stability of Time-Varying Nonlinear Feedback Systems, Part II: Conditions Involving Circles in the Frequency Plane and Sector Nonlinearities," *IEEE Trans. Automatic Control*, vol. AC-11, no. 3, pp. 465–476, 1966.

2

Review of Linear State-Variable Systems

SUMMARY

In this chapter we lay the foundations for modern control by reviewing the linear state-variable formulation. The state-space model provides a powerful technique for describing the dynamics of complicated multi-input/multi-output systems. It is the basis for design of control systems and filters using modern theory. The basics of feedback control using state-space models are covered.

INTRODUCTION

In the early 1960s, R. Kalman introduced linear algebra and the state-variable model into system theory. He showed the power of this model in the control problem [1960a] and the filtering or estimation problem [1960b, 1961]. Sophisticated modern-day systems such as aircraft, satellites, fuel-injected automobiles, and so on are driven by many inputs using measurements from many outputs. The state space is well suited to applications for such multi-input/multi-output systems.

The state-space description is straightforward to derive using either the Lagrangian or the Hamiltonian of a physical system, both of which are found from the kinetic and potential energy. Once this description is available, modern techniques allow the design of a variety of control schemes for many applications. The designs are often based on solving *matrix equations*, for which many commercial packages are available (e.g., ORACLS [Armstrong 1980], [IMSL], MATLAB [Moler et al. 1987], MATRIX$_x$ [1989]). In classical control theory, multivariable systems are treated by closing one control loop at a time. By contrast, in modern control theory the gains are calculated from the matrix design equations so that all the control loops are closed simultaneously.

A major development in the past years has been the small, fast, digital computer. Computers are needed to solve the matrix design equations, as well as to implement modern control systems, which may consist of involved structures with time-varying gains. Thus, the marriage between computers and the state-variable model is a natural one.

In this chapter we review the theory of state-variable systems, covering their solution, computer simulation, properties, and special forms. An introduction to closed-loop control using state feedback and output feedback is given. Both continuous-time and discrete-time systems are treated. We restrict the discussion to linear time-invariant systems.

It would be very useful at this point to examine Appendix B, "Review of Matrix Algebra."

2.1 CONTINUOUS-TIME SYSTEMS

The linear state-variable model

$$\dot{x} = Ax + Bu, \qquad (2.1.1)$$

with dot denoting the time derivative, appears innocuous enough, yet it has opened doors that make controls design for multi-input/multi-output (MIMO) systems straightforward.

The time function $x(t)$ is the *system state*. The state describes where energy is stored in a system. For instance, in a block diagram, the states are the outputs of the integrators. In a circuit, the states might be the capacitor voltages and the inductor currents. The time function $u(t)$ is the *control input*.

The state model is a first-order differential equation. The departure from classical system descriptions lies in the fact that $x(t)$ and $u(t)$ are *vectors*, not scalars. We denote the number of components of x as n, and write $x(t) \in \mathbf{R}^n$ (which denotes the real n-vectors); we denote the number of components $u(t)$ as m, and write $u(t) \in \mathbf{R}^m$. This means that A is an $n \times n$ matrix and B is an $n \times m$ matrix. We call A the *system or plant matrix* and B the *control input matrix*. If A and B are constant, we call the system *time-invariant*.

To solve the first-order vector differential equation (2.1.1) it is necessary to know the control input $u(t)$ as well as the *initial condition* $x(0)$. Kalman's definition of the system state was: the additional information needed at time $t = 0$ which, together with the input $u(t)$, uniquely specifies the trajectory $x(t)$ for $t > 0$.

The *state equation* (2.1.1) is derived from the dynamics of the system using, for instance, Lagrange's or Hamilton's equations of motion [Marion 1965]. See D'Azzo and Houpis [1988], McClamroch [1980], Kailath [1980], and Brogan [1974] for good expositions on deriving state equations.

The *output equation*

$$y = Cx + Du \qquad (2.1.2)$$

describes how the measured output $y(t)$ is derived from the states $x(t)$ and the inputs $u(t)$. The outputs can be selected to a certain extent by the designer, depending on what sensors are available. The number of components of the output vector will be denoted by p, so

2.1 Continuous-Time Systems

that $y(t) \in \mathbf{R}^p$. Matrix C is called the *output or measurement matrix*. Matrix D is the *direct feed* matrix.

We shall denote the state-variable model as (A, B, C, D). This is a convenient *matrix representation* of a multivariable system with m inputs and p outputs. A block diagram of the state-variable model is shown in Fig. 2.1-1. Note that there are really n integrators, since this is a vector block diagram. The output of the integrators are the components of the state vector $x(t)$.

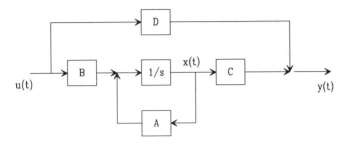

Figure 2.1-1 Block diagram of linear state-variable model

Some examples of state-space systems are now given.

Example 2.1-1: State Model for a Pendulum

This example, though simple, shows all the steps in deriving the state-space model for physical systems using Lagrange's equations of motion. Fig. 2.1-2 shows a pendulum of length ℓ with a mass of m kg at its end.

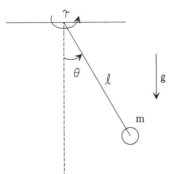

Figure 2.1-2 Pendulum

a. Lagrange's Equation of Motion

The kinetic energy of the system is

$$K = \frac{1}{2} I \dot{\theta}^2 = \frac{1}{2} m \ell^2 \dot{\theta}^2, \tag{1}$$

where we have used the fact that the moment of inertia I of a mass m at a distance ℓ from the center of rotation is $m\ell^2$. The potential energy is

$$P = -mg\ell \cos \theta, \tag{2}$$

whence the *Lagrangian* is given by

$$L = K - P = \frac{1}{2} m\ell^2 \dot{\theta}^2 + mg\ell \cos \theta. \tag{3}$$

According to Lagrange's equation of motion,

$$\frac{d}{dt}\frac{\partial L}{\partial \dot{\theta}} - \frac{\partial L}{\partial \theta} = \tau, \tag{4}$$

with τ the torque input to the pivot point. Computing the quantities needed for Lagrange's equation we obtain

$$\frac{\partial L}{\partial \dot{\theta}} = m\ell^2 \dot{\theta}$$

$$\frac{d}{dt}\frac{\partial L}{\partial \dot{\theta}} = m\ell^2 \ddot{\theta}$$

$$\frac{\partial L}{\partial \theta} = -mg\ell \sin \theta,$$

so that (4) yields

$$m\ell^2 \ddot{\theta} + mg\ell \sin \theta = \tau,$$

or

$$\ddot{\theta} + \frac{g}{\ell} \sin \theta = \frac{\tau}{m\ell^2} \tag{5}$$

with $\tau/m\ell^2$ a normalized torque input.

b. Nonlinear State Equation

Defining the state as

$$x = [\theta \quad \dot{\theta}]^T \tag{6}$$

we may write (5) as

$$\frac{d}{dt}\begin{bmatrix} \theta \\ \dot{\theta} \end{bmatrix} = \begin{bmatrix} \dot{\theta} \\ -(g/\ell) \sin \theta \end{bmatrix} + \begin{bmatrix} 0 \\ \tau/m\ell^2 \end{bmatrix} \tag{7}$$

This is of the form

$$\dot{x} = f(x, u, t), \tag{8}$$

with $u \equiv \tau/m\ell^2$ the control input. This is a *nonlinear state-space equation*.

c. Linearization

To find a linear state equation that is accurate for small angles θ, we may set $\sin \theta \approx \theta$ to obtain the approximation

$$\frac{d}{dt}\begin{bmatrix} \theta \\ \dot{\theta} \end{bmatrix} = \begin{bmatrix} \dot{\theta} \\ -(g/\ell) \theta \end{bmatrix} + \begin{bmatrix} 0 \\ \tau/m\ell^2 \end{bmatrix} \tag{9}$$

2.1 Continuous-Time Systems

This may be more conveniently written as

$$\dot{x} = \begin{bmatrix} 0 & 1 \\ -g/\ell & 0 \end{bmatrix} x + \begin{bmatrix} 0 \\ 1 \end{bmatrix} u \equiv Ax + Bu, \qquad (10)$$

which is the linear state-space model of a pendulum for small θ.

d. Measurements

Placing a potentiometer at the pivot point allows us to measure the angle, so that the output equation is

$$y = \theta(t) = [c_1 \quad 0]x \equiv Cx. \qquad (11)$$

The constant c_1 depends on the voltage applied across the potentiometer and the density of its windings.

∎

Example 2.1-2: State Model for a Circuit

Though this example is simple, it shows all the steps involved in writing a state-space model for an RLC circuit using Kirchhoff's laws. The circuit is shown in Fig. 2.1-3, where $u(t)$ is a voltage source.

a. Kirchhoff's Laws

A voltage balance around the left-hand loop yields

$$u = i_1 R_1 + v_2 \qquad (1)$$

and a current balance at the central node shows that

$$i_1 = i_2 + i_3 = C\dot{v}_2 + i_3. \qquad (2)$$

A voltage balance in the right-hand loop shows that

$$v_2 = i_3 R_2 + L\dot{i}_3. \qquad (3)$$

b. State Equation

The state equation has the state derivatives on the left and only the states and inputs on the right. For a circuit, we generally select the state components as the capacitor voltages and the inductor currents. Therefore

$$x = [v_2 \quad i_3]^\mathrm{T}. \qquad (4)$$

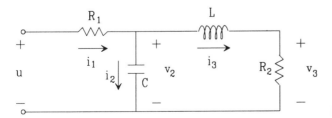

Figure 2.1-3 Electronic circuit

According to (1) and (2)

$$i_1 = \frac{u - v_2}{R_1} = C\dot{v}_2 + i_3,$$

which is rearranged to give

$$\dot{v}_2 = -\frac{1}{R_1 C} v_2 - \frac{1}{C} i_3 + \frac{1}{R_1 C} u. \qquad (5)$$

Rewriting (3) yields

$$\dot{i}_3 = \frac{1}{L} v_2 - \frac{R_2}{L} i_3. \qquad (6)$$

Defining two time constants

$$\tau_C \equiv \frac{1}{R_1 C} \qquad \tau_L \equiv \frac{R_2}{L}, \qquad (7)$$

we may write the state-space equations

$$\frac{d}{dt}\begin{bmatrix} v_2 \\ i_3 \end{bmatrix} = \begin{bmatrix} -\tau_C & -1/C \\ 1/L & -\tau_L \end{bmatrix}\begin{bmatrix} v_2 \\ i_3 \end{bmatrix} + \begin{bmatrix} \tau_C \\ 0 \end{bmatrix} u = Ax + Bu. \qquad (8)$$

c. Measurements

If we want to measure the voltage v_3 across R_2, then

$$v_3 = i_3 R_2,$$

so that the output equation is

$$y \equiv v_3 = [0 \quad R_2] x = Cx. \qquad (9)$$

■

Example 2.1-3: DC Motor with Compliant Coupling

An important practical system is a motor coupled to its load through a shaft that has significant compliance. This system appears in disk drives, antenna and radar pointing, robotics, and elsewhere.

Assuming an armature-coupled DC motor, the electric and mechanical subsystems appear in Fig. 2.1-4. The compliant shaft is modeled as a rotary spring with spring constant k and damping b.

a. Electrical Equation

Refer to the motor armature circuit shown in Fig. 2.1-4a, where R is the armature resistance, L is the armature winding leakage inductance, and $i(t)$ is the armature current. In the case of an armature-controlled DC motor, the field current is held constant and the control input is the armature current $u(t)$ [Franklin, Powell, and Emami-Naeini 1986; de Silva 1989; McClamroch 1980]. Then, the back e.m.f. v_b generated by the motion is given by

$$v_b = k'_m \dot{\theta}_m, \qquad (1)$$

with $\theta_m(t)$ the motor angular position and k'_m the back e.m.f. constant.

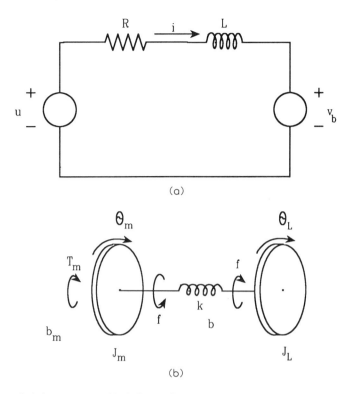

Figure 2.1-4 DC motor with shaft compliance. (a) Electrical subsystem. (b) Mechanical subsystem.

Using standard circuit theory, the electrical equation may be written as

$$L\dot{i} = -Ri - k'_m \dot{\theta}_m + u. \qquad (2)$$

b. Mechanical Equations

Refer to Fig. 2.1-4b, where $\theta_L(t)$ is the angular position of the load, T_m is the torque provided by the motor, b_m is the rotor equivalent damping constant, and J_m (resp. J_L) is the moment of inertia of the rotor (resp. load).

The interaction force exerted by the compliant shaft is given by

$$f = b(\dot{\theta}_m - \dot{\theta}_L) + k(\theta_m - \theta_L). \qquad (3)$$

For an armature-controlled DC motor, the torque provided is equal to

$$T_m = k_m i, \qquad (4)$$

with k_m the motor torque constant.

Using Newton's laws for angular motion, the mechanical equations of motion may be written as

$$J_m \ddot{\theta}_m + b_m \dot{\theta}_m = T_m - b(\dot{\theta}_m - \dot{\theta}_L) - k(\theta_m - \theta_L) \qquad (5)$$

$$J_L \ddot{\theta}_L = b(\dot{\theta}_m - \dot{\theta}_L) + k(\theta_m - \theta_L), \qquad (6)$$

or

$$J_m \ddot{\theta}_m + b_m \dot{\theta}_m + b(\dot{\theta}_m - \dot{\theta}_L) + k(\theta_m - \theta_L) = k_m i \qquad (7)$$

$$J_L \ddot{\theta}_L - b(\dot{\theta}_m - \dot{\theta}_L) - k(\theta_m - \theta_L) = 0. \qquad (8)$$

c. State Equations

The complete dynamical equations for the armature-controlled DC motor with compliant coupling are (2), (7), (8). To place them into state-space form, define the state as

$$x = [i \quad \theta_m \quad \omega_m \quad \theta_L \quad \omega_L]^T, \qquad (9)$$

with $\omega_m = \dot{\theta}_m$ and $\omega_L = \dot{\theta}_L$ the motor and load angular velocities. Then

$$\dot{x} = \begin{bmatrix} \dfrac{-R}{L} & 0 & \dfrac{-k'_m}{L} & 0 & 0 \\ 0 & 0 & 1 & 0 & 0 \\ \dfrac{k_m}{J_m} & \dfrac{-k}{J_m} & \dfrac{-(b+b_m)}{J_m} & \dfrac{k}{J_m} & \dfrac{b}{J_m} \\ 0 & 0 & 0 & 0 & 1 \\ 0 & \dfrac{k}{J_L} & \dfrac{b}{J_L} & \dfrac{-k}{J_L} & \dfrac{-b}{J_L} \end{bmatrix} x + \begin{bmatrix} \dfrac{1}{L} \\ 0 \\ 0 \\ 0 \\ 0 \end{bmatrix} u. \qquad (10)$$

d. Noncompliant Coupling Shaft

Adding (7) and (8) yields

$$J_m \ddot{\theta}_m + J_L \ddot{\theta}_L + b_m \dot{\theta}_m = k_m i. \qquad (11)$$

In the case that the coupling shaft is rigid, we have $k = \infty$ and $\theta_L = \theta_m$. Therefore

$$(J_m + J_L) \ddot{\theta}_m + b_m \dot{\theta}_m = k_m i. \qquad (12)$$

The dynamical equations of motion are now (2) and (12).

Defining the total moment of inertia as $J = J_m + J_L$ and the state as

$$x = [i \quad \omega_m]^T \qquad (13)$$

results in the state-space model for noncompliant coupling given by

$$\dot{x} = \begin{bmatrix} -R/L & -k'_m/L \\ k_m/J & -b_m/J \end{bmatrix} x + \begin{bmatrix} 1/L \\ 0 \end{bmatrix} u. \qquad (14)$$

∎

Frequency-Domain Solution

We shall derive frequency-domain and time-domain solutions for the state equation. The results to be derived are collected for easy reference in Table 2.1-1. Examples are presented at the end of the derivations.

2.1 Continuous-Time Systems

We assume that the plant matrix A, control matrix B, and output matrix C are constant. Taking the one-sided Laplace transform of (2.1.1) yields

$$sX(s) - x_0 = AX(s) + BU(s),$$

with $x(0) = x_0$ the known initial condition. Letting I represent the $n \times n$ identity matrix (e.g., $I = \text{diag}\{1\}$, a matrix of zeros with ones on the diagonal), we may write

$$sIX(s) - AX(s) = x_0 + BU(s)$$

$$(sI - A)X(s) = x_0 + BU(s)$$

$$X(s) = (sI - A)^{-1}x_0 + (sI - A)^{-1}BU(s). \qquad (2.1.3)$$

Now, output equation (2.1.2) yields

$$Y(s) = C(sI - A)^{-1}x_0 + [C(sI - A)^{-1}B + D]U(s). \qquad (2.1.4)$$

The $n \times n$ matrix $(sI - A)^{-1}$ may be expanded (using long division of $sI - A$ into I, for instance) into the Laurent series about infinity

$$(sI - A)^{-1} = s^{-1}I + s^{-2}A + s^{-3}A^2 + s^{-4}A^3 + \ldots$$

$$= s^{-1}\sum_{i=0}^{\infty} A^i s^{-i}. \qquad (2.1.5)$$

To verify this expression, one may multiply the infinite series by $(sI - A)$ to obtain the

---

TABLE 2.1-1 SOLUTION OF THE STATE EQUATION

State Equation:

$\dot{x} = Ax + Bu$

$y = Cx + Du$

Frequency-Domain Solution:

$X(s) = (sI - A)^{-1}x_0 + (sI - A)^{-1}BU(s)$

$Y(s) = C(sI - A)^{-1}x_0 + H(s)U(s)$

where transfer function is $H(s) = C(sI - A)^{-1}B + D$.

System poles are given by the characteristic polynomial $\Delta(S) = |sI - A|$.

Time-Domain Solution:

$$x(t) = e^{A(t-t_0)}x(t_0) + \int_{t_0}^{t} e^{A(t-\tau)}Bu(\tau)\,d\tau$$

$$y(t) = Ce^{A(t-t_0)}x(t_0) + \int_{t_0}^{t} h(t - \tau)u(\tau)\,d\tau$$

where impulse response is $h(t) = Ce^{At}B + D\delta(t)$

identity. The matrix $(sI - A)^{-1}$ is of sufficient importance to grace it with a name, the *resolvent matrix*, as well as a symbol, $\Phi(s)$. Then, we may write

$$X(s) = \Phi(s)x_0 + \Phi(s)BU(s) \tag{2.1.6}$$

$$Y(s) = C\Phi(s)x_0 + [C\Phi(s)B + D]U(s). \tag{2.1.7}$$

These two equations constitute the frequency-domain solution of the state equation. Note that they each consist of two parts—one part depending on the initial conditions $x(0) = x_0$ and one (the forced solution) depending on the control input $u(t)$. Setting $u(t) = 0$ we obtain the initial-condition term, which we therefore call the *zero-input response*. Likewise, the forced solution is called the *zero (initial)-state response*.

Having in mind the classical definition of the system *transfer function*

$$Y(s) = H(s)U(s), \text{ when } x_0 = 0, \tag{2.1.8}$$

we may compare (2.1.7) and (2.1.8) to express the transfer function in terms of A, B, C, and D as

$$H(s) = C(sI - A)^{-1}B + D = C\Phi(s)B + D. \tag{2.1.9}$$

This allows us to compute $H(s)$ given the system (A, B, C, D). It also allows us to write for (2.1.7)

$$Y(s) = C\Phi(s)x_0 + H(s)U(s), \tag{2.1.10}$$

so that, if the initial conditions are not zero, the output is equal to the transfer function term plus an initial condition term.

The resolvent matrix is given by

$$\Phi(s) = (sI - A)^{-1} = \frac{\text{adj}(sI - A)}{|sI - A|}, \tag{2.1.11}$$

with $|sI - A|$ the *determinant* of the matrix $sI - A$ and $\text{adj}(sI - A)$ its *adjoint*, which may be computed by standard means such as Laplace expansion [Gantmacher 1977]. Therefore

$$H(s) = \frac{C\{\text{adj}(sI - A)\}B}{|sI - A|} + D = \frac{C\{\text{adj}(sI - A)\}B + D|sI - A|}{|sI - A|}. \tag{2.1.12}$$

Denote the scalar polynomial $|sI - A|$ by $\Delta(s)$. Recalling now that the system poles are classically defined as the roots of the denominator of the transfer function, let us define the poles of the multivariable system (A, B, C, D) as the roots of the equation

$$\Delta(s) = |sI - A| = 0. \tag{2.1.13}$$

We call $\Delta(s)$ the *characteristic polynomial* and (2.1.13) the *characteristic equation*.

The zeros of a multivariable system are not so easy to define, since $H(s)$ is a $p \times m$ polynomial matrix. The zeros of the individual entries of $H(s)$ mean little from the point

2.1 Continuous-Time Systems

of view of multivariable control. If $p = m$ so that $H(s)$ is square, we may define the zeros as the roots of the polynomial equation

$$\det[C\{\mathrm{adj}(sI - A)\}B + D|sI - A|] = 0. \tag{2.1.14}$$

This means we should set the determinant of the numerator of the transfer function equal to zero and solve for s. If $p \neq m$ the zeros occur where $H(s)$ loses rank.

Time-Domain Solution

Now we express the solutions to the state equation in the time domain. The results to be derived are collected in Table 2.1-1 for reference purposes.

Exactly as in the scalar case, the *matrix exponential* is defined by

$$e^{At} = I + At + \frac{A^2 t^2}{2!} + \frac{A^3 t^3}{3!} + \ldots \tag{2.1.15}$$

Now, a term-by-term Laplace transform yields nothing but (2.1.5). (Note that, to ensure causality we use the one-sided Laplace transform, so that e^{At} should be multiplied by the unit step prior to finding the transform.) Therefore

$$L(e^{At}) = \Phi(s) = (sI - A)^{-1}, \tag{2.1.16}$$

with $L(\ )$ denoting Laplace transform.

Recalling now that the Laplace transform of a convolved pair of time functions is the product of their Laplace transforms, we may examine (2.1.6) from a new point of view. Taking its inverse Laplace transform yields

$$x(t) = e^{At}x(0) + \int_0^t e^{A(t-\tau)}Bu(\tau)\, d\tau, \tag{2.1.17}$$

where the second term is the convolution of e^{At} and $Bu(t)$. Due to linearity of the system, we may select any initial time t_0 so that, in general

$$x(t) = e^{A(t-t_0)}x(t_0) + \int_{t_0}^t e^{A(t-\tau)}Bu(\tau)\, d\tau. \tag{2.1.18}$$

This is the time-domain solution of the state equation. One might recognize it as identical to the solution of a first-order ordinary differential equation first seen in an undergraduate mathematics course. This equation, however, involves *vectors* $x(t)$, $u(t)$, and *matrices* A, B. Therefore, it is not suitable for computation, for which purpose (2.1.6) should be used. See the examples.

According to the output equation (2.1.2) one may now write the solution for the measured output as

$$y(t) = Ce^{A(t-t_0)}x(t_0) + \int_{t_0}^t [Ce^{A(t-\tau)}B + D\delta(t - \tau)]u(\tau)\, d\tau. \tag{2.1.19}$$

In writing this equation, we have used the *sifting property* of the *Dirac delta function* $\delta(t)$, that is

$$\int_0^\infty \delta(t - \tau)u(\tau)\, d\tau = u(t). \tag{2.1.20}$$

Exactly as in the frequency-domain solution, the time-domain solutions (2.1.18), (2.1.19) each have a zero-input response, depending only on the initial conditions, and a zero-state response depending only on the input $u(t)$.

Let us now draw some correspondences between time-domain and frequency-domain quantities. According to (2.1.16) we are justified in denoting

$$\phi(t) \equiv e^{At}. \tag{2.1.21}$$

Note that $L(\phi(t)) = \Phi(s)$. Setting $u(t) = 0$ in (2.1.18) yields the *homogeneous solution*

$$x(t) = e^{A(t-t_0)}x(t_0), \tag{2.1.22}$$

and reveals that the state propagation from time t_0 to any time t is determined by multiplying by $\phi(t - t_0)$. We therefore call $\phi(t)$ the *state transition matrix*.

When $x(t_0) = 0$ in (2.1.19), the output is seen to be obtained by convolution of the input $u(t)$ with the quantity

$$h(t) \equiv Ce^{At}B + D\delta(t), \tag{2.1.23}$$

which we therefore identify as the *impulse response* of system (A, B, C, D). Note that

$$L(h(t)) = H(s), \tag{2.1.24}$$

that is, the transfer function (2.1.9) is the Laplace transform of the impulse response. It should be clear at this point that we are only generalizing well-known classical notions to the vector case.

Natural Modes and Stability

The transfer function may be written in the *partial-fraction expansion (PFE)* form

$$H(s) = \sum_{i=1}^{n} \frac{K_i}{s - p_i}, \tag{2.1.25}$$

where p_i denotes the poles, which are n in number (in writing the PFE, we have assumed for simplicity that the poles are nonrepeated or *simple*). The numerator coefficient K_i is the *residue* of pole p_i. Note that the K_i are $p \times m$ matrices. The partial-fraction expansion may be found by one of many standard techniques [D'Azzo and Houpis 1988]. Then, the impulse response is found by term-by-term inverse transformation to be

$$h(t) = \sum_{i=1}^{n} K_i e^{p_i t} u_{-1}(t). \tag{2.1.26}$$

2.1 Continuous-Time Systems

The *unit step* $u_{-1}(t)$ has been inserted to correspond to the one-sided Laplace transform, since the systems in this book will be causal. The unit step has a value of zero for $t < 0$ and one for $t \geq 0$.

Internal stability

From (2.1.26) it is evident that with each pole p_i there is associated a time function of the form $e^{p_i t}$. These time functions are called the *natural modes* of the system. We say a natural mode is (asymptotically) stable if it decays to zero with time.

The system poles are the roots of the characteristic equation $\Delta(s) = |sI - A| = 0$. A pole is said to be stable if it has a stable natural mode. Write the natural mode as

$$e^{p_i t} = e^{\alpha t} e^{j\beta t}, \qquad (2.1.27)$$

with $p_i = \alpha + j\beta$. Then, $e^{\alpha t}$ is a real part describing the exponential decay or growth of the time function, while $e^{j\beta t}$ is a complex part with a magnitude of one. It is apparent that for stability α must be negative. That is, a stable pole has a negative real part, and so is in the *(open) left-half of the s-plane*. The terminology "open" means that the pole cannot be on the $j\omega$-axis.

A natural mode is said to be *marginally stable* if it does not increase without bound as t increases. Any stable function is marginally stable by definition. Other examples of marginally stable functions include the unit step and the sine function. The unit step $u_{-1}(t)$ has a Laplace transform of

$$U(s) = \frac{1}{s} \qquad (2.1.28)$$

and $u(t) = \sin \beta t$, $t \geq 0$, has a transform of

$$U(s) = \frac{\beta}{s^2 + \beta^2} = \frac{\beta}{(s + j\beta)(s - j\beta)}. \qquad (2.1.29)$$

Both of these have nonrepeated (or simple) poles on the $j\omega$-axis. In fact, a pole is marginally stable (i.e., gives rise to a marginally stable natural mode) if and only if it is either a nonrepeated $j\omega$-axis pole or in the open left-half plane.

The unit ramp $u(t) = t u_{-1}(t)$ has a transform of

$$U(s) = \frac{1}{s^2}, \qquad (2.1.30)$$

and $u(t) = t \sin \beta t$, $t \geq 0$, has a transform of

$$U(s) = \frac{2\beta s}{(s^2 + \beta^2)^2} = \frac{2\beta s}{(s + j\beta)^2 (s - j\beta)^2}. \qquad (2.1.31)$$

Both of these have repeated poles on the $j\omega$-axis and neither is marginally stable. Time functions that are unbounded as time increases are said to be *unstable*. Unstable poles are those that have associated unstable natural modes; poles in the right-half plane and repeated poles on the $j\omega$-axis are unstable.

A system is (asymptotically) stable if all its poles are in the open left-half plane. It is marginally stable if all the poles are in the open left-half plane, except possibly for some nonrepeated poles on the $j\omega$-axis. Otherwise, it is unstable.

External stability

Let us now discuss stability from an external point of view. Assume the initial state $x(0)$ is zero. The input/output relation of a linear time-invariant system is given by $Y(s) = H(s)U(s)$, with $H(s)$ the transfer function, or, in the time domain

$$y(t) = h(t)*u(t), \qquad (2.1.32)$$

with * denoting convolution.

A system is said to be *bounded-input/bounded-output (BIBO)* stable if bounded inputs give rise only to bounded outputs. If $u(t)$ in (2.1.32) is bounded, then $y(t)$ is bounded if and only if the impulse response $h(t)$ decays to zero with t. Thus, a system is BIBO stable if and only if all the poles of the transfer function $H(s)$ are in the open left-half plane.

The transfer function $H(s)$ is given in terms of (A, B, C, D) by (2.1.9), (2.1.12). Clearly, if all the system poles (i.e., the roots of $|sI - A| = 0$) are stable, then $H(s)$ has all its poles in the left-half plane. However, it may be that some cancellation occurs in (2.1.12). Thus, there may be system poles that are canceled by zeros and so do not appear in the transfer function after simplification. The modes associated with such poles can be excited only by the initial condition $x(0)$, and not by the input $u(t)$. Since one can imagine that canceling poles may be unstable, BIBO stability does not imply asymptotic stability. That is, even if the denominator of $H(s)$ is stable, $\Delta(s)$ may have unstable factors if pole-zero cancellation occurs in (2.1.12).

Example 2.1-4: Systems Obeying Newton's Law

Some of the most ubiquitous systems are those obeying Newton's law of motion $F = ma$. Defining the state

$$x = \begin{bmatrix} d \\ v \end{bmatrix} \qquad (1)$$

with $d(t)$ the position and $v(t)$ the velocity, we see that $\dot{d} = v$ and $\dot{v} = a = F/m$. Therefore, Newton's law may be written in state-space form as

$$\dot{x} = \begin{bmatrix} 0 & 1 \\ 0 & 0 \end{bmatrix} x + \begin{bmatrix} 0 \\ 1 \end{bmatrix} u \equiv Ax + Bu, \qquad (2)$$

with $u(t) \equiv F(t)/m$ an input driving acceleration. Assuming that position measurements are taken, we have $y(t) = d(t)$ or

$$y = \begin{bmatrix} 1 & 0 \end{bmatrix} x \equiv Cx. \qquad (3)$$

2.1 Continuous-Time Systems

a. Resolvent Matrix

The first step in solving the state equation is to find the resolvent matrix. Thus

$$\Phi(s) = (sI - A)^{-1} = \left[\begin{bmatrix} s & 0 \\ 0 & s \end{bmatrix} - \begin{bmatrix} 0 & 1 \\ 0 & 0 \end{bmatrix}\right]^{-1} = \begin{bmatrix} s & -1 \\ 0 & s \end{bmatrix}^{-1} \quad (4)$$

$$= \frac{1}{s^2}\begin{bmatrix} s & 1 \\ 0 & s \end{bmatrix} = \begin{bmatrix} 1/s & 1/s^2 \\ 0 & 1/s \end{bmatrix}.$$

b. System Poles

The characteristic polynomial is

$$\Delta(s) = |sI - A| = \frac{1}{s^2}. \quad (5)$$

Therefore, the system has two poles at the origin $s = 0$ and is unstable.

The natural mode of one pole at the origin, which has a term like $1/s$, is the unit step, denoted $u_{-1}(t)$. The natural mode of the pole pair at the origin, corresponding to $1/s^2$, is $tu_{-1}(t)$, the unit ramp. Note that, given an initial velocity $v(0)$, the position will increase linearly as $tv(0)u_{-1}(t)$ (exhibiting the ramp mode), while the velocity will remain constant at $v(0)u_{-1}(t)$ (exhibiting the step mode).

c. Transfer Function

The transfer function from driving acceleration to position is given by

$$H(s) = C\Phi(s)B = \frac{1}{s^2}[1 \quad 0]\begin{bmatrix} s & 1 \\ 0 & s \end{bmatrix}\begin{bmatrix} 0 \\ 1 \end{bmatrix} = \frac{1}{s^2}. \quad (6)$$

d. State Transition Matrix

The state transition matrix $\phi(t) = e^{At}$ is given by the inverse Laplace transform of $\Phi(s)$. Standard transform tables yield

$$\phi(t) = \begin{bmatrix} 1 & t \\ 0 & 1 \end{bmatrix}u_{-1}(t). \quad (7)$$

It is important to note that $\phi(t)$ may be written as

$$\phi(t) = \begin{bmatrix} 1 & 0 \\ 0 & 1 \end{bmatrix}u_{-1}(t) + \begin{bmatrix} 0 & 1 \\ 0 & 0 \end{bmatrix}tu_{-1}(t). \quad (8)$$

Indeed, the matrix exponential can always be written in the form

$$\phi(t) = e^{At} = \sum_{i=1}^{n} M_i e^{p_i t}, \quad (9)$$

(assuming nonrepeated poles p_i) with M_i denoting coefficient matrices. That is, e^{At} is always a linear combination of the natural modes of A.

e. Impulse Response

The impulse response $h(t)$ may be computed using

$$h(t) = C\phi(t)B \tag{10}$$

or by inverse Laplace transformation of $H(s)$. Either technique yields

$$h(t) = tu_{-1}(t). \tag{11}$$

Thus, in response to an impulsive input acceleration, the position increases linearly with t.

f. Solution

Let us determine the solution of Newton's system given initial conditions of

$$x(0) = x_0 = [d_0 \quad v_0]^T \tag{12}$$

and a constant input acceleration of a.

According to (2.1.6), we have

$$X(s) = \Phi(s)x_0 + \Phi(s)BU(s). \tag{13}$$

The input $u(t)$ has a constant value of a applied at time $t = 0$, so that

$$u(t) = au_{-1}(t) \tag{14}$$

$$U(s) = a/s. \tag{15}$$

Therefore

$$X(s) = \frac{1}{s^2}\begin{bmatrix} s & 1 \\ 0 & s \end{bmatrix}\begin{bmatrix} d_0 \\ v_0 \end{bmatrix} + \frac{1}{s^2}\begin{bmatrix} s & 1 \\ 0 & s \end{bmatrix}\begin{bmatrix} 0 \\ 1 \end{bmatrix}\frac{a}{s}$$

$$= \begin{bmatrix} d_0/s + v_0/s^2 \\ v_0/s \end{bmatrix} + \begin{bmatrix} a/s^3 \\ a/s^2 \end{bmatrix}. \tag{16}$$

Inverse Laplace transformation yields

$$x(t) = \begin{bmatrix} d(t) \\ v(t) \end{bmatrix} = \begin{bmatrix} d_0 + v_0 t + at^2/2 \\ v_0 + at \end{bmatrix}, \tag{17}$$

which we may recognize as the laws of motion under constant acceleration from our high school days.

∎

Example 2.1-5: Damped Harmonic Oscillator

In this example we shall gain considerable insight on the complex pole pair $s = \alpha \pm j\beta$ that will be used throughout the book.

Consider the state-variable system given by

$$\dot{x} = \begin{bmatrix} 0 & 1 \\ -\omega_n^2 & -2\alpha \end{bmatrix} x + \begin{bmatrix} 0 \\ 1 \end{bmatrix} u \equiv Ax + Bu. \tag{1}$$

Let us measure the first state component so that

$$y = [1 \quad 0]x \equiv Cx. \tag{2}$$

2.1 Continuous-Time Systems

Notice that this model covers the pendulum of Example 2.1-1 if we set $\alpha = 0$ and define $\omega_n^2 = g/\ell$.

a. Poles

The characteristic equation is

$$\Delta(s) = |sI - A| = \begin{vmatrix} s & -1 \\ \omega_n^2 & s + 2\alpha \end{vmatrix} = s^2 + 2\alpha s + \omega_n^2. \quad (3)$$

Thus, if $\omega_n > \alpha$ the poles are at

$$s = -\alpha \pm j\beta, \quad (4)$$

where the imaginary part is

$$\beta = \sqrt{\omega_n^2 - \alpha^2}. \quad (5)$$

Now we recognize the system as as damped harmonic oscillator with natural frequency of ω_n, oscillation frequency of β, and damping ratio of

$$\zeta = \alpha/\omega_n. \quad (6)$$

b. Characteristic Polynomial of the Complex Pole Pair

It is extremely useful to notice that the characteristic polynomial (3) of the complex pole pair can be written in the equivalent forms

$$\Delta(s) = s^2 + 2\alpha s + \omega_n^2 \quad (7)$$

$$= s^2 + 2\zeta\omega_n s + \omega_n^2 \quad (8)$$

$$= (s + \alpha)^2 + \beta^2 \quad (9)$$

$$= [s + (\alpha + j\beta)][s + (\alpha - j\beta)]. \quad (10)$$

All of these represent the polynomial with roots at the complex pair of poles $s = \alpha \pm j\beta$. We shall use these representations throughout the book. Since it is desirable to avoid the appearance of j, the square root of -1, we shall prefer the formulations (7)–(9).

c. Resolvent Matrix

The resolvent matrix is given by

$$\Phi(s) = (sI - A)^{-1} = \begin{bmatrix} s & -1 \\ \omega_n^2 & s + 2\alpha \end{bmatrix}^{-1} = \frac{1}{\Delta(s)} \begin{bmatrix} s + 2\alpha & 1 \\ -\omega_n^2 & s \end{bmatrix}. \quad (11)$$

d. Transfer Function

The transfer function is

$$H(s) = C\Phi(s)B = \frac{1}{s^2 + 2\alpha s + \omega_n^2} = \frac{1}{(s + \alpha)^2 + \beta^2}. \quad (12)$$

e. State Transition Matrix

The state transition matrix e^{At} is the inverse transform of $\Phi(s)$, found from standard Laplace tables to be

$$\phi(t) = \begin{bmatrix} 1 & 0 \\ 0 & 1 \end{bmatrix} e^{-\alpha t} \cos\beta t \, u_{-1}(t) + \begin{bmatrix} \alpha/\beta & 1/\beta \\ -\omega_n^2/\beta & -\alpha/\beta \end{bmatrix} e^{-\alpha t} \sin\beta t \, u_{-1}(t). \quad (13)$$

It is a linear combination of the natural modes.

f. Impulse Response

The impulse response is found by inverse transformation of $H(s)$ to be

$$h(t) = \frac{1}{\beta} e^{-\alpha t} \sin\beta t \, u_{-1}(t). \quad (14)$$

This is shown in Fig. 2.1-5 for the case $\alpha = 1$, $\beta = 3$.

g. Solution

Let us find the solution $y(t)$ given initial conditions of $x(0) = [1 \quad 0]^T$ and input of $u(t) = e^{-\alpha t} u_{-1}(t)$.

Using the given initial condition x_0 and $U(s) = 1/(s + \alpha)$ in (2.1.10) yields

$$Y(s) = \frac{s + 2\alpha}{(s + \alpha)^2 + \beta^2} + \frac{1}{(s + \alpha)^2 + \beta^2} \cdot \frac{1}{s + \alpha}.$$

Now a partial-fraction expansion yields

$$Y(s) = \frac{s + 2\alpha}{(s + \alpha)^2 + \beta^2} + \frac{-(s + \alpha)/\beta^2}{(s + \alpha)^2 + \beta^2} + \frac{1/\beta^2}{s + \alpha}, \quad (15)$$

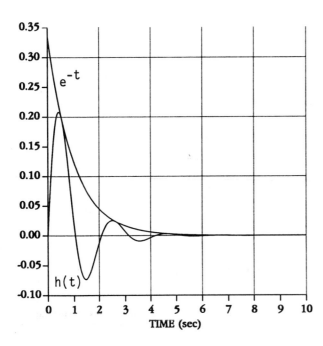

Figure 2.1-5 Impulse response of damped harmonic oscillator

2.1 Continuous-Time Systems

whence inverse Laplace transformation gives the final result

$$y(t) = \frac{(\beta^2 - 1)}{\beta^2} e^{-\alpha t} \cos \beta t + \frac{\alpha}{\beta} e^{-\alpha t} \sin \beta t + \frac{1}{\beta^2} e^{-\alpha t}. \quad (16)$$

It is worth noting that the output of a linear system generally has frequencies corresponding to the natural modes and frequencies corresponding to the input driving terms.

h. System Time Constant

The natural modes of the system with poles at $s = \alpha \pm j\beta$ are $e^{-\alpha t} \cos \beta t$ and $e^{-\alpha t} \sin \beta t$. The *time constant* of an exponential is defined as the quantity τ such that the exponential may be written as $e^{-t/\tau}$. Thus

$$\tau = \frac{1}{\alpha}. \quad (17)$$

The natural behavior of the system is an oscillation at the frequency of β sec^{-1} and a superimposed exponential decay with time constant of τ sec. See Fig. 2.1-5.

The *settling time* t_s is the time it takes the response to settle out to its steady-state value. For exponential responses, we may take it to be

$$t_s = 5\tau, \quad (18)$$

although 4τ is also popular. After t_s sec, the transient exponential portion of the response has, for all practical purposes, vanished. ∎

Computer Simulation of Continuous Systems

We have just seen how to solve continuous-time systems analytically. It is important to be able to *simulate* them on a digital computer to obtain time responses of their trajectories. It is very easy to simulate any system that is in the state-variable formulation. In fact, even systems in the *nonlinear state-variable formulation*

$$\dot{x} = f(x, u, t), \quad (2.1.33)$$

with $x(t)$ the state and $u(t)$ the control input, are easily simulated.

In Appendix A there appears a FORTRAN V driver program called TRESP that contains a Runge-Kutta integrator. This may be used to integrate the dynamics (2.1.33) and obtain time plots of the system trajectories. A general nonlinear output equation is given by

$$y = h(x, u, t); \quad (2.1.34)$$

these output can also be obtained as a function of time using TRESP.

TRESP is written in a modular fashion so that it is only necessary to write a subroutine F(time, x, xp) that computes the state derivative $\dot{x}$ (called xp, or xprime) given the current states $x(t)$ and inputs $u(t)$. The argument "time" is used if the system (2.1.33) is time-varying. Using this subroutine, the Runge-Kutta integrator computes the new state at time $t + T_R$, with T_R the Runge-Kutta step size. The subroutine should also compute the outputs using (2.1.34). The inputs should appear in a common block labeled CONTROL for passing to F(time, x, xp). The outputs should appear in a common block labeled OUPUT for plotting purposes.

The Runge-Kutta integrator calls F(time, x, xp) four times during each Runge-Kutta integration period T_R. During this period, the input $u(t)$ should be held constant. That is, for the numerical integration to give correct results, $u(t)$ should be changed only at the beginning of each integration period T_R; the Runge-Kutta routine then calls F(time, x, xp) four times using the same constant value of $u(t)$.

An alternative for systems simulation is to use commercially available packages such as MATRIX$_X$ [1989], Program CC [Thompson 1985], and SIMNON [Åström and Wittenmark 1984].

Some examples of simulation of systems are now given.

Example 2.1-6: Simulation of van der Pol Oscillator

The equations of the van der Pol oscillator are

$$\dot{x}_1 = x_2$$
$$\dot{x}_2 = -\alpha(x_1^2 - 1)x_2 - x_1. \qquad (1)$$

This is a nonlinear state equation with state $x = [x_1 \ x_2]^T$ and α a parameter. Although nonlinear equations are difficult to solve analytically, they are easy to simulate.

The subroutine f(time, x, xp) needed to simulate (1) using program TRESP in Appendix A appears in Fig. 2.1-6. Note that it is nothing but an exact transcription of (1).

```
C     SUBROUTINE TO SIMULATE VAN DER POL OSCILLATOR
C     FOR USE WITH PROGRAM TRESP

      SUBROUTINE F(t,x,xp)
      REAL x(*), xp(*)
      DATA al/0.8/

      xp(1)= x(2)
      xp(2)= -al*(x(1)**2 - 1)*x(2) - x(1)

      RETURN
      END
```

Figure 2.1-6 Subroutine for simulation of van der Pol oscillator

Using this subroutine, the time response plots shown in Fig. 2.1-7 were obtained with an initial state of $x(0) = (0.1, 0.1)$. Phase-plane portraits of x_2 versus x_1 are shown. Figure 2.1-7a shows the response over 20 sec for $\alpha = 0.1$. Figure 2.1-7b shows the response over 20 sec for $\alpha = 0.8$. The limit cycle is clearly visible. The Runge-Kutta integration period was 0.05 sec.

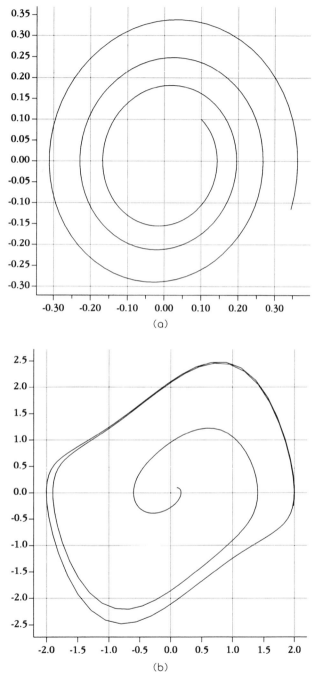

Figure 2.1-7 Phase-plane time portrait for van der Pol oscillator, showing x_2 vs. x_1. (a) $\alpha = 0.1$. (b) $\alpha = 0.8$ showing limit cycle.

Example 2.1-7: Simulation of Damped Harmonic Oscillator

The damped harmonic oscillator of Example 2.1-4 is

$$\dot{x}_1 = x_2 \qquad (1)$$
$$\dot{x}_2 = -\omega_n^2 x_1 - 2\alpha x_2 + \omega_n^2 u$$

with state $x = [x_1 \; x_2]^T$. We have added a control gain of ω_n^2 so that the transfer function becomes

$$H(s) = \frac{\omega_n^2}{s^2 + 2\alpha s + \omega_n^2}. \qquad (2)$$

```
C     SUBROUTINE TO SIMULATE DAMPED HARMONIC OSCILLATOR
C     FOR USE WITH DRIVER PROGRAM TRESP

      SUBROUTINE F(time,x,xp)
      REAL x(*), xp(*)
      COMMON/CONTROL/u
      COMMON/OUTPUT/y
      DATA al, omsq/ 0.5, 16./

      xp(1)=              x(2)
      xp(2)= -omsq*x(1) - 2*al*x(2) + omsq*u
      y    =      x(1)

      RETURN
      END
```

Figure 2.1-8 Subroutine for simulation of damped harmonic oscillator

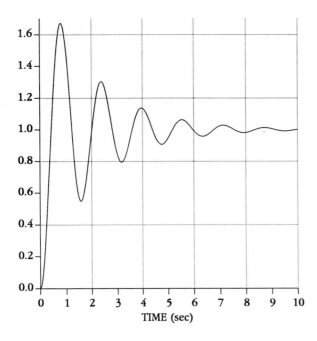

Figure 2.1-9 Step response of damped harmonic oscillator

2.1 Continuous-Time Systems

Setting $s = 0$ it is seen that the DC gain is now 1 for all values of α and ω_n^2, which facilitates scaling a step-response plot.

The subroutine used with TRESP for simulation purposes is shown in Fig. 2.1-8. Note that it contains a duplicate of the state equations. We selected $\alpha = 0.5$, $\omega_n = 4$ and set $x(0) = 0$, $u(t) = 1$ to find the step response. The simulation results appear in Fig. 2.1.9. The Runge-Kutta integration period was 0.01 sec.

∎

Example 2.1-8: Simulation of DC Motor with Compliant Coupling

In Example 2.1-3 we derived the state equations for an armature-controlled DC motor with a flexible coupling shaft. In this example we intend to use computer simulation to study the effects of the shaft compliance on the motor performance.

a. Rigid Coupling Shaft

If there is no compliance in the coupling shaft, the state equations are (see Example 2.1-3d)

$$\dot{x} = \begin{bmatrix} -R/L & -k_m'/L \\ k_m/J & -b_m/J \end{bmatrix} x + \begin{bmatrix} 1/L \\ 0 \end{bmatrix} u \equiv Ax + Bu, \tag{1}$$

where $J = J_m + J_L$. Defining the output as the motor speed gives

$$y = [0 \quad 1]x \equiv Cx.$$

The state is $x = [i \quad \omega]^T$.

The transfer function is computed to be

$$H(s) = C(sI - A)^{-1}B = \frac{k_m}{(Ls + R)(Js + b_m) + k_m k_m'}. \tag{2}$$

Using parameter values of $J_m = J_L = 0.1$ kg.m², $k_m = k_m' = 1$ V.s, $L = 0.5$ H, $b_m = 0.2$ N.m./rad/s, and $R = 5$ Ω yields

$$H(s) = \frac{10}{(s + 2.3)(s + 8.7)}, \tag{3}$$

so that there are two real poles at $s = -2.3$, $s = -8.7$.

Using Program TRESP in Appendix A to perform a simulation yields the step response shown in Fig. 2.1-10.

b. Very Flexible Coupling Shaft

Coupling shaft parameters of $k = 2$ N.m./rad and $b = 0.2$ N.m./rad/s correspond to a very flexible shaft. Using these values, software like PC-MATLAB [Moler et al. 1987] can be employed with the state model of Example 2.1-3c to obtain the two transfer functions

$$\frac{\omega_m}{u} = \frac{20s[(s + 1)^2 + 4.36^2]}{s(s + 3.05)(s + 6.14)[(s + 3.4)^2 + 5.6^2]} \tag{4}$$

$$\frac{\omega_L}{u} = \frac{40s(s + 10)}{s(s + 3.05)(s + 6.14)[(s + 3.4)^2 + 5.6^2]}. \tag{5}$$

The shaft flexible mode has the poles $s = -3.4 \pm j5.6$, and so has a damping ration of $\zeta = 0.52$ and a natural frequency of $\omega = 6.55$ rad/sec.

Note that the system is marginally stable, with a pole at $s = 0$. It is BIBO stable due to pole/zero cancellation.

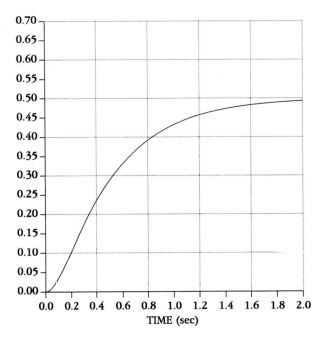

Figure 2.1-10 Step response of DC motor with no shaft flexibility. Motor speed in rad/sec.

Program TRESP yielded the step response shown in Fig. 2.1-11. Several points are worthy of note. Initially the motor speed ω_m rises more quickly than in Fig. 2.1-10, since the shaft flexibility means that only the rotor moment of inertia J_m initially affects the speed. Then, as the load J_L is coupled back to the motor through the shaft, the rate of increase of ω_m slows. Note also that the load speed ω_L exhibits a *delay* of approximately 0.1 sec due to the flexibility in the shaft.

It is extremely interesting to note that the shaft flexibility has the effect of speeding up the slowest motor real pole [compare (3) and (4)], so that ω_L approaches its steady-state value more quickly than in the rigid shaft case. This is due to the "whipping" action of the flexible shaft.

c. Fairly Rigid Coupling Shaft

Coupling shaft parameters of $k = 10$ N.m./rad and $b = 1$ N.m./rad/s correspond to a fairly rigid shaft. Using these values, the transfer functions from armature voltage to the motor and load speeds are

$$\frac{\omega_m}{u} = \frac{20s[(s+5)^2 + 8.66^2]}{s(s+2.4)(s+7.43)[(s+11.1)^2 + 10.1^2]} \tag{6}$$

$$\frac{\omega_L}{u} = \frac{20s(s+10)}{s(s+2.4)(s+7.43)[(s+11.1)^2 + 10.1^2]}. \tag{7}$$

The shaft flexible mode now occurs with poles at $s = -11.1 \pm j10.1$, and so has a damping ration of $\zeta = 0.74$ and a natural frequency of $\omega = 15$ rad/sec. Note that the real poles are similar to the rigid poles in (3).

Program TRESP yielded the step response shown in Fig. 2.1-12. This is much like the rigid response in Fig. 2.1-10, though the load speed ω_L still exhibits some initial delay.

2.1 Continuous-Time Systems

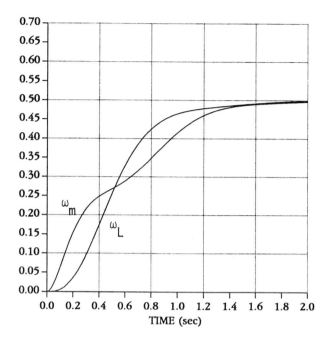

Figure 2.1-11 Step response of motor with very flexible shaft

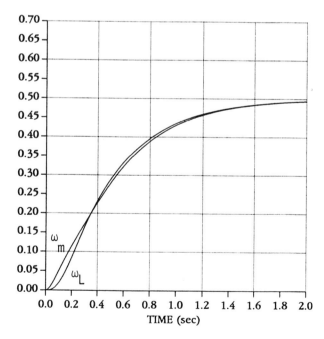

Figure 2.1-12 Step response of motor with fairly rigid shaft

∎

2.2 DISCRETE-TIME SYSTEMS

In controls applications, the systems are generally continuous-time. However, with microprocessors so fast, light, and versatile, it is common to implement control schemes on digital signal processors (DSPs). As we shall see in Chapter 7, digital control laws are conveniently designed using the *linear discrete-time state-space model* given by

$$x_{k+1} = Ax_k + Bu_k \qquad (2.2.1)$$

$$Y_k = Cx_k + Du_k. \qquad (2.2.2)$$

If the matrices (A, B, C, D) are constant, we say the system is *time-invariant*. Since the state at time $k + 1$ is found from the state x_k and control u_k at time k, this is a simple recursive equation which is easily implemented on a digital computer using, for instance, a do loop. Thus, computer simulation of discrete systems is very straightforward.

We have shown how to derive the continuous-time state model from the physical system. In section 7.1, when we discuss digital controllers, we show how to obtain the discrete state model from the continuous-time state model. Let us review the analysis of discrete-time systems.

Solution of Discrete State Equation

This development will be very similar to that for continuous systems in section 2.1; therefore, it will be brief.

Frequency-domain solution

The (one-sided) z-transform of a discrete time function u_k is defined by [Oppenheim and Schafer 1975]

$$U(z) = \sum_{k=0}^{\infty} u_k z^{-k}. \qquad (2.2.3)$$

The z-transform of (2.2.1) is

$$zX(z) - zx_0 = AX(z) + BU(z),$$

with x_0 the given initial condition. We may write

$$zIX(z) - AX(z) = zx_0 + BU(z)$$

$$X(z) = (zI - A)^{-1}zx_0 + (zI - A)^{-1}BU(z), \qquad (2.2.4)$$

which is the frequency-domain solution for the state. According to (2.2.2), the output solution is

$$Y(z) = C(zI - A)^{-1}zx_0 + [C(zI - A)^{-1}B + D]U(z). \qquad (2.2.5)$$

Defining the *discrete-time resolvent matrix* as

$$\Phi(z) = (zI - A)^{-1} \qquad (2.2.6)$$

2.2 Discrete-Time Systems

we have

$$X(z) = \Phi(z)zx_0 + \Phi(z)BU(z) \tag{2.2.7}$$

$$Y(z) = C\Phi(z)zx_0 + [C\Phi(z)B + D]U(z). \tag{2.2.8}$$

The discrete *transfer function* is

$$H(z) = C(zI - A)^{-1}B + D = C\Phi(z)B + D, \tag{2.2.9}$$

whence

$$Y(z) = C\Phi(z)zx_0 + H(z)U(z). \tag{2.2.10}$$

All of these solutions have a *zero-input* portion—the first term which depends only on the initial state, and a *zero-state* portion—the second term which depends only on the input. It is interesting to note the appearance of an extra z in the zero-input solution that does not occur in the continuous-time case.

Since

$$H(z) = \frac{C\{\text{adj}(zI - A)\}B}{|zI - A|} + D = \frac{C\{\text{adj}(zI - A)\}B + D|zI - A|}{|zI - A|}, \tag{2.2.11}$$

the system poles are the roots of the characteristic equation

$$\Delta(z) \equiv |zI - A| = 0, \tag{2.2.12}$$

where $\Delta(z)$ is the characteristic polynomial.

Time-domain solution

The Laurent series expansion about infinity of $z\Phi(z)$ is

$$z\Phi(z) = z(zI - A)^{-1} = I + z^{-1}A + z^{-2}A^2 + \ldots = \sum_{k=0}^{\infty} A^k z^{-k} \tag{2.2.13}$$

whence it is evident that

$$\mathbf{Z}(A^k) = z\Phi(z) = z(zI - A)^{-1}, \tag{2.2.14}$$

with $\mathbf{Z}(\ )$ denoting the z-transform. Note the role of the extra term z. We call A^k the *discrete matrix exponential*.

Recalling that the z-transform of a convolved pair of time functions is the product of their z-transforms, we may perform an inverse z-transformation on (2.2.7) to obtain

$$x_k = A^k x_0 + \sum_{i=0}^{k-1} A^{k-i-1} B u_i, \tag{2.2.15}$$

where the second term is the convolution of A^{k-1} and Bu_k.

Setting u_k to zero, the homogeneous solution is seen to be

$$x_k = A^k x_0, \tag{2.2.16}$$

so that, in the absence of an input, the state propagates from time 0 to time k by multiplication by A^k. We therefore call A^k the *discrete state transition matrix*.

Now, the output is given as

$$y_k = CA^k x_0 + \sum_{i=0}^{k-1} CA^{k-i-1} Bu_i + Du_k. \qquad (2.2.17)$$

The output is the convolution of the input and the pulse response, so that the pulse response is given in terms of (A, B, C, D) by

$$h_k = \begin{cases} D, & k = 0 \\ CA^{k-1}B, & k \geq 0. \end{cases} \qquad (2.2.18)$$

Using (2.2.13) in (2.2.9) shows that $H(z) = \mathbf{Z}(h_k)$. Using the definition of h_k, the output may be written as

$$y_k = CA^k x_0 + \sum_{i=0}^{k} h_{k-i} u_i. \qquad (2.2.19)$$

Discrete Natural Modes and Stability

The transfer function may be written in the PFE form

$$H(z) = \sum_{i=1}^{n} \frac{zK_i}{z - p_i} = \sum_{i=1}^{n} \frac{K_i}{1 - p_i z^{-1}}, \qquad (2.2.20)$$

where we have assumed for simplicity that all n of the poles p_i are nonrepeated. The residues K_i are generally matrices. Now, a term-by-term inverse z-transform yields

$$h_k = \sum_{i=1}^{n} K_i p_i^k, \quad k \geq 0. \qquad (2.2.21)$$

From this expression it is evident that with each pole, p_i is associated a discrete exponential time function of the form p_i^k. These are the *natural modes* of the poles.

Internal stability

A discrete pole is (asymptotically) stable if its natural mode p_i^k decays with k. Writing

$$p_i^k = |p_i|^k e^{jk \angle p_i}, \qquad (2.2.22)$$

with $|p_i|$ the magnitude and $\angle p_i$ the angle, we see that a pole is stable if and only if it is strictly within the unit circle in the z-plane, for then its magnitude is less than one.

A discrete-time system is (asymptotically) stable if all of its poles are within the unit circle.

The z-transform of the discrete unit step, which is defined as zero for $k < 0$ and one for $k \geq 0$, is

2.3 System Properties

$$U(z) = \frac{1}{1 - z^{-1}}. \qquad (2.2.23)$$

It is a good exercise to verify this by long division of $1 - z^{-1}$ into 1 and then interpreting the quotient in light of definition (2.2.3). The unit step is marginally stable as it neither decays nor grows without bound as a function of k. Its pole is at $z = 1$ on the unit circle. In fact, a discrete system is marginally stable if and only if its poles are within the unit circle, with possibly some nonrepeated poles on the unit circle.

A discrete system is unstable if it has poles outside the unit circle, or repeated poles on the unit circle. For example, note that the discrete unit ramp $u_k = k$, $k \geq 0$, has the transform

$$U(z) = \frac{z^{-1}}{(1 - z^{-1})^2}, \qquad (2.2.24)$$

which has two poles at $z = 1$. The ramp is unstable.

External stability

A discrete system is BIBO stable if its pulse response h_k decays with k, or equivalently if its transfer function $H(z)$ has all its poles inside the unit circle. Asymptotic stability implies BIBO stability, but the reverse statement is not true unless no pole/zero cancellation occurs in (2.2.11).

2.3 SYSTEM PROPERTIES

In this section we shall extend our ability to analyze systems by covering some of their fundamental properties. It is important to know the basic nature of a system before one attempts to design a control system for it. We have discussed one system property, namely, stability. Stable systems behave fundamentally differently than unstable systems, as we have seen. This has an impact on the sort of control system one should design.

Other properties that reveal the nature of a system are the minimum-phase property, reachability, and observability. There are fundamental limitations or our ability to control nonminimum-phase systems. Reachability has to do with our ability to control a state-variable system using the given inputs. Observability has to do with our ability to determine the internal states by measuring the given outputs.

Minimum-Phase Systems

The system

$$G(s) = \frac{s + 1}{s + 5} \qquad (2.3.1)$$

has the Bode phase plot shown in Fig. 2.3-1. The system

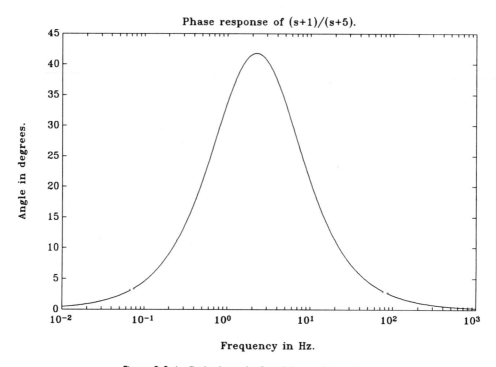

Figure 2.3-1 Bode phase plot for minimum-phase system

$$G_1(s) = \frac{s-1}{s+5} \quad (2.3.2)$$

has the Bode phase plot shown in Fig. 2.3-2. The first system has both its pole and zero in the left-half plane, while the second has a zero in the right-half plane. The right-half-plane zero has the effect of increasing the phase for low frequencies. Indeed, note that the DC gain of the latter system is $G_1(0) = -0.2$, which has a phase of 180°.

Systems with both poles and zeros in the left-half plane are called *minimum-phase systems*. Note that if the system $G(s)$ is of minimum phase, then both it and its inverse $G^{-1}(s)$ are stable. The minimum-phase property is a very important one in feedback design, for nonminimum-phase zeros (i.e., those in the right-half plane) can cause serious problems, as we shall now see.

Nonminimum-phase zeros in closed-loop control

There are some fundamental limitations in what may be achieved using closed-loop control for systems with nonminimum-phase zeros.

Consider the system $G(s)$ connected in the feedback configuration of Fig. 2.3-3.

2.3 System Properties

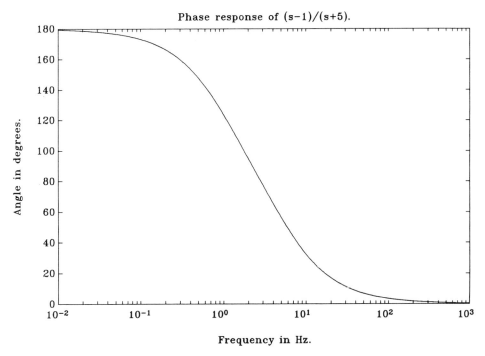

Figure 2.3-2 Bode phase plot for nonminimum-phase system

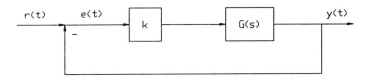

Figure 2.3-3 System $G(s)$ in tracker feedback configuration

This is a tracking system which, with proper design, will cause the system output $y(t)$ to follow the reference command $r(t)$. The feedback gain is k.

Having in mind the root locus theory, recall that as the feedback gain increases, the closed-loop poles approach the open-loop zeros. A root locus plot versus k when

$$G(s) = \frac{s+5}{s-5} \qquad (2.3.3)$$

is shown in Fig. 2.3-4. $G(s)$ has an unstable pole that is easily stabilized using high-gain feedback, for the closed-loop pole approaches the stable zero as k increases.

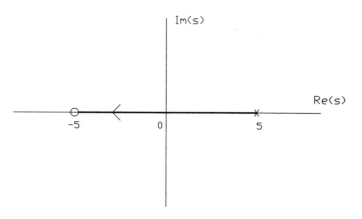

Figure 2.3-4 Root locus for system with stable zero

A root locus plot when

$$G(s) = \frac{s-5}{s+5} \quad (2.3.4)$$

is shown in Fig. 2.3-5. Although this system is open-loop stable, as k increases the stability is lost since the closed-loop pole approaches the unstable open-loop zero. In view of the fact that the gain k must be high to achieve a small steady-state error (note that the system is of Type 0), this is not a good situation. That is, for this system with a nonminimum-phase zero, there is a design trade-off between a small steady-state error and a fast transient response, for, as k increases the closed-loop system slows down, and finally passes into instability.

Design limitations in feedback control are common for systems with nonminimum-phase zeros. One should realize that systems with unstable poles do not suffer such problems, since, if all the zeros are stable, the poles may be stabilized using high-gain negative feedback.

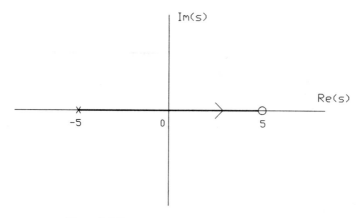

Figure 2.3-5 Root locus for system with unstable zero

2.3 System Properties

Bode gain-phase relation

Using classical techniques the properties of closed-loop systems like the one in Fig. 2.3-3 can be investigated in terms of the *open-loop* quantity $kG(s)$, which is the loop gain. A small steady-state error and robustness to low-frequency disturbances are guaranteed by making the loop gain large at low frequencies. Stability is guaranteed by correct selection of the crossover frequency (e.g., the frequency at which $|kG(j\omega)| = 1$). Rejection of high-frequency noise is guaranteed by making the loop gain small at high frequencies.

To verify stability, one may examine the Bode magnitude and phase plots of $kG(s)$ and compute the gain and phase margins [Franklin, Powell, and Emami-Naeini 1986]. However, for minimum-phase systems, there is a simplification in this procedure. The *Bode gain-phase relation* says that, for minimum-phase systems, there is a definite relation between magnitude and phase. In fact, given the magnitude versus ω, one can reconstruct the phase versus ω. An approximate statement of the Bode relation is

$$\angle kG(j\omega) \approx 90°n, \qquad (2.3.5)$$

where n is the slope of $|kG(j\omega)|$ in units of decade of amplitude per decade of frequency. For instance, a slope of 20 dB/decade = 6 dB/octave corresponds to $n = 1$.

Given the Bode relation, it is possible to deal with only the magnitude Bode plot for design in minimum-phase systems. The phase plot is not needed. Specifically, suppose that the magnitude slope is $n = -1$, or -20 dB/decade, persisting for about a decade in frequency prior to the crossover point. Then, according to (2.3.5), the phase at crossover is $-90°$, yielding a phase margin of $90°$. On the other hand, if the magnitude slope is $n = -2$, then the phase at crossover is $-180°$, resulting in no phase margin. Thus, for a good stability margin it is only necessary to ensure that the slope of the loop gain magnitude plot is $n = -1$ in the vicinity of crossover.

This ability to design control systems using only the magnitude portion of the Bode plot is particularly important in Chapter 8, where we show how to perform frequency-domain analysis for multi-input/multi-output systems. This is because the multivariable magnitude plot is not difficult to compute, relying on the notion of singular values, or to interpret. However, the multivariable phase plot is more involved to construct, relying on the notion of principal phases [Postlethwaite et al. 1981], and also to use. Thus, multivariable frequency-domain analysis is easy for minimum-phase systems, but more complicated for nonminimum-phase systems.

Undershoot and delay

A property common in systems with nonminimum-phase zeros is that of undershoot, which causes a delay in the system response.

The step response for the system

$$G(s) = \frac{-(s-2)}{(s+1)^2 + 1} = \frac{-s+2}{s^2 + 2s + 2} \qquad (2.3.6)$$

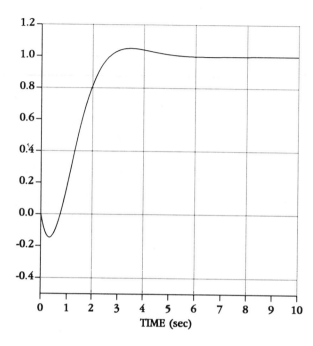

Figure 2.3-6 Nonminimum-phase step response

is shown in Fig. 2.3-6. The effect of the nonminimum-phase zero is clearly seen in the initial undershoot of the response. This causes a delay in the response.

Note that the DC gain of this system is $G(0) = 1$, so that the steady-state error in following a unit step is zero. The coefficient of the highest power of s in the transfer function numerator is known as the *static loop sensitivity*. Due to the unstable zero, it is necessary to make the static loop sensitivity of (2.3.6) equal to -1 to obtain a positive DC gain.

To see why the undershoot occurs, one may examine the high-frequency performance of the system by using the initial-value theorem. Using a step input so that $U(s) = 1/s$, one obtains an initial output value of

$$\lim_{s \to \infty} sG(s)U(s) = \frac{-1}{s},$$

which has a phase of 180°. The high-frequency phase characterizes the initial transient response of the system, which is therefore negative in the step-response plot, leading to the undershoot.

The undershoot leads to a delay in the system response. Thus, systems with nonminimum-phase zeros exhibit *delay phenomena* which are very difficult to eliminate using feedback control. Note that the zeros are not influenced using feedback, so that the closed-loop system will also be nonminimum-phase.

Using the notion of static loop sensitivity, it is not difficult to show that systems with an odd number of nonminimum-phase zeros on the real axis exhibit undershoot.

2.3 System Properties

Reachability

In this subsection we shall examine the input-coupling structure of the state-variable model given in continuous form as

$$\dot{x} = Ax + Bu \tag{2.3.7}$$

and discrete form as

$$x_{k+1} = Ax_k + Bu_k. \tag{2.3.8}$$

Of concern is the following question: how much control effectiveness does one have in manipulating the internal state x using the given input u? If some of the components of x cannot be influenced by u, there is a limit on what can be achieved using feedback control. We shall take $x \in \mathbf{R}^n$ and $u \in \mathbf{R}^m$.

For either a continuous or a discrete system, we say a given state vector $x \in \mathbf{R}^n$ is reachable if there exists a control input u and a final time such that, starting with an initial state x_0 of zero, the final state of the system is x.

We say a continuous or discrete system is reachable if all n-vectors x are reachable, that is, if one can find a control input to drive the system from zero to any desired state x at some final time.

Let us now discuss reachability for discrete-time systems and then for continuous-time systems, deriving a simple test for reachability in terms of the system matrices (A, B).

Reachability for discrete systems

The recursive nature of discrete systems makes it easy to solve the reachability problem in discrete time. The discrete state solution (2.2.15) can be written conveniently as

$$x_k = A^k x_0 + \sum_{i=0}^{k-1} A^{k-i-1} B u_i \tag{2.3.9}$$

$$= A^k x_0 + [B \quad AB \quad A^2 B \quad \ldots \quad A^{k-1} B] \begin{bmatrix} u_{k-1} \\ u_{k-2} \\ \vdots \\ u_1 \\ u_0 \end{bmatrix}$$

or

$$x_k = A^k x_0 + U_k \bar{u}_{0,k}, \tag{2.3.10}$$

where we have defined the $n \times mk$ *reachability matrix*

$$U_k = [B \quad AB \quad A^2 B \quad \ldots \quad A^{k-1} B], \tag{2.3.11}$$

and the vector of inputs $\bar{u}_{0,k} \in \mathbf{R}^{mk}$ as the right-hand vector in (2.3.9).

The reachability problem is now easily confronted. Suppose we desire to reach a given state x_N at a specified time N. Setting $x_0 = 0$, as required by the definition, and $k = N$ in (2.3.10) we obtain

$$x_N = U_N \bar{u}_{0,N}. \tag{2.3.12}$$

In order for the state x_N to be reachable, this equation should have a solution for the control input $\bar{u}_{0,N}$. This occurs if and only if the given n-vector x_N is in the column space, or *range*, of the reachability matrix U_N, for then the columns of U_N may be combined to yield x_N by appropriate selection of the entries of the vector $\bar{u}_{0,N}$.

If x_N is not in the range of the reachability matrix, there is no solution $\bar{u}_{0,N}$ and hence no control that drives the system from $x_0 = 0$ to x_N on the specified time interval $[0, N]$. In this event, the range of the reachability matrix U_N can sometimes be increased by adding more columns of the form $A^k B$. According to (2.3.11), this amounts to increasing the time N. That is, the given final state might be reachable over a longer time interval.

In order for the system to be reachable, all n-vectors x_N should fall in the range of U_N for some N. That is, U_N should have full rank of n for some N. If this occurs, then not only is any x_N in the range of U_N, but so is $(x_N - A^N x_0)$ for any x_N and x_0. That is, if the system is reachable then *any* final state may be reached from *any* initial state.

Testing the system for reachability on the interval $[0, N]$ may not be easy. Generally, the system is multi-input so that $m > 1$, then the reachability matrix U_N is an $n \times mN$ matrix, generally nonsquare. To test it for full rank n one could attempt to find an $n \times n$ submatrix that is nonsingular. However, there is an easier alternative. A theorem of linear algebra states that

$$\text{rank}(U_N) = \text{rank}(U_N U_N^T), \tag{2.3.13}$$

so that the rank of the reachability matrix equals the rank of the *discrete reachability gramian*

$$P_N \equiv U_N U_N^T, \tag{2.3.14}$$

which is a square $n \times n$ matrix. Therefore, to test the system for reachability one may find the determinant of P_N, which may be accomplished using standard numerical routines (e.g., IMSL, ORACLS [Armstrong 1980], MATLAB [Moler et al. 1987]). Note that the gramian can be written as

$$P_N = \sum_{k=0}^{N-1} A^{N-k-1} B B^T (A^T)^{N-k-1} \tag{2.3.15}$$

Finally, note that the rank of the reachability matrix U_k is low for small k. In fact $U_1 = B$, which is an $n \times m$ matrix. Since the number m of controls is usually smaller than the number n of states, U_1 rarely has full rank n. The rank of U_k may increase as more columns $A^k B$ are added. Therefore, if the system is not reachable on a given interval, one may attempt to reach a given final state on a longer interval. Testing for reachability for increasing values of the final time N could become tedious after a few tries.

2.3 System Properties

There is, fortunately, a limit on how much the rank of U_N can be increased by appending additional powers of $A^k B$. It is based on the *Cayley-Hamilton Theorem*.

The Cayley-Hamilton Theorem states that

$$\Delta(A) = 0, \tag{2.3.16}$$

or, every matrix satisfies its own characteristic equation. An example is necessary at this point.

Example 2.3-1: Cayley-Hamilton Theorem

The Cayley-Hamilton Theorem applies for both continuous and discrete systems. Thus, consider the damped harmonic oscillator of Example 2.1-5

$$\dot{x} = \begin{bmatrix} 0 & 1 \\ -\omega_n^2 & -2\alpha \end{bmatrix} x + \begin{bmatrix} 0 \\ 1 \end{bmatrix} u \equiv Ax + Bu. \tag{1}$$

which has characteristic equation of

$$\Delta(s) = s^2 + 2\alpha s + \omega_n^2. \tag{2}$$

To verify the Cayley-Hamilton Theorem, it is easy to compute that

$$\Delta(A) = A^2 + 2\alpha A^1 + \omega_n^2 A^0 = A^2 + 2\alpha A + \omega_n^2 I \tag{3}$$

$$= \begin{bmatrix} -\omega_n^2 & -2\alpha \\ 2\alpha\omega_n^2 & 4\alpha^2 - \omega_n^2 \end{bmatrix} + \begin{bmatrix} 0 & 2\alpha \\ -2\alpha\omega_n^2 & -4\alpha^2 \end{bmatrix} + \begin{bmatrix} \omega_n^2 & 0 \\ 0 & \omega_n^2 \end{bmatrix} = 0.$$

■

Suppose the characteristic polynomial is

$$\Delta(s) = s^n + a_1 s^{n-1} + \ldots + a_n. \tag{2.3.17}$$

Then, the Cayley-Hamilton Theorem says that $\Delta(A) = 0$ or

$$A^n = -\sum_{i=0}^{n-1} a_{n-1} A^i, \tag{2.3.18}$$

that is, A^n *is a linear combination of lower powers of A*. Note that $A^0 = I$, the identity matrix.

This fact has important ramifications in terms of the reachability matrix, for it implies that $A^n B$ is a linear combination of $B, \ldots, A^{n-1}B$. That is

$$U_n = [B \quad AB \ldots A^{n-1}B] \tag{2.3.19}$$

has maximal rank over all U_k, since adding another column $A^n B$ will not increase the rank of U_n. What this means is that if the system is not reachable in n time steps, with n the dimension of the system, then it is not reachable for any $N > n$. It follows that the system is reachable if and only if

$$\text{rank}(U_n) = n. \tag{2.3.20}$$

Therefore, we call U_n the *system reachability matrix*.

Reachability for continuous systems

The reachability issue for continuous systems is not as simple as for discrete systems due to the integral form of the solution in section 2.1, which, over the time interval $[t_0, T]$, is given by

$$x(T) = e^{A(T-t_0)}x_0 + \int_{t_0}^{T} e^{A(T-\tau)}Bu(\tau)\, d\tau. \tag{2.3.21}$$

with $x(t_0) = x_0$ the initial condition. Setting $x_0 = 0$ as required by the definition of reachability, the final state $x(T)$ is seen to be reachable if it is in the range of the integral operator; that is, if there exists a function $u(t)$ such that the integral is equal to $x(T)$.

The system is reachable if this is true for any choice of final state $x(T)$, or equivalently if the integral operator has full rank n. Then, it is not necessary for the initial condition to be zero, for if the operator has full rank then $[x(T) - e^{A(T-t_0)}x_0]$ is in its range for all $x(T)$ and x_0. That is, if the system is reachable, then *any* initial condition can be driven to any final state for some $u(t)$.

We call the integral operator in (2.3.21) the *continuous-time input-coupling operator*. By continuity, it is nonsingular either for all $T - t_0$ or for no $T - t_0$. That is, in the continuous-time case, if the system is reachable the final state can be reached over a time interval *of any length*. This is in contrast to discrete-time control, where the rank of the reachability matrix U_N can continue to increase as columns like $A^k B$ are added until $N = n$, meaning that the system may not be reachable for small intervals $[0, N]$ even though it is reachable on $[0, n]$.

In section 3.2 we show how to compute the control input $u(t)$ that drives the system from a given state x_0 to a desired final state $x(T)$. In the rest of this subsection we show how to conveniently test a continuous-time system for reachability.

It is difficult to determine the rank of the input-coupling operator. However, its rank is equal to that of the *continuous reachability gramian*

$$P(t_0, T) = \int_{t_0}^{T} e^{A(T-\tau)}BB^T e^{A^T(T-\tau)}\, d\tau, \tag{2.3.22}$$

which is the input-coupling operator postmultiplied by its adjoint [Brogan 1974]. Given a fixed t_0 and T, $P(t_0, T)$ may be computed and is just an $n \times n$ constant matrix, whose determinant can be evaluated using standard software. The system is reachable if and only if $|P(t_0, T)| \neq 0$. Compare $P(t_0, T)$ to the discrete reachability gramian (2.3.15).

Checking a continuous system for reachability by finding $P(t_0, T)$ and examining its determinant is far more inconvenient than the simple test for the reachability of discrete-time systems in terms of the easily constructed reachability matrix U_n. Let us now show that the nonsingularity of $P(t_0, T)$ is equivalent to the full rank n of the reachability matrix. Thus, this matrix provides the test for reachability for *both discrete and continuous systems*.

To show this, recall that

$$e^{At} = I + At + \frac{A^2 t^2}{2!} + \ldots \tag{2.3.23}$$

2.3 System Properties

and, by the Cayley-Hamilton Theorem, A^n may be expressed in terms of lower powers of A using (2.3.18). Therefore, the infinite series for e^{At} can be expressed as a finite series involving A^{n-1} and lower powers of A. Write this series as

$$e^{At} = \sum_{i=0}^{n-1} \alpha_i(t) A^i, \qquad (2.3.24)$$

where the coefficients $\alpha_i(t)$ depend on the coefficients a_i of $\Delta(A)$ as well as the time-varying coefficients of A^i in (2.3.23).

Substituting now (2.3.24) into (2.3.21) (with $x_0 = 0$) yields

$$x(T) = \int_{t_0}^{T} \sum_{i=0}^{n-1} \alpha_i(T - \tau) A^i B u(\tau) \, d\tau$$

$$= \sum_{i=0}^{n-1} A^i B \int_{t_0}^{T} \alpha_i(T - \tau) u(\tau) \, d\tau = \sum_{i=0}^{n-1} A^i B v_i(t_0, T),$$

where we have defined n auxiliary signals derived from the input $u(t)$ as

$$v_i(t_0, T) = \int_{t_0}^{T} \alpha_i(T - \tau) u(\tau) \, d\tau; \qquad i = 0, \ldots, n - 1. \qquad (2.3.25)$$

Therefore, in terms of v_i we may write the final state as

$$x(T) = [B \quad AB \quad A^2B \ldots A^{n-1}B] \begin{bmatrix} v_0 \\ v_1 \\ \vdots \\ v_{n-1} \end{bmatrix}. \qquad (2.3.26)$$

This solution has exactly the form of the discrete-time solution (2.3.9). Therefore, all the remarks made there in terms of the range of the reachability matrix U_n apply here as well. That is, the continuous-time system is reachable if and only if U_n has full rank n. [To prove sufficiency, one must show that the map (2.3.25) is onto, that is, that for every choice of $v_i(t_0, T)$, $i = 0, 1, \ldots n - 1$, there exists a control $u(t)$, $t_0 \leq t \leq T$.]

We can now say that a system, continuous or discrete, is reachable if and only if the reachability matrix U_n has full rank n.

Observability

We shall now concern ourselves with the output-coupling structure of the homogeneous state model given in continuous form as

$$\dot{x} = Ax \qquad (2.3.27)$$

$$y = Cx \qquad (2.3.28)$$

and discrete form as

$$x_{k+1} = Ax_k \tag{2.3.29}$$

$$y_k = Cx_k. \tag{2.3.30}$$

Of interest is the following question: when can the internal state x by reconstructed from measurements of the available outputs y? If some of the state components do not influence y, then the measurements are somehow deficient and there are some limitations on what can be achieved using feedback of the output y. We shall take $x \in \mathbf{R}^n$ and $y \in \mathbf{R}^p$.

For either a continuous or a discrete system, we say the initial condition x_0 is *unobservable* if it yields an output of $y = 0$. If $x_0 \neq 0$ is unobservable, then a nonzero state trajectory results in a zero output; this indicates that our measurements are badly designed since they cannot sense all motions in the state-space.

We say a continuous or discrete system is observable if there are no nonzero unobservable initial conditions; that is, if the output is nonzero whenever the state trajectory is nonzero.

In Chapter 9 we show how to design observers that provide estimates of the state given the measured outputs. Let us now discuss some basic notions of observability for discrete-time systems and then for continuous-time systems, providing a simple test for observability in terms of the system matrices (A, C).

Observability for discrete systems

Setting the input equal to zero, (2.2.17) shows that the complete output solution sequence over an interval $[0, k]$ may be written as

$$\begin{bmatrix} y_{k-1} \\ \vdots \\ y_1 \\ y_0 \end{bmatrix} = \begin{bmatrix} CA^{k-1} \\ \vdots \\ CA \\ C \end{bmatrix} x_0, \tag{2.3.31}$$

or

$$\bar{y}_{0,k} = V_k x_0, \tag{2.3.32}$$

where we have defined the *pk $\times$ n observability matrix*

$$V_k \equiv \begin{bmatrix} CA^{k-1} \\ \vdots \\ CA \\ C \end{bmatrix} \tag{2.3.33}$$

and the vector of outputs $\bar{y}_{0,k} \in \mathbf{R}^{pk}$. As is traditional, we have not used the last measurement y_k.

The observability problem is now easily solved, for according to standard results in linear algebra x_0 yields a zero output sequence $\bar{y}_{0,k}$ if and only if x_0 lies in the nullspace of the matrix V_k. If the nullspace of V_k is zero for some k, then only $x_0 = 0$ yields the

2.3 System Properties

zero output sequence. Consequently, the system is observable if and only if V_k has full rank n for some k.

If x_0 is in the nullspace of V_k for a given k, then $y_0, y_1, \ldots, y_{k-1}$ is a zero sequence. The nullspace of V_k may be reduced by adding more rows of the form CA^k; according to the definition of V_k this can be achieved by increasing the length of the measurement interval. Thus, for a given initial state x_0, although $y_0, y_1, \ldots, y_{k-1}$ are all zero, one may have nonzero vectors among $y_k, y_{k+1}, \ldots$, so that x_0 is not unobservable if a longer measurement interval is used.

According to the Cayley-Hamilton Theorem, however, the rows of CA^n are spanned by the rows of CA^k, $k < n$. That is, it does no good to increase the measurement interval beyond $[0, n]$ since the data beyond this point do not provide new information in terms of increasing the rank of V_k. Thus, the system is observable if and only if

$$\text{rank}(V_n) = n, \tag{2.3.34}$$

and we call V_n the *system observability matrix*.

To check the full rank of the $pk \times n$ matrix V_n, which is generally nonsquare, one could attempt to find an $n \times n$ nonsingular submatrix. However, since

$$\text{rank}(V_n) = \text{rank}(V_n^T V_n), \tag{2.3.35}$$

the system is observable if and only if the $n \times n$ *discrete observability gramian*

$$Q_n \equiv V_n^T V_n \tag{2.3.36}$$

is nonsingular. This may easily be checked using standard software [IMSL] to evaluate the determinant of Q_n. Note that the gramian may be written as

$$Q_n = \sum_{k=0}^{n-1} (A^T)^{n-k-1} C^T C A^{n-k-1}. \tag{2.3.37}$$

Observability for continuous systems

The homogeneous output solution for the continuous system is given by

$$y(t) = Ce^{A(t-t_0)} x_0, \tag{2.3.38}$$

with $x(t_0) = x_0$ the initial condition. Thus, x_0 is unobservable over a fixed interval $[t_0, T]$ if it lies in the nullspace of the *output coupling operator* $Ce^{A(t-t_0)}$ over $[t_0, T]$; that is, if $y(t) = 0$ for $t_0 \leq t \leq T$. The system is observable if the nullspace of this operator is zero.

Such conditions are difficult to check directly in terms of the output-coupling operator. However, the rank over $[t_0, T]$ of this operator is the same as the rank of the *continuous observability gramian*

$$Q(t_0, T) = \int_{t_0}^{T} e^{A^T(\tau-t_0)} C^T C e^{A(\tau-t_0)} \, d\tau, \tag{2.3.39}$$

which is constructed by premultiplying the output-coupling operator by its adjoint [Brogan 1974]. Compare $Q(t_0, T)$ to (2.3.37). For a fixed interval $[t_0, T]$ this is a constant $n \times n$ matrix whose determinant can be evaluated by standard means.

By continuity, $Q(t_0, T)$ has full rank n for all $(T - t_0)$ or for no $(T - t_0)$. That is, in contrast to the discrete case where observability might be improved by taking measurements over a longer time interval (up to one of length n), in the continuous case the initial condition of an observable system can be determined using the output function measured over a nonzero time interval of *any* length, no matter how small.

The test for observability in terms of the continuous observability gramian is inconvenient to perform. It may be shown, exactly as for reachability, that this condition for the observability of continuous systems is equivalent to the full rank of the observability matrix V_n. Thus, for both discrete and continuous systems, the system is observable if and only if rank$(V_n) = n$.

2.4 REALIZATION AND CANONICAL FORMS

In classical control theory the system was generally described by a transfer function. In modern controls the system is generally described by a state-variable model. It is important to be able to convert between these two system descriptions. We have seen how to compute the transfer function from the system matrices (A, B, C, D) in both the continuous and discrete cases. In this section we shall show how to find a state-space *realization* (A, B, C, D) for a given transfer function. This allows us to use modern controls techniques to control any system for which we know the transfer function.

As another application of state-space realization theory, transfer function or differential equation descriptions of the *actuators and sensors* in a control system are often available. For modern controls design, it is necessary to find a state-space model.

Yet another application of realization theory arises in the implementation of digital controllers. As we shall see in Chapter 6, it is very important how a digital controller is implemented on a microprocessor with finite wordlength. The *real Jordan form*, introduced in Example 2.4-3, affords a convenient way to find a numerically stable implementation of a controller in terms of simple first- and second-order subsystems in series and parallel.

We shall take two approaches to realization theory. First, we will introduce some very simple forms of the state equations which are called *canonical forms*. In the SISO case it is easy to write these forms down by inspection given a transfer function; this gives the required state-space realization. Then, we shall present Gilbert's method for realizing MIMO systems. A fundamental concept in connection with canonical forms is the *state-space transformation*, which we now review.

All the results in this section apply for both continuous and discrete-time systems. Some of the key results are collected in Table 2.4-1 for reference purposes.

2.4 Realization and Canonical Forms

TABLE 2.4-1 STATE-SPACE TRANSFORMATION AND CANONICAL FORMS

State Equation:

$$\dot{x} = Ax + Bu$$
$$y = Cx$$

State-Space Transformation:

If $\bar{x} = Tx$, then

$$\dot{\bar{x}} = (TAT^{-1})x + (TB)u$$
$$y = (CT^{-1})\bar{x}$$

Modal Decomposition: (for A diagonalizable)

$$y(t) = \sum_{i=1}^{n} (w_i^T x_0) e^{\lambda_i t} C v_i + \sum_{i=1}^{n} C v_i \int_0^t e^{\lambda_i (t-\tau)} w_i^T B u(\tau) \, d\tau$$

where
$$(A - \lambda_i I) v_i = 0$$
$$w_i^T (A - \lambda_i I) = 0$$

Transfer Function:

$$H(s) = \frac{Y(s)}{U(s)} = \frac{b_1 s^3 + b_2 s^2 + b_3 s + b_4}{s^4 + a_1 s^3 + a_2 s^2 + a_3 s + a_4}$$

Reachable Canonical Form:

$$\dot{x} = \begin{bmatrix} 0 & 1 & 0 & 0 \\ 0 & 0 & 1 & 0 \\ 0 & 0 & 0 & 1 \\ -a_4 & -a_3 & -a_2 & -a_1 \end{bmatrix} x + \begin{bmatrix} 0 \\ 0 \\ 0 \\ 1 \end{bmatrix} u = Ax + Bu$$

$$y = [b_4 \quad b_3 \quad b_2 \quad b_1] x = Cx$$

Observable Canonical Form:

$$\dot{x} = \begin{bmatrix} -a_1 & 1 & 0 & 0 \\ -a_2 & 0 & 1 & 0 \\ -a_3 & 0 & 0 & 1 \\ -a_4 & 0 & 0 & 0 \end{bmatrix} x + \begin{bmatrix} b_1 \\ b_2 \\ b_3 \\ b_4 \end{bmatrix} u = Ax + Bu$$

$$y = [1 \quad 0 \quad 0 \quad 0] x = Cx$$

State-Space Transformations

Suppose a system has the given state-space description

$$\dot{x} = Ax + Bu \tag{2.4.1}$$

$$y = Cx + Du \tag{2.4.2}$$

with state $x(t) \in \mathbf{R}^n$, control input $u(t) \in \mathbf{R}^m$, and measured output $y(t) \in \mathbf{R}^p$. It is often useful to be able to determine the state description when the state is redefined by transforming it to other coordinates.

Define the *state-space transformation (SST)*

$$\bar{x} = Tx, \tag{2.4.3}$$

where $\bar{x}(t)$ is the state vector described in the new coordinates and T is an $n \times n$ transformation matrix that relates the new coordinates to the old coordinates. It is now required to find the state-space model

$$\dot{\bar{x}} = \bar{A}\bar{x} + \bar{B}u \tag{2.4.4}$$

$$y = \bar{C}\bar{x} + \bar{D}u \tag{2.4.5}$$

which describes the dynamics of the new state $\bar{x}(t)$ expressed in the new coordinates. Note that the original system (A, B, C, D) and the transformed system $(\bar{A}, \bar{B}, \bar{C}, \bar{D})$ have the same input $u(t)$ and output $y(t)$. This is because all we have done is redefine the way we describe the state vector. Clearly, if we walk up to a circuit and tell it that we have decided to express the state in different coordinates, the input and output will not be affected.

To find the new system description $(\bar{A}, \bar{B}, \bar{C}, \bar{D})$ in terms of the original (A, B, C, D) and the transformation matrix T, we may write

$$x = T^{-1}\bar{x} \tag{2.4.6}$$

so that, according to (2.4.1)

$$\dot{x} = T^{-1}\dot{\bar{x}} = AT^{-1}\bar{x} + Bu$$

or

$$\dot{\bar{x}} = TAT^{-1}x + TBu. \tag{2.4.7}$$

Also, using (2.4.2) it is seen that

$$y = CT^{-1}\bar{x} + Du. \tag{2.4.8}$$

Comparing the last two equations to (2.4.4), (2.4.5), it is evident that

$$(\bar{A}, \bar{B}, \bar{C}, \bar{D}) = (TAT^{-1}, TB, CT^{-1}, D) \tag{2.4.9}$$

Thus, the transformed state $\bar{x}(t)$ satisfies the state equations (2.4.4), (2.4.5) with transformed system matrices given by (2.4.9).

Graphical analysis

It is instructive to examine the SST from a pictorial point of view. Figure 2.4-1 shows the state-space $\mathbf{X} \equiv \mathbf{R}^n$, the input space $\mathbf{U} \equiv \mathbf{R}^m$, and the output space $\mathbf{Y} \equiv \mathbf{R}^p$. Matrix B is a transformation from controls in $\mathbf{U}$ to the state space $\mathbf{X}$ (note that B is an $m \times n$ matrix so that it maps from $\mathbf{R}^m$ to $\mathbf{R}^n$). Likewise, A maps from $\mathbf{X}$ to $\mathbf{X}$, and C

2.4 Realization and Canonical Forms

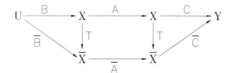

Figure 2.4-1 Graphical depiction of the state-space transformation T

maps from **X** to **Y**. Since the direct feed matrix D maps from **U** to **Y** and is not changed by a SST, we have not shown it in the figure.

The SST T acts as shown to map the state-space **X** into a new state-space $\bar{\mathbf{X}}$ with new coordinates. The state $\bar{x}(t)$ resides in $\bar{\mathbf{X}}$. In a fashion similar to that just described, we may draw in the matrices $\bar{A}$, $\bar{B}$, $\bar{C}$, and $\bar{D}$, showing their respective domains and co-domains.

In terms of these constructions, it is now straightforward to express $\bar{A}$, $\bar{B}$, $\bar{C}$, and $\bar{D}$ in terms of A, B, C, D, and T. Consider first $\bar{B}$. In the figure, there are two ways to get from **U** to $\bar{\mathbf{X}}$. One may either go via $\bar{B}$, or via B and then T. Recalling that concatenation of operators is performed from right to left (that is, given a u in **U**, first B operates on u to give Bu, then T operates on Bu to yield TBu), it follows that $\bar{B} = TB$, exactly as in (2.4.9).

In similar fashion, there are two ways to go from the left-hand $\bar{\mathbf{X}}$ to the right-hand $\bar{\mathbf{X}}$. One may either go via $\bar{A}$, or via T^{-1} (i.e. backwards along the arrow labeled by the left-hand T), then A, then down along T. Concatenating these three operators from right to left yields $\bar{A} = TAT^{-1}$, as in (2.4.9). Similar reasoning yields $\bar{C} = CT^{-1}$.

Example 2.4-1: State-Space Transformation in the 2-Body Problem

We shall consider a simplified version of the 2-body problem, where two bodies interact through a spring/damper system instead of through gravitational attraction. See Fig. 2.4-2. For simplicity we assume each body has a mass of $m = 1$ kg. The spring constant is k and the damping coefficient is b. The control input is a force of $u(t)$ nt applied to the second mass.

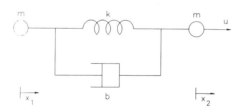

Figure 2.4-2 Two-body system with spring/damper interaction

a. State Equations

The equations of motion are

$$\ddot{x}_1 = k(x_2 - x_1) + b(\dot{x}_2 - \dot{x}_1) \qquad (1)$$
$$\ddot{x}_2 = k(x_1 - x_2) + b(\dot{x}_1 - \dot{x}_2) + u$$

Defining the state as

$$x = [x_1 \quad \dot{x}_1 \quad x_2 \quad \dot{x}_2]^T \qquad (2)$$

we may write the state equations

$$\frac{d}{dt}\begin{bmatrix} x_1 \\ \dot{x}_1 \\ x_2 \\ \dot{x}_2 \end{bmatrix} = \begin{bmatrix} 0 & 1 & 0 & 0 \\ -k & -b & k & b \\ 0 & 0 & 0 & 1 \\ k & b & -k & -b \end{bmatrix} \begin{bmatrix} x_1 \\ \dot{x}_1 \\ x_2 \\ \dot{x}_2 \end{bmatrix} + \begin{bmatrix} 0 \\ 0 \\ 0 \\ 1 \end{bmatrix} u. \tag{3}$$

b. State-Space Transformation

The coordinates chosen for the state vector are inconvenient. More insight may be gained by shifting to a center-of-mass frame of reference. The position of the center of mass of two bodies of equal masses is given by $z_1 = (x_1 + x_2)/2$. An independent variable is the distance between the masses, $z_2 = x_1 - x_2$. Therefore, let us define a SST by

$$z = Tx \tag{4}$$

with $z = [z_1 \ \dot{z}_1 \ z_2 \ \dot{z}_2]^T$ and nonsingular transformation matrix given by

$$T = \begin{bmatrix} 1/2 & 0 & 1/2 & 0 \\ 0 & 1/2 & 0 & 1/2 \\ 1 & 0 & -1 & 0 \\ 0 & 1 & 0 & -1 \end{bmatrix} \tag{5}$$

Note that $\dot{z}_1 = (\dot{x}_1 + \dot{x}_2)/2$, and $\dot{z}_2 = \dot{x}_1 - \dot{x}_2$.

Computing

$$T^{-1} = \begin{bmatrix} 1 & 0 & 1/2 & 0 \\ 0 & 1 & 0 & 1/2 \\ 1 & 0 & -1/2 & 0 \\ 0 & 1 & 0 & -1/2 \end{bmatrix} \tag{6}$$

the system matrices in the new coordinates are given by (2.4.9). Performing these operations yields the transformed system description

$$\frac{d}{dt}\begin{bmatrix} z_1 \\ \dot{z}_1 \\ z_2 \\ \dot{z}_2 \end{bmatrix} = \begin{bmatrix} 0 & 1 & 0 & 0 \\ 0 & 0 & 0 & 0 \\ 0 & 0 & 0 & 1 \\ 0 & 0 & -2k & -2b \end{bmatrix} \begin{bmatrix} z_1 \\ \dot{z}_1 \\ z_2 \\ \dot{z}_2 \end{bmatrix} + \begin{bmatrix} 0 \\ 1/2 \\ 0 \\ -1 \end{bmatrix} u = \bar{A}z + \bar{B}u. \tag{7}$$

c. Analysis of Transformed System

The importance of the SST is that it allows us to see the structure of the system. According to (7) the new A matrix is block diagonal. Thus, the center-of-mass subsystem, described by z_1 and $\dot{z}_1$, is decoupled from the subsystem describing the distance between the masses and described by z_2 and $\dot{z}_2$. The control input $u(t)$ influences both subsystems.

The center-of-mass subsystem has the characteristic equation

$$\Delta_1(s) = \begin{vmatrix} s & -1 \\ 0 & s \end{vmatrix} = s^2 = 0. \tag{8}$$

Thus, under the influence of the external force $u(t)$ the center of mass moves according to Newton's law $u = 2m\ddot{z}_1$.

The distance-between-the-masses subsystem has the characteristic equation

$$\Delta_2(s) = \begin{vmatrix} s & -1 \\ 2k & s + 2b \end{vmatrix} = s^2 + 2bs^2 + 2k = 0. \tag{9}$$

2.4 Realization and Canonical Forms

That is, under the influence of the external force $u(t)$ the distance between the masses satisfies the equation of a damped harmonic oscillator

$$\ddot{z}_2 + 2b\dot{z}_2 + 2kz_2 = -u \tag{10}$$

with natural frequency and damping ratio of

$$\omega_n = \sqrt{2k}, \qquad \zeta = \frac{b}{\sqrt{2k}}. \tag{11}$$

∎

Characteristic polynomial and transfer function

The characteristic polynomials and transfer function are not changed by a SST. To show this, use properties of determinants to write

$$\bar{\Delta}(s) = |sI - \bar{A}| = |sTT^{-1} - TAT^{-1}| = |T| \cdot |sI - A| \cdot |T^{-1}|$$
$$= |sI - A| = \Delta(s).$$

For the transfer function one has (setting $D = 0$ for convenience)

$$\bar{H}(s) = \bar{C}(sI - \bar{A})^{-1}\bar{B} = CT^{-1}(sTT^{-1} - TAT^{-1})^{-1}TB$$
$$= CT^{-1}T(sI - A)^{-1}T^{-1}TB = C(sI - A)^{-1}B = H(s).$$

We have used the fact that, for any two nonsingular matrices M and N, $(MN)^{-1} = N^{-1}M^{-1}$.

These results only say that by a mathematical redefinition of coordinates, we do not change the physical properties of the system. Thus, the system poles and input/output transmission properties do not change.

SST and reachability

We have seen that the SST T transforms the system (A, B, C, D) to the new system description $(TAT^{-1}, TB, CT^{-1}, D)$. It is interesting to determine the effect of T on the reachability matrix.

The reachability matrix of the transformed system $(\bar{A}, \bar{B})$ is

$$\bar{U}_n = [\bar{B} \quad \bar{A}\bar{B} \quad \bar{A}^2\bar{B} \ldots \bar{A}^{n-1}\bar{B}]$$
$$= [TB \quad TAT^{-1}TB \quad (TAT^{-1})^2TB \ldots]$$
$$= T[B \quad AB \quad A^2B \ldots A^{n-1}B]$$

so that

$$\bar{U}_n = TU_n. \tag{2.4.10}$$

That is, the reachability matrix transforms by multiplication by the SST T.

This means several things. First, since T is nonsingular

$$\text{rank}(\bar{U}_n) = \text{rank}(U_n), \tag{2.4.11}$$

so that rank($\bar{U}_n$) = n if and only if rank (U_n) = n. Thus, reachability is preserved by a SST. Clearly, the control effectiveness in a system is not changed by simply redefining the states in a mathematical sense.

If the system is reachable, the relation (2.4.10) offers a way of finding the SST that relates two system descriptions (A, B, C, D) and ($\bar{A}$, $\bar{B}$, $\bar{C}$, $\bar{D}$). To determine T, one need only find the reachability matrix U_n of the original system and the reachability matrix $\bar{U}_n$ of the transformed system. This may be accomplished knowing only A, B, $\bar{A}$, and $\bar{B}$. Then, T is obtained from (2.4.10) by

$$T = \bar{U}_n U_n^+, \qquad (2.4.12)$$

with U_n^+ the Moore-Penrose inverse [Rao and Mitra 1971] of U_n. Since U_n has full row rank n, U_n^+ is the right inverse, which is given by

$$U_n^+ = U_n^T(U_n U_n^T)^{-1}. \qquad (2.4.13)$$

Therefore

$$T = \bar{U}_n U_n^T(U_n U_n^T)^{-1}. \qquad (2.4.14)$$

If the system has only one input, then U_n is square and $U_n^+ = U_n^{-1}$ the usual matrix inverse. Then

$$T = \bar{U}_n U_n^{-1}. \qquad (2.4.15)$$

SST and observability

In a similar fashion to that just presented, one may show (see the problems) that

$$\bar{V}_n T = V_n, \qquad (2.4.16)$$

with V_n the observability matrix in the original basis and $\bar{V}_n$ the observability matrix in the new coordinates. Since T is nonsingular, rank($\bar{V}_n$) = rank(V_n), so that the original system (A, B, C, D) is observable if and only if the transformed system ($\bar{A}$, $\bar{B}$, $\bar{C}$, $\bar{D}$) is observable.

If (A, C) is observable, the SST may be determined from (A, C) and the transformed ($\bar{A}$, $\bar{C}$) by finding V_n and $\bar{V}_n$ and then using

$$T = (\bar{V}_n^T \bar{V}_n)^{-1} \bar{V}_n^T V_n. \qquad (2.4.17)$$

If the system has only one output so that V_n is square, then

$$T = \bar{V}_n^{-1} V_n. \qquad (2.4.18)$$

Jordan Normal Form

We have seen that using a state-space transformation it is often possible to place the system in a more convenient form that reveals its structure. Here, we explore this notion further.

2.4 Realization and Canonical Forms

The notions to be explored in this subsection are very important in implementing digital control systems, for the Jordan normal form consists of simple subsystems connected in series and parallel. Thus, it affords a stable way to implement control systems on a digital computer. We illustrate this in Example 2.4-3 and discuss it further in section 6.4.

A vector space $\mathbf{R}^n$ has no structure. It is homogeneous, that is, invariant with respect to translation, and isotropic, that is, invariant with respect to rotation. Once a linear operator $A \in \mathbf{R}^{n \times n}$ is defined on the space, however, the situation changes, for A induces a *structure* on the space. Our objective now is to examine the structure induced on $\mathbf{R}^n$ by a square linear operator A.

Eigenvalues and eigenvectors

Define the *eigenvalues* of A as those values λ_i of λ for which

$$\Delta(\lambda) \equiv |\lambda I - A| = 0. \tag{2.4.19}$$

These are just the poles of the system having A as system matrix; they are always n in number. Since the matrices $[A - \lambda_i I]$ are singular by the definition of λ_i, one can now find vectors v_i^1 in their nullspaces, so that

$$[A - \lambda_i I] v_i^1 = 0, \tag{2.4.20}$$

or

$$A v_i^1 = \lambda_i v_i^1. \tag{2.4.21}$$

This is an interesting statement. It says that the vectors v_i^1 are privileged in the sense that they are not rotated by A, but only scaled by the factor λ_i. One may imagine a sheet of rubber that is stretched in two directions: vectors in those two directions are not rotated, but a vector in any other direction will be rotated toward the stretching axes.

The vector v_i^1 is called the *rank 1 eigenvector associated with* λ_i. There are n eigenvalues. Supposing that there are n rank 1 eigenvectors, we may use (2.4.21) to write

$$A[v_1^1 \; v_2^1 \ldots v_n^1] = [v_1^1 \; v_2^1 \ldots v_n^1] \begin{bmatrix} \lambda_1 & & & \\ & \lambda_2 & & \\ & & \ddots & \\ & & & \lambda_n \end{bmatrix}. \tag{2.4.22}$$

By defining the *modal matrix*

$$M = [v_1^1 \; v_2^1 \ldots v_n^1] \tag{2.4.23}$$

and the diagonal matrix

$$J = \mathrm{diag}\{\lambda_i\} \tag{2.4.24}$$

this becomes

$$AM = MJ. \qquad (2.4.25)$$

If the n rank 1 eigenvectors are linearly independent, then M is nonsingular and one may write

$$J = M^{-1}AM. \qquad (2.4.26)$$

Therefore, $T = M^{-1}$ is a *state-space transformation* that converts the matrix A to a particularly convenient form, namely, the form J that is diagonal with entries equal to the eigenvalues.

Rank k eigenvectors and Jordan form

The reader is no doubt wondering about the superscript "1" appended to the vectors v_i. Unfortunately, it is not always possible to find n linearly independent rank 1 eigenvectors that satisfy (2.4.20) for some λ_i. If it is possible for a given A, the matrix A is said to be *simple*. Consider the case where A has nonrepeated (i.e., distinct) eigenvalues. Then, clearly, the dimension of the nullspace of $[A - \lambda_i I]$ is equal to at least 1 for each i so that there do exist n linearly independent rank 1 eigenvectors. (In fact, since there cannot be more than n independent eigenvectors, for a matrix with distinct eigenvalues the dimension of the nullspace of $[A - \lambda_i I]$ is *equal* to one.) Therefore, if the eigenvalues are distinct, then A is simple. (The converse is not true—that is, A may have repeated eigenvalues and still be simple.)

If there are not n linearly independent rank 1 eigenvectors, then the modal matrix M defined in (2.4.23) does not have n columns, and we are faced with finding some suitable additional columns to make M nonsingular. These columns should result in a transformation of A to a convenient form like the diagonal form J just given.

Define the rank k eigenvector associated with λ_i by

$$(A - \lambda_i I)v_i^{k+1} = v_i^k. \qquad (2.4.27)$$

To incorporate the definition of rank 1 eigenvector into this, we define $v_i^0 = 0$. This relation says that

$$Av_i^{k+1} = \lambda_i v_i^{k+1} + v_i^k. \qquad (2.4.28)$$

That is, the rank $k + 1$ eigenvector is not simply scaled by λ_i on multiplication with A (unless $k = 0$), but it is transformed in an elementary way. In fact, it is scaled, but with a component of v_i^k also added in. We call the sequence $\{v_i^1, v_i^2, \ldots\}$ an *eigenvector chain* for λ_i.

Define now the modal matrix by

$$M = [v_1^1 \quad v_1^2 \ldots v_2^1 \quad v_2^2 \ldots v_r^1 \quad v_r^2 \ldots], \qquad (2.4.29)$$

with r the number of independent rank 1 eigenvectors. That is, in defining M, the chains

2.4 Realization and Canonical Forms

for the λ_i are kept together. Then, a little thought on the definition (2.4.28) shows that (2.4.25) holds with the *Jordan matrix* of the form

$$J = \left[\begin{array}{ccc|cc|c} \lambda_1 & 1 & & & & \\ & \lambda_1 & 1 & & & \\ & & \lambda_1 & & & \\ \hline & & & \lambda_2 & 1 & \\ & & & & \lambda_2 & \\ \hline & & & & & \lambda_3 \end{array}\right] \equiv \begin{bmatrix} J_1 & & \\ & J_2 & \\ & & J_3 \end{bmatrix}. \quad (2.4.30)$$

The blocks J_i have been defined to correspond to a partition of J so that J_1 includes the entries with λ_1, and so on. We call the submatrices J_i the *Jordan blocks* corresponding to A. The structure they reveal is known as the *eigenstructure of A*. The structure illustrated here has three eigenvalue chains, of lengths three, two, and one. If A is simple, then J has no superdiagonal ones and is diagonal.

There is more to the study of the Jordan form than we are able to cover here. For more details see Brogan [1974] and Kailath [1980].

There are many commercially available routines for finding the eigenvalues and eigenvectors ([IMSL], LINPACK [Dongarra et al. 1979]). Therefore, suppose that the system (A, B, C, D) is given and that the modal matrix M has been found using one of these techniques. Then the *Jordan Normal Form* (*JNF*) of the system is given by

$$\dot{\bar{x}} = J\bar{x} + B^J u \quad (2.4.31)$$
$$y = C^J \bar{x} + Du$$

which is derived from (A, B, C, D) using the SST $T = M^{-1}$, so that

$$J = M^{-1}AM, \quad B^J = M^{-1}B, \quad C^J = CM. \quad (2.4.32)$$

The Jordan matrix J is diagonal with possibly some superdiagonal ones.

Left eigenvectors

According to (2.4.25) we may write

$$M^{-1}A = JM^{-1}. \quad (2.4.33)$$

Let us consider the case of simple A so that $J = \text{diag}\{\lambda_i\}$ and define the rows of M^{-1} as vectors w_i^T. Then

$$\begin{bmatrix} w_1^T \\ w_2^T \\ \vdots \\ w_n^T \end{bmatrix} A = \begin{bmatrix} \lambda_1 & & & \\ & \lambda_2 & & \\ & & \ddots & \\ & & & \lambda_n \end{bmatrix} \begin{bmatrix} w_1^T \\ w_2^T \\ \vdots \\ w_n^T \end{bmatrix} \quad (2.4.34)$$

or
$$w_i^T A = w_i^T \lambda_i, \tag{2.4.35}$$

which means that
$$w_i^T (A - \lambda_i I) = 0. \tag{2.4.36}$$

We call the vectors w_i the (rank 1) left eigenvectors of A. For clarity, we sometimes call v_i the *right* eigenvectors of A. (We eliminate the superscript "1" here since all the v_i are of rank 1 in this subsection.) Transposing (2.4.36) yields

$$(A - \lambda_i I)^T w_i = 0, \tag{2.4.37}$$

so that the left eigenvectors of A are seen to be the (right) eigenvectors of A^T. Since $M^{-1} M = I$, we see that

$$w_i^T v_j = \delta_{ij}, \tag{2.4.38}$$

with δ_{ij} the Kronecker delta. That is, the sets $\{w_i\}$ and $\{v_i\}$ are orthogonal. They are said to be *reciprocal bases*.

Parallel form realization

The transfer function of (A, B, C, D) is not changed by a SST, so that it equals the transfer function of the Jordan system. To conveniently find the transfer function of the Jordan system, partition it into *block diagonal subsystems* corresponding to the Jordan blocks as in (2.4.30). Illustrating for the case of three blocks, we may write

$$\frac{d}{dt}\begin{bmatrix} x_1 \\ x_2 \\ x_3 \end{bmatrix} = \begin{bmatrix} J_1 & & \\ & J_2 & \\ & & J_3 \end{bmatrix} \begin{bmatrix} x_1 \\ x_2 \\ x_3 \end{bmatrix} + \begin{bmatrix} B_1 \\ B_2 \\ B_3 \end{bmatrix} u \tag{2.4.39}$$

$$y = [C_1 \ C_2 \ C_3] \begin{bmatrix} x_1 \\ x_2 \\ x_3 \end{bmatrix} \tag{2.4.40}$$

The direct feed matrix D has been set to zero for notational ease. Since J is block diagonal, so is $(sI - J)$, which may therefore be inverted block by block to obtain the transfer function

$$H(s) = C_1(sI - J_1)^{-1} B_1 + C_2(sI - J_2)^{-1} B_2 + C_3(sI - J_3)^{-1} B_3.$$

Therefore, it is seen that the Jordan form partitions the system into subsystems connected in parallel corresponding to the Jordan blocks. We call this a *parallel form realization of the transfer function*. To implement this system using integrators, we may use three parallel channels. If J has the form in (2.4.30), one channel will have three integrators, one will have two, and one will have a single integrator.

2.4 Realization and Canonical Forms

JNF from partial fraction expansion

For a system (A, B, C), the Jordan form is given by $(M^{-1}AM, M^{-1}B, CM)$, with M the modal matrix. (Set $D = 0$ for convenience.) Therefore, the transfer function may be written as

$$H(s) = C(sI - A)^{-1}B = CMM^{-1}(sI - A)^{-1}MM^{-1}B = CM(sI - J)^{-1}M^{-1}B$$

Assume that the A matrix is simple for notational ease. Then

$$H(s) = C[v_1 \ v_2 \ldots v_n] \begin{bmatrix} s - \lambda_1 & & & \\ & s - \lambda_2 & & \\ & & \ddots & \\ & & & s - \lambda_n \end{bmatrix}^{-1} \begin{bmatrix} w_1^T \\ w_2^T \\ \vdots \\ w_n^T \end{bmatrix} B$$

or

$$H(s) = \sum_{i=1}^{n} \frac{C v_i w_i^T B}{s - \lambda_i}. \qquad (2.4.41)$$

(We suppress the superscript "1" on the v_i.) This gives a partial fraction expansion (PFE) in terms of the modal structure of A. The simplicity of A means precisely that the PFE has no terms involving $(s - \lambda_i)^k$ for $k > 1$.

The residue of the pole λ_i is equal to $Cv_i w_i^T B$, so that if either Cv_i or $w_i^T B$ is zero, then λ_i does not contribute to the input/output response. Note that for a MIMO system with m inputs and p outputs, the residues are $p \times m$ *matrices* and not scalars.

It is not difficult to show that if $w_i^T B = 0$ for any w_i satisfying (2.4.36), then rank$(U_n) \neq n$ so that the system is not reachable (see the problems). Then, we say that the eigenvalue λ_i is unreachable. Similarly, if $Cv_i = 0$ for any rank 1 right eigenvector v_i, then the system is unobservable, since rank$(V_n) \neq n$. Then, λ_i is said to be unobservable. In fact, these statements are necessary and sufficient; they are known as the Popov-Belevitch-Hautus (PBH) eigenvector tests for reachability and observability [Kailath 1980].

Therefore, a pole that is not both reachable and observable makes no contribution to the transfer function. This means that the associated natural mode plays no role in the input/output behavior of the system. Algebraically, this means that the transfer function has a zero that cancels out that pole.

The PFE gives us a convenient way to find a parallel state-space realization for SISO systems. Thus, given a transfer function $H(s)$, one may perform a PFE to obtain

$$H(s) = \sum_{i=1}^{n} \frac{K_i}{s - \lambda_i}. \qquad (2.4.42)$$

Then, a JNF state-space system with transfer function $H(s)$ is given by

$$\dot{x} = \begin{bmatrix} \lambda_1 & & & \\ & \lambda_2 & & \\ & & \ddots & \\ & & & \lambda_n \end{bmatrix} x + \begin{bmatrix} b_1 \\ b_2 \\ \vdots \\ b_3 \end{bmatrix} u$$

$$y = [c_1 \quad c_2 \quad c_n] x \qquad (2.4.43)$$

where b_i, c_i are any scalars such that $c_i b_i = K_i$.

Example 2.4-2: Pole/Zero Cancellation and JNF Realization

Consider the SISO transfer function

$$H(s) = \frac{s^2 - 1}{s^3 + 6s^2 + 11s + 6}. \qquad (1)$$

a. Pole/Zero Cancellation

A partial fraction expansion yields

$$H(s) = \frac{s^2 - 1}{(s+1)(s+2)(s+3)} = \frac{0}{s+1} + \frac{-3}{s+2} + \frac{4}{s+3}. \qquad (2)$$

Notice, however, that

$$H(s) = \frac{(s+1)(s-1)}{(s+1)(s+2)(s+3)} = \frac{s-1}{(s+2)(s+3)}, \qquad (3)$$

which has the same PFE. Thus, for SISO systems a zero term in the PFE means exactly that there is a pole/zero cancellation in $H(s)$.

b. JNF Realization

From the PFE, one may write down a JNF state-space realization of $H(s)$. One realization is

$$\dot{x} = \begin{bmatrix} -2 & 0 \\ 0 & -3 \end{bmatrix} x + \begin{bmatrix} 1 \\ 1 \end{bmatrix} u \qquad (4)$$

$$y = [-3 \quad 4] x. \qquad (5)$$

One should verify that the transfer function of this system is the given $H(s)$.

Other realizations are possible that have different B and C matrices $B = [b_1 \quad b_2]^T$ and $C = [c_1 \quad c_2]$, as long as $b_1 c_1 = -3$, $b_2 c_2 = 4$.

c. Time Response Simulation

Using the state-space realization (4)–(5), computer simulation may be used to plot the system response given any input $u(t)$, as discussed in section 2.1.

■

2.4 Realization and Canonical Forms

Example 2.4-3: Real Jordan Form in Controller Implementation

As we shall see in section 6.4, due to computer roundoff errors it is important to implement digital controllers using small subsystems in series and parallel. This may be accomplished using the *real Jordan form*.

a. Real Jordan Form

The real Jordan form has the structure

$$J = \text{diag}(J_i), \tag{1}$$

with Jordan blocks of the form

$$J_i = \begin{bmatrix} \Lambda & I & & & \\ & \Lambda & \cdot & & \\ & & \cdot & \cdot & I \\ & & & \cdot & \Lambda & I \\ & & & & & \Lambda \end{bmatrix} \tag{2}$$

where, for complex eigenvalues $\lambda_R + j\lambda_I$, I is the 2×2 identity matrix and

$$\Lambda = \begin{bmatrix} \lambda_R & \lambda_I \\ -\lambda_I & \lambda_R \end{bmatrix} \tag{3}$$

and for real eigenvalues λ, $I = 1$ and $\Lambda = \lambda$. The real Jordan form is easily determined from the (complex) Jordan form.

The real Jordan form consists of first- and second-order systems in series and parallel. The Jordan blocks are in parallel, and the first- and second-order subsystems within each Jordan block are in series.

b. Digital Controller Implementation

A digital controller can be expressed using the discrete-time state equations

$$x_{k+1} = Ax_k + Bu_k \tag{4}$$

$$y_k = Cx_k + Du_k. \tag{5}$$

If the dimension of x_k is large, severe numerical problems can result if the controller is implemented directly on a digital computer with finite wordlength. The real Jordan form can be used to minimize these numerical problems.

Suppose a two-input one-output digital controller is designed using the techniques to be discussed in Part III of the book. Let the dynamical description of the controller be

$$x_{k+1} = \begin{bmatrix} 0.875 & 0.1 & -0.025 \\ -0.025 & 0.9 & 0.075 \\ 0.075 & -0.1 & 0.975 \end{bmatrix} x_k + \begin{bmatrix} 1 & -1 \\ 1 & 1 \\ 1 & 1 \end{bmatrix} u_k \tag{6}$$

$$y_k = [0.5 \quad 0.5 \quad 0] x_k. \tag{7}$$

Using the state-space transformation

$$T = \begin{bmatrix} 1 & 1 & -1 \\ 1 & 1 & 1 \\ 1 & -1 & 1 \end{bmatrix} \qquad (8)$$

the system may be brought to the real Jordan form

$$\bar{x}_{k+1} = \begin{bmatrix} 0.95 & 0 & 0 \\ 0 & 0.9 & 0.1 \\ 0 & -0.1 & 0.9 \end{bmatrix} \bar{x}_k + \begin{bmatrix} 1 & 0 \\ 0 & 0 \\ 0 & 1 \end{bmatrix} u_k \qquad (9)$$

$$y_k = [1 \ 1 \ 0] x_k. \qquad (10)$$

It is now seen that the controller poles are at $z = 0.95$, $z = 0.9 \pm j0.1$.

The form (9), (10) of the controller consists of a first-order subsystem in parallel with a second-order subsystem. Thus, it will not suffer as many numerical problems on a finite wordlength machine as a direct implementation of the third-order controller (6).

Defining the state components by $x_k = [x_k^1 \ x_k^2 \ x_k^3]^T$ and the control input components by $u_k = [u_k^1 \ u_k^2]^T$, a numerically stable difference equation implementation of this controller is now given by

$$x_{k+1}^1 = 0.95 x_k^1 + u_k^1$$

$$x_{k+1}^2 = 0.9 x_k^2 + 0.1 x_k^3 \qquad (11)$$

$$x_{k+1}^3 = -0.1 x_k^2 + 0.9 x_k^3 + u_k^2$$

$$y_k = x_k^1 + x_k^2.$$

These simple equations are easily programmed on a microprocessor for controls implementation purposes. See Chapter 6.

■

Modal decomposition

The Jordan form reveals a great deal about the structure of the system in the time domain as well as the frequency domain. Let us now investigate this structure.

Since $A = MJM^{-1}$, and e^{At} is a polynomial in A, we may write

$$e^{At} = Me^{Jt}M^{-1}.$$

Therefore, the solution to the continuous state equation from section 2.1 may be written as

$$x(t) = e^{At}x_0 + \int_0^t e^{A(t-\tau)}Bu(\tau) \, d\tau$$

$$= Me^{Jt}M^{-1}x_0 + \int_0^t Me^{J(t-\tau)}M^{-1}Bu(\tau) \, d\tau.$$

2.4 Realization and Canonical Forms

For notational ease, let us take A simple so that $J = \text{diag}\{\lambda_i\}$. The superscript "1" denoting rank 1 eigenvectors will be suppressed. Focusing on the first term, one has

$$x(t) = [v_1 \; v_2 \; \ldots \; v_n] \begin{bmatrix} e^{\lambda_1 t} & & & \\ & e^{\lambda_2 t} & & \\ & & \ddots & \\ & & & e^{\lambda_n t} \end{bmatrix} \begin{bmatrix} w_1^T \\ w_2^T \\ \vdots \\ w_n^T \end{bmatrix} x_0$$

$$= \sum_{i=1}^{n} v_i e^{\lambda_i t} w_i^T x_0 \tag{2.4.44}$$

or, since $w_i^T x_0$ is a scalar

$$x(t) = \sum_{i=1}^{n} (w_i^T x_0) e^{\lambda_i t} v_i. \tag{2.4.45}$$

In similar fashion, one may deal with the forced solution to obtain the complete state solution

$$x(t) = \sum_{i=1}^{n} (w_i^T x_0) e^{\lambda_i t} v_i + \sum_{i=1}^{n} v_i \int_0^t e^{\lambda_i(t-\tau)} w_i^T B u(\tau) \, d\tau. \tag{2.4.46}$$

The output (with $D = 0$ for simplicity) is now seen to be equal to

$$y(t) = \sum_{i=1}^{n} (w_i^T x_0) e^{\lambda_i t} C v_i + \sum_{i=1}^{n} C v_i \int_0^t e^{\lambda_i(t-\tau)} w_i^T B u(\tau) \, d\tau. \tag{2.4.47}$$

These important expressions yield a great deal of insight. To understand why, note that these solutions are a sum of motions in the directions of the eigenvectors v_i. Each motion is an exponential of the form $e^{\lambda_i t}$. That is, the solution has been decomposed into a sum of the natural modes $e^{\lambda_i t}$. We call this the *modal decomposition* of the solution of the state equation.

Since $\{w_i\}$ is a reciprocal basis for $\{v_i\}$, the scalar coefficient $(w_i^T x_0)$ is the component of the initial condition x_0 in the direction of eigenvector v_i. The row vector $w_i^T B$ maps the input $u(t)$ into the direction v_i. That is, $w_i^T B u(t)$ represents the influence of the input on the natural mode of λ_i. The influence of the natural mode $e^{\lambda_i t}$ on the output $y(t)$ is described by the term $C v_i$. Therefore, the modal decomposition decomposes the input and the initial conditions into terms that separately excite the natural modes, and then recombines them to reconstruct the output response $y(t)$.

Reachability means that $w_i^T B \neq 0$ for all i, and guarantees that the input independently influences all the modes. Observability means that $C v_i \neq 0$ for all i, and guarantees that all modes appear independently in the output.

Reachable Canonical Form

We have just seen how to use a partial fraction expansion of a given transfer function to find a JNF parallel state-space realization. In this section and the next we show how to find two series state-variable realizations of a differential equation or transfer function [Kailath 1980]. In many applications where realization theory is needed, the transfer function is SISO. For instance, we may have a first-order model of an electric motor that actuates a plant we are interested in controlling. For modern controls design, the actuator must be incorporated into a state model of the plant. Therefore, we shall restrict our attention to SISO systems in this subsection and the next.

Suppose there is available an ordinary differential equation (ODE) describing the relation between the input $u(t)$ and output $y(t)$ of a plant. Write it as

$$y^{(n)} + a_1 y^{(n-1)} + \ldots + a_n y = b_1 u^{(n-1)} + \ldots + b_n u, \qquad (2.4.48)$$

where superscript (i) denotes the i-th derivative and a_i, b_i are known real numbers. Taking the Laplace transform, with initial conditions of zero, yields the frequency-domain description

$$(s^n + a_1 s^{n-1} + \ldots + a_n) Y(s) = (b_1 s^{n-1} + \ldots + b_n) U(s). \qquad (2.4.49)$$

Therefore, the transfer function is given by

$$H(s) = \frac{Y(s)}{U(s)} = \frac{b_1 s^{n-1} + \ldots + b_n}{s^n + a_1 s^{n-1} + \ldots + a_n}. \qquad (2.4.50)$$

RCF block diagram

It is not difficult to draw a block diagram corresponding to this transfer function. Such a diagram is given in Fig. 2.4-3 for the case $n = 4$. A number, such as a_i, next to a path represents the path's transmission, so that signals in that path are multiplied by a_i. Using Mason's rule [Franklin, Powell, and Emami-Naeini 1986], it is straightforward to verify that the transfer function of this block diagram is nothing but $H(s)$.

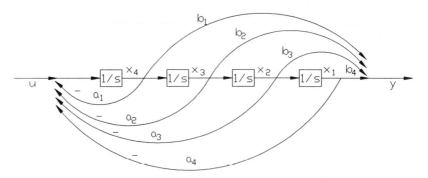

Figure 2.4-3 Reachable canonical form block diagram

2.4 Realization and Canonical Forms

Note that the diagram is in series form. For reasons that will soon become clear, we call this specific series arrangement the *reachable canonical form* (*RCF*) block diagram of $H(s)$. If the system is discrete-time, then the integrators $1/s$ in the block diagram are replaced by delay elements z^{-1}, so that the diagram shows a shift register of length n.

RCF state-space form

To find a state-space realization, number the states from right to left as shown in the figure; each state corresponds to an integrator output since the integrators are where energy is stored in the system. Now the state equation can be written down by inspection by considering the summer at the input of each integrator. The result, shown for $n = 4$, is

$$\dot{x} = \begin{bmatrix} 0 & 1 & 0 & 0 \\ 0 & 0 & 1 & 0 \\ 0 & 0 & 0 & 1 \\ -a_4 & -a_3 & -a_2 & -a_1 \end{bmatrix} x + \begin{bmatrix} 0 \\ 0 \\ 0 \\ 1 \end{bmatrix} u = Ax + Bu \quad (2.4.51)$$

$$y = [b_4 \ b_3 \ b_2 \ b_1] = Cx, \quad (2.4.52)$$

where $x = [x_1 \ x_2 \ x_3 \ x_4]^T$.

We call this state-space realization the *reachable canonical form* (RCF) realization of $H(s)$. The A matrix is called a *bottom companion matrix* to the characteristic equation

$$\Delta(s) = s^n + a_1 s^{n-1} + \cdots + a_n. \quad (2.4.53)$$

It is a good exercise to show that $|sI - A|$ is indeed given by (2.4.53) (see the problems) and that the transfer function $C(sI - A)^{-1}B$ is given by (2.4.50).

Except for the bottom row, the A matrix consists of zeros with superdiagonal ones. This means precisely that all the integrators are connected in series (i.e., $\dot{x}_i = x_{i+1}$). The B matrix is all zero except for the bottom entry, meaning exactly that the input is hard-wired to the left-most integrator.

Note that the RCF state equations can be written down directly from $H(s)$ without drawing the block diagram; the C matrix is just composed of the numerator coefficients in reverse order, and the last row of the A matrix consists of the negatives of the denominator coefficients in reverse order.

The *relative degree* of the transfer function (2.4.50) is defined as the denominator degree minus the numerator degree. The relative degree is one if $b_1 \neq 0$, otherwise it is greater than one. To find the RCF of a transfer function with relative degree of zero, one must perform an initial decomposition of $H(s)$ so that

$$H(s) = H'(s) + D$$

with D a constant and $H'(s)$ having a relative degree of one or more. Then, the RCF for $H(s)$ is found by writing the RCF (2.4.51) for $H'(s)$, and adding the direct feed matrix D so that (2.4.52) becomes $y = Cx + Du$.

The states x_i could have been numbered in a different order from the one chosen. Then, the matrices A, B, and C would have different forms, but the transfer function is still $H(s)$. In fact, two different possible systems (A, B, C) obtained by different orderings of the states are related by a state-space transformation.

Reachability of the RCF

By direct computation using A and B in (2.4.51), the reachability matrix of the RCF is found to be

$$U_4 = \begin{bmatrix} 0 & 0 & 0 & 1 \\ 0 & 0 & 1 & -a_1 \\ 0 & 1 & -a_1 & a_1^2 - a_2 \\ 1 & -a_1 & a_1^2 - a_2 & -a_1^3 + 2a_1 a_2 - a_3 \end{bmatrix}, \quad (2.4.54)$$

where we have taken $n = 4$. For any choice of a_i this is nonsingular, so that the *RCF is always reachable*. Hence, its name.

Example 2.4-4: Analysis and Simulation Using the RCF

The differential equation describing the linearized input/output behavior of a hydraulic transmission coupled through a spring to its load is given by D'Azzo and Houpis [1988]:

$$y''' + 11y'' + 39y' + 29y = 29u'' + 29u, \quad (1)$$

where $u(t)$ is the control stroke (between ± 1) and $y(t)$ is the motor output angle.

a. Transfer Function

The transfer function is written by inspection as

$$H(s) = \frac{29s^2 + 29}{s^3 + 11s^2 + 39s + 29} \quad (2)$$

or

$$H(s) = \frac{29(s^2 + 1)}{(s + 1)[(s + 5)^2 + 2^2]}. \quad (3)$$

Therefore, the system has a real pole at $s = -1$, which corresponds to the hydraulic motor time constant of 1 sec, and a complex pole pair at $s = -5 \pm j2$, corresponding to the high-frequency behavior of the spring coupler. There are two zeros at $s = \pm j$.

b. RCF Realization

This system could be analyzed and simulated using the theory of ordinary differential equations. However, given the availability of digital computers, the state-space theory is more convenient.

The RCF is written by inspection as

$$\dot{x} = \begin{bmatrix} 0 & 1 & 0 \\ 0 & 0 & 1 \\ -29 & -39 & -11 \end{bmatrix} x + \begin{bmatrix} 0 \\ 0 \\ 1 \end{bmatrix} u = Ax + Bu \quad (4)$$

$$y = [29 \quad 0 \quad 29]x. \quad (5)$$

2.4 Realization and Canonical Forms

Now, the theory of section 2.1 could be used to analytically find the time response $y(t)$ given any control input $u(t)$ and initial conditions.

c. Time Response Simulation

Using program TRESP in Appendix A, a computer simulation of the step response of the system is easily obtained as described in section 2.1. It is shown in Fig. 2.4-4. The natural modes corresponding to both the real motor pole and the complex spring coupler poles are clearly visible.

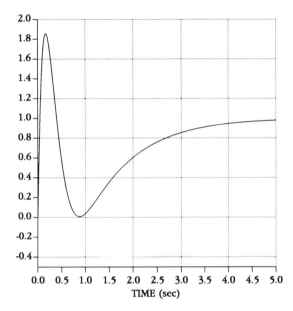

Figure 2.4-4 Step response of hydraulic transmission system

When the step command is applied, the motor angle initially increases. As the spring couples the motion to the inertial load, it begins to move, and due to the snap provided by the spring it pulls the motor angle to a value that is too high. Then, the load angle decreases as the spring oscillates; this retards the motor angle. Finally, after the spring oscillation dies out (at a rate given by the decay term e^{-5t}), the motor angle settles t the desired value of 1 rad. The final motion is governed by the motor natural mode of e^{-t}.

∎

Observable Canonical Form

There are many block diagrams that correspond to the differential equation (2.4.48) and associated transfer function (2.4.50). Another one is given in Fig. 2.4-5. Use Mason's rule to verify it has the same transfer function as Fig. 2.4-3.

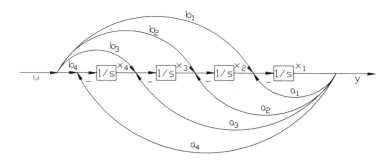

Figure 2.4-5 Observable canonical form block diagram

Labeling the states as shown results in the state-variable equations

$$\dot{x} = \begin{bmatrix} -a_1 & 1 & 0 & 0 \\ -a_2 & 0 & 1 & 0 \\ -a_3 & 0 & 0 & 1 \\ -a_4 & 0 & 0 & 0 \end{bmatrix} x + \begin{bmatrix} b_1 \\ b_2 \\ b_3 \\ b_4 \end{bmatrix} u = Ax + Bu \qquad (2.4.55)$$

$$y = [1 \ 0 \ 0 \ 0]x = Cx, \qquad (2.4.56)$$

where $x = [x_1 \ x_2 \ x_3 \ x_4]^T$.

We call this state-space realization the *observable canonical form* (OCF) realization of $H(s)$. The A matrix is called a *left companion matrix* to the characteristic equation (2.4.53). It is a good exercise to show that $|sI - A|$ is indeed given by (2.4.53) (see the problems) and that the transfer function $C(sI - A)^{-1}B$ is given by (2.4.50).

Except for the first column, the A matrix consists of zeros with superdiagonal ones. This means precisely that all the integrators are connected in series [i.e., $\dot{x}_i = x_{i+1}$ plus feedback from x_1 and feedforward from $u(t)$]. The C matrix is all zero except for the first entry, meaning exactly that the output is hard-wired to the right-most integrator.

Note that the OCF state equations can be written down directly from $H(s)$ without drawing the block diagram; the B matrix is just composed of the numerator coefficient vector stood on end, and the first column of the A matrix consists of the negatives of the denominator coefficient vector stood on end.

The remarks made in connection with the RCF on ordering the states differently and handling a transfer function with relative degree of zero hold here also.

By direct computation using A and C in the OCF state-space form, the observability matrix of the OCF is found to be

$$V_4 = \begin{bmatrix} -a_1^3 + 2a_1a_2 - a_3 & a_1^2 - a_2 & -a_1 & 1 \\ a_1^2 - a_2 & -a_1 & 1 & 0 \\ -a_1 & 1 & 0 & 0 \\ 1 & 0 & 0 & 0 \end{bmatrix}, \qquad (2.4.57)$$

where we have taken $n = 4$. For any choice of a_i this is nonsingular, so that *the OCF is always observable*, hence its name.

2.4 Realization and Canonical Forms

Duality

The RCF and the OCF are related in quite an interesting way. Note that Fig. 2.4-3, when held to the light and viewed through the back of the page, has the same form as Fig. 2.4-5. To make these two views identical, one should reverse the arrows along all the paths of the RCF block diagram, and reverse the numbering of the state components x_i. In circuit theory, this relation is called *duality*. The reversal of the arrows corresponds to exchanging the roles of the input $u(t)$ and the output $y(t)$.

In state-space terms, the dual to a system (A, B, C, D) is given by writing $(A^D, B^D, C^D, D^D) = (A^T, C^T, B^T, D^T)$. That is, A and D are replaced by their transposes, and B (resp. C) is replaced by C^T (resp. B^T). One may verify that by performing these replacements on the RCF state-variable form, one obtains the OCF equations. To obtain a perfect match, it will also be necessary to renumber the states in reverse order. This corresponds to reversing the rows of A and B, and the columns of A and C.

Minimality and Gilbert's Realization Method

We have shown several methods for obtaining state-space realizations for SISO transfer functions. We now show how to find a realization for MIMO systems using Gilbert's method [Brogan 1974]. This method works for many systems of interest, and finds a *minimal* state-space realization; that is, one of the smallest order n.

While discussing (2.4.41) we saw that if the system is unreachable, so that $w_i^T B = 0$ for some rank 1 left eigenvector w_i, or unobservable, so that $Cv_i = 0$, for some rank 1 right eigenvector v_i, then a PFE of a SISO transfer function will have some zero terms. The next example is relevant.

Example 2.4-5: Realization and Minimality for SISO Systems

From the ordinary differential equation

$$y''' + 6y'' + 11y' + 6y = u'' - u, \tag{1}$$

the transfer function can be written down. It is

$$H(s) = \frac{s^2 - 1}{s^3 + 6s^2 + 11s + 6}. \tag{2}$$

This is the same transfer function that was used in Example 2.4-2.

a. RCF

From $H(s)$ one may directly write down the RCF realization

$$\dot{x} = \begin{bmatrix} 0 & 0 & 1 \\ 0 & 1 & 0 \\ -6 & -11 & -6 \end{bmatrix} x + \begin{bmatrix} 0 \\ 0 \\ 1 \end{bmatrix} u$$

$$y = \begin{bmatrix} -1 & 0 & 1 \end{bmatrix} x. \tag{3}$$

Recall that a PFE of $H(s)$ yields

$$H(s) = \frac{s^2 - 1}{(s+1)(s+2)(s+3)} = \frac{0}{s+1} + \frac{-3}{s+2} + \frac{4}{s+3}. \tag{4}$$

Since this has a zero residue for the pole at $s = -1$, the system cannot be both reachable and observable. One may easily check U_3 to see that the system is reachable. (Indeed, since (3) is in RCF, it must be reachable.) However,

$$V_3 = \begin{bmatrix} C \\ CA \\ CA^2 \end{bmatrix} = \begin{bmatrix} -1 & 0 & 1 \\ -6 & -11 & -7 \\ 42 & 66 & 36 \end{bmatrix} \tag{5}$$

which has a determinant of zero, so that (A, C) is not observable. Therefore, system (3) is deficient in its output coupling so that $y(t)$ does not convey all the information stored in the state $x(t)$.

b. Minimal State-Space Realization

If a system (A, B, C) is not both reachable and observable, we say it is not *minimal*. In the SISO case, this means that there is pole/zero cancellation in its transfer function. Notice that

$$H(s) = \frac{(s+1)(s-1)}{(s+1)(s+2)(s+3)} = \frac{s-1}{(s+2)(s+3)} = \frac{s-1}{s^2 + 5s + 6}, \tag{6}$$

which also has the PFE (4). A RCF realization of the reduced transfer function is given by

$$\dot{x} = \begin{bmatrix} 0 & 1 \\ -6 & -5 \end{bmatrix} x + \begin{bmatrix} 0 \\ 1 \end{bmatrix} u$$
$$y = [-1 \quad 1]x. \tag{7}$$

The reader should verify that this system is both reachable and observable and hence minimal.

The differential equation corresponding to the reduced transfer function (6) may be directly written down. It is

$$y'' + 5y' + 6 = u' - u. \tag{8}$$

This ODE has the same input/output behavior as (1); that is, if the initial conditions are zero in each equation, their outputs $y(t)$ are identical for the same inputs $u(t)$.

c. Time Response

Using either (3) or (7), the time response program TRESP in Appendix A (see section 2.1) may be used to plot the step response, or indeed the response $y(t)$ for any input $u(t)$. This is considerably easier than solving the ODE or simulating it on a digital computer in the form of a higher-order ODE.

∎

We say that any system (A, B, C) that is not both reachable and observable is not *minimal*. If a system is not minimal, then one can find a state-space description *of smaller dimension* that has the same transfer function.

2.4 Realization and Canonical Forms

Minimality can be important from several points of view. Suppose, for instance, that one wants to simulate the ODE (1) of the previous example on an analog computer. Suppose, however, that the only available analog computer has only four integrators. Then, the equation (1) cannot be simulated. However, the minimal system (7) can easily be simulated using four integrators—one for the integrator associated with each state, one for the summer to manufacture $y(t)$, and one for the summer to provide the feedback. See Fig. 2.4-3.

If the system is SISO, then minimality corresponds to the fact that *the transfer function has no pole/zero cancellations*. However, in the multivariable case, this is not so. That is, the transfer function may have no apparent pole/zero cancellations yet still be nonminimal.

To illustrate this consider the next example, which shows how to obtain a minimal realization of a given MIMO transfer function as long as it has only linear terms in its PFE.

Example 2.4-6: Gilbert's Method for Multivariable Minimal Realizations

Consider the state-space system

$$\dot{x} = \begin{bmatrix} -1 & 0 & 0 & 0 & 0 & 0 \\ 0 & 0 & 1 & 0 & 0 & 0 \\ 0 & -3 & -4 & 0 & 0 & 0 \\ 0 & 0 & 0 & 0 & 1 & 0 \\ 0 & 0 & 0 & -2 & -3 & 0 \\ 0 & 0 & 0 & 0 & 0 & -3 \end{bmatrix} x + \begin{bmatrix} 1 & 0 \\ 0 & 0 \\ 1 & 0 \\ 0 & 0 \\ 0 & 1 \\ 0 & 1 \end{bmatrix} u \quad (1)$$

$$y = \begin{bmatrix} 1 & 0 & 0 & 2 & 0 & 0 \\ 0 & 1 & 0 & 0 & 0 & 1 \end{bmatrix} x. \quad (2)$$

Since this system has four diagonal blocks of orders 1, 2, 2, and 1, all in RCF, it is easy to analyze. Defining $u = [u_1 \; u_2]^T$ and $y = [y_1 \; y_2]^T$, the block diagram in Fig. 2.4-6 may easily be drawn by inspection; it is just four systems like the RCF in Fig. 2.4-3 placed in parallel.

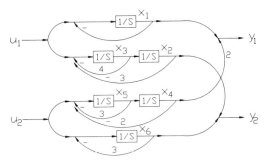

Figure 2.4-6 Block diagram of MIMO system

a. Nonmimality of System

Using our knowledge of the RCF, the transfer function may be written down by inspection. It is

$$H(s) = \begin{bmatrix} \dfrac{1}{s+1} & \dfrac{2}{s^2+3s+2} \\ \dfrac{1}{s^2+4s+3} & \dfrac{1}{s+3} \end{bmatrix} \tag{3}$$

To understand how this is written down, note, for instance, that the second subsystem is

$$A_2 = \begin{bmatrix} 0 & 1 \\ -3 & -4 \end{bmatrix}, \quad B_2 = \begin{bmatrix} 0 & 0 \\ 1 & 0 \end{bmatrix}, \quad C_2 = \begin{bmatrix} 0 & 0 \\ 1 & 0 \end{bmatrix}, \tag{4}$$

which has the second column of B_2 and the first row of C_2 equal to zero. Therefore, this subsystem is in RCF and maps u_1 to y_2. It thus corresponds to the (2, 1) entry of the 2 × 2 transfer function matrix $H(s)$,

There is no evident pole/zero cancellation in $H(s)$, as each term is irreducible. However, one may find the reachability and observability matrices to convince oneself that the system is not minimal.

b. Gilbert's Method for Minimal Realizations

We shall now demonstrate how to find a minimal state-space realization of $H(s)$ that works in many situations. First, perform a PFE on $H(s)$ to obtain

$$H(s) = \begin{bmatrix} \dfrac{1}{s+1} & \dfrac{2}{(s+1)(s+2)} \\ \dfrac{1}{(s+1)(s+3)} & \dfrac{1}{s+3} \end{bmatrix}$$

$$= \begin{bmatrix} \dfrac{1}{s+1} & \dfrac{2}{s+1} - \dfrac{2}{s+2} \\ \dfrac{1/2}{s+1} - \dfrac{1/2}{s+3} & \dfrac{1}{s+3} \end{bmatrix} \tag{5}$$

$$= \dfrac{\begin{bmatrix} 1 & 2 \\ 1/2 & 0 \end{bmatrix}}{s+1} + \dfrac{\begin{bmatrix} 0 & -2 \\ 0 & 0 \end{bmatrix}}{s+2} + \dfrac{\begin{bmatrix} 0 & 0 \\ -1/2 & 1 \end{bmatrix}}{s+3}.$$

Note that in the MIMO case the residues of the poles are *matrices* and not scalars.

The second step is to factor the residue matrices into a *minimal number* of vector outer-products of the form vz^T. The result is

$$H(s) = \dfrac{\begin{bmatrix} 1 \\ 0 \end{bmatrix}[1 \ 2] + \begin{bmatrix} 0 \\ 1 \end{bmatrix}[1/2 \ 0]}{s+1} + \dfrac{\begin{bmatrix} 1 \\ 0 \end{bmatrix}[0 \ -2]}{s+2} + \dfrac{\begin{bmatrix} 0 \\ 1 \end{bmatrix}[-1/2 \ 1]}{s+3} \tag{6}$$

It should be understood that this step is not unique. That is, other definitions of the outer products are possible.

2.5 Feedback Control

The third step is to pull the column vectors out to the left and the row vectors out to the right to yield

$$H(s) = \begin{bmatrix} 1 & 0 & 1 & 0 \\ 0 & 1 & 0 & 1 \end{bmatrix} \begin{bmatrix} (s+1)^{-1} & & & \\ & (s+1)^{-1} & & \\ & & (s+2)^{-1} & \\ & & & (s+3)^{-1} \end{bmatrix} \quad (7)$$

$$\times \begin{bmatrix} 1 & 2 \\ 1/2 & 0 \\ 0 & -2 \\ -1/2 & 1 \end{bmatrix}$$

We are now done, for compare this to $H(s) = C(sI - A)^{-1}B$ to see that it is just four scalar subsystems in parallel. The middle diagonal matrix corresponds to $(sI - A)^{-1}$, so that the state-space system with transfer function $H(s)$ is seen to be

$$\dot{x} = \begin{bmatrix} -1 & & & \\ & -1 & & \\ & & -2 & \\ & & & -3 \end{bmatrix} x + \begin{bmatrix} 1 & 2 \\ 1/2 & 0 \\ 0 & -2 \\ -1/2 & 1 \end{bmatrix} u = Ax + Bu \quad (8)$$

$$y = \begin{bmatrix} 1 & 0 & 1 & 0 \\ 0 & 1 & 0 & 1 \end{bmatrix} x. \quad (9)$$

This procedure has resulted in a realization of order 4 with the same transfer function as the realization (1), (2) which has order 6. The reachability and observability matrices may be found to verify that this system is both reachable and observable, and hence minimal.

The outer-product factors in (6) are not unique. However, it is important that the *minimum number* of outer-products be used, since to each one there will correspond a state.

Gilbert's method works, and is guaranteed to produce a minimal realization as long as the factorization (6) is performed with the smallest possible number of outer products, whenever they are only linear terms in the PFE of $H(s)$. For instance, no terms like $(s + p_i)^2$ are allowed in the PFE. ∎

2.5 FEEDBACK CONTROL

We have seen how to derive the state equations for systems, and how to analyze state-space models and perform some manipulations with them. In this section we shall use the state model for *design* of control systems, introducing the concept of *feedback control* for state-space systems. In classical control, the loops of a multi-input/multi-output or multiloop system are closed *one at a time* using SISO techniques like root locus design. A basic notion in modern control, by contrast, is that of closing *all the feedback loops simultaneously*.

Using state-space techniques, all the feedback gains in a multiloop system are computed simultaneously by solving *matrix design equations* using digital computers. Different design equations apply for different sorts of control schemes. One objective of this book is to derive the matrix design equations for different situations and show how to use them.

Some important notions of classical control carry over to multivariable state-space control. In this section we shall mention the multivariable loop gain and return difference. In Chapter 8 we shall further discuss these, while also introducing the multivariable sensitivity and Bode plot, which is based on the singular value decomposition (see Appendix B).

In this section we shall discuss *state-variable feedback*, where all the internal states are available for feedback purposes, and *output feedback*, where only the measured outputs are available for feedback.

State-Variable Feedback

Figure 2.5-1 shows a basic control system consisting of the plant or system to be controlled

$$\dot{x} = Ax + Bu \qquad (2.5.1)$$

$$y = Cx \qquad (2.5.2)$$

and the feedback control

$$u = -Kx + v, \qquad (2.5.3)$$

where $v(t)$ is an auxillary input signal. By substituting (2.5.3) into (2.5.1) the state equation for the closed-loop system is found to be

$$\dot{x} = (A - BK)x + Bv \equiv A_c x + Bv. \qquad (2.5.4)$$

The state-feedback design problem is to select the feedback gain K so that the closed-loop system has desired properties. The convenience of our matrix notation conceals the fact that this can be a complicated issue, for the system can have *multiple inputs*, corresponding to the multiple components of the input vector $u(t)$. Let us take $x \in \mathbf{R}^n$, $u(t) \in \mathbf{R}^m$, so that there are m inputs. Then K is an $m \times n$ matrix, so that it is necessary to select mn scalar gains. This corresponds to closing mn feedback loops to obtain good

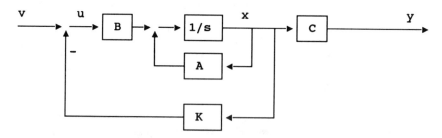

Figure 2.5-1 State-variable feedback system

2.5 Feedback Control

closed-loop performance. Clearly, this would not be easy to achieve using classical one-loop-at-a-time techniques.

Loop gain and closed-loop transfer function

Defining the open-loop transfer function

$$G(s) = (sI - A)^{-1}B \qquad (2.5.5)$$

from $u(t)$ to the state $x(t)$, we may draw the feedback control system as in Fig. 2.5-2. Then, we may write in terms of Laplace transforms

$$X = GU = -GKX + GV$$

$$(I + GK)X = GV$$

$$X(s) = (I + GK)^{-1}GV(s) \qquad (2.5.6)$$

$$Y(s) = C(I + GK)^{-1}GV(s). \qquad (2.5.7)$$

We call $-G(s)K$ the (multivariable) *loop gain* and $I + G(s)K$ the multivariable *return difference*. Since $I + GK$ is an $n \times n$ matrix, its inverse must be used in (2.5.6).

Using the definition of $G(s)$, the closed-loop transfer function relation (2.5.7) is seen to be

$$Y(s) = C[I + (sI - A)^{-1}BK]^{-1}(sI - A)^{-1}BV(s). \qquad (2.5.8)$$

According to (2.5.4) and (2.5.2), we may also write the closed-loop transfer relation as

$$Y(s) = C(sI - A_c)^{-1}B = C[sI - (A - BK)]^{-1}BV(s). \qquad (2.5.9)$$

That these two expressions are the same may be verified by some simple manipulations involving the property $(MN)^{-1} = N^{-1}M^{-1}$ for any nonsingular matrices M and N.

The closed-loop transfer relations may alternatively be formulated as follows. Write

$$U = -KX + V = -KGU + V$$

$$(I + KG)U = V$$

$$U(s) = (I + KG)^{-1}V(s)$$

so that

$$X(s) = G(I + KG)^{-1}V(s) \qquad (2.5.10)$$

$$Y(s) = CG(I + KG)^{-1}V(s). \qquad (2.5.11)$$

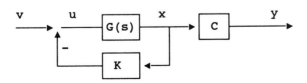

Figure 2.5-2 State-variable feedback system in frequency domain

These relations are equivalent to those just derived, and correspond to breaking the loop at the input $u(t)$. The previous relations corresponded to breaking the loop at the state $x(t)$. Note that $I + KG$ is an $m \times m$ matrix, and so is generally of lower dimension than $I + GK$ since m is usually less than n. We call $-KG(s) = -K(sI - A)^{-1}B$ the loop gain referred to the input and $I + KG(s)$ the return difference referred to the input.

One can now appreciate the difference between the scalar case and the multivariable case, for $I + GK$ is not equal to $I + KG$. However, evidently

$$(I + GK)^{-1}G = G(I + KG)^{-1}. \tag{2.5.12}$$

In Chapter 8 we shall use such frequency-domain transfer relations to study robustness of the closed-loop system to noise, disturbances, and inaccurate modeling of the plant dynamics. We will discover that, exactly as in classical control, the multivariable loop gain should be large at low frequencies for disturbance rejection, but small at high frequencies for stability in the presence of unmodeled dynamics. We show that the natural generalization of the loop gain Bode magnitude plot to MIMO systems is based on the singular-value decomposition (see Appendix B) of $KG(s)$ [or of $G(s)K$].

Pole placement by Ackermann's formula

In the remainder of this subsection we discuss methods for choosing K to assign the closed-loop poles. Thus, we want to choose K so that the closed-loop characteristic equation

$$\Delta_c(s) = |sI - A_c| = |sI - (A - BK)| = 0 \tag{2.5.13}$$

has desired roots. This is called the *pole placement* or *pole assignment problem*. First, let us cover the single-input case $m = 1$.

Given desired closed-loop poles, one may compute the desired closed-loop characteristic polynomial

$$\Delta^D(s) = s^n + \alpha_1 s^{n-1} + \cdots + \alpha_n. \tag{2.5.14}$$

Then, the state feedback K required to place the closed-loop poles at the desired locations is given by *Ackermann's formula* [Franklin and Powell 1980]

$$K = e_n^T U_n^{-1} \Delta^D(A), \tag{2.5.15}$$

where U_n is the reachability matrix, $e_n^T = [0 \cdots 0 \ 1]$ is the last row of the $n \times n$ identity matrix and

$$\Delta^D(A) = A^n + \alpha_1 A^{n-1} + \cdots + \alpha_n I. \tag{2.5.16}$$

$\Delta^D(A)$ is the desired characteristic polynomial evaluated at A. It is an $n \times n$ *matrix polynomial*.

Ackermann's formula is a *matrix design equation* that finds all the mn feedback gains, thus closing all the control loops simultaneously to obtain the desired closed-loop poles.

2.5 Feedback Control

A fundamental result by W. M. Wonham is the fact that *all the closed-loop poles may be arbitrarily assigned using state-variable feedback if and only if the system is reachable*. Indeed, note that reachability is needed to apply Ackermann's formula since the reachability matrix U_n must be invertible.

To demonstrate the usefulness of Ackermann's formula, let us consider a design example.

Example 2.5-1: Control of Inverted Pendulum Using Ackermann's Formula

Figure 2.5-3 shows a rod attached to a cart through a pivot. A force $u(t)$ is applied to the cart through a motor attached to an axle. The control objective is to use $u(t)$ to balance the pendulum upright while simultaneously keeping the horizontal position $p(t)$ of the cart small. This is known as the *inverted pendulum* problem, and is often used to demonstrate control principles. Note that it is the same problem as balancing a stick on the end of one's finger. Although this may not be so exciting, one may note that it is also related to the problem of balancing a rocket during liftoff, which one might consider quite exciting.

The inverted pendulum is a notoriously difficult system for which to design a good controller. Here, we shall use pole-placement using state-variable feedback to stabilize the pendulum.

a. Actuators and Sensors

The motor attached to the axle is the actuator that provides the control force $u(t)$. We shall ignore the motor dynamics, assuming that the armature inductance is small so that the transfer function from the motor voltage to $u(t)$ is merely a gain constant. This corresponds to assuming that the motor response is much faster than the cart response. In a more detailed design, the motor dynamics could easily be included (by augmenting the system dynamics) once it is seen how to approach the problem.

For state feedback purposes, we shall need to measure the rod angle $\theta(t)$, the rod angular velocity $\dot{\theta}(t)$, the cart position $p(t)$, and the cart velocity $\dot{p}(t)$. To accomplish this,

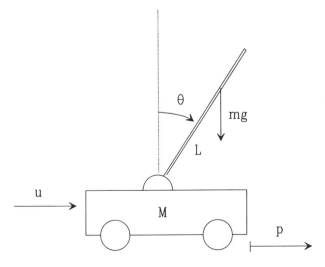

Figure 2.5-3 Inverted pendulum on a cart

we may place potentiometers at the rod pivot point [for $\theta(t)$] and on one wheel [for $p(t)$]. To measure $\dot{\theta}(t)$ and $\dot{p}(t)$ it is necessary to place tachometers at the rod pivot point and on one wheel. This involves much instrumentation, and the apparatus might be complicated to build. In section 4.2 we shall see how to design an *output feedback* control system that only needs measured outputs of $\theta(t)$ and $p(t)$.

b. Inverted Pendulum Dynamics

Assume that the rod mass m is concentrated at its center of gravity, which is at a distance of L from the pivot. The mass of the cart is M. Then, the dynamics of the inverted pendulum, linearized about the equilibrium position of $\theta = \dot{\theta} = p = \dot{p} = 0$, are given by Friedland [1986] (see the problems for section 2.1):

$$\dot{x} = \begin{bmatrix} 0 & 1 & 0 & 0 \\ \frac{(M+m)g}{ML} & 0 & 0 & 0 \\ 0 & 0 & 0 & 1 \\ -\frac{mg}{M} & 0 & 0 & 0 \end{bmatrix} x + \begin{bmatrix} 0 \\ -\frac{1}{ML} \\ 0 \\ \frac{1}{M} \end{bmatrix} u, \qquad (1)$$

with $g = 9.8$ m/sec^2 the acceleration due to gravity. The state is

$$x = [\theta \ \dot{\theta} \ p \ \dot{p}]^T \qquad (2)$$

Friction effects have been neglected, but they could easily be added; they make the control problem easier since their efforts are passive.

Assuming that $M = 5$ kg, $m = 0.5$ kg, $L = 1$ m, we obtain

$$\dot{x} = \begin{bmatrix} 0 & 1 & 0 & 0 \\ 10.78 & 0 & 0 & 0 \\ 0 & 0 & 0 & 1 \\ -0.98 & 0 & 0 & 0 \end{bmatrix} x + \begin{bmatrix} 0 \\ -0.2 \\ 0 \\ 0.2 \end{bmatrix} u = Ax + Bu. \qquad (3)$$

To gain some insight on the system, we may determine its open-loop poles to be at

$$s = 0, 0, \pm 3.283. \qquad (4)$$

The poles at ± 3.283 arise from the angle subsystem and are involved with θ and $\dot{\theta}$. The two poles at $s = 0$ evolve from the Newton's system relating to the horizontal position, and involve p and $\dot{p}$.

Thus, with no control input the rod will clearly fall over due to the unstable pole at $s = 3.283$ (unless it is balanced exactly at $\theta = 0$ and there are no disturbances).

Let us mention that the linearized dynamics describe the system behavior near the equilibrium point $x = 0$. If we can provide a stabilizing control, then the control will hold x in the vicinity of the origin, so that the accuracy of the linearized model will be enhanced.

c. Control Input Design

We shall take as the control input $u(t)$ the state feedback

$$u = -Kx = -[k_1 \ k_2 \ k_3 \ k_4]x = -k_1\theta - k_2\dot{\theta} - k_3 p - k_4\dot{p}. \qquad (5)$$

2.5 Feedback Control

The desired closed-loop poles are selected as

$$s = -1 \pm j1 \atop -2 \pm j2. \qquad (6)$$

This yields the desired characteristic polynomial of

$$\Delta^D(s) = [(s + 1)^2 + 1][(s + 2)^2 + 2^2] = s^4 + 6s^3 + 18s^2 + 24s + 16. \qquad (7)$$

A FORTRAN V program was written to implement Ackermann's formula (2.5.15). It is called ACKERM and appears in Appendix A. It computes the reachability matrix U_n and the 4×4 matrix polynomial

$$\Delta^D(A) = A^4 + 6A^3 + 18A^2 + 24A + 16I. \qquad (8)$$

For the closed-loop poles in (6), the resulting feedback gain is

$$K = [-152.06 \quad -42.24 \quad -8.16 \quad -12.24]. \qquad (9)$$

d. Computer Simulation

A computer simulation on the closed-loop system $(A - BK, B)$ was performed using program TRESP in Appendix A. The initial conditions were $\theta(0) = 0.1$ rad, $p(0) = 0.1$ m, and the closed-loop response is shown in Fig. 2.5-4. After about 6 sec the system comes to rest with the rod balanced at $\theta = 0$ and the cart at the position $p = 0$.

The rod response $\theta(t)$ is quite good, and it remains small so that the linear approximation holds. (Note that the nonlinearities involve terms like $\sin \theta$, $\cos \theta$, $\dot{\theta}^2$.) However, the position $p(t)$ increases to 0.43 m before it returns to zero. This is necessary since the cart must run up under the rod to balance it before coming back to the origin $p = 0$.

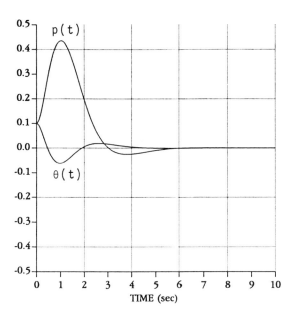

Figure 2.5-4 Inverted pendulum closed-loop response. (a) Rod angle $\theta(t)$ (rads) and cart position $p(t)$ (m). (b) Control input $u(t)$.

In section 4.2 we shall perform this design using output feedback based on measurements of only $\theta(t)$ and $p(t)$. There, the response will be quite a bit better. Pole-placement design using full state feedback is sometimes unsuitable, since it does not take advantage of the full design freedom in a problem. Thus, although the poles in (6) appear quite stable, the problem here is that the zeros in the SISO transfer functions from $u(t)$ to $\theta(t)$ and from $u(t)$ to $p(t)$ are not manipulated in the pole-placement design algorithm. The large excursion in $p(t)$ is because of unsuitable zeros in the transfer function from $u(t)$ to $p(t)$. Output feedback design using optimal techniques (section 4.2) corrects this problem.

■

Multi-input eigenstructure assignment

Since Ackermann's formula relies on finding U_n^{-1}, it does not apply for the multi-input (MI) case where $m > 1$ and U_n, which is an $n \times mn$ matrix, is not square. There is no easy generalization of this formula to the MI case. Thus, we shall take a different approach for pole placement in MI systems that is due to B. C. Moore [1976].

The basic issue is that for MI systems *we can do more than simply place the closed-loop poles* λ_i. In fact, we may to a certain extent assign also the closed-loop eigenvectors v_i. This is important in view of the discussion in section 2.4 on modal decomposition, for the eigenvectors v_i define the directions in $\mathbf{R}^n$ that respond according to the natural modes $e^{\lambda_i t}$. They also help determine the numerator of the transfer function (see the PFE in section 2.4). Thus, the eigenvectors help determine the shape of the closed-loop response.

For ease of presentation we shall assume that B has full rank m. Our discussion will be based on the input-coupling polynomial matrix

$$R(s) = [sI - A \quad B], \qquad (2.5.17)$$

with s a complex variable and A and B the plant system and input matrices.

In the MI case, it will be necessary to discuss the desired closed-loop poles as well as the desired closed-loop eigenvectors. Suppose it is required to select K in (2.5.3) so that a desired eigenvalue λ_i and a desired associated eigenvector v_i are assigned to the closed-loop system (2.5.4). Suppose moreover that we can find a vector $u_i \in \mathbf{R}^m$ to satisfy the equation

$$[\lambda_i I - A \quad B]\begin{bmatrix} v_i \\ u_i \end{bmatrix} = 0. \qquad (2.5.18)$$

Now, choose the feedback gain K to satisfy

$$Kv_i = u_i. \qquad (2.5.19)$$

Using the last two equations we may write

$$0 = (\lambda_i I - A)v_i + Bu_i \qquad (2.5.20)$$

$$0 = [\lambda_i I - (A - BK)]v_i, \qquad (2.5.21)$$

2.5 Feedback Control

so that using this feedback v_i is assigned as an eigenvector of the closed-loop system matrix A_c for eigenvalue λ_i.

As Cv_i was shown in section 2.4 to be a direction in the output space associated with v_i, so u_i is the associated direction in the input space $\mathbf{R}^m$. That is, motions of $u(t)$ in the direction of u_i will cause motions of $x(t)$ in the direction of v_i, resulting in motions of $y(t)$ in the direction of Cv_i.

To complete the picture, suppose that n desired closed-loop eigenvalues λ_i and associated eigenvectors v_i are chosen, and that in each case we have found a vector u_i that satisfies (2.5.18). Then, we may define K by

$$K[v_1 \quad v_2 \ldots v_n] = [u_1 \quad u_2 \ldots u_n] \qquad (2.5.22)$$

or by appropriate definition of the matrices V and U

$$KV = U. \qquad (2.5.23)$$

Then, for each value of $i = 1, \ldots, n$ equation (2.5.21) will hold, so that each λ_i will be assigned as a closed-loop pole with associated eigenvector v_i.

In terms of the matrices U and V and the diagonal matrix of desired closed-loop eigenvectors

$$J = \text{diag}\{\lambda_i\} \qquad (2.5.24)$$

we may write down a *matrix design equation for eigenstructure assignment* for MI systems. According to (2.5.18)/(2.5.20)

$$0 = \lambda_i v_i - Av_i + Bu_i$$

or

$$AV - VJ = BU. \qquad (2.5.25)$$

If, given desired closed-loop poles J and eigenvectors V, this *matrix Sylvester equation* can be solved for U, then the feedback gain that yields the desired closed-loop structure is given by (2.5.23).

Some notions on solving the Sylvester equation are given in Chen [1984]. Note that if $U = 0$, then the feedback gain in (2.5.23) is zero and $AV = VJ$, which means that J is the Jordan matrix for A, consisting of the open-loop system poles.

We have shown how to select the feedback gain K to place the poles in MI systems using the Sylvester matrix design equation. It remains only to discuss a few points.

Since, by definition, the closed-loop eigenvectors must be linearly independent, it is necessary to select v_i as linearly independent vectors. Then (2.5.23) may be solved for K to give

$$K = UV^{-1} \qquad (2.5.26)$$

Another issue is that the closed-loop system and feedback gain must be real and not complex. Thus, if a complex closed-loop pole λ_i is selected, it is also necessary to select as a closed-loop pole its complex conjugate λ_i^*. Moreover, if v_i is to be the closed-loop eigenvector associated with a complex pole λ_i, then in order for (2.5.22) to have a real solution K, v_i^* (i.e., the complex-conjugate of v_i) must be selected as the eigenvector for λ_i^*.

To see that under these circumstances (2.5.23) indeed has a real solution K, note first that if u_i solves (2.5.20) for a given λ_i and v_i, then u_i^* solves the equation for their complex conjugates. Therefore, if $u_i = u_R + ju_I$ and $v_i = v_R + jv_I$, then to assign the desired eigenstructure, K must satisfy

$$K[v_R + jv_I \quad v_R - jv_I] = [u_R + ju_I \quad u_R - ju_I]. \tag{2.5.27}$$

Postmultiplying both sides of this equation by

$$M = \frac{1}{2}\begin{bmatrix} 1 & -j \\ 1 & j \end{bmatrix}$$

It is seen that this equation is equivalent to the real equation

$$K[v_R \quad v_I] = [u_R \quad u_I], \tag{2.5.28}$$

which clearly has as a solution a real gain matrix K. Thus, if v_i is complex, then to obtain a real value for K it is only necessary to use not v_i and v_i^* (resp. u_i and u_i^*) in (2.5.22), but the real and imaginary parts of v_i (resp. u_i).

Finally, we must investigate the conditions for existence of a solution to (2.5.18)/(2.5.25). It can be shown that J may be selected with arbitrary poles λ_i if and only if (A, B) is reachable. Thus, arbitrary closed-loop pole placement is equivalent to system reachability. That is, the *open-loop property* of reachability indicates what can be achieved in the closed-loop system. This investigation of closed-loop properties in terms of open-loop properties is in the spirit of classical control analysis.

Eigenstructure assignment can still be used if a system is unreachable. If λ_i is an unreachable pole, then it cannot be moved using feedback. However, the reachable poles can be moved using feedback. Thus, it is only necessary to select the unreachable pole λ_i as one of the closed-loop poles, and choose the remaining closed-loop poles as desired. Then, the technique just described may be applied. It is remarkable that, although an unreachable pole cannot be moved, feedback can be used to modify its closed-loop eigenvector. Thus, the contribution of the unreachable pole to the system response can be changed.

It is not usually possible to specify independently both an arbitrary λ_i and v_i and obtain a solution u_i to (2.5.18). Indeed, assuming that λ_i is not an open-loop pole, we have

$$v_i = -(\lambda_i I - A)^{-1} B u_i. \tag{2.5.29}$$

Thus, for the existence of a solution u_i, the desired v_i must be a linear combination of the m columns of the linear operator

$$L_i = (\lambda_i I - A)^{-1} B. \tag{2.5.30}$$

2.5 Feedback Control

Since B has full rank m by assumption, the matrix L_i also has rank m. Thus, v_i must lie in an m-dimensional subspace of $\mathbf{R}^n$ that depends on the choice of λ_i. This means we have *m degrees of freedom in selecting the closed-loop eigenvector* v_i once λ_i has been selected.

This last point is the crucial difference between classical SISO design and multi-variable eigenstructure assignment. If $m = 1$, which corresponds to the single-input case, then eigenvector v_i has only one degree of freedom once the desired eigenvalue λ_i has been selected; that is, there is no additional freedom to choose the eigenvector. However, in the multi-input case where $m > 1$, we can have additional freedom to specify the *internal structure of the closed-loop system* by selecting m degrees of freedom of v_i arbitrarily.

The successive loop closure approach of classical control, where only one feedback gain is selected at a time, obscures the extra design freedom arising from multiple inputs. In modern control, where all gains are selected simultaneously, this freedom is clearly revealed.

Example 2.5-2: Multivariable Decoupling Using Eigenstructure Assignment

A primary use of eigenstructure assignment design is in the decoupling of multivariable systems. A system is *decoupled* if certain inputs affect only certain groups of the states or outputs, and have little influence on the remaining states. Let us demonstrate decoupling control using a two-input example.

The circuit in Fig. 2.5-5 has the state-variable representation [Brogan 1974]

$$\dot{x} = \begin{bmatrix} -1/CR_1 & 0 & 1/C & 0 \\ 0 & -1/RC_1 & 0 & -1/C \\ -1/L & 0 & -R_2/L & -R_2/L \\ 0 & 1/L & -R_2/L & -R_2/L \end{bmatrix} x + \begin{bmatrix} 0 & 0 \\ 0 & 0 \\ 1/L & 0 \\ 0 & 1/L \end{bmatrix} u \quad (1)$$

$$y = \begin{bmatrix} 1 & 1 & 0 & 0 \\ 0 & 1 & 0 & 0 \end{bmatrix} x = Cx. \quad (2)$$

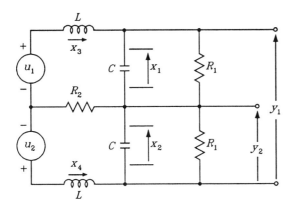

Figure 2.5-5 Multi-input electric circuit

Assuming values of 1 for all components yields

$$\dot{x} = \begin{bmatrix} -1 & 0 & 1 & 0 \\ 0 & -1 & 0 & -1 \\ -1 & 0 & -1 & -1 \\ 0 & 1 & -1 & -1 \end{bmatrix} x + \begin{bmatrix} 0 & 0 \\ 0 & 0 \\ 1 & 0 \\ 0 & 1 \end{bmatrix} u = Ax + Bu. \qquad (3)$$

a. Open-Loop Response

The poles of this system are at

$$\begin{aligned} s &= -0.5 \pm j0.866 \\ &\quad -1.5 \pm j0.866. \end{aligned} \qquad (4)$$

The open-loop response of this circuit is shown in Fig. 2.5-6. Part (a) shows the states x_1 through x_4 when $u_1 = 1$ and $u_2 = 0$, and part (b) shows the response for $u_1 = 0$, $u_2 = 1$. Clearly, each input excites all four states. The responses attain their steady-state values in about 8 sec.

b. Decoupling Design

We should like to design a state-feedback control law

$$u = -Kx + w, \qquad (5)$$

with $w(t)$ the new system input. Then, the closed-loop system is

$$\dot{x} = (A - BK)x + Bw. \qquad (6)$$

The controller should decouple the system so that input $w_1(t)$ only influences states x_1 and x_3, and has little effect on components x_2 and x_4. Similarly, control input $w_2(t)$ should only excite x_2 and x_4, and not x_1 and x_3. To speed up the response, it is also desired to place the closed-loop poles at

$$\begin{aligned} s &= -1.5 \pm j1.5 \\ s &= -1.0 \pm j0.5. \end{aligned} \qquad (7)$$

To achieve these multiple objectives, one may use eigenstructure assignment design.

Let us define the desired closed-loop poles $\lambda_1 = -1.5 + j1.5$, $\lambda_2 = -1.5 - j1.5$. To associate this pole pair with the state components x_1 and x_3, and decouple this mode from components x_2 and x_4, we may select the desired closed-loop eigenvectors

$$v_1 = \begin{bmatrix} 1 \\ 0 \\ -1 \\ 0 \end{bmatrix} + j \begin{bmatrix} 1 \\ 0 \\ 1 \\ 0 \end{bmatrix} \qquad (8)$$

and $v_2 = v_1^*$. Note that this choice is not unique, except that for decoupling, the second and fourth components of v_1 should be zero. Then, the motion of this natural mode is restricted to a state-space direction that does not involve x_2 and x_4. See the discussion on natural modes in section 2.4.

2.5 Feedback Control

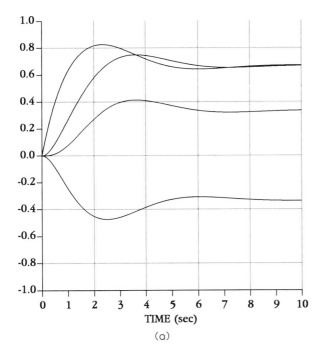

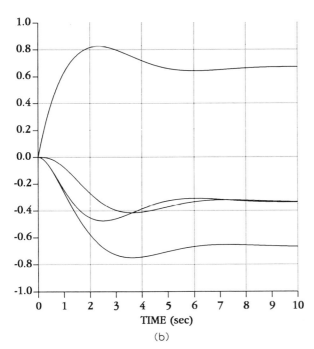

Figure 2.5-6 Open-loop response. (a) Response for $u_1 = 1$, $u_2 = 0$. (b) Response for $u_1 = 0$, $u_2 = 1$.

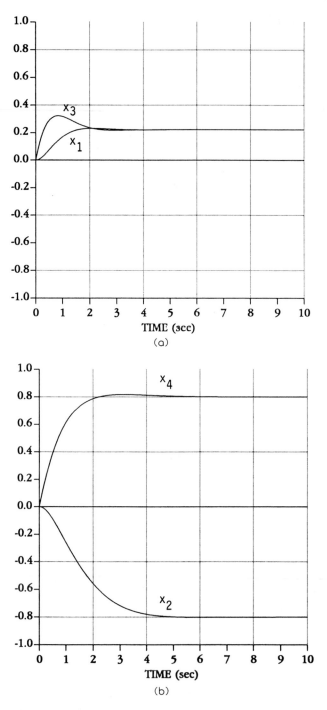

Figure 2.5-7 Closed-loop response showing decoupling. (a) Response for $w_1 = 1$, $w_2 = 0$. (b) Response for $w_1 = 0$, $w_2 = 1$.

2.5 Feedback Control

Similarly, define the desired closed-loop poles $\lambda_3 = -1.0 + j0.5$, $\lambda_4 = -1.0 - j0.5$. To associate this pole pair with the state components x_2 and x_4, and decouple this mode from components x_1 and x_3, select the desired closed-loop eigenvectors

$$v_3 = \begin{bmatrix} 0 \\ -1 \\ 0 \\ 1 \end{bmatrix} + j\begin{bmatrix} 0 \\ 1 \\ 0 \\ 1 \end{bmatrix} \qquad (9)$$

and $v_4 = v_3^*$. Again, the choice is not unique, but for decoupling, the first and third components of v_3 should be zero.

A program was written in FORTRAN V to solve the matrix design equation (2.5.25) with J equal to the diagonal matrix of desired eigenvalues and V the matrix of desired eigenvectors. Recognizing that there is not always a solution U given J and V, the program fixed J and found the solution for U and V^a, where V^a is as close as possible to the desired V in a least-squares sense. This was achieved using subroutine LLBQF in [IMSL].

The columns of V^a are the *achievable closed-loop eigenvectors*. A solution U was found for the Sylvester equation (2.5.25) with the given J and achievable eigenvectors of

$$v_1^a = \begin{bmatrix} 0.29 \\ 0 \\ -0.96 \\ 0 \end{bmatrix} + j\begin{bmatrix} 0.54 \\ 0 \\ 0.16 \\ 0 \end{bmatrix}, \quad v_3^a = \begin{bmatrix} 0 \\ -1 \\ 0 \\ 0 \end{bmatrix} + j\begin{bmatrix} 0 \\ 0 \\ 0 \\ 0.5 \end{bmatrix} \qquad (10)$$

The important point to note is that the zeros are in the correct place to guarantee the required decoupling.

To solve for the feedback gain, the program used the real and imaginary portions of the v_i^a and the u_i [see (2.5.28)]. The gain so determined was

$$K = \begin{bmatrix} 1.5 & 0 & 1 & -1 \\ 0 & 0.75 & -1 & 0 \end{bmatrix} \qquad (11)$$

c. Simulation

To verify the performance of the controller, a computer simulation was carried out on the closed-loop system (6) using program TRESP in Appendix A. The result is shown in Fig. 2.5-7. In part (a) of the figure, x_2 and x_4 are zero, while in part (b) x_1 and x_3 are zero. Clearly, the decoupling behavior of the closed-loop system is excellent. Input $w_1(t)$ has no influence on x_2 and x_4, while $w_2(t)$ has no influence on x_1 and x_3. Moreover, the response has been sped up so that the steady-state values are reached in about 4 sec. ∎

Stabilizability and Detectability

We have seen that (A, B) is reachable if and only if all the poles may be placed arbitrarily using state-variable feedback. A weaker condition than reachability that is often useful in modern control is *stabilizability*. We say (A, B) is *stabilizable* if there exists a state-feedback gain K such that the closed-loop system matrix $(A - BK)$ is stable.

Although there is no simple test for stabilizability in terms of the reachability matrix U_n, a test can be given in terms of the eigenstructure of A.

A necessary and sufficient condition for reachability is given by the PBH test in section 2.4. Thus, (A, B) is reachable if and only if $w_i^T(A - \lambda_i I) = 0$ for some λ_i and $w_i^T B = 0$ imply that $w_i = 0$. Similarly, (A, B) is stabilizable if and only if $w_i^T(A - \lambda_i I) = 0$ and $w_i^T B = 0$ imply that either $w_i = 0$ or λ_i is stable [Kailath 1980]. That is, stabilizability is equivalent to the reachability of all the unstable modes.

A pole is reachable if and only if it can be moved using state feedback. Thus, stabilizability means that all poles are either stable or can be moved to stable locations using state feedback.

It can be shown that (A, C) is observable if and only if all the poles of $(A - LC)$ may be placed arbitrarily by appropriate choice of the matrix L. The matrix $(A - LC)$ is important in the observer design problem (see Chapter 9), and L is called an *output-injection matrix*. A weaker condition than observability that is often useful in modern control is *detectability*. We say (A, C) is *detectable* if there exists an output injection L such that the matrix $(A - LC)$ is stable.

Although there is no simple test for detectability in terms of the observability matrix V_n, a test can be given in terms of the eigenstructure of A.

A necessary and sufficient condition for observability is given by the PBH test in section 2.4. Thus, (A, C) is observable if and only if $(A - \lambda_i I)v_i = 0$ for some λ_i and $Cv_i = 0$ imply that $v_i = 0$. Similarly, (A, C) is detectable if and only if $(A - \lambda_i I)v_i = 0$ and $Cv_i = 0$ imply that either $v_i = 0$ or λ_i is stable [Kailath 1980]. That is, detectability is equivalent to the observability of all the *unstable modes*.

A pole is observable if and only if it can be moved using output injection. Thus, detectability means that all poles are either stable or can be moved to stable locations using output injection.

There is an important duality relation between reachability-related notions and observability-related notions. Note that

$$(A - LC)^T = A^T - C^T L^T \tag{2.5.31}$$

which has the free design matrix L^T to the right exactly as in the state-feedback problem $(A - BK)$. Note further that the poles of $(A - LC)$ are equal to the poles of $(A - LC)^T$. Therefore, we can say that (A, B) is reachable if and only if (A^T, B^T) is observable. Similarly, (A, B) is stabilizable if and only (A^T, B^T) is detectable. In view of the fact that the left eigenvectors of A are the right eigenvectors of A^T, this makes a great deal of sense.

Output Feedback

We have seen that if (A, B) is reachable, then all the closed-loop poles may be exactly assigned to any locations desired in the s-plane using state-variable feedback. This is quite a powerful result, especially in view of the fact that matrix design equations were given for finding the feedback gain matrix K that places the poles as desired. By solving these

2.5 Feedback Control

design equations, which is easy using modern-day computing facilities, one computes K, thus closing *all the feedback loops simultaneously* to obtain the specified closed-loop behavior. This amounts to finding mn feedback gains simultaneously, with n the number of states and m the number of inputs. The determination of this number of feedback gains using classical techniques of closing one loop at a time would be very difficult.

Given the state-variable system

$$\dot{x} = Ax + Bu \tag{2.5.32}$$

$$y = Cx, \tag{2.5.33}$$

measurements of all the states $x(t)$ are, unfortunately, seldom available. Some state components may simply be too expensive or even impossible to measure. Therefore, full state-variable feedback is usually impossible to implement in practice and only the measured outputs $y(t)$ should be used for feedback purposes.

There are basically two techniques for control system design using the available outputs. In one technique, a state feedback controller $u = -Kx$ is first designed. Then, the measured output $y(t)$ is used to *estimate* the state. Finally, the state estimate $\hat{x}(t)$ is used in the state-variable feedback controller *as if* it were exactly equal to $x(t)$; that is, we set $u = -K\hat{x}$. In Chapter 9 we show how to estimate the state from $y(t)$ by using a *dynamic observer*. Then, in Chapter 10 we show how to use the state-feedback law and an observer to design a compensator for suitable closed-loop performance.

The other technique for output-feedback design is to restrict oneself from the outset to dealing only with the available outputs. Several forms of control law are common. We shall use the constant or *static* output-feedback control

$$u = -Ky + v, \tag{2.5.34}$$

with K the constant output-feedback gain matrix. This control law only uses the available measured outputs $y(t)$.

Taking $x \in \mathbf{R}^n$, $u \in \mathbf{R}^m$, $y \in \mathbf{R}^p$, the gain matrix K is $m \times p$, so that the output-feedback design problem is to select the mp entries of K to guarantee suitable closed-loop performance. This amounts to closing mp control loops. Using (2.5.34) in (2.5.32) yields the closed-loop system

$$\dot{x} = (A - BKC)x + Bv. \tag{2.5.35}$$

The selection of K for good closed-loop performance is not an easy problem. There are few results for output feedback that are as convenient as the results derived for state feedback in the previous subsection. For instance, if the system is reachable then all the closed-loop poles can be placed exactly as desired using full state feedback, and there are convenient matrix design equations for computing the required feedback gain. However, even reachability plus observability do not guarantee that all the poles can be placed using output feedback.

We say that (A, B, C) is *output-feedback stabilizable* if there exists a K such that $(A - BKC)$ is stable. Unfortunately, it is difficult to provide convenient tests for this property, even using the eigenstructure of A. Moreover, it is difficult to find an output-feedback gain K that guarantees closed-loop stability.

It is worth mentioning a partial result for output feedback design, for it sometimes gives good results or at least allows one to select a gain K that stabilizes the system. Suppose that p closed-loop poles λ_i and their associated eigenvectors v_i have been chosen, and that vectors u_i have been found that satisfy (2.5.18). Then, if the output-feedback gain in (2.5.34) is computed according to

$$KC[v_1 \quad v_2 \ldots v_p] = [u_1 \quad u_2 \ldots u_p], \quad (2.5.36)$$

we have $KCv_i = u_i$ for $i = 1, 2, \ldots, p$, so that, according to (2.5.20)

$$[\lambda_i I - (A - BKC)]v_i = 0. \quad (2.5.37)$$

This equation reveals that the λ_i have been assigned as closed-loop poles with eigenvectors v_i for the system (2.5.35).

This technique is easy to apply. Defining $U = [u_1 \ldots u_p]$, $V = [v_1 \ldots v_p]$ allows us to write for (2.5.36)

$$KCV = U.$$

To obtain a unique solution for any U, the eigenvectors v_i must be selected so that CV is invertible. Then the output-feedback gain is given by

$$K = U(CV)^{-1}, \quad (2.5.38)$$

where U and V are determined using the matrix design equation (2.5.25) for a given $p \times p$ matrix J of desired closed-loop poles. Note that the invertibility requirement on CV means that V can have no more than p columns, so that only p poles can be assigned using this approach.

Therefore, using an eigenstructure assignment approach to output-feedback design allows us to assign p closed-loop poles, with p the number of measured outputs. As seen in the previous subsection, each eigenvector can be assigned with m degrees of freedom, with m the number of control inputs.

Unfortunately, this is at best a partial result, for although it assigns p poles precisely, there is no guarantee as to what happens to the other $n - p$ poles in the closed-loop system. They may be stable or unstable. There are ways to extend the eigenstructure assignment approach so that $m + p$ poles can often be assigned [Kwon and Youn 1987], but they are not pleasing from a design point of view.

The output-feedback design problem is still the focus of research. It has been shown that, while the state feedback problem is linear, the output feedback problem is *nonlinear*.

Chapter 4 will be devoted to determining output-feedback gains K that provide closed-loop performance that is in some sense optimal. We shall see that output feedback provides far more design flexibility than does full state feedback, although the design equations are fairly complicated. Fortunately, they are not difficult to solve using digital computers.

PROBLEMS FOR CHAPTER 2

Problems for Section 2.1

2.1-1 For the pendulum system in Example 2.1-1, find the poles, resolvent matrix, transfer function, state transition matrix, impulse response, and solution for $x(0) = [0 \ 1]^T$, $u = \cos t \, u_{-1}(t)$. Do not use the results of Example 2.1-5, but work the problem from scratch.

2.1-2 Repeat Problem 2.1-1 for the circuit in Example 2.1-2.

2.1-3 For the system

$$\dot{x} = \begin{bmatrix} -3 & 2 \\ 1 & -2 \end{bmatrix} x + \begin{bmatrix} 1 \\ b \end{bmatrix} u, \ y = [1 \ 0]x$$

a. Find the poles, resolvent matrix, transfer function, state transition matrix, and impulse response.

b. Find b so that the input excites only the most stable pole.

c. Let $u(t) = 2e^t u_{-1}(t)$ and $x(0) = [1 \ -1]^T$. Find b so that the output is stable. For this value of b, find the output.

2.1-4 Inverted Pendulum Dynamic Equations. An inverted pendulum attached to a cart is shown in Fig. P.2-1. The rod can pivot about its point of attachment, and has a mass of m with a center of gravity at a distance of L from the pivot point. The cart mass is M and $g = 9.8$ m/sec² the acceleration due to gravity. The rod angle is $\theta(t)$ and the cart position is $p(t)$. A force input $u(t)$ is applied to the cart through a motor that drives the wheels. All motion occurs in the plane.

a. Show that the Lagrangian is given by

$$L = \frac{1}{2}(M + m)\dot{p}^2 + mL\dot{p}\dot{\theta} \cos \theta + \frac{1}{2} mL^2 \dot{\theta}^2 - mgL \cos \theta$$

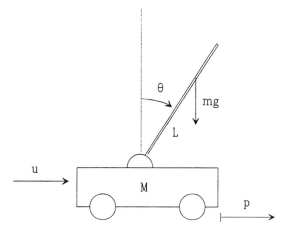

Figure P.2-1 Inverted pendulum on a cart

b. Show that the equations of motion are

$$\ddot{\theta} = \frac{mL\dot{\theta}^2 \sin\theta \cos\theta - (M+m)g \sin\theta + u\cos\theta}{mL\cos^2\theta - (M+m)L}$$

$$\ddot{p} = \frac{mg \sin\theta \cos\theta - mL\dot{\theta}^2 \sin\theta - u}{m\cos^2\theta - (M+m)}$$

c. To find a linearized state-space model that is good for small θ, set $\theta \approx 0$, so that $\sin\theta \approx \theta$, $\cos\theta = 1$, and $\dot{\theta}^2 \approx 0$. Then, show that

$$\dot{x} = \begin{bmatrix} 0 & 1 & 0 & 0 \\ \frac{(M+m)g}{ML} & 0 & 0 & 0 \\ 0 & 0 & 0 & 1 \\ \frac{-mg}{M} & 0 & 0 & 0 \end{bmatrix} x + \begin{bmatrix} 0 \\ \frac{-1}{ML} \\ 0 \\ \frac{1}{M} \end{bmatrix} u$$

where the state is $x = [\theta \ \dot{\theta} \ p \ \dot{p}]^T$. Note that this model consists of a 2×2 angle subsystem and a 2×2 cart position subsystem in block triangular form.

2.1-5 For the angle subsystem in Problem 2.1-4, find the poles, resolvent matrix, transfer function, state transition matrix, and impulse response.

2.1-6 For the inverted pendulum in Problem 2.1-4 find the transfer function and draw a block diagram, labeling the states. Notice the triangularly decoupled form. That is, the angle subsystem does not directly depend on the cart position or velocity.

2.1-7 Motor with Flexible Coupling. For the system in Example 2.1-3, find the poles. Find the transfer functions from $u(t)$ to $\omega_m(t)$ and from $u(t)$ to $\omega_L(t)$.

2.1-8 Ball Balancer Dynamic Equations. A ball rolling on two parallel bars is shown in Fig. P.2-2. The bars have a length of $2R$ and a combined mass of M. They are driven by a motor attached to their centers, which can change their angle of inclination θ. The distance of the center of the ball from the pivot point of the bars is r. The ball has a radius of ρ and a mass of m, and $g = 9.8$ m/sec² is the acceleration due to gravity.

a. Show that the Lagrangian is given by

$$L = \frac{1}{2}m[(1+k)\dot{r}^2 + r^2\dot{\theta}^2] + \frac{1}{2}I\dot{\theta}^2 - mgr \sin\theta,$$

where $I = MR^2/3$ and $k = 2c_r^2/5$, with c_r the coefficient of rotation for the ball. For pure rotation of the ball $c_r = 1$, while for pure sliding $c_r = 0$. (Hint: the angular velocity of the ball is $\omega_b = c_r\dot{r}/\rho$.)

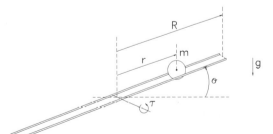

Figure P.2-2 Ball rolling on two parallel bars

b. Show that the equations of motion are

$$\ddot{\theta} = \frac{-mr(2\dot{r}\dot{\theta} + g\cos\theta) + \tau}{mr^2 + I} \tag{1}$$

$$\ddot{r} = \frac{r\dot{\theta}^2 - g\sin\theta}{1 + k}, \tag{2}$$

with τ the torque input supplied by the motor in the direction of positive θ.

c. A control objective might be to balance the ball at a distance of r_0 from the central pivot point by tilting the bars back and forth using the motor. To find a linearized state-space model that is good for small motions about $\theta = 0$ and $r = r_0$, proceed as follows.

First, define the state as $X = [\theta \;\; \dot{\theta} \;\; r \;\; \dot{r}]^T$ and write (1), (2) in the nonlinear state-space form

$$\dot{X} = f(X, \tau).$$

This defines the nonlinear 4-vector function f.

The desired equilibrium point is $X_0 = [0 \;\; 0 \;\; r_0 \;\; 0]^T$. Substituting these values into (1) yields the required equilibrium torque of $\tau_0 = mgr_0$.

Now the linearized state-space model is given by

$$\dot{x} = Ax + Bu,$$

with $x = \Delta X$ and $u = \Delta\tau$ the incremental changes in X and τ. To find the constant plant matrices A and B, compute the jacobians $A(X) = \partial f/\partial X$, $B(X) = \partial f/\partial \tau$ as defined in Appendix B, and evaluate them at the equilibrium value of $\theta = 0$, $\dot{\theta} = 0$, $r = r_0$, $\dot{r} = 0$, $\tau_0 = mgr_0$. The result should be

$$\dot{x} = \begin{bmatrix} 0 & 1 & 0 & 0 \\ 0 & 0 & \frac{-mg}{I + mr_0^2} & 0 \\ 0 & 0 & 0 & 1 \\ \frac{-g}{1+k} & 0 & 0 & 0 \end{bmatrix} x + \begin{bmatrix} 0 \\ \frac{1}{I + mr_0^2} \\ 0 \\ 0 \end{bmatrix} u.$$

(This problem was contributed by P. Panecki.)

Problems for Section 2.2

2.2-1 A discrete-time system is

$$x_{k+1} = \begin{bmatrix} 0 & 1 \\ -1/8 & 3/4 \end{bmatrix} x_k + \begin{bmatrix} 0 \\ 1 \end{bmatrix} u_k$$

$$y_k = [-1/8 \;\; -1/4] x_k.$$

a. Using iteration, find y_0, y_1, y_2 given $x_0 = 0$ and $u_k = 1$ for all k.
b. Find the system poles and natural modes. Find the resolvent matrix.
c. Find an analytic expression for the discrete state-transition matrix A^k.
d. Find the transfer function $H(z)$.
e. Find an analytic expression for the impulse response h_k.
f. Find an analytic expression for the output y_k if $x_0 = 0$ and $u_k = 1$. Check with the answer to part a.

2.2-2 Consider the system

$$x_{k+1} = \begin{bmatrix} 0 & 1 \\ -1 & 5/2 \end{bmatrix} x_k + \begin{bmatrix} 0 \\ 1 \end{bmatrix} u_k$$

$$y_k = [-2 \quad 1] x_k.$$

a. Find the poles and natural modes. Is the system asymptotically stable?
b. Is the system BIBO stable?

2.2-3 Compound Interest Formula. If x_k is the balance of a bank account at the beginning of the k-th period, β is the interest rate, and u_k is the amount deposited at the end of the k-th period, then

$$x_{k+1} = \beta x_k + u_k.$$

Assume that the initial balance is x_0 and that the deposit is an annuity, that is u_k has the constant value of a. Using the time-domain solution for a state-equation, derive the compound interest formula

$$x_k = \beta^k x_0 + \frac{(1 - \beta^k)}{1 - \beta} a.$$

You will need to use the formula for the sum of a geometric series

$$\sum_{i=0}^{k-1} c^i = \frac{1 - c^k}{1 - c}.$$

Problems for Section 2.3

2.3-1 Inversion Using High-Gain Feedback. Given a plant $G(s)$, a precompensator arrangement for forcing the plant output $y(t)$ to follow a given reference input $r(t)$ is shown in Fig. P.2-3. Show that, as the feedback gain k becomes large, the transfer function from $r(t)$ to $w(t)$ tends to $G^{-1}(s)$, so that $y(t)$ tends to $r(t)$. What happens with such high-gain feedback if the plant has nonminimum-phase zeros?

2.3-2 Reachability. A discrete-time system is

$$x_{k+1} = \begin{bmatrix} 0 & 1 \\ -1 & 5/2 \end{bmatrix} x_k + \begin{bmatrix} 0 \\ 1 \end{bmatrix} u_k.$$

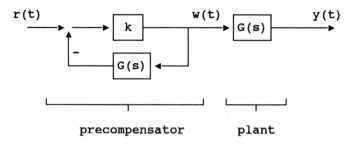

Figure P.2-3 Precompensator based on inverse system

Problems for Chapter 2

a. Is the system stable? Find x_2 if $x_0 = [1 \ 0]^T$ and $u_k = 0$.
b. Find the reachability matrix U_n. Is the system reachable?
c. It is desired to drive $x_0 = [1 \ 0]^T$ to the origin in two time steps. That is, we want to select u_0 and u_1 so that $x_2 = 0$. Using (2.3.10) and the fact that U_2 is nonsingular, find the required u_0 and u_1.
d. Find the state trajectory x_0, x_1, x_2 for the u_0 and u_1 determined in part c.

2.3-3 Observability. Consider the system

$$x_{k+1} = \begin{bmatrix} 0 & 1 \\ -1/4 & -1 \end{bmatrix} x_k, \quad y_k = [1 \ 1] x_k.$$

a. Is the system observable?
b. For some initial condition x_0 the measurements $y_0 = 1$, $y_1 = 0$, $y_2 = -1/4$ are taken. Find x_0 using (2.3.32).
c. If the system is observable, then x_0 can be found from y_k. Moreover, the state trajectory can be reconstructed from y_k. For the x_0 found in part b, find x_k for $k = 1, 2, 3, 4$. Find y_k for $k = 3, 4$.

2.3-4 MIMO Systems. The prescribed system is

$$\dot{x} = \begin{bmatrix} -3 & 4 \\ -2 & 3 \end{bmatrix} x + \begin{bmatrix} 1 & 1 \\ 1 & 1 \end{bmatrix} u, \quad y = \begin{bmatrix} -1 & 2 \\ 1 & -2 \end{bmatrix}.$$

a. Is the system reachable? Is it observable?
b. Now the control inputs and measurements are changed so that

$$\dot{x} = \begin{bmatrix} -3 & 4 \\ -2 & 3 \end{bmatrix} x + \begin{bmatrix} 1 & 1 \\ 2 & 1 \end{bmatrix} u, \quad y = \begin{bmatrix} 1 & 2 \\ 1 & -2 \end{bmatrix}.$$

Is the new system reachable and observable?

2.3-5 Pole/Zero Cancellation. Consider the system

$$\dot{x} = \begin{bmatrix} -1 & 1 \\ 2 & 0 \end{bmatrix} x + \begin{bmatrix} 1 \\ -1 \end{bmatrix} u, \quad y = [1 \ 0] x.$$

a. Is the system asymptotically stable?
b. Find the poles and zeros. Is the system BIBO stable?
c. Is the system reachable and observable?

Problems for Section 2.4

2.4-1 Redo Example 2.4-1 if the masses of the two bodies are M and m, generally not equal.

2.4-2 Using $\bar{A} = TAT^{-1}$, $\bar{C} = CT^{-1}$, demonstrate (2.4.16).

2.4-3 PBH Eigenvector Test. To demonstrate the PBH reachability eigenvector test for single-input systems, consider the simple single-input system in Jordan form

$$\dot{x} = \begin{bmatrix} \lambda_1 & & \\ & \lambda_2 & \\ & & \lambda_3 \end{bmatrix} x + \begin{bmatrix} b_1 \\ b_2 \\ b_3 \end{bmatrix} u$$

with all the λ_i distinct. Find the reachability matrix U_3, and demonstrate that $|U_3| \neq 0$ if and only if $b_i \neq 0$ for all i.

To draw the connection with the eigenvector test, note that the left eigenvector associated with λ_i is e_i, the i-th column of the 3×3 identity matrix. Therefore, $w_i^T B = 0$ is equivalent to $b_i = 0$.

2.4-4 PBH Eigenvector Test. To examine the PBH tests from a different point of view, show that if $w_i^T A = \lambda w_i^T$ and $w_i^T B = 0$ for some λ and $w_i \neq 0$, then $w_i^T [B \quad AB \ldots A^{n-1}B] = 0$, so that U_n does not have full rank and the system is not reachable.

2.4-5 Jordan Form. Consider the system

$$\dot{x} = \begin{bmatrix} -1 & 1 & 1 \\ 0 & 0 & 1 \\ 0 & -2 & -3 \end{bmatrix} x + \begin{bmatrix} 1 & 0 \\ 0 & 0 \\ 0 & 1 \end{bmatrix} u, \quad y = \begin{bmatrix} 1 & 0 & 0 \\ 0 & 2 & 0 \end{bmatrix} u. \tag{1}$$

a. Find the eigenvalues and eigenvectors of A.
b. Find the transformation to Jordan form and the Jordan form of the system.
c. Draw a block diagram of (1) and also of its Jordan form.
d. Find the transfer function of (1) and of the Jordan form.
e. Find the left eigenvectors of A.
f. Using the eigenstructure of A, find a PFE of $H(s)$. Now, find a PFE directly from $H(s)$.
g. Verify reachability using the reachability matrix of (1), and also using the PBH eigenvector test.
h. Verify observability using the observability matrix of (1), and also using the PBH eigenvector test.

2.4-6 Rank 2 Eigenvector. For the system

$$\dot{x} = \begin{bmatrix} 3 & -1 & 1 \\ 0 & 1 & 1 \\ -2 & 1 & 1 \end{bmatrix} x + \begin{bmatrix} 1 & -1 \\ 2 & 0 \\ 0 & 1 \end{bmatrix} u, \quad y = [3 \quad -1 \quad 3] u. \tag{1}$$

Find the Jordan form. Draw a block diagram of the Jordan form. Is the system reachable and observable?

2.4-7 Show that (2.4.53) is the characteristic equation for the system (2.4.51), (2.4.52), and that the transfer function of the system is (2.4.50). Use the case $n = 3$.

2.4-8 Show that (2.4.53) is the characteristic equation for the system (2.4.55), (2.4.56), and that the transfer function of the system is (2.4.50). Use the case $n = 3$.

2.4-9 Reachable Canonical Form. For the system

$$\dot{x} = \begin{bmatrix} -1 & 0 & 0 \\ 0 & -2 & 0 \\ 0 & 0 & -3 \end{bmatrix} x + \begin{bmatrix} 1 \\ 1 \\ 1 \end{bmatrix} u, \quad y = [1 \quad 1 \quad 1] u \tag{1}$$

find the reachable canonical form. Find the transformation that takes (1) to the RCF.

2.4-10 Observable Canonical Form. For the system in Problem 2.4-9, find the observable canonical form. Find the transformation that takes (1) to the OCF.

2.4-11 Minimal Realization. A system is described by the differential equation

$$y''' + y'' - 4y' - 4y = u' - 2u. \tag{1}$$

a. Find the poles. Is this a minimal system? Is it stable?
b. Find a minimal state-variable realization. Is the minimal realization stable? Using program TRESP in Appendix A, write a subroutine to plot the step response of (1).

2.4-12 MIMO Minimal Realization. For the system

$$\dot{x} = \begin{bmatrix} 0 & 1 & 0 & 0 & 0 & 0 \\ -2 & -3 & 0 & 0 & 0 & 0 \\ 0 & 0 & 0 & 1 & 0 & 0 \\ 0 & 0 & -3 & -4 & 0 & 0 \\ 0 & 0 & 0 & 0 & -2 & 0 \\ 0 & 0 & 0 & 0 & 0 & -3 \end{bmatrix} x + \begin{bmatrix} 0 & 0 \\ 1 & 0 \\ 0 & 0 \\ 0 & 1 \\ 1 & 0 \\ 0 & 1 \end{bmatrix} u$$

$$y = \begin{bmatrix} 3 & 2 & 7 & 3 & 0 & 0 \\ 0 & 0 & 0 & 0 & 3 & -1 \end{bmatrix} x$$

find the transfer function. Find a minimal realization. Draw block diagrams of the original system and its minimal realization.

Problems for Section 2.5

2.5-1 Ackermann's Formula. An unstable harmonic oscillator has the system description

$$\dot{x} = \begin{bmatrix} 0 & 1 \\ -2 & 2 \end{bmatrix} x + \begin{bmatrix} 0 \\ 1 \end{bmatrix} u.$$

Find the poles. Using Ackermann's formula, determine a state-feedback gain K in $u = -Kx$ to place the poles at $s = -1 \pm j$. Using program TRESP in Appendix A, simulate the closed-loop system to find the step response.

2.5-2 Analytic Computation of Feedback Gain. As an alternative to Ackermann's formula, set $u = -Kx = -[k_1 \quad k_2]x$ in Problem 2.5-1. In terms of k_1 and k_2 find the closed-loop system $(A - BK)$ and the closed-loop characteristic polynomial $\Delta_c(s)$. Find the desired characteristic polynomial $\Delta^D(s)$ that has poles at $s = -1 \pm j$. Set $\Delta_c(s)$ equal to $\Delta^D(s)$ to compute the gain elements k_1 and k_2 required to place the closed-loop poles as desired.

2.5-3 Multi-input Feedback Using Eigenstructure Assignment. Consider the two-input system

$$\dot{x} = \begin{bmatrix} 1 & 0 \\ 0 & 2 \end{bmatrix} x + \begin{bmatrix} 1 & 1 \\ 1 & -1 \end{bmatrix} u.$$

Using Moore's technique, find the state-feedback gain to place the poles at $\lambda_1 = -1$, $\lambda_2 = -2$, with respective closed-loop eigenvectors of $v_1 = [0 \quad 1]^T$, $v_2 = [1 \quad 0]^T$.

2.5-4 Output Feedback. Consider output feedback of the form $u = -ky$ applied to the system in Problem 2.5-1 with output measurements given by $y = [1 \quad 1]x$.

  a. In terms of the scalar gain k, find the closed-loop characteristic polynomial $\Delta_c(s)$. Put $\Delta_c(s)$ into the form $\Delta(s)[1 + kn(s)/\Delta(s)]$, with $\Delta(s)$ the open-loop characteristic polynomial. Now, draw a root locus for $\Delta_c(s)$ versus k. Can output feedback stabilize this system?

  b. Now, compute the transfer function $H(s)$ of the open-loop system. Use standard classical control theory to plot a root locus versus k. Notice that this is exactly what you did in part a, from a different point of view. The advantage of the approach in part a is that it can be generalized to MIMO systems.

  c. Now, measurements of the form $y = [-1 \quad 1]x$ are taken. Repeat part a. Can output feedback with positive k stabilize this new system? Find the new transfer function. Note that it has a nonminimum-phase zero.

2.5-5 MIMO Output Feedback Using Eigenstructure Assignment. Consider the 2-input 1-output system

$$\dot{x} = \begin{bmatrix} 1 & 0 \\ 0 & -2 \end{bmatrix} x + \begin{bmatrix} 1 & 1 \\ 1 & -1 \end{bmatrix} u, \quad y = [1 \quad 1]x.$$

  a. Using eigenstructure assignment, find the output-feedback gain K in $u = -Ky$ to place one closed-loop pole at $\lambda_1 = -1$ with an eigenvector of $v_1 = [1 \quad 0]^T$. Compute the closed-loop poles using this K—is the system stable?
  b. Repeat part a with $v_1 = [0 \quad 1]^T$. Note that this choice for v_1 does not yield a stable closed-loop system.

PART II

CONTINUOUS-TIME CONTROL

In this part of the book we shall discuss the design of continuous-time control systems. In Part III we cover digital control. Our objective is to present practical design techniques that take advantage of the intuition of classical control theory yet employ the powerful methods of modern control theory in dealing with the multiple control gains in multi-input/multi-output and multiloop systems.

To lay a firm theoretical infrastructure, in Chapter 3 we cover the foundations of standard continuous-time optimal control using state-variable feedback. Then, in Chapter 4 we show how to design practical control systems using output-feedback design. Output feedback provides a powerful approach to control systems design. It allows designers to use their engineering intuition to select the structure of a multivariable controller, and then employ the matrix design equations of modern control to tune multiple control gains so that the closed-loop system has optimal time responses. In contrast to classical controls design, this amounts to closing *all the feedback loops simultaneously* to obtain good performance.

3

Optimal Control of Continuous-Time Systems

SUMMARY

In this chapter, controls design equations are developed assuming that all the states are measurable and available for feedback. The fundamentals of modern optimal control are derived—namely, the solution to the linear quadratic regulator problem. Some aspects of minimum-time control are covered.

INTRODUCTION

In this chapter we shall introduce the foundations of optimal control for continuous-time dynamical systems. To make the discussion as direct as possible we assume that all internal variables, or states, of the system are known. This will give us a feel for the best that is possible using optimal control. In Chapter 4 we discuss optimal control where only some of the states are measurable as outputs; this will allow us to design practical control systems using compensators with any desired structure. The fundamental design techniques used in this book are introduced in sections 4.1 and 4.2.

In section 3.1 we present the general design equations for nonlinear systems, showing how they may be solved numerically. In section 3.2 we solve perhaps the most basic problem of modern control theory—the selection of state-feedback controls that minimize quadratic performance indices for linear systems. This is termed the linear quadratic regulator (LQR) problem. The LQR optimal control gains are time-varying. In section 3.3 we show how to derive time-invariant control gains, which are easier to implement and adequate for most applications. In section 3.4 we show some techniques for controls design where one of the objectives is to complete a task in minimum time.

For the student, all parts of each section are important. Those who are more interested in the design equations and their use may skip the subsections labeled "Derivation." Some notions of matrix algebra that will be quite useful through the chapter are introduced in Appendix B.

3.1 THE GENERAL CONTINUOUS-TIME OPTIMAL CONTROL PROBLEM

A state-variable model for a nonlinear time-varying dynamical system is given by

$$\dot{x} = f(x, u, t) \tag{3.1.1}$$

where $x(t) \in \mathbf{R}^n$ is the vector of internal states and $u(t) \in \mathbf{R}^m$ is the vector of control inputs. This is the plant to be controlled.

As we shall see, a broad range of performance objectives may be achieved by selecting the control $u(t)$ to minimize a *performance index* (PI) or *cost* given by

$$J(t_0) = \phi(x(T), T) + \int_{t_0}^{T} L(x(t), u(t), t) \, dt, \tag{3.1.2}$$

with t_0 the initial time and T the final time of interest.

The *final-state weighting* function $\phi(x(T), T)$ and weighting function $L(x, u, t)$ are selected depending on the performance objectives, as we shall subsequently demonstrate in several examples.

The *optimal control problem* is to determine a control input $u(t)$ for the system that minimizes the PI and also ensures that the *final state constraint*

$$\Psi(x(T), T) = 0 \tag{3.1.3}$$

is satisfied for a given function $\Psi \in \mathbf{R}^p$.

The roles of the final weighting function ϕ and the final constraint Ψ should not be confused. The former is a function we would like to make small, such as the final energy $x^T(T)P(T)x(T)$, with $P(T)$ a specified weighting matrix. On the other hand, $\Psi(x(T), T)$ is required to be exactly equal to zero.

For instance, if it is desired that the two angles θ_1 and θ_2 of a 2-link robot arm be driven exactly to specified values of r_1 and r_2 radians at the final time T, then $p = 2$ and the final state constraint is

$$\Psi(x(T), T) = \begin{bmatrix} \theta_1(T) - r_1 \\ \theta_2(T) - r_2 \end{bmatrix} = 0.$$

On the other hand, if it is sufficient merely to have the final angles *near* r_1, r_2, then we might select a final state weighting like

$$\phi(x(T), T) = (\theta_1(T) - r_1)^2 + (\theta_2(T) - r_2)^2.$$

Generally speaking, it is more convenient to derive practical (i.e., feedback) control laws using performance objective formulations in terms of ϕ and not Ψ.

Solution of the Nonlinear Optimal Control Problem

First we derive the design equations in Table 3.1-1. Then, we discuss the results and provide some examples to illustrate their use. In the following we denote partial derivatives by subscripts; for example, Ψ_x represents $\partial \Psi / \partial x$. A review of the matrix calculus we shall use is provided in Appendix B.

Derivation

To solve the optimal control problem we shall use Lagrange multipliers to adjoin the constraints (3.1.1) and (3.1.3) to the performance index (3.1.2). Since the system equation (3.1.1) is an equality constraint that must hold at each time, we require an associated multiplier $\lambda(t) \in \mathbf{R}^n$ that is a function of time. Since the final state constraint only holds at the final time T, we use a constant associated multiplier $\nu \in \mathbf{R}^p$. The augmented PI is therefore

$$J' = \phi(x(T), T) + \nu^T \Psi(x(T), T)$$
$$+ \int_{t_0}^{T} [L(x, u, t) + \lambda^T(t)(f(x, u, t) - \dot{x})] \, dt. \quad (3.1.4)$$

Defining the *Hamiltonian function*

$$H(x, u, t) = L(x, u, t) + \lambda^T(t) f(x, u, t) \quad (3.1.5)$$

we may write this as

$$J' = \phi(x(T), T) + \nu^T \Psi(x(T), T)$$
$$+ \int_{t_0}^{T} [H(x, u, t) - \lambda^T(t)\dot{x}] \, dt. \quad (3.1.6)$$

According to the theory of Lagrange multipliers, the problem of determining the control function $u(t)$ that minimizes (3.1.2) subject to the constraints (3.1.1) and (3.1.3) has now been converted to the problem of minimizing (3.1.6) without constraints, which we shall next accomplish.

Using Leibniz's rule (see the problems), the increment in J' as a function of increments in x, λ, ν, u, and t is

$$dJ' = (\phi_x + \Psi_x^T \nu)^T dx|_T + (\phi_t + \Psi_t^T \nu) dt|_T + \Psi^T|_T d\nu$$
$$+ (H - \lambda^T \dot{x}) dt|_T - (H - \lambda^T \dot{x}) dt|_{t_0} \quad (3.1.7)$$
$$+ \int_{t_0}^{T} [H_x^T \delta x + H_u^T \delta u - \lambda^T \delta \dot{x} + (H_\lambda - \dot{x})^T \delta \lambda] \, dt,$$

where the vertical bar means that the quantity to its left should be evaluated at the time indicated to its lower right. (Note that we define the gradient, e.g., $\partial \phi / \partial x$, as a *column* vector.)

The state differential is represented by dx, while δx represents the state variation. The relationship between these at a fixed time T is given by

$$dx(T) = \delta x(T) + \dot{x}(T) dT, \quad (3.1.8)$$

since $dx(T)$ is the overall change in $x(T)$ including small changes in T, while $\delta x(T)$ is defined as a small change in $x(T)$ for a fixed T. See Bryson and Ho [1975] and Lewis [1986].

To eliminate the variation in $\dot{x}$ from dJ', integrate by parts to see that

$$-\int_{t_0}^{T} \lambda^T \delta \dot{x} \, dt = -\lambda^T \delta x|_T + \lambda^T \delta x|_{t_0} + \int_{t_0}^{T} \dot{\lambda}^T \delta x \, dt. \quad (3.1.9)$$

Using this and (3.1.8) in (3.1.7) yields

$$\begin{aligned} dJ' &= (\phi_x + \Psi_x^T \nu - \lambda)^T dx|_T + (\phi_t + \Psi_t^T \nu + H) dt|_T \\ &\quad + \Psi^T|_T d\nu - H dt|_{t_0} + \lambda^T dx|_{t_0} \\ &\quad + \int_{t_0}^{T} [(H_x + \dot{\lambda})^T \delta x + H_u^T \delta u + (H_\lambda - \dot{x})^T \delta \lambda] \, dt. \end{aligned} \quad (3.1.10)$$

A minimum value of J' is attained when $dJ' = 0$ for all independent increments in its arguments. Setting to zero the coefficients of $d\nu$, δx, δu, and $\delta \lambda$ yields the necessary conditions for a minimum which appear in Table 3.1-1. These equations must be solved to yield the optimal control input $u(t)$ that minimizes the PI. We shall always assume that

TABLE 3.1-1 CONTINUOUS NONLINEAR OPTIMAL CONTROLLER

System model:

$$\dot{x} = f(x, u, t), \qquad t \geq t_0, \qquad t_0 \text{ fixed}$$

Performance index:

$$J(t_0) = \phi(x(T), T) + \int_{t_0}^{T} L(x, u, t) \, dt$$

Final state constraint:

$$\Psi(x(T), T) = 0$$

Optimal Controller:
  Hamiltonian:

$$H(x, u, t) = L(x, u, t) + \lambda^T f(x, u, t)$$

State equation:

$$\dot{x} = \frac{\partial H}{\partial \lambda} = f, \qquad t \geq t_0$$

Costate equation:

$$-\dot{\lambda} = \frac{\partial H}{\partial x} = \frac{\partial f^T}{\partial x} \lambda + \frac{\partial L}{\partial x}, \qquad t \leq T$$

Stationarity condition:

$$0 = \frac{\partial H}{\partial u} = \frac{\partial L}{\partial u} + \frac{\partial f^T}{\partial u} \lambda$$

Boundary conditions:

$$x(t_0) \text{ given}$$
$$(\phi_x + \Psi_x^T \nu - \lambda)^T|_T dx(T) + (\phi_t + \Psi_t^T \nu + H)|_T dT = 0 \quad (3.1.11)$$

the initial time t_0 and the initial state $x(t_0)$ are both known and fixed, so that dt_0 and $dx(t_0)$ are both zero and need not have zero coefficients.

Discussion

Necessary conditions for the solution of the nonlinear optimal control problem are given in Table 3.1-1. That is, any control $u(t)$ that results in a minimum value of the PI when it is applied to the system must satisfy the equations given there. Conditions under which these equations are sufficient as well are addressed in Bryson and Ho [1975] and Lewis [1986].

We shall use these as design equations for determining the control $u(t)$ that minimizes the PI. The structure of these design equations is worth discussing.

Note that, according to the table, $\lambda(t)$ is a dynamical variable that satisfies its own dynamical equation. It is called the *costate*. It is defined by a differential equation that develops *backward* in time (by defining a backward time variable $\tau = T - t$, we see that $d\tau = -dt$), with the final condition $\lambda(T)$ determined by (3.1.11). Thus, to determine the optimal control it is necessary to solve a *two-point boundary-value problem*, namely the state equation with initial condition $x(t_0)$ and the costate equation with final condition $\lambda(T)$. The optimal control $u(t)$, as we shall see, is then generally determined in terms of $x(t)$ and $\lambda(t)$ by using the stationarity condition [so named because this is the condition that guarantees a minimum or stationary point with respect to changes in $u(t)$]. We shall discuss the numerical solution of two-point boundary-value problems in the next subsection.

We do not really care about $\lambda(t)$, but it is an intermediate variable which must evidently be determined to solve for the optimal control $u(t)$ that minimizes the PI $J(t_0)$ while ensuring that the constraints (3.1.1) and (3.1.3) are satisfied.

The dynamical state and costate equations, along with the control as specified by the stationarity condition, are called the *Hamiltonian system*. These equations may be used to derive Lagrange's and Hamilton's equations of motion in physics [see, e.g., Lewis 1986]. The costate equation and stationarity condition are called *Euler's equations*.

The final condition (3.1.11) in Table 3.1-1 needs further discussion. Generally, dT and $dx(T)$ are not independent [see (3.1.8)]. Therefore, their individual coefficients are not necessarily equal to zero, but the entire expression must be considered.

It is worth noting a fundamental property of optimal control for time-invariant systems. The time derivative of the Hamiltonian is

$$\dot{H} = H_t + H_x^T \dot{x} + H_u^T \dot{u} + \dot{\lambda}^T f = H_t + H_u^T \dot{u} + (H_x + \dot{\lambda})^T f. \quad (3.1.12)$$

If $u(t)$ is an optimal control, then

$$\dot{H} = H_t. \quad (3.1.13)$$

In the time-invariant case, f and L are not explicit functions of t, so that neither is H. In this situation

$$\dot{H} = 0. \quad (3.1.14)$$

Thus, for time-invariant systems and cost functions, the Hamiltonian is a *constant* on the optimal trajectory. This is a quite general statement of the principle of conservation of energy.

The general nonlinear optimal controller in Table 3.1-1 is not the focus of this book. In section 3.2 we shall derive the Linear Quadratic Regulator, which will prove more useful to us from a practical controls engineering point of view. In spite of this, it is worth looking at a few examples to develop a feel for the optimal controller of Table 3.1-1.

Example 3.1-1: Optimal Control of a Nonlinear Steering System

A boat traveling with a fixed velocity of V has a variable steering angle of θ. It is desired to find the steering angle $\theta(t)$ which will allow the boat to cross a stream with a constant current of h and arrive at a specified point on the opposite side in the minimum time T.

This situation is depicted in Fig. 3.1-1, where the boat starts at the origin and the desired destination on the opposite bank of the stream is $(x(T), y(T)) = (x_T, y_T)$. The equations of motion are nonlinear:

$$\dot{x} = V\cos\theta + h$$
$$\dot{y} = V\sin\theta, \tag{1}$$

and we shall denote the state by $X = [x \quad y]^T$. The control input is $\theta(t)$.

In this example the final time T is unknown, but to make the journey in minimum time we could determine $\theta(t)$ to minimize the PI

$$J = \int_0^T 1 \, dt = T. \tag{2}$$

According to Table 3.1-1, the Hamiltonian is

$$H = 1 + [\lambda_x \quad \lambda_y]\begin{bmatrix} V\cos\theta + h \\ V\sin\theta \end{bmatrix} \tag{3}$$

where $\lambda = [\lambda_x \quad \lambda_y]^T$ is the Lagrange multiplier. Note that $\lambda \in \mathbf{R}^n$, so that there is one component of λ associated with each state component, as we have denoted using subscripts.

The costate equations are given by $-\dot{\lambda} = \partial H/\partial X$, or

$$-\dot{\lambda}_x = \frac{\partial H}{\partial x} = 0 \tag{4}$$

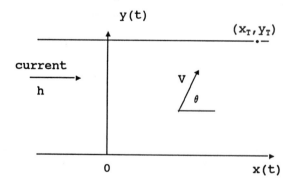

Figure 3.1-1 Boat crossing a region of constant currents

3.1 The General Continuous-Time Optimal Control Problem

$$-\dot{\lambda}_y = \frac{\partial H}{\partial y} = 0, \tag{5}$$

so that the costate is a constant.

The stationarity condition is

$$0 = \frac{\partial H}{\partial \theta} = -\lambda_x V \sin\theta + \lambda_y \cos\theta, \tag{6}$$

which may be solved for the control to obtain

$$\tan\theta = \frac{\lambda_y}{\lambda_x}. \tag{7}$$

This is called the *linear tangent control law*. Since the costate is constant, we see that the optimal steering angle θ is a constant.

Integrating (1) assuming that $\theta(t)$ is a constant results in

$$x = (V\cos\theta + h)t \tag{8}$$
$$y = Vt\sin\theta.$$

Evaluating this at the final time T gives

$$x(T) = x_T = (V\cos\theta + h)T \tag{9}$$
$$y(T) = y_T = VT\sin\theta,$$

whence

$$\sin\theta = y_T/VT \tag{10}$$
$$\cos\theta = (x_T - hT)/VT.$$

Dividing the first equation by the second shows that

$$\tan\theta = \frac{y_T}{x_T - hT}. \tag{11}$$

It is now necessary to solve for the unknown final time T, for then the optimal steering angle is given by (11) in terms of the desired final position. To accomplish this, note that $\sin^2\theta + \cos^2\theta = 1$, so that (10) implies

$$\frac{y_T^2}{V^2 T^2} + \frac{(x_T - hT)^2}{V^2 T^2} = 1, \tag{12}$$

or

$$(h^2 - V^2)T^2 - 2x_T hT + (x_T^2 + y_T^2) = 0. \tag{13}$$

This is typical of many problems in that we have a nonlinear equation to solve for the minimum time T.

The solutions of (13) are given by

$$T = \frac{x_T h \pm \sqrt{V^2(x_T^2 + y_T^2) - h^2 y_T^2}}{(h^2 - V^2)} \tag{14}$$

Only one of these roots, the positive real one, will make physical sense. For there to exist a positive real root, it is necessary that the term under the square root be positive, that is

$$\frac{V}{h} \geq \frac{y_T}{\sqrt{x_T^2 + y_T^2}}. \tag{15}$$

The quantity on the right may be understood from Fig. 3.1-1 as the sine of the desired path angle, that is, the angle of the straight line connecting the origin to (x_T, y_T). The path angle should be distinguished from the steering angle θ, which is given by (11) in terms of T, which we have just determined.

The procedure for finding the optimal steering angle θ, then, is to solve (13) for the minimum time T and then use (11). In the special case that the current h is zero, then the steering angle and the path angle are the same, and the travel time is just

$$T = \frac{\sqrt{x_T^2 + y_T^2}}{V}. \tag{16}$$

∎

Example 3.1-2: Shortest Distance Between Two Points

The length of a curve $x(t)$ dependent on a parameter t between $t = a$ and $t = b$ is given by

$$J = \int_a^b \sqrt{1 + \dot{x}^2(t)} \, dt. \tag{1}$$

In order to specify that the curve join two points (a, A), (b, B), in the plane, we need to impose the boundary conditions

$$x(a) = A, \tag{2}$$

$$x(b) = B. \tag{3}$$

See Schultz and Melsa [1967].

It is desired to find the curve $x(t)$ joining (a, A) and (b, B) that minimizes (1).

To put this into the form of an optimal control problem, define the "input" by

$$\dot{x} = u. \tag{4}$$

This is the "plant." Then (1) becomes

$$J = \int_a^b \sqrt{1 + u^2} \, dt. \tag{5}$$

The Hamiltonian is

$$H = \sqrt{1 + u^2} + \lambda u \tag{6}$$

Now, Table 3.1-1 yields the conditions

$$\dot{x} = H_\lambda = u, \tag{7}$$

$$-\dot{\lambda} = H_x = 0, \tag{8}$$

3.1 The General Continuous-Time Optimal Control Problem

$$0 = H_u = \lambda + \frac{u}{\sqrt{1 + u^2}}. \tag{9}$$

To solve these for the optimal slope u, note that by (9)

$$u = \frac{\lambda}{\sqrt{1 - \lambda^2}}, \tag{10}$$

but according to (8), λ is constant. Hence

$$u = \text{const} \tag{11}$$

is the optimal "control." Now use (7) to get

$$x(t) = c_1 t + c_2. \tag{12}$$

To determine c_1 and c_2, use the boundary conditions (2) and (3) to see that

$$x(t) = \frac{(A - B)t + (aB - bA)}{a - b}. \tag{13}$$

The optimal trajectory (13) between two points is thus a straight line.

∎

Two-Point Boundary-Value Problems

The solution for the optimal control $u(t)$ in Table 3.1-1 depends on solving two coupled differential equations, the state and costate equations, each of which is of order n. These two dynamical equations

$$\dot{x} = f(x, u, t), \qquad t \geq t_0 \tag{3.1.15}$$

$$-\dot{\lambda} = \frac{\partial H}{\partial x} = \frac{\partial f^T}{\partial x}\lambda + \frac{\partial L}{\partial x}, \qquad t \leq T \tag{3.1.16}$$

comprise the Hamiltonian system once the stationarity condition has been used to eliminate $u(t)$. The boundary conditions are:

n conditions: $\quad x(t_0)$ given $\tag{3.1.17}$

p conditions: $\quad \Psi(x(T), T) = 0 \tag{3.1.18}$

$n - p$ conditions: $\quad (\phi_x + \Psi_x^T \nu - \lambda)^T|_T dx(T) = 0, \tag{3.1.19}$

where we have assumed for simplicity that the final time T is specified and hence fixed, so that $dT = 0$ in condition (3.1.11).

Since n boundary conditions are specified at the initial time t_0 and n conditions are specified at the final time T, this is a two-point boundary-value problem. There are many methods available for solving such problems [Press et al. 1986], but here we shall only discuss two approaches.

Shooting-point method

One good numerical routine for solving the two-point boundary-value problem is DTPTB in the IMSL library of subroutines [IMSL], which uses the *shooting-point* method. The next example shows how to use it.

Example 3.1-3: Numerical Solution of Hamiltonian System for Armature-Controlled DC Motor

In Example 2.1-3 we derived the state equations for an armature-controlled DC motor. If there is no shaft flexibility, the state equations are

$$L\dot{i} = -Ri - v_b + u \tag{1}$$

$$J_m\dot{\omega} = T_m - T_L - b_m\omega, \tag{2}$$

where the motor torque and back e.m.f. are

$$T_m = k_m i \tag{3}$$

$$v_b = k'_m \omega \tag{4}$$

with $i(t)$ the armature current, $\omega(t)$ the motor speed, and control input $u(t)$ the armature voltage. The various constants were defined in Example 2.1-3. We shall assume that the load torque T_L is zero.

By appropriate definition of variables we may write

$$\dot{i} = -ai - k'\omega + bu \tag{5}$$

$$\dot{\omega} = -\alpha\omega + ki, \tag{6}$$

with $1/a = L/R$ the electrical time constant and $1/\alpha = J_m/b_m$ the mechanical time constant. Let us define the state as $x = [i \quad \omega]^T \in \mathbf{R}^n$ with $n = 2$.

It is desired to drive the motor from rest ($i = 0$, $\omega = 0$) to a specified final value of motor speed ω_T at a fixed final time T while minimizing control energy. A suitable PI for this control objective is

$$J = \frac{1}{2} s(\omega(T) - \omega_T)^2 + \frac{1}{2} \int_0^T u^2 \, dt. \tag{7}$$

The final-state weighting s is a constant design parameter. If it is selected large, then if J is finite it will be necessary for the final state component $\omega(T)$ to be very near the desired final speed ω_T. Thus, s is a design parameter which may be selected to achieve a suitable trade-off between the closeness of $\omega(t)$ to ω_T and the smallness of the control energy u^2. We shall illustrate this point in this example.

To apply the shooting-point method to compute the optimal control we only need to find the Hamiltonian system and the boundary conditions. According to Table 3.1-1, the Hamiltonian is

$$H = \frac{1}{2}u^2 + [\lambda_i \quad \lambda_\omega]\begin{bmatrix} -ai - k'\omega + bu \\ -\alpha\omega + ki \end{bmatrix}, \tag{8}$$

with $\lambda = [\lambda_i \quad \lambda_\omega]^T \in \mathbf{R}^2$ an undetermined Lagrange multiplier.

3.1 The General Continuous-Time Optimal Control Problem

The costate equations are

$$\dot{\lambda}_i = \frac{-\partial H}{\partial i} = a\lambda_i - k\lambda_\omega \tag{9}$$

$$\dot{\lambda}_\omega = \frac{-\partial H}{\partial \omega} = \alpha\lambda_\omega + k'\lambda_i \tag{10}$$

and the stationarity condition is

$$0 = \frac{\partial H}{\partial u} = u + b\lambda_i, \tag{11}$$

so that if we can determine $\lambda_i(t)$, then the optimal control may be found since

$$u = -b\lambda_i. \tag{12}$$

Using (12) in (5) yields the Hamiltonian system

$$\dot{i} = -ai - k'\omega - b^2\lambda_i \tag{13}$$

$$\dot{\omega} = -\alpha\omega + ki \tag{14}$$

$$\dot{\lambda}_i = \frac{-\partial H}{\partial i} = a\lambda_i - k\lambda_\omega \tag{15}$$

$$\dot{\lambda}_\omega = \frac{-\partial H}{\partial \omega} = \alpha\lambda_\omega + k'\lambda_i. \tag{16}$$

The initial conditions are

$$i(0) = 0 \tag{17}$$

$$\omega(0) = 0. \tag{18}$$

Since T is fixed but the final state is free, $dT = 0$ but $dx(T) \neq 0$ in (3.1.11), so it is required that the costate satisfy the final conditions

$$\lambda_i(T) = \left.\frac{\partial \phi}{\partial i}\right|_T = 0 \tag{19}$$

$$\lambda_\omega(T) = \left.\frac{\partial \phi}{\partial \omega}\right|_T = s(\omega(T) - \omega_T). \tag{20}$$

Routine DTPTB in IMSL requires three subroutines. Subroutine FCNI($N, TIME, X, XP$) provides the dynamics (13)–(16), where the overall state of the Hamiltonian system is $X = [i \ w \ \lambda_i \ \lambda_\omega]^T$. XP denotes the derivative of X, which must be computed by FCNI. $N = 2n$ is the dimension of the Hamiltonian system. The argument TIME will not be required in this example since the Hamiltonian system is time-invariant. This subroutine is shown in Fig. 3.1-2.

Subroutine FCNB($N, X0, XF, F$) provides the boundary conditions; it is also shown in Fig. 3.1-2. X0 is the value of X at the initial time, XF is the value of X at the final time,

```
C     ARMATURE-CONTROLLED DC MOTOR
C     SUBROUTINES FOR USE WITH IMSL SUBROUTINE DTPTB

C     STATE AND COSTATE EQUATIONS

      SUBROUTINE FCN1(N,TIME,X,XP)
      REAL X(*),XP(*),J,K,KM,KMP,KP,L
      COMMON/JACOB/A,AL,K,KP,B
      DATA J,KM,KMP,L,BM,R/0.1,1.,1.,.5,0.2,5./

      A= R/L
      AL= BM/J
      KP= KMP/L
      K= KM/J
      B= 1/L

      XP(1)= -A*X(1) - KP*X(2) - B**2*X(3)
      XP(2)= -AL*X(2) + K*X(1)
      XP(3)=  A*X(3) - K*X(4)
      XP(4)=  AL*X(4) + KP*X(3)

      RETURN
      END

C     BOUNDARY CONDITIONS

      SUBROUTINE FCNB(N,XO,XF,F)
      REAL XO(*),XF(*),F(*)
      DATA R,S/50,100/

      F(1)= XO(1)
      F(2)= XO(2)
      F(3)= XF(3)
      F(4)= XF(4) - S*(XF(2)-R)

      RETURN
      END

C     JACOBIAN

      SUBROUTINE FCNJ(N,TIME,X,JAC)
      REAL X(*),JAC(N,N),K,KP
      COMMON/JACOB/A,AL,K,KP,B

      JAC(1,1)= -A
      JAC(1,2)= -KP
      JAC(1,3)= -B**2
      JAC(2,1)=  K
      JAC(2,2)= -AL
      JAC(3,3)=  A
      JAC(3,4)= -K
      JAC(4,3)=  KP
      JAC(4,4)=  AL

      RETURN
      END
```

Figure 3.1-2 Subroutines for use with IMSL routine DTPTB

3.1 The General Continuous-Time Optimal Control Problem

and F is an N-vector function that is equal to zero when the boundary conditions are satisfied. It is, therefore, defined as

$$F(1) = i(0) = X0(1)$$
$$F(2) = \omega(0) = X0(2)$$
$$F(3) = \lambda_i(T) = XF(3) \tag{21}$$
$$F(4) = \lambda_\omega(T) - s(\omega(T) - \omega_T) = XF(4) - s(XF(2) - \omega_T).$$

Subroutine FCNJ(N,TIME,X,JAC) supplies the $N \times N$ Jacobian [$JAC(i, j)$] of the Hamiltonian system with respect to its state X. Differentiating (13)–(16) (Appendix B) it is found that

$$\dot{X} = \begin{bmatrix} -a & -k' & -b^2 & 0 \\ k & -\alpha & 0 & 0 \\ 0 & 0 & a & -k \\ 0 & 0 & k' & \alpha \end{bmatrix} X \equiv JAC \cdot X. \tag{22}$$

This subroutine is also shown in the figure.

For a sample run we used values for the motor parameters of $J_m = 0.1$ kg.m^2, $k_m = k'_m = 1$ V.s, $L = 0.5$ H, $b_m = 0.2$ N.m./rad/s, and $R = 5\Omega$. The desired final speed was $\omega_T = 50$ rad/s and the final time was $T = 2$ sec. DTPTB was used, with a driver program (provided in Lewis 1986), to find the initial conditions $X(0)$ that are *equivalent* to the split boundary conditions (17)–(20). Then, a Runge-Kutta integrator (TRESP in Appendix A) was used to integrate the Hamiltonian system forward in time using these initial conditions.

Figure 3.1-3 shows the plots of the state and costate trajectories for the value $s = 100$ of the final-state weighting. The final value of motor speed $\omega(T)$ was 48.828 rad/s. The optimal control voltage $u(t)$ is simply given by (12) in terms of the costate component $\lambda_i(t)$. It may be stored in computer memory and applied to the motor during the control run if these plots are deemed satisfactory.

Figure 3.1-4 shows the trajectories of $\omega(t)$ for the three values of final-state weighting $s = 10, 100, 1000$. The final speeds are respectively 40.32 rad/s, 48.82 rad/s, and 49.88 rad/s. Note that the final value of $\omega(t)$ approaches the desired value of $\omega_T = 50$ rad/s more closely as the value of the design parameter s is increased. This design parameter reflects a trade-off between the two performance objectives of closeness of $\omega(T)$ to ω_T and the desire to conserve control energy. That is, as s is increased, $\omega(T)$ approaches ω_T more closely, but at the expense of greater control energy $u^2(t)$. See (7).

One final word is in order. Note that optimal control design makes no reference to what occurs after the time interval of interest $[0, T]$. Indeed, note that $\lambda_i(T)$, and hence $u(T)$, is equal to zero. However, according to (5), (6) at steady-state we have $di/dt = 0$ and $d\omega/dt = 0$ so that

$$i = \frac{\alpha}{k} \omega \tag{23}$$

$$u = \frac{kk' + a\alpha}{bk} \omega. \tag{24}$$

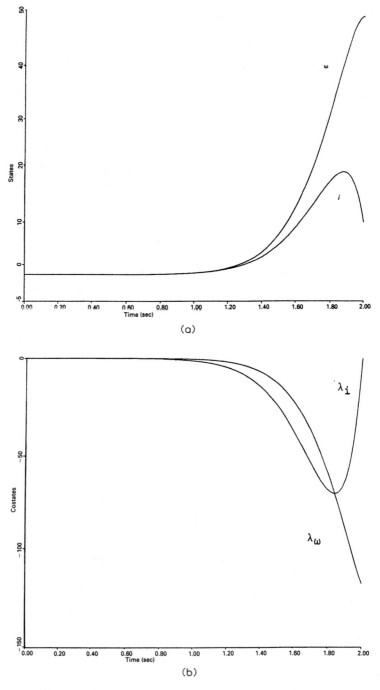

Figure 3.1-3 Optimal trajectories for $s = 100$ (a) State. (b) costate.

3.1 The General Continuous-Time Optimal Control Problem

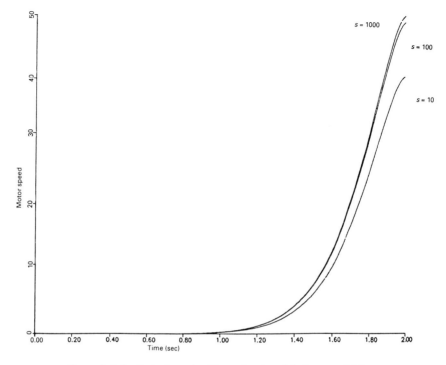

Figure 3.1-4 Optimal speed $\omega(t)$ as a function of final-state weighting s

That is, once the speed has been driven to 50 rad/s over the time interval $[0.\ T]$, to hold the speed at this value after time T it is necessary to switch in the armature voltage (24), which evaluates to $u = 2\omega = 100\ V$ with the parameters we have used.

Note that, since in Fig. 3.1-3 (a) $\dot\omega$ is approximately equal to zero, $i(T)$ should be approximately given by (23) so that $i(T) \approx 0.2\omega$. In fact, referring to the graph, this is the case. ∎

Unit solution method

Another way to solve the two-point boundary-value problem in the case of linear systems is to solve several initial-value problems and then solve a system of simultaneous equations. This *unit solution* method is of interest because it yields some intuition. It proceeds as follows:

1. Integrate the Hamiltonian system using as initial conditions $\lambda(t_0) = 0$ and $x(t_0) = r_0$, where r_0 is the given initial state. Call the resulting solutions $x_0(t)$, $\lambda_0(t)$.
2. Suppose $\lambda \in \mathbf{R}^n$ and let e_i represent the *i*th column of the $n \times n$ identity matrix. Determine *n unit solutions* by integrating the Hamiltonian system n times, using as initial conditions

$$x(t_0) = 0 \qquad (3.1.20)$$
$$\lambda(t_0) = e_i, \ i = 1, \ldots, n.$$

Call the resulting unit solutions $x_i(t)$, $\lambda_i(t)$ for $i = 1, \ldots, n$.

3. General initial conditions can be expressed as

$$x(t_0) = r_0 \text{ given,} \qquad (3.1.21)$$
$$\lambda(t_0) = \sum_{i=1}^{n} c_i e_i$$

for constants c_i. By superposition, the overall solutions for these general initial conditions are

$$x(t) = x_0(t) + \sum_{i=1}^{n} c_i x_i(t) \qquad (3.1.22)$$
$$\lambda(t) = \lambda_0(t) + \sum_{i=1}^{n} c_i \lambda_i(t).$$

Evaluate these solutions at the final time $t = T$, and then solve for the initial costate values c_i required to ensure that the terminal conditions (3.1.18) and (3.1.19) are satisfied.

This technique is important because it shows clearly that there exist *initial conditions* $x(t_0)$, $\lambda(t_0)$ that are *equivalent* to the specified split boundary conditions, in the sense that they yield the same solution to the system equations.

Example 3.1-4: Unit Solution Method for Scalar System

For the scalar plant

$$\dot{x} = ax + bu \qquad (1)$$

let us select the PI

$$J = \frac{s(x(T) - r_T)^2}{2} + \frac{1}{2} \int_0^T (qx^2 + ru^2) \, dt. \qquad (2)$$

Thus, our control objectives are to drive $x(T)$ to a given desired final state of r_T, while also keeping $x(t)$ and $u(t)$ small over the interval $[0, T]$. The initial state $x(0) = r_0$ is known and the final time T is fixed. The final state $x(t)$ is free, since we have not fixed it at r_T, but only weight the difference $(x(T) - r_t)$ to make it small. To determine the optimal control $u(t)$ on $[0, T]$ by the method of unit solutions, we proceed as follows.

3.1 The General Continuous-Time Optimal Control Problem

From Table 3.1-1, the Hamiltonian is

$$H = \frac{qx^2}{2} + \frac{ru^2}{2} + \lambda(ax + bu). \tag{3}$$

The Euler equations are

$$\dot{\lambda} = -\frac{\partial H}{\partial x} = -qx - a\lambda, \tag{4}$$

$$0 = \frac{\partial H}{\partial u} = ru + b\lambda. \tag{5}$$

Therefore, the optimal control is

$$u = -\frac{b}{r}\lambda. \tag{6}$$

Eliminating $u(t)$ in (1) yields the Hamiltonian system

$$\begin{bmatrix} \dot{x} \\ \dot{\lambda} \end{bmatrix} = \begin{bmatrix} a & -b^2/r \\ -q & -a \end{bmatrix} \begin{bmatrix} x \\ \lambda \end{bmatrix}. \tag{7}$$

The split boundary conditions are

$$x(0) = r_0 \text{ given}, \tag{8}$$

$$\lambda(T) = s(x(T) - r_T). \tag{9}$$

Instead of solving the split boundary-value problem (7)–(9), we solve two (i.e., $n + 1$) initial-value problems, one with initial conditions $x(0) = r_0$, $\lambda(0) = 0$, and one with $x(0) = 0$, $\lambda(0) = 1$. Then we solve for the $\lambda(0)$ required to make (9) hold.

If $x(0) = r_0$, $\lambda(0) = 0$, then the solution can be found by Laplace transforms to be

$$\begin{bmatrix} x_0(t) \\ \lambda_0(t) \end{bmatrix} = \frac{r_0}{2\alpha} \begin{bmatrix} \alpha - a \\ q \end{bmatrix} e^{-\alpha t} + \frac{r_0}{2\alpha} \begin{bmatrix} \alpha + a \\ -q \end{bmatrix} e^{\alpha t}, \quad t \geq 0, \tag{10}$$

where $\alpha = (a^2 + qb^2/r)^{1/2}$.

If $x(0) = 0$, $\lambda(0) = 1$, the unit solution is

$$\begin{bmatrix} x_1(t) \\ \lambda_1(t) \end{bmatrix} = \frac{1}{2\alpha} \begin{bmatrix} b^2/r \\ \alpha + a \end{bmatrix} e^{-\alpha t} + \frac{1}{2\alpha} \begin{bmatrix} b^2/r \\ \alpha - a \end{bmatrix} e^{\alpha t}, \quad t \geq 0. \tag{11}$$

Now consider the general initial condition $x(0) = r_0$, $\lambda(0) = c$ for some constant c. The solution with these initial conditions is

$$\begin{bmatrix} x(t) \\ \lambda(t) \end{bmatrix} = \begin{bmatrix} x_0(t) \\ \lambda_0(t) \end{bmatrix} + c\begin{bmatrix} x_1(t) \\ \lambda_1(t) \end{bmatrix}$$

$$= \frac{1}{2\alpha}\begin{bmatrix} r_0(\alpha - a) + \frac{cb^2}{r} \\ r_0 q + c(\alpha + a) \end{bmatrix} e^{-\alpha t} + \frac{1}{2\alpha}\begin{bmatrix} r_0(\alpha + a) - \frac{cb^2}{r} \\ -r_0 q + c(\alpha - a) \end{bmatrix} e^{\alpha t} \tag{12}$$

Now it remains only to determine the initial costate value c so that boundary condition (9) holds.

Evaluating (12) at $t = T$, and substituting $x(T)$ and $\lambda(T)$ into (9), yields the required initial-costate value of

$$\lambda(0) = c = \frac{r_0[(q + sa)\sinh \alpha T + s\alpha \cosh \alpha T] - r_T s\alpha}{(sb^2/r - a)\sinh \alpha T + \alpha \cosh \alpha T}. \tag{13}$$

Note that the initial costate is a linear combination of the initial and final states.

Using this value of c in (12) yields the optimal state and costate trajectories. Then (6) yields the optimal control. This method yields the optimal control as an open-loop control law, that is, as a function of time, not of the current state.

■

3.2 CONTINUOUS-TIME LINEAR QUADRATIC REGULATOR

The optimal controller for general nonlinear systems is given in Table 3.1-1. Through several examples we have hinted at the power of the equations in the table; however, it has taken some insight and experience to complete the examples. Indeed, no systematic design approach has yet been suggested.

In this section we will discuss the design of optimal controls for linear systems with quadratic performance indices; this is the so-called *linear quadratic (LQ) problem*. For two cases we shall derive methodical techniques for controls design, obtaining first an open-loop controller and then a closed-loop controller.

The results to be presented in this section form the basis of modern controls design. This is due to the fact that many systems are linear to begin with, while many nonlinear systems may be considered as linear when they are operating near an equilibrium point.

For those whose interest is in the final design equations and their use, the sections labeled "Derivation" may be skipped.

Open-Loop Control

We should now like to develop a methodical design technique for optimal controllers for linear time-invariant systems of the form

$$\dot{x} = Ax + Bu, \tag{3.2.1}$$

with $x \in \mathbf{R}^n$ and control input $u \in \mathbf{R}^m$, and an associated quadratic PI

$$J(t_0) = \frac{1}{2} \int_{t_0}^{T} u^T R u \, dt. \tag{3.2.2}$$

The initial time is t_0 and the final time is T. The symmetric *control weighting matrix R* is chosen by the designer depending on the control objectives, as we shall see.

We shall assume that R is positive definite ($R > 0$); that is, R has positive eigenvalues so that $u^T R u > 0$ for all $u(t) \neq 0$. In this case, J is always bounded below by zero, so that a sensible minimization problem results. Since the squares of the control inputs

3.2 Continuous-Time Linear Quadratic Regulator

occur in (3.2.2), we are trying to minimize a generalized control energy (consider for illustration the case when some of the control components are currents and voltages).

We shall determine the control $u(t)$ that minimizes $J(t_0)$ and drives the system from a given initial state

$$x(t_0) = x_0 \qquad (3.2.3)$$

to a *fixed* final reference value of r_T specified at a fixed value of the final time T. That is, we require that

$$x(T) = r_T \qquad (3.2.4)$$

for a given value of T.

Since the system (3.2.1) is linear and the PI (3.2.2) is quadratic, we call this the *fixed-final-state linear quadratic (LQ) problem*. The open-loop LQ controller appears in Table 3.2-1—let us now derive it.

Derivation

According to Table 3.1-1, the solution to the fixed-final-state control problem may be derived as follows. The Hamiltonian is

$$H(t) = \frac{1}{2} u^T R u + \lambda^T (Ax + Bu), \qquad (3.2.5)$$

where $\lambda(t) \in \mathbf{R}^n$ is an undetermined multiplier. The state and costate equations are

$$\dot{x} = \frac{\partial H}{\partial \lambda} = Ax + Bu \qquad (3.2.6)$$

$$-\dot{\lambda} = \frac{\partial H}{\partial x} = A^T \lambda, \qquad (3.2.7)$$

and the stationarity condition is

$$0 = \frac{\partial H}{\partial u} = Ru + B^T \lambda. \qquad (3.2.8)$$

Solving the last equation yields the optimal control in terms of the costate

$$u(t) = -R^{-1} B^T \lambda(t), \qquad (3.2.9)$$

and using this in (3.2.6) yields

$$\dot{x} = Ax - BR^{-1} B^T \lambda. \qquad (3.2.10)$$

To determine the optimal control, we must now solve the Hamiltonian system (3.2.10) and (3.2.7) taking into account the boundary conditions. Since both the final time T and final state $x(t)$ are fixed, $dT = 0$ and $dx(T) = 0$ so that the boundary condition in Table 3.1-1 is automatically satisfied. Thus, the boundary conditions are the given x_0 and (3.2.4).

Integrating (3.2.7) we obtain its solution

$$\lambda(t) = e^{A^T(T-t)}\lambda(T), \qquad (3.2.11)$$

where $\lambda(T)$ is still unknown. Using this in (3.2.10) yields

$$\dot{x} = Ax - BR^{-1}B^T e^{A^T(T-t)}\lambda(T), \qquad (3.2.12)$$

whose solution is

$$x(t) = e^{A(t-t_0)}x_0 - \int_{t_0}^{t} e^{A(t-\tau)}BR^{-1}B^T e^{A^T(T-\tau)}\lambda(T)\, d\tau. \qquad (3.2.13)$$

To find $\lambda(T)$, evaluate this at T to obtain

$$x(T) = e^{A(T-t_0)}x_0 - G(t_0, T)\lambda(T) \qquad (3.2.14)$$

where the *continuous reachability gramian* is the symmetric matrix

$$G(t_0, t) \equiv \int_{t_0}^{t} e^{A(t-\tau)}BR^{-1}B^T e^{A^T(T-\tau)}\, d\tau. \qquad (3.2.15)$$

Due to the requirement (3.2.4), we may solve for the final costate to obtain

$$\lambda(T) = -G^{-1}(t_0, T)[r_T - e^{A(T-t_0)}x_0]. \qquad (3.2.16)$$

Note that $\lambda(T)$ is expressed in terms of the given initial state and desired final state.

Finally, the optimal control is found by using this result and (3.2.11) in (3.2.9) to be

$$u(t) = R^{-1}B^T e^{A^T(T-t)}G^{-1}(t_0, T)[r_T - e^{A(T-t_0)}x_0]. \qquad (3.2.17)$$

This is the minimum-energy control that drives the state from a given x_0 to a desired value of r_T at a specified final time T.

There is a better way to compute the reachability gramian in practice than the integration (3.2.15). Using Leibniz's rule, the solution to the equation

$$\dot{P} = AP + PA^T + BR^{-1}B^T \qquad (3.2.18)$$

may be shown (see the problems) to be

$$P(t) = e^{A(t-t_0)}P(t_0)e^{A^T(t-t_0)} + \int_{t_0}^{t} e^{A(t-\tau)}BR^{-1}B^T e^{A^T(t-\tau)}\, d\tau. \qquad (3.2.19)$$

Thus, if (3.2.18) is solved using $P(t_0) = 0$, then $G(t_0, T) = P(T)$. Equation (3.2.18) for P is called a *Lyapunov equation*; it is linear in P. Note that P is symmetric as long as $P(t_0)$ is, since the Lyapunov equation is equal to its own transpose.

The optimal state trajectory may be determined by using (3.2.16) in (3.2.13). It would be useful to know as well the optimal value of the PI using the proposed control. Defining the *final-state difference* as

$$d(t_0, T) = r_T - e^{A(T-t_0)}x_0 \qquad (3.2.20)$$

3.2 Continuous-Time Linear Quadratic Regulator

we may use (3.2.17) in (3.2.2) to write the optimal PI as

$$J(t_0) = \frac{1}{2} \int_{t_0}^{T} d^T G^{-1} e^{A(T-t)} BR^{-1} B^T e^{A^T(T-t)} G^{-1} d \, dt. \quad (3.2.21)$$

Using the definition of the (symmetric) gramian yields the optimal cost

$$\begin{aligned} J(t_0) &= \frac{1}{2} d^T(t_0, T) G^{-1}(t_0, T) d(t_0, T) \\ &= \frac{1}{2} d^T(t_0, T) P^{-1}(T) d(t_0, T). \end{aligned} \quad (3.2.22)$$

TABLE 3.2-1 OPEN-LOOP LINEAR QUADRATIC CONTROLLER

System model:

$$\dot{x} = Ax + Bu, \qquad t \geq t_0, \qquad x(t_0) = x_0 \text{ given}$$

Desired final state:

$$x(T) = r_T, \qquad r_T \text{ given}$$

Performance index:

$$J(t_0) = \frac{1}{2} \int_{t_0}^{T} u^T Ru \, dt, \qquad R > 0$$

Optimal Open-Loop Control:
Lyapunov equation:

$$\dot{P} = AP + PA^T + BR^{-1}B^T, \qquad P(t_0) = 0$$

Open-loop control:

$$u(t) = R^{-1} B^T e^{A^T(T-t)} P^{-1}(T) d(t_0, T)$$

where $d(t_0, T) = r_T - e^{A(T-t_0)} x_0$

Optimal cost:

$$J(t_0) = \frac{1}{2} d^T(t_0, T) P^{-1}(T) d(t_0, T)$$

Discussion

In Table 3.2-1 we summarize the design equations for the fixed-final-state LQ controller that drives the system from a known initial state of x_0 to a fixed desired final state of r_T while minimizing the generalized control energy $J(t_0)$. Let us discuss them to get a feel for the controller.

According to the table, the optimal control $u(t)$ that minimizes the PI $J(t_0)$ while ensuring that $x(T) = r_T$ is found as follows. First, integrate the Lyapunov equation from

t_0 to T with $P(t_0) = 0$ to find $P(T) = G(t_0, T)$. This may be accomplished *off-line* before the control run. The integration may easily be accomplished numerically using a Runge-Kutta integrator [IMSL 1980]; it is only necessary to write a driver program (e.g., program TRESP in Appendix A). Then, the optimal control is given in terms of $P^{-1}(T)$.

Since the control $u(t)$ does not depend on the state $x(t)$ at time t, but only on the specified end points of the state trajectory, it is an *open-loop control*. If for some reason $x(t)$ is perturbed off of the optimal path, the control has no way of sensing this, so then $x(T)$ will not generally equal the desired value r_T. That is, open-loop controls are not *robust* to disturbances or uncertain parameters in the system matrices.

In the next subsection we shall derive the closed-loop LQ regulator.

Since the control is open-loop, if desired, $u(t)$ may be precomputed and stored in computer memory *before the control run*. Then it may be applied to the plant to achieve the control objectives during the actual implementation phase.

The homogeneous solution to the state equation at time T is

$$x(T) = e^{A(T-t_0)}x_0, \qquad (3.2.23)$$

so that $d(t_0, T)$ is just the difference between the desired final state r_T and the final state that the system would arrive at on its own. It makes sense that the optimal control should be proportional to this difference. If $d(t_0, T) = 0$, then no control is required to make the state $x(t)$ go to r_T at $t = T$.

The optimal control exists if and only if $G(t_0, T)$ is nonsingular. Since R is assumed nonsingular, this corresponds to *reachability* of the plant (see Chapter 2). Thus, if (A, B) is reachable, then there exists a minimum-energy control that drives any given initial state to any desired final state.

Note that the optimal value of the PI depends only on x_0 and r_T. Thus, given the initial state and the desired final state, the required control energy can be calculated off line before the optimal control $u(t)$ is applied to the system. If it is too large, then too much control energy will be required. In this event, $u(t)$ should be redesigned, selecting a larger time interval $(T - t_0)$. According to (3.2.15) this will make the gramian $G(t_0, T)$ larger, so that according to (3.2.17) the control magnitude will be smaller.

Let us now consider some examples of open-loop LQ control. These examples will be analytical in nature as their purpose is to demonstrate the computational aspects of the design equations in Table 3.2-1.

Example 3.2-1: Open-Loop Control of a DC Motor

Assuming zero load torque, the transfer relation for an armature-controlled DC motor is

$$\omega = \frac{k_m}{(Ls + R)(J_m s + b_m) + k_m k'_m} u \qquad (1)$$

with $\omega(t)$ the output speed, control input $u(t)$ the armature voltage, k_m the torque constant, k'_m the back e.m.f. constant, L the armature winding leakage inductance, R the armature resistance, J_m the rotor moment of inertia, and b_m the rotor equivalent damping constant. See Examples 2.1-3 and 2.1-8.

Neglecting the electrical time constant L/R, which is usually at least an order of magnitude smaller than the mechanical time constant J_m/b_m, we may write [de Silva 1989]

3.2 Continuous-Time Linear Quadratic Regulator

$$\omega = \frac{k}{1 + s\tau} u \qquad (2)$$

where

$$k = \frac{k_m}{Rb_m + k_m k'_m}, \qquad \tau = \frac{RJ_m}{Rb_m + k_m k'_m}. \qquad (3)$$

A state-variable model of this is given by

$$\dot{x} = -ax + bu \qquad (4)$$

with $x(t) = \omega(t)$, $a = 1/\tau$, and $b = k/\tau$.

The initial speed $x(0) = \omega_0$ is known. The objective is to drive the motor to a desired final speed $x(T) = \omega_T$ at a specified final time T. In order to use minimum control energy we want to accomplish this while minimizing the PI

$$J = \frac{1}{2} \int_0^T r u^2(t) \, dt. \qquad (5)$$

The Lyapunov equation in Table 3.2-1 is

$$\dot{p} = -2ap + b^2/r, \qquad (6)$$

whose solution when $p(0) = 0$ is given by

$$p(t) = G(0, t) = \int_0^t e^{-2a(t-\tau)} \frac{b^2}{r} \, d\tau$$
$$= \frac{b^2}{2ar}(1 - e^{-2at}). \qquad (7)$$

According to Table 3.2-1 this yields the optimal control

$$u(t) = \frac{a(\omega_T - \omega_0 e^{-aT})}{b \sinh aT} e^{at}. \qquad (8)$$

It is interesting to note that the result is independent of the control weighting r. This is because $u(t)$ is a scalar.

Recall (Chapter 2) that the solution to the state equation (4) is given by

$$x(t) = e^{-at}x(0) + \int_0^t e^{-a(t-\tau)} bu(\tau) \, d\tau. \qquad (9)$$

Using (8) in (9) and simplifying shows that, under the influence of the open-loop control, the speed as a function of time is given by

$$x(t) = \omega_0 e^{-at} + (\omega_T - \omega_0 e^{-aT}) \frac{\sinh at}{\sinh aT}, \qquad (10)$$

which indeed has a final value of $x(T) = \omega_T$.

It is easy to write a FORTRAN subroutine to simulate applying $u(t)$ to the system (4) using a Runge-Kutta integrator. This procedure will be illustrated in subsequent examples.

Our purpose here has been to show how the equations of Table 3.2-1 are used to compute the control $u(t)$. For actual implementation purposes, we shall use digital controls, as we describe in Part III of the book.

■

Example 3.2-2: Open-Loop Control of Systems Obeying Newton's Laws

A particle obeying Newton's Laws satisfies

$$\dot{x} = \begin{bmatrix} 0 & 1 \\ 0 & 0 \end{bmatrix} x + \begin{bmatrix} 0 \\ 1 \end{bmatrix} u \qquad (1)$$

where $x = [d \quad v]^T$ with $d(t)$ the position and $v(t)$ the velocity, and $u(t)$ is an acceleration input. See Example 2.1-4.

It is easy to use the equations in Table 3.2-1 to find an analytic expression for the control required to drive any given $x(0)$ to any desired $x(T)$, while minimizing

$$J(0) = \frac{1}{2} \int_0^T r u^2 \, dt. \qquad (2)$$

To find the reachability gramian, we solve the Lyapunov equation. P is symmetric, so let

$$P(t) = \begin{bmatrix} p_1(t) & p_2(t) \\ p_2(t) & p_3(t) \end{bmatrix}. \qquad (3)$$

Then the Lyapunov equation is

$$\dot{P} = \begin{bmatrix} 0 & 1 \\ 0 & 0 \end{bmatrix} P + P \begin{bmatrix} 0 & 0 \\ 1 & 0 \end{bmatrix} + \begin{bmatrix} 0 & 0 \\ 0 & 1/r \end{bmatrix}, \qquad (4)$$

which yields the scalar equations

$$\dot{p}_1 = 2p_2, \qquad (5)$$

$$\dot{p}_2 = p_3, \qquad (6)$$

$$\dot{p}_3 = 1/r. \qquad (7)$$

For the gramian, we integrate (7), (6), and then (5) with $P(0) = 0$ to get

$$p_3 = \frac{t}{r}, \qquad (8)$$

$$p_2 = \frac{t^2}{2r}, \qquad (9)$$

$$p_1 = \frac{t^3}{3r}, \qquad (10)$$

so that

$$G(0, t) = P(t) = \begin{bmatrix} \dfrac{t^3}{3r} & \dfrac{t^2}{2r} \\ \dfrac{t^2}{2r} & \dfrac{t}{r} \end{bmatrix}. \qquad (11)$$

The state-transition matrix is

$$e^{At} = \begin{bmatrix} 1 & t \\ 0 & 1 \end{bmatrix}. \qquad (12)$$

3.2 Continuous-Time Linear Quadratic Regulator

According to Table 3.2-1, the optimal control becomes

$$u(t) = \frac{1}{r}[T-t \quad 1]\begin{bmatrix} 12r/T^3 & -6r/T^2 \\ -6r/T^2 & 4r/T \end{bmatrix}\left(x(T) - \begin{bmatrix} 1 & T \\ 0 & 1 \end{bmatrix}x(0)\right), \quad (13)$$

or

$$u(t) = \begin{bmatrix} \dfrac{6T - 12t}{T^3} & \dfrac{-2T + 6t}{T^2} \end{bmatrix}\left(x(T) - \begin{bmatrix} 1 & T \\ 0 & 1 \end{bmatrix}x(0)\right). \quad (14)$$

Once again, since $u(t)$ is a scalar, it is independent of r. Note also that the control magnitude decreases as the control interval $[0, T]$ increases. More control is required to move the system more quickly from one state to another.

■

Closed-Loop Control

The results to be presented in this subsection are perhaps the most widely used of any we shall discuss. This is because they apply to a large class of systems that frequently appear, namely linear systems, while providing for *feedback* control. This is in contrast with the open-loop control of the previous subsection. The advantages of feedback are well known in terms of sensitivity reduction, self-regulation, robustness to disturbances, and so on; such issues will be discussed in Chapters 8 and 10.

Consider again the linear system

$$\dot{x} = Ax + Bu, \quad (3.2.24)$$

with $x \in \mathbf{R}^n$ and control input $u \in \mathbf{R}^m$. The plant may be time-varying, though for notational convenience we shall not show this dependence explicitly.

Instead of demanding a fixed final state, we shall now only require that the final state $x(T)$ be *near* zero at a specified final time T. Thus, the final state is *free* and we are interested in choosing the control that minimizes the quadratic PI

$$J(t_0) = \frac{1}{2}x^T(T)S(T)x(T) + \frac{1}{2}\int_{t_0}^{T}(x^T Q x + u^T R u)\, dt. \quad (3.2.25)$$

This formulation will result in a radically different control law than the one in Table 3.2-1.

The *control weighting* R, *state weighting* Q, and *final state weighting* $S(T)$ are symmetric matrices chosen by the designer depending on the control objectives, as we shall see. As a preview, if the elements of $S(T)$ are selected larger, then the final state $x(T)$ must be smaller to keep the PI small.

We shall assume that Q and $S(T)$ are positive-semidefinite ($Q \geq 0$, $S(T) \geq 0$). Thus, Q and $S(T)$ have nonnegative eigenvalues so that $x^T Q x$ and $x^T(T)S(T)x(T)$ are nonnegative for all $x(t)$. Likewise, we shall assume that R is positive definite ($R > 0$); that is, R has positive eigenvalues so that $u^T R u > 0$ for all $u(t) \neq 0$. In this case, J is always bounded below by zero, so that a sensible minimization problem results. Since the squares of the states and control inputs occur in (3.2.25), we are trying to minimize a generalized energy (consider the case when some of the state components are velocities, or currents and voltages).

Since we are using a quadratic PI to regulate the state of a linear system to zero, but do not require the final state to take on any fixed value, we call this the *free-final-state linear quadratic regulator* (LQR) problem. The design equations for the optimal LQR controller appear in Table 3.2-2, which we now derive.

Derivation

According to Table 3.1-1 the solution to the free-final-state optimal control problem may be derived as follows. The Hamiltonian is

$$H(t) = \frac{1}{2}(x^T Q x + u^T R u) + \lambda^T (Ax + Bu), \qquad (3.2.26)$$

where $\lambda(t) \in \mathbf{R}^n$ is an undetermined multiplier. The state and costate equations are

$$\dot{x} = \frac{\partial H}{\partial \lambda} = Ax + Bu \qquad (3.2.27)$$

$$-\dot{\lambda} = \frac{\partial H}{\partial x} = Qx + A^T \lambda, \qquad (3.2.28)$$

and the stationarity condition is

$$0 = \frac{\partial H}{\partial u} = Ru + B^T \lambda. \qquad (3.2.29)$$

Solving the last equation yields the optimal control in terms of the costate

$$u(t) = -R^{-1} B^T \lambda(t). \qquad (3.2.30)$$

The controller implied by these equations is shown in Fig. 3.2-1. It is a feedback compensator with dynamics given by the costate equation. Unfortunately, it cannot be implemented in this fashion since the costate equation is noncausal [because conditions on the final costate $\lambda(T)$ are specified—see (3.2.33)].

Using (3.2.30) in (3.2.27) yields

$$\dot{x} = Ax - BR^{-1} B^T \lambda, \qquad (3.2.31)$$

which may be combined with the costate equation into the homogeneous Hamiltonian system

$$\begin{bmatrix} \dot{x} \\ \dot{\lambda} \end{bmatrix} = \begin{bmatrix} A & -BR^{-1}B^T \\ -Q & -A^T \end{bmatrix} \begin{bmatrix} x \\ \lambda \end{bmatrix}. \qquad (3.2.32)$$

The coefficient matrix is called the *Hamiltonian matrix*. We shall see in section 3.3 that its eigenvalues and eigenvectors are of major importance in the analysis of the time-invariant LQR.

3.2 Continuous-Time Linear Quadratic Regulator

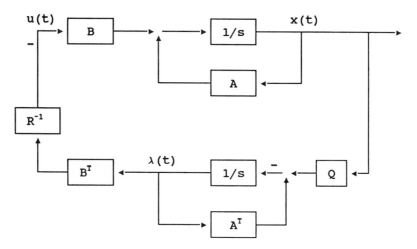

Figure 3.2-1 State-costate formulation of the LQR

The initial condition is the known value x_0 of $x(t_0)$. The final time T is fixed. Since the final state $x(T)$ is free, $dx(T)$ is nonzero, so that the final condition from Table 3.1-1 is

$$\lambda(T) = \left.\frac{\partial \phi}{\partial x}\right|_T = S(T)x(T). \tag{3.2.33}$$

To find the optimal control, we shall solve this two-point boundary-value problem using the *sweep method* [Bryson and Ho 1975]. Thus, assume that $x(t)$ and $\lambda(t)$ satisfy a linear relation like (3.2.33) for all times in the interval of interest $[t_0, T]$ so that

$$\lambda(t) = S(t)x(t) \tag{3.2.34}$$

for some as yet unknown $n \times n$ matrix function $S(t)$.

To find the auxiliary function $S(t)$, differentiate the costate to get

$$\dot{\lambda} = \dot{S}x + S\dot{x} = \dot{S}x + S(Ax - BR^{-1}B^T Sx) \tag{3.2.35}$$

where we have used the state equation. Now, the costate equation shows that, for all t

$$-\dot{S}x = (A^T S + SA - SBR^{-1}B^T S + Q)x. \tag{3.2.36}$$

Since this holds for any initial condition x_0 and hence all state trajectories, we must have

$$-\dot{S} = A^T S + SA - SBR^{-1}B^T S + Q, \quad t \leq T. \tag{3.2.37}$$

This is a matrix *Riccati* equation; it is bilinear in S. If $S(t)$ is the solution with final condition $S(T)$, then (3.2.34) holds for all $t \leq T$. Thus, our assumption was valid.

In terms of the Riccati solution $S(t)$, the optimal control is given by

$$u(t) = -R^{-1}B^T S(t)x(t). \tag{3.2.38}$$

Defining the *Kalman gain* by

$$K(t) = R^{-1}B^T S(t), \tag{3.2.39}$$

we may write the *state-feedback control law*

$$u(t) = -K(t)x(t). \tag{3.2.40}$$

To determine the optimal cost using this controller, note that

$$\frac{1}{2}\int_{t_0}^{T} \frac{d}{dt}(x^T S x)\, dt = \frac{1}{2} x^T(T)S(T)x(T) - \frac{1}{2} x^T(t_0)S(t_0)x_0. \tag{3.2.41}$$

Therefore, add zero, in the form of the left-hand side of this relation minus its right-hand side to (3.2.25), to obtain

$$J(t_0) = \frac{1}{2} x_0^T S(t_0) x_0$$
$$+ \frac{1}{2}\int_{t_0}^{T} (x^T Q x + u^T R u + \dot{x}^T S x + x^T \dot{S} x + x^T S \dot{x})\, dt. \tag{3.2.42}$$

It is important to note at this point that the dependence on the (unknown) final state is gone. Now, the state equation (3.2.24) allows us to write

$$J(t_0) = \frac{1}{2} x_0^T S(t_0) x_0$$
$$+ \frac{1}{2}\int_{t_0}^{T} [x^T(\dot{S} + Q + A^T S + SA)x + x^T S B u + u^T B^T S x + u^T R u]\, dt. \tag{3.2.43}$$

However, if $S(t)$ satisfies the Riccati equation, then

$$J(t_0) = \frac{1}{2} x_0^T S(t_0) x_0$$
$$+ \frac{1}{2}\int_{t_0}^{T} [x^T S B R^{-1} B^T S x + x^T S B u + u^T B^T S + u^T R u]\, dt. \tag{3.2.44}$$

Denoting $\|v\|_R^2 = v^T R v$ as the R-norm of a vector v (for $R > 0$), this may be written as

$$J(t_0) = \frac{1}{2} x_0^T S(t_0) x_0$$
$$+ \frac{1}{2}\int_{t_0}^{T} \|R^{-1} B^T S x + u\|_R^2\, dt. \tag{3.2.45}$$

This is an important result. We should first understand clearly that the function of the Riccati equation is to allow us to write the integrand as a perfect square. Thus, it performs for the matrix case exactly the same function as *completing the square*. Next, if we select the control given by (3.2.38), then $J(t_0)$ is minimized, and its optimal value is

$$J(t_0) = \frac{1}{2} x_0^T S(t_0) x_0. \tag{3.2.46}$$

Thus, this is an alternative derivation of the LQR.

3.2 Continuous-Time Linear Quadratic Regulator

TABLE 3.2-2 CONTINUOUS LINEAR QUADRATIC REGULATOR

System model:

$$\dot{x} = Ax + Bu, \qquad t \geq t_0, \qquad x(t_0) = x_0 \text{ given}$$

Performance index:

$$J(t_0) = \frac{1}{2} x^T(T) S(T) x(T) + \frac{1}{2} \int_{t_0}^{T} (x^T Q x + u^T R u) \, dt.$$

with:

$$S(T) \geq 0, \qquad Q \geq 0, \qquad R > 0$$

Optimal feedback control:
Riccati equation:

$$-\dot{S} = A^T S + SA - SBR^{-1}B^T S + Q, \qquad t \leq T, \qquad S(T) \text{ given}$$

Kalman gain:

$$K = R^{-1} B^T S$$

Time-varying feedback:

$$u = -K(t)x$$

Optimal cost:

$$J(t_0) = \frac{1}{2} x_0^T S(t_0) x_0$$

Discussion

The optimal LQ regulator is summarized in Table 3.2-2. A block diagram is shown in Fig. 3.2-2—it is a *feedback control system*. Let us discuss the control law we have just derived.

The optimal LQR is determined by solving the Riccati equation *backward* in time for the auxiliary matrix function $S(t)$, using as final condition the value of $S(T)$ selected for the PI. Then the optimal feedback gain is given by the Kalman gain $K(t)$. Even if the system (A, B) is time-invariant, the optimal control $u(t)$ is a *time-varying state feedback*. This is why the optimal LQ controller may not be determined using classical frequency-domain techniques.

Feedback or closed-loop control of this sort is more useful in practice than open-loop control, for it is robust to uncertainties in the plant parameters as well as to unaccounted-for disturbances. That is, even if the system model is not an exact description of the plant, the LQR performs well if it is a close description. In Chapter 10 we shall see that the LQR has important *guaranteed robustness properties*.

The initial state of the plant is known. Therefore, the expression in Table 3.2-2 allows us to compute the optimal cost before we ever apply the control to the plant. If it

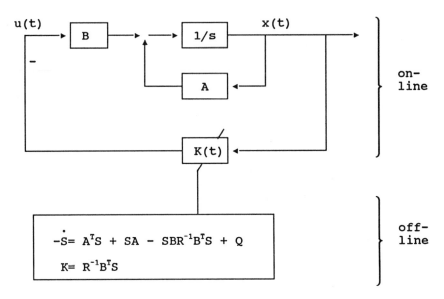

Figure 3.2-2 Kalman gain formulation of the LQR

is too high, we may select different weighting matrices Q, R, and $S(T)$ in the performance index and try another design.

Note that, in terms of the Kalman gain, the Riccati equation may be written as

$$-\dot{S} = A^T S + SA - K^T R K + Q. \qquad (3.2.47)$$

In terms of the closed-loop plant matrix (see the problems) it may be written in the *Joseph stabilized formulation*

$$-\dot{S} = (A - BK)^T S + S(A - BK) + K^T R K + Q. \qquad (3.2.48)$$

In contrast to the fixed-final-state controller of the previous subsection, reachability of the plant is *not* required in the LQR problem. Regardless of whether this property holds, the LQR will do its best to minimize the performance index. However, we shall see in section 3.3 that reachability of the plant results in some very nice properties of the closed-loop system as the final time T goes to infinity.

The Riccati equation may be solved off-line for $S(t)$, and the Kalman gain $K(t)$ computed and stored in a microprocessor. Then, during the implementation or control run, the states may be measured and the feedback control $u(t)$ applied to the plant. Thus, we have decomposed the LQR problem into two stages: off-line computation of the optimal gains using a backward differential equation, and the actual control of the plant using feedback. Such *hierarchical control schemes*, consisting of an inner linear feedback loop whose gain is computed by an outer quadratic design equation, are typical of modern control schemes.

To solve the Riccati equation we may use a fourth-order Runge-Kutta integrator. To simulate the action of the plant under the influence of the designed control, we may also use a Runge-Kutta integrator to integrate the closed-loop system

3.2 Continuous-Time Linear Quadratic Regulator

$$\dot{x} = (A - BK)x \qquad (3.2.49)$$

with $K(t)$ the time-varying Kalman gain. A driver program may be written to accomplish this along the lines of the block diagram in Fig. 3.2-3. Such a driver program (CTLQR) is given in Lewis [1986] and is a small modification of program TRESP in Appendix A. In the examples, we shall show how simple the simulation of the LQR is by providing some subroutines for use with the program in Fig. 3.2-3 that implement the Riccati equation solution and the closed-loop plant.

Since most Runge-Kutta integrators work forward in time, we convert the Riccati equation into an equation that is integrated forward in time as follows. Define a change of variables by

$$\tau = T - t, \qquad (3.2.50)$$

so that $d\tau = -dt$ and the Riccati equation may be written as

$$\dot{S}_b = A^T S_b + S_b A - S_b B R^{-1} B^T S_b + Q, \qquad t \leq T. \qquad (3.2.51)$$

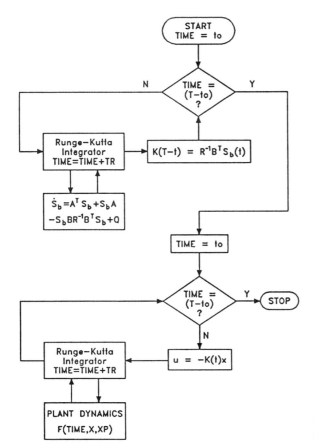

Figure 3.2-3 Continuous-time LQR simulation procedure

where

$$S(t) = S_b(T - t). \quad (3.2.52)$$

Thus, to solve the Riccati equation it is only necessary to integrate (3.2.51) forward in time with the given initial condition $S_b(0) = S(T)$ and then reverse the resulting solution and shift it to $t = T$.

It is important to notice that none of our derivation has depended on the fact that the system and PI weighting matrices are constant. That is, the equations of Table 3.2-2 hold as well for the case that $A(t)$, $B(t)$, $Q(t)$, and $R(t)$ are time-varying.

Example 3.2-3: LQR for DC Motor Using Scalar Model

In Example 3.2-1 we derived the optimal open-loop control for a DC motor assuming the scalar model

$$\dot{x} = -ax + bu, \quad (1)$$

with $x(t)$ the motor speed. Here, we shall find the optimal *feedback* control that minimizes the PI

$$J = \frac{1}{2} s_T x^2(T) + \frac{1}{2} \int_0^T (qx^2 + ru^2) \, dt \quad (2)$$

for specified final time T and weights s_T, q, and r.

According to Table 3.2-2 the Riccati equation is

$$-\dot{s} = -2as - \frac{b^2}{r} s^2 + q, \quad t \le T \quad (3)$$

with final condition of $s(T) = s_T$. The Kalman gain is

$$k = \frac{b}{r} s \quad (4)$$

and the optimal feedback control is

$$u = -kx. \quad (5)$$

This control scheme is shown in Fig. 3.2-4. It is a *two-level hierarchical controller* in which the Riccati solution $s(t)$ is manufactured by a sort of "doubled" or "squared" version of the plant. Then, the optimal feedback loop in the plant is built using a *multiplier*. This nonlinear hierarchical configuration is typical of advanced controllers that are designed by modern control techniques.

Were it not for the requirement to reverse the Riccati solution in time, the controller of Fig. 3.2-4 could easily be implemented using analog circuitry. In the next example we shall show how to simulate such control schemes using digital computers.

To find an explicit formula for the optimal control we may proceed as follows.

Using separation of variables to solve for $s(t)$ we have

$$\int_{s(t)}^{s_T} \frac{ds}{(b^2/r)s^2 + 2as - q} = \int_t^T dt, \quad (6)$$

3.2 Continuous-Time Linear Quadratic Regulator

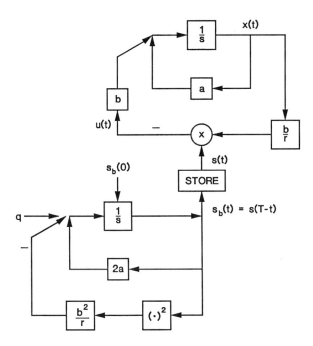

Figure 3.2-4 Scalar continuous LQ regulator

whence integrating yields

$$s(t) = s_2 + \frac{s_1 + s_2}{[(s_T + s_1)/(s_T - s_2)]e^{2\beta(T-t)} - 1} \tag{7}$$

where

$$\beta \equiv \sqrt{a^2 + \frac{b^2 q}{r}}, \tag{8}$$

$$s_1 \equiv \frac{r}{b^2}(\beta + a), \qquad s_2 \equiv \frac{r}{b^2}(\beta - a). \tag{9}$$

We see that even for the scalar case the optimal LQ control is a complicated time-varying feedback given by (4) and (7). For implementation purposes, however, it is only necessary to compute the Kalman gain $K(t)$ and store it in computer memory for application to the plant. This may easily be accomplished, as we show in the next example.

It is interesting to examine the steady-state behavior of the optimal control; that is, the behavior as the control interval of interest $[0, T]$ becomes large. As $(T - t)$ becomes large, the steady-state value s_∞ of $s(t)$ is given by s_2, so that

$$s_\infty = \frac{ar}{b^2}(\sqrt{1 + \gamma} - 1) \tag{10}$$

where the "control effectiveness ratio" is

$$\gamma \equiv \frac{b^2 q}{a^2 r}. \tag{11}$$

The steady-state closed-loop plant is given by

$$\dot{x} = (-a - bk_\infty)x = -\left(a + \frac{b^2}{r}s_\infty\right)x, \quad (12)$$

or

$$\dot{x} = -a\sqrt{1 + \gamma}\, x. \quad (13)$$

Now it may be seen that as the ratio q/r increases, the closed-loop system becomes more stable. Thus, either an increase in the state weighting q or a decrease in the control weighting r will tend to *speed up* the optimal closed-loop response. This is because as q/r increases, the PI weights $x(t)$ more heavily, tending to make it vanish more quickly for large values of t, as well as weighting $u(t)$ less heavily, thus allowing greater control effort in an attempt to keep $x(t)$ small.

■

Example 3.2-4: LQR for Armature-Controlled DC Motor

In Example 3.1-3 we showed how to obtain an open-loop control for an armature-controlled DC motor by numerically solving the Hamiltonian system two-point boundary-value problem. Here, we shall show how to obtain a *feedback* control law using the LQR in Table 3.2-2 and computer software to solve the Riccati equation.

a. Control Problem Formulation

The system equations from Examples 2.1-3 and 3.1-3 are

$$\dot{i} = -ai - k'\omega + bu \quad (1)$$
$$\dot{\omega} = -\alpha\omega + ki$$

with $i(t)$ the armature current, $\omega(t)$ the motor speed, control input $u(t)$ the armature voltage, $1/a$ the electrical time constant, $1/\alpha$ the mechanical time constant, and the remaining parameters as previously defined.

Defining the state as $x = [i \quad \omega]^T$ we may write

$$\dot{x} = \begin{bmatrix} -a & -k' \\ k & -\alpha \end{bmatrix} x + \begin{bmatrix} b \\ 0 \end{bmatrix} u \equiv Ax + Bu. \quad (2)$$

We shall be interested in determining $u(t)$ to minimize the PI

$$J = \frac{1}{2} x^T(T) \begin{bmatrix} s_i & 0 \\ 0 & s_\omega \end{bmatrix} x(T) + \frac{1}{2} \int_0^T \left[x^T \begin{bmatrix} q_i & 0 \\ 0 & q_\omega \end{bmatrix} x + ru^2 \right] dt \quad (3)$$

with s_i, s_ω the final state weights, q_i, q_ω the (intermediate) state weights, and r the control weight. These are design parameters that may be adjusted or *tuned* using computer simulations to yield suitable closed-loop behavior, as we shall see in this example.

The minimization of J corresponds to the control objective of driving the motor to a speed of zero from any initial speed while keeping the control energy small.

b. Optimal Control

Since the Riccati equation solution $S(t)$ is symmetric, we may assume that

$$S = \begin{bmatrix} s_1 & s_2 \\ s_2 & s_3 \end{bmatrix} \quad (4)$$

3.2 Continuous-Time Linear Quadratic Regulator

where the scalars $s_i(t)$ are to be determined. Substituting A, B from the state equation and $S(T)$, Q, R from the PI into the Riccati equation in Table 3.2-2 yields the three nonlinear scalar coupled differential equations

$$-\dot{s}_1 = -2as_1 + 2ks_2 - \beta s_1^2 + q_i \tag{5a}$$

$$-\dot{s}_2 = -(a + \alpha)s_2 - k's_1 + ks_3 - \beta s_1 s_2 \tag{5b}$$

$$-\dot{s}_3 = -2\alpha s_3 - 2k's_2 - \beta s_2^2 + q_\omega, \tag{5c}$$

where

$$\beta \equiv b^2/r. \tag{6}$$

The reader should derive these equations.

Writing the feedback gain as $K(t) = [k_i \; k_\omega]$, the table shows that $K = R^{-1}B^T S$, so that

$$k_i = \frac{b}{r} s_1, \quad k_\omega = \frac{b}{r} s_2. \tag{7}$$

Then, the optimal control is given by the time-varying feedback

$$u = -k_i i - k_\omega \omega. \tag{8}$$

c. Computer Simulation

Although the equations for $s_i(t)$ are difficult to solve analytically, this is fortunately not required for implementation of the control law. Indeed, we may write a driver program that uses a standard Runge-Kutta integrator to implement Fig. 3.2-3 [Lewis 1986]. Such a program is a minor modification of program TRESP in Appendix A. According to Fig. 3.2-3, the program is structured to use four subroutines. They appear in Fig. 3.2-5.

The first two subroutines are needed for the off-line phase of Fig. 3.2-3 (i.e., the top loop). Subroutine FRIC(TIME, S, SP) provides the Riccati equation dynamics (5) for the preliminary backward integration. To use a forward Runge-Kutta routine, the Riccati equation should be converted to the forward equation (3.2.51) (i.e. ignore the minus signs on the left-hand sides of (5)). Then, the replacement (3.2.52) is accomplished by the driver program. The argument S represents the elements s_i of the Riccati solution arranged into a vector, and SP represents the derivative of S. TIME is an argument needed for controlling time-varying systems; it is not used here. Given S, the subroutine must compute SP.

The second subroutine, FBGAIN(K, S, AK) computes the optimal feedback gains $K(t)$ using (7). Argument K keeps track of the Runge-Kutta iteration number (i.e., the time), and the subroutine must compute the feedback gain AK given the Riccati solution S. The resulting time-varying gains are stored in memory.

The last two subroutines are needed during the on-line simulation phase of Fig. 3.2-3 (the bottom loop). The control input U is computed by subroutine CONUP(K, X, AK) using the stored feedback gains AK, the state vector X, and (8). Control U is passed through a common block to subroutine F(TIME, X, XP), which contains the plant dynamics (1). This subroutine must compute the state derivative XP given the state vector X and U.

Using this software, the optimal state trajectories and control voltage were plotted for $r = 1$ and several values of $q = q_i = q_\omega$. The results are displayed in Fig. 3.2-6. Note that the states go to zero more quickly as q increases, while the controls become larger. Also shown are the elements $s_i(t)$ of $S(t)$ for two values of q. Note that, as q increases, their steady-state values are obtained more quickly as we move backward from the final time T.

```
C      SUBROUTINES FOR USE WITH CTLQR
C      ARMATURE-CONTROLLED DC MOTOR
C
C      SUBROUTINE TO COMPUTE RICCATI EQUATION SOLUTION
C
       SUBROUTINE FRIC(TIME,S,SP)
       PARAMETER (NX=20)
       REAL S(*), SP(*), KP, KS
       COMMON/PLANT/A,KP,KS,ALPHA,B
       COMMON/WEIGHTS/Q(NX,NX),R
C
       BETA= B**2/R
       SP(1) = -2*A*S(1) + 2*KS*S(2) - BETA*S(1)**2 + Q(1,1)
       SP(2) = -(ALPHA+A)*S(2) - KP*S(1) + KS*S(3) - BETA*S(1)*S(2)
       SP(3) = -2*ALPHA*S(3) - 2*KP*S(2) - BETA*S(2)**2 + Q(2,2)
C
       RETURN
       END
C
C
C      SUBROUTINE TO COMPUTE OPTIMAL FEEDBACK GAINS
C
       SUBROUTINE FBGAIN(K,S,AK)
       PARAMETER (NX=20)
       REAL S(*), AK(NX,1), KP, KS
       COMMON/PLANT/A,KP,KS,ALPHA,B
       COMMON/WEIGHTS/Q(NX,NX),R
C
       AK(1,K) = B*S(1)/R
       AK(2,K) = B*S(2)/R
C
       RETURN
       END
C
C
C      SUBROUTINE TO COMPUTE OPTIMAL CONTROL INPUT
C
       SUBROUTINE CONUP(K,X,AK)
       PARAMETER (NX=20)
       REAL X(*), AK(NX,1)
       COMMON/COMMAND/U(1)
C
       U(1) = -AK(1,K)*X(1) - AK(2,K)*X(2)
C
       RETURN
       END
C
C
C      MOTOR STATE EQUATIONS
C
       SUBROUTINE F(TIME,X,XP)
       PARAMETER (NX=20)
       REAL X(*), XP(*), KP, KS
       COMMON/COMMAND/U(1)
       COMMON/PLANT/A,KP,KS,ALPHA,B
       COMMON/WEIGHTS/Q(NX,NX), R
C      DATA ((Q(I,J), J= 1,2), I= 1,2)/1.,0.,0.,1./
       DATA R/1./
       DATA A,KP,KS,ALPHA,B/10.,2.,10.,2.,2./
C
       XP(1) = -A*X(1) -KP*X(2) +B*U(1)
       XP(2) = KS*X(1) -ALPHA*X(2)
C
       RETURN
       END
```

Figure 3.2-5 Subroutines for use with LQR driver program

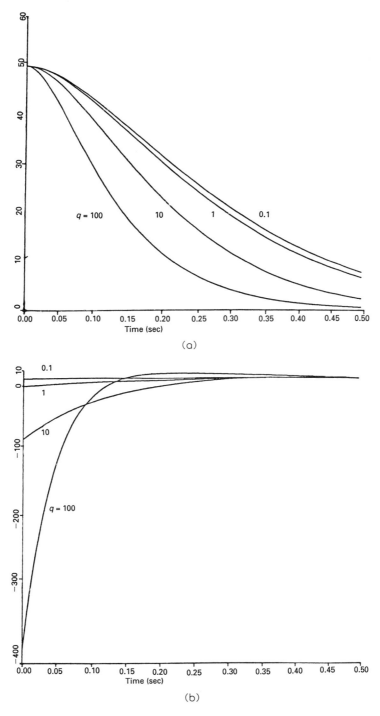

Figure 3.2-6 Results of DC motor simulation. (a) Motor speed. (b) Optimal control voltage. (c) Riccati solution for $q = 0.1$. (d) Riccati solution for $q = 100$.

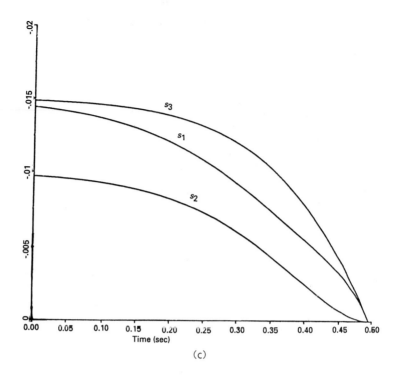

(c)

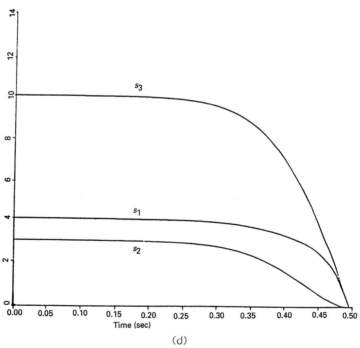

(d)

Figure 3.2-6 (*continued*)

We used final weights of $s_t = s_\omega = 0$; it is quite instructive to perform this experiment with a small value for q and various values of the final state weights. Basically, for a given control interval $[0, T]$, as s_t and s_ω increase, the final states are closer to zero.

Based on the simulation results we may select suitable values for the PI weights. Then, the associated $K(t)$ may be stored in memory and applied to the actual motor during the control implementation run.

■

3.3 STEADY-STATE AND SUBOPTIMAL CONTROL

We have seen that, even for time-invariant plants the optimal LQ control is a *time-varying* state-variable feedback. Such feedbacks are inconvenient to implement, since they require the storage in computer memory of time-varying gains. In this section we shall examine an alternative control scheme in which the time-varying Kalman gain $K(t)$ is replaced by its constant steady-state (e.g., $t \to \infty$) value. In many applications, this use of the steady-state feedback gain is adequate.

Steady-State Control

Suppose the plant to be controlled has the linear description

$$\dot{x} = Ax + Bu, \tag{3.3.1}$$

with $x \in \mathbf{R}^n$ and control input $u \in \mathbf{R}^m$. For this section it will be necessary to assume that the plant is time-invariant.

Now, we are interested in choosing the control that minimizes the quadratic PI

$$J(t_0) = \frac{1}{2} \int_0^\infty (x^T Q x + u^T R u)\, dt, \tag{3.3.2}$$

with $Q \geq 0$ and $R > 0$. Since the integration interval is infinite, we call this an *infinite horizon* performance index. Our performance objectives are now with reference to an infinite control interval $[0, \infty)$.

The control law of Table 3.2-2 still applies; however, it is now necessary to integrate the Riccati equation backwards over an infinite interval. Supposing that the Riccati equation has a limiting solution so that $\dot{S}(t)$ vanishes for large values of $(T - t)$, we may write

$$0 = A^T S + SA - SBR^{-1}B^T S + Q. \tag{3.3.3}$$

This is called the *algebraic Riccati equation* (ARE); it is a symmetric matrix quadratic equation. The limiting solution S_∞ to the Riccati differential equation, if it exists, is a solution to the ARE.

The converse may not be true. That is, positive definite solutions to the ARE may not be limiting solutions to the Riccati equation for any value of $S(T)$. Moreover, the ARE may have solutions that are not symmetric and, just as the scalar quadratic equation, it may have real or complex solutions.

If the limiting solution S_∞ to the Riccati equation does exist, then the optimal infinite-horizon Kalman gain is the *constant* matrix given by

$$K_\infty = R^{-1}B^T S_\infty. \qquad (3.3.4)$$

Thus, the optimal steady-state control is the constant state-variable feedback

$$u(t) = -K_\infty x(t). \qquad (3.3.5)$$

Moreover, the optimal cost is given in terms of the initial state by

$$J = \frac{1}{2} x^T(0) S_\infty x(0). \qquad (3.3.6)$$

Under the influence of the steady-state control the closed-loop plant has the time-invariant dynamics

$$\dot{x} = (A - BK_\infty)x \equiv A_c x. \qquad (3.3.7)$$

The advantages of this simplified control that uses a constant feedback are clear. Therefore, we should like to answer several questions to determine the usefulness of the proposed scheme. Specifically:

1. When does there exist a bounded limiting solution S_∞ to the Riccati equation for all choices of the final state weighting $S(T)$?
2. In general, S_∞ will depend on $S(T)$. However, our new formulation no longer involves $S(T)$. Thus, when does there exist a unique S_∞ that does not depend on the choice of $S(T)$?
3. When is the closed-loop plant A_c asymptotically stable?

The answers to these questions are so fundamentally important to the LQ theory that we present them as two theorems (for proofs see Lewis [1986]). They rely on some of the system properties discussed in Chapter 2.

Theorem 1. Let (A, B) be stabilizable. Then for every $S(T)$ there exists a bounded limiting solution S_∞ to the Riccati equation. Furthermore, S_∞ is a symmetric positive semi-definite solution to the ARE.

Theorem 2. Let C be any square root of Q so that $Q = C^T C$. Suppose (C, A) is observable. Then (A, B) is stabilizable if and only if:

1. There is a unique symmetric positive-definite limiting solution S_∞ to the Riccati equation. Furthermore, S_∞ is the unique positive-definite solution of the ARE.
2. The closed-loop plant A_c is asymptotically stable.

The power of these results cannot be overemphasized. They mean that, as long as the system and PI satisfy certain basic requirements, the steady-state LQ regulator will

3.3 Steady-State and Suboptimal Control

yield gains that stabilize the system. Considering the difficulty encountered by classical control techniques in stabilizing multi-input systems, this is a remarkable property.

As may be seen by comparing the two theorems, the observability of ($\sqrt{Q}$, A) considerably strengthens the results. This property means roughly that all the plant modes should be weighted in the PI. If J is bounded, then the integrand in (3.3.2) vanishes with time. If, in addition, all states are observable in the PI, this will in turn ensure that $x(t)$ vanishes with time, thereby guaranteeing closed-loop stability.

The closed-loop poles will depend on the selection of the design matrices Q and R; however, the poles will always be stable as long as we select $R > 0$ and $Q \geq 0$ with ($\sqrt{Q}$, A) observable. Thus, the elements of Q and R may be varied during an interactive computer-aided design procedure to obtain suitable closed-loop performance. That is, the optimal gain K is found for given values of Q and R, and the closed-loop time responses are found by simulation. If these responses are unsuitable, new values for Q and R are selected and the design is repeated. Given good software to solve for K, this procedure is quite convenient. Such software is available, for instance, in ORACLS [Armstrong 1980], MATLAB [Moler et al. 1987], and MATRIX$_x$ [1989].

The next example will illustrate the dependence of the closed-loop poles on Q and R.

It is not actually required for ($\sqrt{Q}$, A) to be observable to guarantee a stable closed-loop system. All that is required is the weaker condition of *detectability*, which corresponds to the observability of the unstable modes of A. However, if only detectability holds, then S_∞ will generally be only positive semidefinite.

Example 3.3-1: Steady-State LQ Design for Systems Obeying Newton's Laws

In Example 3.2-2 we discussed open-loop control for systems obeying Newton's laws

$$\dot{x} = \begin{bmatrix} 0 & 1 \\ 0 & 0 \end{bmatrix} x + \begin{bmatrix} 0 \\ 1 \end{bmatrix} u = Ax + Bu, \quad (1)$$

where the state is $x = [p \ v]^T$, with $p(t)$ the position and $v(t)$ the velocity, and control $u(t)$ is an acceleration input. In this example, we want to illustrate steady-state LQ design for the regulator problem for this system, thus obtaining a closed-loop feedback control.

To regulate the state to zero, we may select the PI

$$J = \frac{1}{2} \int_0^\infty (x^T Q x + u^2) \, dt \quad (2)$$

with $Q = \text{diag}\{q_p^2, q_v\}$. Note that it is not useful to include a separate control weighting r, since only the ratios q_p^2/r and q_v/r are important in J.

In this example, we are interested in determining some analytical results to see the effect of the design parameters q_p and q_v.

a. Steady-State Kalman Gain

Note that (A, B) is reachable and ($\sqrt{Q}$, A) is observable (since Q and hence also any root of Q are nonsingular). Therefore, Theorem 2 guarantees that the ARE has a unique positive definite solution, and that the steady-state Kalman gain will stabilize the system. Let us find the stabilizing gain K.

Since the solution to the ARE (3.3.3) is symmetric, we may as well assume that

$$S = \begin{bmatrix} s_1 & s_2 \\ s_2 & s_3 \end{bmatrix} \tag{3}$$

for some constant scalars s_i. Substituting A, B, Q, and $R = I$ into the ARE and simplifying (check!) yields the three nonlinear scalar coupled equations

$$0 = -s_2^2 + q_p^2 \tag{4}$$

$$0 = s_1 - s_2 s_3 \tag{5}$$

$$0 = 2s_2 - s_3^2 + q_v. \tag{6}$$

These are easily solved in the order (4), (6), (5) to obtain

$$s_2 = q_p \tag{7}$$

$$s_3 = \sqrt{q_v + 2q_p} \tag{8}$$

$$s_1 = q_p \sqrt{q_v + 2q_p}. \tag{9}$$

We have selected the signs that result in a positive definite S, which should be verified by the reader.

According to (3.3.4), the steady-state Kalman gain is given by

$$K = R^{-1}B^T S = [s_2 \quad s_3] \tag{10}$$
$$= [q_p \quad \sqrt{q_v + 2q_p}].$$

b. Analysis of Closed-Loop Plant

The closed-loop plant is given by

$$A_c = A - BK = \begin{bmatrix} 0 & 1 \\ -q_p & -\sqrt{q_v + 2q_p} \end{bmatrix} \tag{11}$$

so that the closed-loop characteristic polynomial is

$$\Delta_c = s^2 + s\sqrt{q_v + 2q_p} + q_p. \tag{12}$$

This yields a closed-loop natural frequency and damping ratio of

$$\omega_n = \sqrt{q_p} \tag{13}$$

$$\zeta = \frac{1}{\sqrt{2}} \sqrt{1 + q_v/2q_p}. \tag{14}$$

The effect of the design parameters on ω_n and ζ is now clear, so that the weights q_p and q_v may be selected for suitable performance. Moreover, the strength of modern LQ design is clear, for the closed-loop plant is stable for any admissible choice of q_p and q_v. A naive approach to design would involve selecting the elements of the gain matrix K directly. However, stability is not guaranteed for all values of K. On the other hand, no matter what weights we select in the PI, as long as $(\sqrt{Q}, A)$ is observable and $Q \geq 0$, the closed-loop system is stable by Theorem 2.

The closed-loop stability is guaranteed even in complex plants with multiple inputs and outputs, where the single-input/single-output classical techniques cannot easily be applied. This is a very powerful result.

Note that, if the velocity weight q_v is zero, then the damping ratio becomes the familiar

3.3 Steady-State and Suboptimal Control

$1/\sqrt{2}$. On the other hand, $q_p = 0$ is not allowed since then $(\sqrt{Q}, A)$ is not observable. (Then, $\sqrt{Q} = [0 \quad \sqrt{q_v}]$ is a root of Q.)

∎

Example 3.3-2: Receding Horizon Control—An Easy Way to Stabilize a System

Here, we shall discuss an easy way to stabilize a linear system using time-invariant feedback gains that is based on the *receding horizon* concept [Kleinman 1970, Kwon and Pearson 1977].

It is desired to drive the system

$$\dot{x} = Ax + Bu \tag{1}$$

from a given state $x(t)$ at the current time t to a final state of zero at time $t + T$, where T is fixed. That is, we require

$$x(t + T) = 0. \tag{2}$$

To attain this control objective we may select $u(t)$ to minimize the performance index

$$J = \frac{1}{2} x^T(t + T) S_T x(t + T) + \frac{1}{2} \int_t^{t+T} (x^T Q x + u^T R u) \, dt, \tag{3}$$

with $S_T = \infty$, $Q \geq 0$, $R > 0$.

The required control $u(t)$ is given by the feedback gain in Table 3.2-2. To find it, we must integrate

$$-\dot{S} = A^T S + SA - SBR^{-1}B^T S + Q \tag{4}$$

backward from $t + T$ to t using $S(t + T) = \infty$. However, note that this integration interval has the *constant* length of T for all t; therefore, the $n \times n$ matrix $S(t)$ required to find the Kalman gain is the same for all t; this results in a constant feedback gain.

To avoid using infinity as the terminal condition for (4), we may define $P = S^{-1}$ and use $d(P^{-1})/dt = -P^{-1}\dot{P}P^{-1}$ to obtain

$$\dot{P} = AP + PA^T - BR^{-1}B^T + PQP, \quad t \leq T, \quad P(T) = 0. \tag{5}$$

We have removed the constant shift of t in the integration interval.

Now, the optimal control is given by the constant feedback

$$u = -Kx \tag{6}$$

where

$$K = R^{-1}B^T P^{-1}(0). \tag{7}$$

Thus, by using the current time as the lower limit of integration in the PI and a *receding horizon* as the upper limit, we have obtained a *constant* state feedback law.

It may be shown that, as long as (A, B) is controllable the control law (6), (7) stabilizes the system. This controller should be compared to Table 3.2-1.

∎

Suboptimal Control

We have seen when the infinite horizon LQ problem has a steady-state solution, namely, when (A, B) is stabilizable and $(\sqrt{Q}, A)$ is observable. Then, the ARE (3.3.3) has a unique positive definite solution that yields the steady-state Kalman gain K_∞ through (3.3.4). This steady-state gain always stabilizes the plant.

We now suggest the following. Even if the control interval [0, T] is not infinite, we may decide to use the steady-state Kalman gain K_∞ instead of the optimal time-varying gain $K(t)$ given in Table 3.2-2. On a finite interval [0, T] the constant gain K_∞ is suboptimal, but the convenience gained by not having to implement a time-varying gain may more than make up for the loss of optimality. Moreover, as T becomes large, the optimal gain $K(t)$ tends to K_∞ so that the decision to use the steady-state gain makes more and more sense.

The use of the steady-state gain K_∞ on a finite control interval amounts to a *suboptimal control strategy*. In addition to the ease of implementation of constant feedback gains, this suboptimal controller has another important advantage: there are efficient numerical routines available for the solution of the ARE (e.g., ORACLS [Armstrong 1980], MATLAB [Moler et al. 1987], and MATRIX$_x$ [1989]).

We should like to know the cost associated with using a gain that is not the optimal gain yielding the cost in Table 3.2-2. This will give us some feel for what is being lost by deciding to use K_∞.

Thus, suppose we use the feedback law

$$u = -Fx \qquad (3.3.8)$$

where F is *any* fixed feedback gain that yields a stable closed-loop system

$$\dot{x} = (A - BF)x. \qquad (3.3.9)$$

Using (3.3.8) in the performance index in Table 3.2-2 yields

$$J = \frac{1}{2} x^T(T) S(T) x(T) + \frac{1}{2} \int_{t_0}^{T} x^T(Q + F^T R F) x \, dt. \qquad (3.3.10)$$

Now, if there exists a matrix S such that

$$\frac{d}{dt} x^T S x = -x^T(Q + F^T R F) x \qquad (3.3.11)$$

then the integrand is a perfect differential so that

$$J = \frac{1}{2} x^T(t_0) S(t_0) x(t_0). \qquad (3.3.12)$$

Differentiating the left-hand side of (3.3.11), using (3.3.9), and canceling out the state trajectory [since the equation holds for all $x(t_0)$] yields

$$-\dot{S} = (A - BF)^T S + S(A - BF) + Q + F^T R F. \qquad (3.3.13)$$

Thus, for any feedback gain, whether or not it is optimal, we may determine the associated value of J as follows. First, solve (3.3.13) backwards in time using the given final weighting $S(T)$. Then, the cost associated with using F is given by (3.3.12). This value may be compared with the *optimal* cost attainable by using the Kalman gain, which is obtained by solving the Riccati equation in Table 3.2-2. On this basis, we may decide whether to use the suboptimal F or the optimal time-varying Kalman gain.

3.3 Steady-State and Suboptimal Control

It is important to note clearly that (3.3.13) is a linear *Lyapunov* equation for S since F is a fixed gain. In particular, it is *not* the Joseph stabilized formulation (3.2.48) of the Riccati equation unless F is the optimal gain given by Table 3.2-2.

If we decide to use steady-state control on a finite control interval, we may select $F = K_\infty$ and evaluate (3.3.12) to see what we are losing in return for the simplicity of using a constant gain.

Example 3.3-3: Suboptimal Control of DC Motor Using a Scalar Model

Let the plant be the scalar model for a DC motor that was considered in Examples 3.2-1 and 3.2-3:

$$\dot{x} = -ax + bu \tag{1}$$

with cost

$$J(t_0) = \frac{1}{2} s(T) x^2(T) + \frac{1}{2} \int_{t_0}^{T} (qx^2 + ru^2) \, dt. \tag{2}$$

The optimal control is a time-varying feedback

$$u^* = -K^*(t)x, \tag{3}$$

where

$$K^*(t) = \frac{b}{r} s^*(t) \tag{4}$$

and $s^*(t)$ satisfies the Riccati equation

$$-\dot{s} = -2as - \frac{b^2}{r} s^2 + q. \tag{5}$$

This equation can be solved analytically to get the optimal solution $s^*(t)$ given by (7) in Example 3.2-3. The optimal value of the cost on any interval $[t, T]$ is

$$J^*(t) = \frac{1}{2} s^*(t) x^2(t). \tag{6}$$

Suppose that we do not want to go to the trouble of storing the time-varying optimal gain sequence $K^*(t)$. Instead, let us try to use its steady-state value (see Example 3.2-3) of

$$K_\infty = \frac{b}{r} s_\infty = \frac{a}{b} (\sqrt{1 + \gamma} - 1) \tag{7a}$$

with

$$\gamma = \frac{b^2 q}{a^2 r} \tag{7b}$$

in the constant feedback law

$$u = -K_\infty x. \tag{8}$$

To see the ramifications of using this suboptimal but simple feedback, let us solve (3.3.13) for the suboptimal cost kernel. This Lyapunov equation is just

$$-\dot{s} = 2a_c s + K_\infty^2 r + q, \tag{9}$$

where the closed-loop matrix is

$$a_c = -a - bK_x = -a\sqrt{1+\gamma}. \tag{10}$$

Note that a_c is always stable regardless of the value of a (note that $a > 0$).

The first-order linear equation (9) is easily solved to give

$$s(t) = e^{2a_c(T-t)}s(T) + \frac{K_x^2 r + q}{2a_c}(e^{2a_c(T-t)} - 1). \tag{11}$$

The solutions $s^*(t)$ and (11) are sketched in Fig. 3.3-1 for $a = 4$, $b = 2$, $r = 1$, $q = 10$, $s(T) = 10$. Note that the suboptimal $s(t)$ given by (11) is an upper bound for $s(t)$, so the suboptimal cost

$$J(t) = \frac{1}{2} s(t) x^2(t) \tag{12}$$

is always greater than or equal to the optimal $J(t)$ in (6).

Note that the limiting solutions $s(\infty)$ and $s^*(\infty)$ are both given by

$$s(\infty) = \frac{ar}{b^2}(\sqrt{1+\gamma} - 1), \tag{13}$$

so that the performance index

$$J(0) = \frac{1}{2} \int_0^\infty (x^T q x + u^T r u)\, dt \tag{14}$$

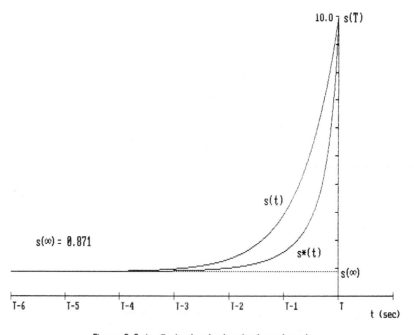

Figure 3.3-1 Optimal and suboptimal cost kernels.

3.3 Steady-State and Suboptimal Control

over the infinite interval $[0, \infty]$ takes on the same value whether we use the optimal feedback (4) or the suboptimal (7). This means that as the control interval $[t_0, T]$ becomes longer, it makes more sense to use the constant feedback K_∞. This is also apparent from Fig. 3.3-1.

■

Example 3.3-4: Suboptimal Control for Systems Obeying Newton's Laws

Let us demonstrate suboptimal control for the Newton's system of Example 3.3-1 given by

$$\dot{x} = \begin{bmatrix} 0 & 1 \\ 0 & 0 \end{bmatrix} x + \begin{bmatrix} 0 \\ 1 \end{bmatrix} u = Ax + Bu, \quad (1)$$

where the state is $x = [p \quad v]^T$, with $p(t)$ the position and $v(t)$ the velocity, and control $u(t)$ is an acceleration input. To regulate the state to zero, we may select the PI

$$J = \frac{1}{2} \int_0^\infty (x^T Q x + u^2) \, dt \quad (2)$$

with $Q = \text{diag}\{q_p^2, q_v\}$.

Although we shall only be interested in a control period of 10 sec, let us use the steady-state Kalman gain

$$K = R^{-1} B^T = [q_p \quad \sqrt{q_v + 2q_p}]. \quad (3)$$

found in Example 3.3-1.

A subroutine for use with the time response program TRESP in Appendix A is shown in Fig. 3.3-2. It simulates a feedback controller—in this case using the steady-state Kalman gain. Note how direct it is to write the subroutine given the system equations.

```
C     SUBROUTINE TO SIMULATE STEADY-STATE LQR FOR NEWTON'S SYSTEM
C     FOR USE WITH TIME RESPONSE PROGRAM TRESP
C
      SUBROUTINE F(T,X,XP)
      REAL X(*), XP(*)
      DATA QP,QV/.1,.1/
C
C     STEADY-STATE KALMAN GAIN
C
      AK1= QP
      AK2= SQRT(QV + 2*QP)
C
C     FEEDBACK CONTROL
C
      U= -AK1*X(1) - AK2*X(2)
C
C     STATE DERIVATIVES
C
      XP(1)= X(2)
      XP(2)=         U
C
      RETURN
      END
```

Figure 3.3-2 Subroutine for use with TRESP to simulate feedback control of Newton's system.

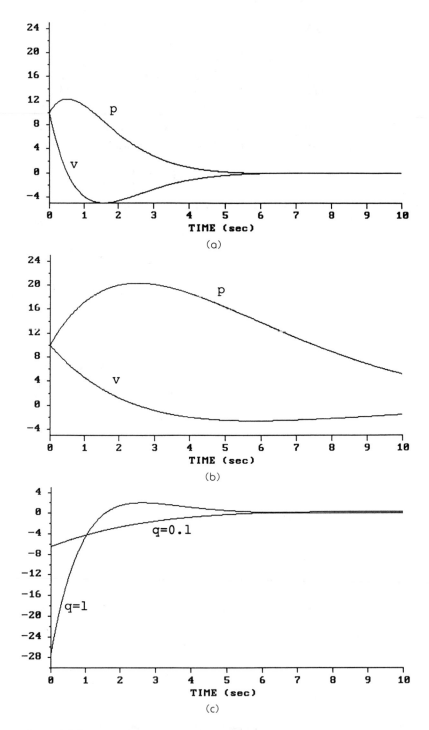

Figure 3.3-3 Newton's system simulation. (a) Position and velocity for $q_p = q_v = 1$. (b) Position and velocity for $q_p = q_v = 0.1$. (c) Control inputs.

3.3 Steady-State and Suboptimal Control

Some simulations are shown in Fig. 3.3-3 for various values of the state weights q_p and q_v. The initial condition was $p(0) = 10$ ft, $v(0) = 10$ ft/sec.

Note that as the weights increase, the state approaches zero more quickly and more control effort is required. Consider this while thinking about the form of the PI (2). We see also that as the state weights increase, the steady-state LQR performs better on any given interval.

To appreciate the difference between implementing the steady-state LQR and the optimal time-varying LQR of Table 3.2-2, note that in this example it was unnecessary to integrate the Riccati equation as required by Fig. 3.2-3.

■

Eigenstructure LQR Design

We now discuss an alternative approach to steady-state controls design that does not involve solving the ARE (3.3.3). Suppose that (A, B) is stabilizable and $(\sqrt{Q}, A)$ is observable so that the conditions of Theorem 2 hold. We want to demonstrate that, instead of solving the ARE (3.3.3) for S_∞ and then using (3.3.4) to find the steady-state gain K_∞, we may determine S_∞, K_∞, and the closed-loop poles directly from the Hamiltonian matrix H in the Hamiltonian system

$$\begin{bmatrix} \dot{x} \\ \dot{\lambda} \end{bmatrix} = \begin{bmatrix} A & -BR^{-1}B^T \\ -Q & -A^T \end{bmatrix} \begin{bmatrix} x \\ \lambda \end{bmatrix} \equiv H \begin{bmatrix} x \\ \lambda \end{bmatrix}. \quad (3.3.14)$$

In fact, we shall show that the poles of the optimal closed-loop system

$$\dot{x} = (A - BK_\infty)x \equiv A_c x, \quad (3.3.15)$$

with K_∞ the steady-state Kalman gain, are exactly the stable eigenvalues of H.

Derivation

First, we shall show H has a special property. Indeed, if μ is an eigenvalue of H, then so is $-\mu$. To wit, let

$$J = \begin{bmatrix} 0 & I \\ -I & 0 \end{bmatrix}. \quad (3.3.16)$$

Then it is easy to show that

$$H = JH^T J.$$

If μ is an eigenvalue of H with eigenvector v, then

$$Hv = \mu v,$$

so that (note $J^{-1} = -J$)

$$JH^T Jv = \mu v$$

$$H^T Jv = -\mu Jv.$$

Hence

$$(Jv)^T H = -\mu(Jv)^T$$

and $-\mu$ is an eigenvalue of H with left eigenvector Jv. This means that H has n stable eigenvalues along with n unstable eigenvalues which are just their reflections in the origin of the complex plane.

The formulations (3.3.14) and (3.3.15) are equivalent ways to characterize the closed-loop dynamics under the influence of the optimal control on the infinite time interval $[0, \infty)$. Theorem 2 guarantees the stability of (3.3.15). Therefore, we may show that the eigenvalues of the optimal closed-loop system are none other than the n stable eigenvalues of H. To accomplish this, suppose that μ_i is an eigenvalue of the optimal closed-loop system [Kailath 1980]. Then if only the mode corresponding to μ_i is excited, since $\lambda(t)$ and $u(t)$ are both linear functions of $x(t)$ [see (3.2.34) and (3.3.5)], we may write

$$x(t) = X_i e^{\mu_i t} \tag{3.3.17}$$

$$u(t) = U_i e^{\mu_i t} \tag{3.3.18}$$

$$\lambda(t) = \Lambda_i e^{\mu_i t} \tag{3.3.19}$$

for some vectors X_i, U_i, Λ_i. Using this in $\dot{x} = Ax + Bu$ yields

$$(\mu_i I - A)X_i = BU_i.$$

According to (3.3.5)

$$U_i = -K_\infty X_i, \tag{3.3.20}$$

so that

$$(\mu_i I - (A - BK_\infty))X_i = 0 \tag{3.3.21}$$

and μ_i is an eigenvalue of the closed-loop system with eigenvector X_i.

Now, let us examine the Hamiltonian system (3.3.14). According to (3.3.17) and (3.3.19) there results

$$\mu_i \begin{bmatrix} X_i \\ \Lambda_i \end{bmatrix} = H \begin{bmatrix} X_i \\ \Lambda_i \end{bmatrix}.$$

That is, μ_i is also an eigenvalue of H with eigenvector $[X_i^T \ \Lambda_i^T]^T$.

Theorem 2 says that the closed-loop system is stable, so that the optimal closed-loop eigenvalues are the n stable eigenvalues of H.

The optimal feedback gain may be determined from the eigenstructure of H. To see this, suppose that the optimal closed-loop eigenvalues are simple (i.e., have no generalized eigenvectors—see section 2.4). Then (3.2.30) shows that, for each i

$$U_i = -R^{-1}B^T \Lambda_i.$$

Now, (3.3.20) shows that

3.3 Steady-State and Suboptimal Control

$$K_\infty X_i = R^{-1}B^T\Lambda_i.$$

Letting X be an $n \times n$ matrix whose columns are X_i and Λ be an $n \times n$ matrix whose columns are Λ_i, we may write

$$K_\infty = R^{-1}B^T\Lambda X^{-1}. \qquad (3.3.22)$$

This gives the steady-state Kalman gain in terms of the eigenstructure of H.

If μ_i is complex, then the gain obtained by this technique is also complex and thus not implementable. In this event μ_i^* is also an eigenvalue and according to (3.3.22) the gain must satisfy

$$K_\infty[X_i \ X_i^*] = R^{-1}B^T[\Lambda_i \ \Lambda_i^*].$$

Postmultiplying this equation by

$$L = \frac{1}{2}\begin{bmatrix} 1 & -j \\ 1 & j \end{bmatrix}$$

results in

$$K_\infty[Re(X_i) \ Im(X_i)] = R^{-1}B^T[Re(\Lambda_i) \ Im(\Lambda_i)], \qquad (3.3.23)$$

where $Re(.)$ and $Im(.)$ denote the real and imaginary parts of a complex variable. Thus, if μ_i is complex, it is only necessary to use in (3.3.22) the real and imaginary parts of the associated vectors X_i and Λ_i instead of the complex vectors themselves. This will yield a real K_∞.

Discussion

We have shown that the poles of the optimal closed-loop system (3.3.15) are the n stable poles of the Hamiltonian system (3.3.14). Moreover, the eigenvector of the closed-loop plant corresponding to μ_i is given by X_i, the "top half" of the eigenvector of H.

In addition, instead of solving the ARE for S_∞ and then using (3.3.4), we may determine the steady-state optimal feedback gain by finding the stable eigenvalues μ_i of the Hamiltonian matrix H, along with their associated eigenvectors $[X_i^T \ \Lambda_i^T]^T$, and then compute K_∞ using (3.3.22). This equation may profitably be compared with Moore's eigenstructure assignment technique in section 2.5.

Comparing (3.3.22) and (3.3.4) it is evident that the ARE solution is given in terms of the eigenstructure of H by

$$S_\infty = \Lambda X^{-1}. \qquad (3.3.24)$$

If a closed-loop pole μ_i is complex, then in (3.3.22) and (3.3.24) we should use the real and imaginary parts of the associated vectors Λ_i and X_i to obtain real K_∞ and S_∞. If the optimal closed-loop eigenvalues are not simple, then it is necessary to use generalized eigenvectors in computing K_∞ and S_∞.

It is worth remarking that if the state-weighting matrix Q is zero, then

$$H = \begin{bmatrix} A & -BR^{-1}B^T \\ 0 & -A^T \end{bmatrix}. \tag{3.3.25}$$

In this case, the eigenvalues of H are just the poles of A plus the poles of $-A$. Thus, to find the optimal steady-state eigenvalues one need only take the stable poles of A along with the negatives of the unstable poles of A. These are also the optimal closed-loop poles in the limit as the control-weighting matrix R goes to infinity.

It is also possible to express the solution to the Riccati differential equation in Table 3.2-2 in terms of the eigenstructure of H. For details, see Lewis [1986].

Example 3.3-5: Eigenstructure LQR Design for Systems Obeying Newton's Laws

In Example 3.3-1 we solved the ARE to find the steady-state Kalman gain for systems obeying Newton's laws

$$\dot{x} = \begin{bmatrix} 0 & 1 \\ 0 & 0 \end{bmatrix} x + \begin{bmatrix} 0 \\ 1 \end{bmatrix} u, \tag{1}$$

where the state is $x = [p \quad v]^T$, with $p(t)$ the position and $v(t)$ the velocity, and control $u(t)$ an acceleration input. In this example, we want to illustrate eigenstructure LQ design for this system.

Let the PI be

$$J = \frac{1}{2} \int_0^\infty (x^T Q x + u^2) \, dt \tag{2}$$

with $Q = \text{diag}\{q_p^2, q_v\}$.

The Hamiltonian matrix is

$$H = \begin{bmatrix} 0 & 1 & 0 & 0 \\ 0 & 0 & 0 & -1 \\ -q_p^2 & 0 & 0 & 0 \\ 0 & -q_v & -1 & 0 \end{bmatrix}, \tag{3}$$

whence we may easily compute the characteristic equation

$$|sI - H| = s^4 - q_v s^2 + q_p^2 = 0. \tag{4}$$

Since this is an even polynomial, if s is a root then so is $-s$, as we have already shown. The standard root-locus theory may be used to plot root loci versus q_v (holding q_p fixed) and versus q_p (holding q_v fixed). See Fig. 3.3-4. According to the theory we have developed, the optimal closed-loop poles for a given value of (q_p, q_v) are the stable poles in the figure.

To gain some more insight, let $\bar{s} = s^2$ so that

$$\bar{s}^2 - q_v \bar{s} + q_p^2 = 0. \tag{5}$$

Assuming that $q_p > q_v/2$, this has a pair of unstable complex roots $\bar{s}_1, \bar{s}_2$ with natural frequency and damping ratio of

$$\bar{\omega} = q_p, \qquad \bar{\zeta} = -q_v/2q_p. \tag{6}$$

3.3 Steady-State and Suboptimal Control

Thus

$$\bar{s}_1 = \bar{\omega}e^{j\bar{\theta}_1}, \qquad \bar{s}_2 = \bar{\omega}e^{j\bar{\theta}_2} \tag{7}$$

with

$$-\bar{\theta}_2 = \bar{\theta}_1 = -\cos^{-1}\bar{\zeta}. \tag{8}$$

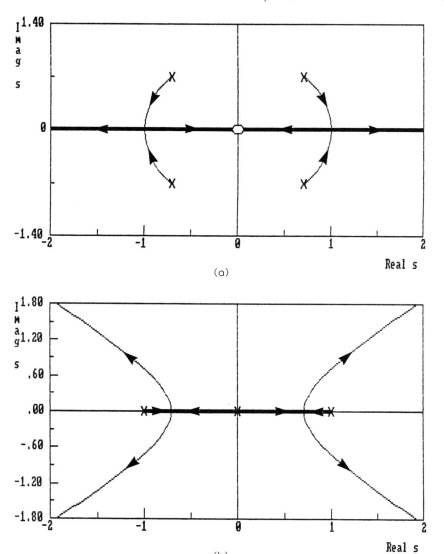

Figure 3.3-4 Root loci of Hamiltonian system for Newton's Laws. (a) vs. q_v holding $q_p = 1$. (b) vs. q_p holding $q_v = 1$.

The poles of the Hamiltonian system are at $\pm\sqrt{\bar{s}_1}$, $\pm\sqrt{\bar{s}_2}$, or

$$\sqrt{\bar{\omega}}e^{\pm j\bar{\theta}_1/2}, \qquad \sqrt{\bar{\omega}}e^{\pm j\bar{\theta}_2/2}. \tag{9}$$

Since

$$\cos(\alpha/2) = \frac{\sqrt{1+\cos\alpha}}{\sqrt{2}} \tag{10}$$

we therefore see that the Hamiltonian system has four poles, corresponding to two complex pairs, with natural frequency and damping ratios of

$$\omega_n = \sqrt{q_p}, \qquad \zeta = \pm\frac{1}{\sqrt{2}}\sqrt{1+q_v/2q_p}. \tag{11}$$

We have seen that the optimal closed-loop poles are nothing but the stable poles of H, so that they correspond to the stable complex pair described by (11) (i.e., positive damping ratio). Compare with Example 3.3-1. It is now clear how selection of the weights in the PI affects the closed-loop behavior. Note that, if no velocity weighting is used, the damping ratio becomes the familiar $1/\sqrt{2}$.

To determine the optimal feedback gains we could find the eigenvectors of H without too much trouble and use (3.3.22). However, since the plant is single-input, there is a simple alternative; we know the closed-loop poles so that Ackermann's formula (section 2.5) may be used. Thus

$$K = e_n^T U_n^{-1}\Delta^D(A), \tag{12}$$

where e_n^T is the last row of the $n\times n$ identity matrix, $\Delta^D(s)$ is the desired closed-loop polynomial, and the reachability matrix U_n is given by

$$U_2 = [B \quad AB] = \begin{bmatrix} 0 & 1 \\ 1 & 0 \end{bmatrix}. \tag{13}$$

The desired closed-loop polynomial is

$$\Delta^D(s) = s^2 + 2\zeta\omega_n s + \omega_n^2, \tag{14}$$

with ζ and ω_n given by (11). Substituting A^0, A^1, and A^2 in place of s^0, s^1, s^2 yields

$$\Delta^D(A) = A^2 + 2\zeta\omega_n A + \omega_n^2 I = \begin{bmatrix} \omega_n^2 & 2\zeta\omega_n \\ 0 & \omega_n^2 \end{bmatrix}, \tag{15}$$

Therefore, the optimal gain is given by

$$K = [\omega_n^2 \quad 2\zeta\omega_n] = [q_p \quad \sqrt{q_v + 2q_p}], \tag{16}$$

exactly as determined in Example 3.3-1. ∎

3.4 MINIMUM-TIME AND CONSTRAINED-INPUT DESIGN

An important class of control problems is concerned with achieving the performance objectives in *minimum time*. A suitable performance index for these problems is

3.4 Minimum-Time and Constrained-Input Design

$$J = \int_{t_0}^{T} 1 \, dt = T - t_0. \qquad (3.4.1)$$

Let us discuss several sorts of minimum-time problems.

Nonlinear Minimum-Time Problems

Suppose our objective is to drive the system

$$\dot{x} = f(x, u) \qquad (3.4.2)$$

from a given initial state $x(t_0) \in \mathbf{R}^n$ to a specified final state $x(T)$ in minimum time. Then, from Table 3.1-1 the Hamiltonian is

$$H = 1 + \lambda^T f \qquad (3.4.3)$$

and the Euler equations are

$$-\dot{\lambda} = \frac{\partial f^T}{\partial x} \lambda \qquad (3.4.4)$$

$$0 = \frac{\partial f^T}{\partial u} \lambda. \qquad (3.4.5)$$

Since the final state is fixed (so that $dx(T) = 0$) but the final time is free, the final condition in Table 3.3-1 says that

$$0 = H(T) = 1 + \lambda^T(T) f(x(T), u(T)). \qquad (3.4.6)$$

If $f(x, u)$ is not an explicit function of time, then according to the conservation principle (3.1.14), this means that $H(t)$ is zero for all time.

The stationarity condition (3.4.5) may usually be used to solve for $u(t)$ in terms of $\lambda(t)$. Then, $u(t)$ may be eliminated in the state and costate equations to obtain the Hamiltonian system. To solve this, we require the n initial conditions ($x(t_0)$ given) and n final conditions ($x(T)$ specified). However, the final time T is now unknown. The function of (3.4.6) is to provide one more equation so that T can be solved for.

We have already solved a minimum-time problem for a nonlinear system in Example 3.1-2, which should be reviewed at this point.

Linear Quadratic Minimum-Time Design

We shall now concern ourselves with finding an optimal control for the linear system

$$\dot{x} = Ax + Bu \qquad (3.4.7)$$

that minimizes the performance index

$$J = \frac{1}{2} x^T(T) S_T x(T) + \frac{1}{2} \int_{t_0}^{T} (1 + x^T Q x + u^T R u) \, dt \qquad (3.4.8)$$

with $S_T \geq 0$, $Q \geq 0$, $R > 0$, and the final time T free. There is no constraint on the final state; thus, the control objective is to make the final state sufficiently small. Due to the term $\frac{1}{2}(T - t_0)$ arising from the integral, there is a concern to accomplish this in a short time period.

This is a general sort of PI that allows for a trade-off between the minimum-time objective and a desire to keep small the states and the controls. Thus, if we select smaller Q and R, the term $\frac{1}{2}(T - t_0)$ in the PI dominates, and the control tries to make the transit time smaller. We call this the *linear quadratic (LQ) minimum-time problem*.

We show here that the optimal time T may be determined using dS/dt, with $S(t)$ the solution to the Riccati equation in Table 3.2-2.

Derivation

Using Table 3.1-1, the Hamiltonian is

$$H = \frac{1}{2} + \frac{1}{2} x^T Q x + \frac{1}{2} u^T R u + \lambda^T (Ax + Bu), \qquad (3.4.9)$$

with $\lambda(t)$ the costate. The Euler equations are

$$-\dot{\lambda} = A^T \lambda + Qx \qquad (3.4.10)$$

$$0 = Ru + B^T \lambda, \qquad (3.4.11)$$

whence

$$u = -R^{-1} B^T \lambda. \qquad (3.4.12)$$

In (3.1.11) both $dx(T)$ and dT are nonzero; however, they are independent in this situation so that the final conditions are

$$\lambda(T) = S_T x(T) \qquad (3.4.13)$$

$$H(T) = 0. \qquad (3.4.14)$$

Indeed, since the system and PI are not explicitly dependent on t, (3.1.14) shows that, for all t

$$H(t) = 0. \qquad (3.4.15)$$

We now remark that, with the exception of (3.4.15), this is the same boundary-value problem we solved in the closed-loop LQR problem in section 3.2. That is, Table 3.2-2 still provides the optimal solution. The difficulty, of course, is that the final time T is unknown.

To find the final time T, recall that for all times t

$$\lambda = Sx \qquad (3.4.16)$$

$$u = -R^{-1} B^T S x. \qquad (3.4.17)$$

3.4 Minimum-Time and Constrained-Input Design

Using these at $t = t_0$ in (3.4.9) and taking into account (3.4.15) we have

$$0 = H(t_0) \qquad (3.4.18)$$
$$= \frac{1}{2} + \frac{1}{2} x^T(t_0)[SBR^{-1}B^TS + Q + (SA + A^TS) - 2SBR^{-1}B^TS]x(t_0),$$

Therefore

$$0 = 1 + x^T(t_0)[A^TS + SA + Q - SBR^{-1}B^TS]x(t_0), \qquad (3.4.19)$$

or, taking into account the Riccati equation from Table 3.2-2

$$x^T \dot{S} x \big|_{t_0} = 1. \qquad (3.4.20)$$

Discussion

We have shown that the solution procedure for the LQ minimum-time problem is to integrate the Riccati equation

$$-\dot{S} = A^TS + SA + Q - SBR^{-1}B^TS \qquad (3.4.21)$$

backwards from some time τ using as the final condition $S(\tau) = S_T$. At each time t, the left-hand side of (3.4.20) is computed using the known initial state and $\dot{S}(t)$. Then, the minimum interval $(T - t_0)$ is equal to $(\tau - t)$ where t is the time for which (3.4.20) first holds. This specifies the minimum final time T.

Finally, the Kalman gain and the optimal control are given using exactly the design equations from Table 3.3-2, namely

$$K = R^{-1}B^TS \qquad (3.4.22)$$
$$u = -K(t)x. \qquad (3.4.23)$$

It is interesting to note that $\dot{S}$ is used to determine the optimal time interval, while S is used to determine the optimal feedback gain.

More details on this control scheme may be found in Verriest and Lewis [1991]. We mention that condition (3.4.20) may never hold. Then, the optimal solution is $T - t_0 = 0$; that is, the PI is minimized by using *no* control. Roughly speaking, if $x(t_0)$ and/or Q and $S(T)$ are large enough, then it makes sense to apply a nonzero control $u(t)$ to make $x(t)$ decrease. On the other hand, if Q and $S(T)$ are too small for the given initial state $x(t_0)$, then it is not worthwhile to apply any control to decrease $x(t)$, for both a nonzero control and a nonzero time interval will increase the PI.

Constrained-Input Design

Up to this point we have discussed minimum-time controls based on the conditions of Table 3.1-1, which were derived using the calculus of variations. Under some smoothness assumptions on $f(x, u, t)$ and $L(x, u, t)$, the resulting controls are also smooth. Here, we shall discuss a fundamentally different sort of control strategy.

If the linear system

$$\dot{x} = Ax + Bu \qquad (3.4.24)$$

with $x \in \mathbf{R}^n$, $u \in \mathbf{R}^m$ is prescribed, there are problems with using the pure minimum-time PI

$$J(t_0) = \int_{t_0}^{T} 1 \, dt, \qquad (3.4.25)$$

where T is free. Indeed, the Hamiltonian is

$$H = 1 + \lambda^T(Ax + Bu), \qquad (3.4.26)$$

whence the stationarity condition is

$$0 = \frac{\partial H}{\partial u} = B^T \lambda. \qquad (3.4.27)$$

This, however, does not involve $u(t)$, so that it may not be used to express the control in terms of $\lambda(t)$.

The problem is that $H(t)$ is *linear* in $u(t)$. To minimize it, we should select $u(t)$ to make $\lambda^T Bu$ as small as possible ("small" means to the left on the real number line). That is, we must select $u(t)$ of infinite magnitude so that $\lambda^T Bu$ is equal to $-\infty$. Obviously, the way to minimize the time is to use infinite control energy!

Since this optimal strategy is not acceptable, we must find a way to reformulate the minimum-time problem for linear systems.

Let us, therefore, discuss the linear minimum-time problem with *constrained input magnitude*. Thus, we shall use the performance index (3.4.25), and the final state will be required to satisfy

$$\Psi(x(T), T) = 0 \qquad (3.4.28)$$

for a prescribed function $\Psi \in \mathbf{R}^p$. This general final condition includes the cases where the final state is required to be equal to a certain value, or on a moving target set, and so on.

However, we shall also require the control to satisfy the magnitude constraint

$$|u(t)| \leq 1 \qquad (3.4.29)$$

for all $t \in [t_0, T]$. This constraint means that *each component* of the m-vector $u(t)$ must have magnitude no greater than 1. Thus, the control is constrained to an *admissible region* (in fact, a hypercube) of $\mathbf{R}^m$. If the constraints on the components of $u(t)$ have a value different than 1, then we may appropriately scale the corresponding columns of the B matrix to obtain the constraints in the form of (3.4.29).

A requirement like (3.4.29) arises in many problems where the control magnitude is limited by physical considerations. For example, the thrust of a rocket certainly has a maximum possible value, as has the armature voltage of a DC motor.

The optimal control problem posed here is to find a control $u(t)$ that minimizes $J(t_0)$, satisfies (3.4.29) at all times, and drives a given $x(t_0)$ to a final state $x(T)$ satisfying (3.4.28) for a given function Ψ.

3.4 Minimum-Time and Constrained-Input Design

Intuitively, to minimize the time, the optimal control strategy appears to be to apply maximum effort (i.e., plus or minus 1) over the entire time interval. We shall now formalize this feeling. When a control component takes on a value at the boundary of its admissible region (i.e., ± 1), it is said to be *saturated*.

The Hamiltonian is

$$H(x, u, \lambda, t) = L + \lambda^T f = 1 + \lambda^T(Ax + Bu). \tag{3.4.30}$$

The objective is to determine $u(t)$ such that $H(t)$ is minimized subject to the constraint (3.4.29). We cannot simply use the stationarity condition $0 = \partial H/\partial u$ due to the fact that the minimum of $H(t)$ with respect to $u(t)$ may occur outside the admissible region.

Pontryagin and co-workers [1962] have shown that, in the case of constrained controls, Table 3.1-1 still applies if the stationarity condition is replaced by the more general condition, known as *Pontryagin's Minimum Principle*

$$H(x^*, u^*, \lambda^*, t) \leq H(x^*, u^* + \delta u, \lambda^*, t), \qquad \text{all admissible } \delta u, \tag{3.4.31}$$

with δu the variation in u, and starred quantities denoting the optimal values. This may also be written as

$$H(x^*, u^*, \lambda^*, t) \leq H(x^*, u, \lambda^*, t), \qquad \text{all admissible } u. \tag{3.4.32}$$

This is an extremely powerful result which we will now employ to solve the linear constrained-input minimum-time problem.

According to Pontryagin's minimum principle, the optimal control $u^*(t)$ must satisfy

$$1 + (\lambda^*)^T(Ax^* + Bu^*) \leq 1 + (\lambda^*)^T(Ax^* + Bu).$$

Now we see the importance of having the optimal state and costate on both sides of the inequality, for this means we can say that for optimality the control $u^*(t)$ must satisfy

$$(\lambda^*)^T Bu^* \leq (\lambda^*)^T Bu \tag{3.4.33}$$

for all admissible $u(t)$. This condition allows us to express $u^*(t)$ in terms of the costate. To see this, let us first discuss the single-input case.

Let $u(t)$ be a scalar, and let b represent the input vector. In this case it is easy to choose $u^*(t)$ to minimize the value of $\lambda^T(t)bu(t)$. (Note: Minimize means that we want $\lambda^T(t)bu(t)$ to take on a value as close to $-\infty$ as possible.)

If $\lambda^T(t)b$ is positive, we should select $u(t) = -1$ to get the largest possible negative value of $\lambda^T(t)bu(t)$. On the other hand, if $\lambda^T(t)b$ is negative, we should select $u(t)$ as its maximum admissible value of 1 to make $\lambda^T(t)bu(t)$ as negative as possible. If $\lambda^T(t)b$ is zero at a single point t in time, then $u(t)$ can take on any value at that time, since then $\lambda^T(t)bu(t)$ is zero for all values of $u(t)$.

This relation between the optimal control and the costate can be expressed in a concise way by defining the *signum function* as

$$\text{sgn}(w) = \begin{cases} 1, & w > 0 \\ \text{indeterminate}, & w = 0 \\ -1, & w < 0. \end{cases} \tag{3.4.34}$$

Then the optimal control is given by

$$u^*(t) = -\text{sgn}(b^T\lambda(t)). \tag{3.4.35}$$

This expression for u^* in terms of the costate should be compared to the expression (3.4.12), which holds for linear systems with quadratic performance indices.

The quantity $b^T\lambda(t)$ is called the *switching function*. A sample switching function and the optimal control it determines are shown in Fig. 3.4-1. When the switching function changes sign, the control switches from one of its extreme values to another. The control in the figure switches four times. The optimal linear minimum-time control is always saturated since it switches back and forth between its extreme values, so it is called *bang-bang* control.

If the control is an m-vector, then according to the minimum principle (3.4.33) we need to select $u^*(t)$ to make $\lambda^T(t)Bu(t)$ take on a value as close to $-\infty$ as possible. To do this, we should select component $u_i(t)$ to be 1 if component $b_i^T\lambda(t)$ is negative, and to be -1 if $b_i^T\lambda(t)$ is positive, where b_i is the ith column of B. This control strategy makes the quantity

$$\lambda^T(t)Bu(t) = \sum_{i=1}^{m} u_i(t)b_i^T\lambda(t) \tag{3.4.36}$$

as small as possible for all $t\varepsilon\ [t_0T]$.

Thus, we can write

$$u^*(t) = -\text{sgn}(B^T\lambda(t)) \tag{3.4.37}$$

if we define the signum function for a vector w as

$$v = \text{sgn}(w) \quad \text{if } v_i = \text{sgn}(w_i) \quad \text{for each } i, \tag{3.4.38}$$

where v_i, w_i are the components of v and w.

It is possible for a component $b_i^T\lambda(t)$ of the switching function $B^T\lambda(t)$ to be zero over a finite time interval. If this happens, component $u_i(t)$ of the optimal control is not well defined by (3.4.33). This is called a *singular condition*. If this does not occur, the time-optimal problem is called *normal*.

If the plant is time-invariant, then we can present some simple results on existence and uniqueness of the minimum-time control. First, we present a test for normality. See Athans and Falb [1966] and Kirk [1970] for more detail on the following results.

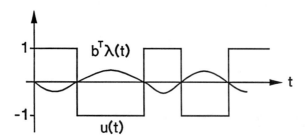

Figure 3.4-1 Sample switching function and associated optimal control

3.4 Minimum-Time and Constrained-Input Design

The time-invariant plant (3.4.24) is reachable if and only if the reachability matrix

$$U_n = [B \quad AB \quad \cdots \quad A^{n-1}B] \tag{3.4.39}$$

has full rank n. If b_i is the ith column of $B \in \mathbf{R}^{n \times m}$, then the plant is *normal* if

$$U_i = [b_i \quad Ab_i \quad \cdots \quad A^{n-1}b_i] \tag{3.4.40}$$

has full rank n for each $i = 1, 2, \ldots, m$; that is, if the plant is reachable by each separate component u_i of $u \in \mathbf{R}^m$. Normality of the plant and normality of the minimum-time control problem are equivalent.

The next results are due to Pontryagin et al. [1962].

Let the plant be normal (and hence reachable), and suppose we want to drive a given $x(t_0)$ to a desired fixed final state $x(T)$ in minimum time with a control satisfying (3.4.29). Then:

1. If the desired final state $x(T)$ is equal to zero, then a minimum-time control exists if the plant has no poles with positive real parts (i.e., no poles in the open right half-plane).
2. For any fixed $x(T)$, if a solution to the minimum-time problem exists, then it is unique.
3. Finally, if the n plant poles are all real and if the minimum-time control exists, then each component $u_i(t)$ of the time-optimal control can switch at most $n - 1$ times.

In both its computation and its final appearance, bang-bang control is fundamentally different from the smooth controls we have seen previously. The minimum principle leads to the expression (3.4.37) for $u^*(t)$, but it is difficult to solve explicitly for the optimal control. Instead, we shall see that (3.4.37) specifies several different control laws, and that we must then select which among these is the optimal control. Thus, the minimum principle keeps us from having to examine all possible control laws for optimality, giving a small subset of potentially optimal controls to be investigated.

To demonstrate these notions and show that $u^*(t)$ can still be expressed as a state feedback control law, let us consider a two-dimensional example, since the two-dimensional plane is easy to draw.

Example 3.4-1: Minimim-Time Control of Systems Obeying Newton's Laws

Let us consider Newton's system

$$\dot{y} = v, \tag{1}$$

$$\dot{v} = u, \tag{2}$$

with $y(t)$ the position, $v(t)$ the velocity, and $u(t)$ the input acceleration. The state is $x = [y \quad v]^T$.

Let the acceleration input u be constrained in magnitude by

$$|u(t)| \leq 1. \tag{3}$$

180   Chap. 3   Optimal Control of Continuous-Time Systems

If a magnitude constraint of, for instance, $\mu \neq 1$ is required, then we may simply redefine $u(t)$ as $u(t)/\mu$, which must satisfy (3).

The control objective is to bring the state from any initial point (y_0, v_0) to the desired final state of $(y_T, 0)$ in the minimum time T. Define the deviation in the position as

$$\tilde{y} = y - y_T, \tag{4}$$

and note that

$$\dot{\tilde{y}} = \dot{y} = v. \tag{5}$$

That is, we may simply redefine the origin of the (y, v) plane to be $(y_T, 0)$; therefore, it is sufficient to determine the optimal constrained control that drives the initial state (y_0, v_0) to the origin in minimum time. Then, in implementing the control law to be derived, we need only replace $y(t)$ everywhere by $y(t) - y_T$.

The final state is thus fixed at

$$\Psi(x(T), T) = \begin{bmatrix} y(T) \\ v(T) \end{bmatrix} = 0. \tag{6}$$

a. Form of the Optimal Control

The Hamiltonian is

$$H = 1 + \lambda_y v + \lambda_v u, \tag{7}$$

where $\lambda = [\lambda_y \;\; \lambda_v]^T$ is the costate. Thus, according to Table 3.1-1, the costate equations are

$$\dot{\lambda}_y = 0, \tag{8}$$

$$\dot{\lambda}_v = -\lambda_y. \tag{9}$$

Since $dT \neq 0$, final condition (3.1.11) requires that

$$0 = H(T) = 1 + \lambda_y(T)v(T) + \lambda_v(T)u(T), \tag{10}$$

or, using (6)

$$\lambda_v(T)u(T) = -1. \tag{11}$$

Pontryagin's minimum principle requires (3.4.37), or (note that $B^T = [0 \;\; 1]$)

$$u(t) = -\text{sgn}(\lambda_v(t)) \tag{12}$$

$$= \begin{cases} 1, & \lambda_v(t) < 0 \\ \text{undetermined}, & \lambda_v(t) = 0 \\ -1, & \lambda_v(t) > 0; \end{cases}$$

so that costate component $\lambda_v(t)$ is the switching function. To determine the optimal control, we need only determine $\lambda_v(t)$.

Solving (8) and (9) with respect to the final time T yields

$$\lambda_y(t) = \text{const} \equiv \lambda_y, \tag{13}$$

$$\lambda_v(t) = \lambda_v(T) + (T - t)\lambda_y. \tag{14}$$

3.4 Minimum-Time and Constrained-Input Design

Using (11) and the fact that $u^*(t)$ is saturated at 1 or -1 requires either

$$u^*(T) = 1 \quad \text{and} \quad \lambda_v^*(T) = -1 \tag{15}$$

or

$$u^*(T) = -1 \quad \text{and} \quad \lambda_v^*(T) = 1. \tag{16}$$

There are several possibilities for the switching function $\lambda_v^*(t)$, depending on the values of $\lambda_v^*(T)$ and λ_y. Some possibilities are shown in Fig. 3.4-2. The actual $\lambda_v^*(t)$ depends on the initial state (y_0, v_0). Note, however, that since $\lambda_v^*(t)$ is linear, it crosses the axis at most one time, so that there is at most one control switching. This agrees with our result on the maximum number of switchings when the plant poles are all real ($n - 1 = 1$). Therefore, the optimal control is one of these choices:

1. -1 for all t.
2. -1 switching to $+1$.
3. $+1$ switching to -1.
4. $+1$ for all t.

These control policies correspond to the switching functions of the forms shown in Fig. 3.4-2a, b, c, and d, respectively.

It remains to determine which of the choices 1 through 4 is the correct optimal control. We must also find the *switching time* t_s (see Fig. 3.4-2) if applicable. We can find a state feedback control law that tells us all this by working in the *phase plane*.

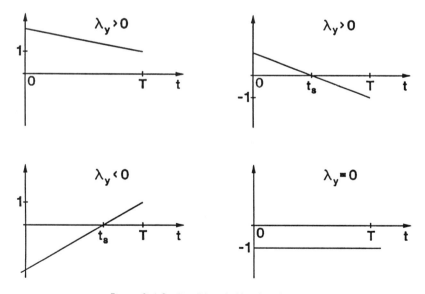

Figure 3.4-2 Possible switching functions $\lambda_v(t)$

b. Phase-Plane Trajectories

Let us determine the state trajectories under the influence of the two possible control inputs: $u(t) = 1$ for all t, and $u(t) = -1$ for all t.

Since in either of these two cases the input u is constant, we can easily integrate state equations (2) and (1) to get

$$v(t) = v_0 + ut, \tag{17}$$

$$y(t) = y_0 + v_0 t + \frac{1}{2} ut^2. \tag{18}$$

To eliminate the time variable, use (17) to say $t = (v(t) - v_0)/u$ and then substitute into (18) to see that

$$u(y - y_0) = v_0(v - v_0) + \frac{1}{2}(v - v_0)^2. \tag{19}$$

This is a parabola passing through (y_0, v_0) and, as the initial state varies, a family of parabolas is defined.

The *phase plane* is a coordinate system whose axes are the state variables. Phase-plane plots of several members of the family of state trajectories (19) are shown for $u = 1$ and for $u = -1$ in Fig. 3.4-3. The arrows indicate the direction of increasing time. Hence, for example, if the initial state (y_0, v_0) is as shown in Fig. 3.4-3a, then under the influence of the control $u = 1$, the state will develop along the parabola with velocity passing through zero and then increasing linearly, and position decreasing to zero and then increasing quadratically. For the particular initial condition shown, the state will be brought exactly to the origin by the control $u = 1$. For the same initial state, if the control $u = -1$ is applied, the trajectory will move along the parabola in Fig. 3.4-3b.

c. Bang-Bang Feedback Control

Since the control input is saturated at 1 or -1, the parabolas in Fig. 3.4-3 are minimum-time paths in the phase plane. Unfortunately, they do not go through the origin for all initial states, so in general (6) is not satisfied. We can construct minimum-time paths to the origin by superimposing the two parts of Fig. 3.4-3. See Fig. 3.4-4. We shall now demonstrate that this figure represents a *state-feedback control law*, which brings any state to the origin in minimum time.

We have argued using the minimum principle and the costate trajectories (13), (14) that at any given time t there are only two control alternatives: $u(t) = 1$ or $u(t) = -1$. Futhermore, at most one control switching is allowed.

Suppose the initial state is as shown in Fig. 3.4-4. Then the only way to arrive at the origin while satisfying these conditions is to apply $u = -1$ to move the state along a parabola to the dashed curve. At this point (labeled "a"), the control is switched to $u = 1$ to drive the state into the origin. Hence, the resulting seemingly roundabout trajectory is in fact a minimum-time path to the origin.

The dashed curve is known as the *switching curve*. For initial states on this curve, a control of $u = 1$ (if $v_0 < 0$) or $u = -1$ (if $v_0 > 0$) for the entire control interval will bring the state to zero. For initial states off this curve, the state must first be driven onto the switching curve, and then the control must be switched to its other extreme value to bring the final state to zero.

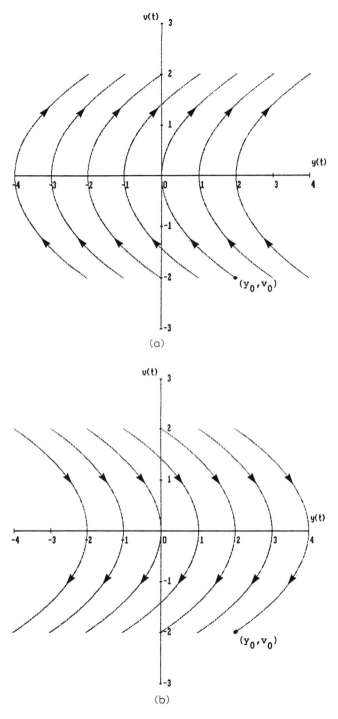

Figure 3.4-3 Phase-plane trajectories. (a) $u = 1$. (b) $u = -1$.

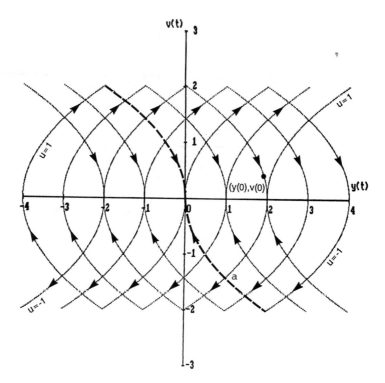

Figure 3.4-4 Bang-bang feedback control law

By setting $v_0 = 0$, $y_0 = 0$ in (19), we can see that the equation of the switching curve is

$$y = \begin{cases} \dfrac{1}{2} v^2, & v < 0 \\ -\dfrac{1}{2} v^2, & v > 0 \end{cases}$$

or

$$y = -\frac{1}{2} v|v|. \tag{20}$$

Simply put, for initial states above the switching curve, the optimal control is $u = -1$, followed by $u = 1$, with the switching occurring when $y(t) = \frac{1}{2}v^2(t)$. For initial states below the switching curve, the optimal control is $u = 1$, followed by $u = -1$, with the switching occurring when $y(t) = -\frac{1}{2}v^2(t)$. Since the control at each time t is completely determined by the state (i.e., by the phase-plane location), Fig. 3.4-4 yields a feedback control law.

3.4 Minimum-Time and Constrained-Input Design

This feedback law, which is represented graphically in the figure, can be stated as

$$u = \begin{cases} -1 & \text{if } y > -\frac{1}{2}v|v| \\ & \text{or if } y = -\frac{1}{2}v|v| \quad \text{and} \quad y < 0 \\ 1 & \text{if } y < -\frac{1}{2}v|v| \\ & \text{or if } y = -\frac{1}{2}v|v| \quad \text{and} \quad y > 0. \end{cases} \quad (21)$$

This control sequence should be contrasted to the continuous control laws developed in Examples 3.2-2 and 3.3-4.

d. Computer Simulation

It is easy to implement the feedback control law (21) and simulate its application on a digital computer.

Subroutine D(IT, X) in Fig. 3.4-5 implements (21). It is based on the function

$$SW = y + \frac{1}{2}v|v|. \quad (22)$$

```
C    SIMULATION OF BANG-BANG CONTROL FOR NEWTON'S SYSTEM
C    SUBROUTINES FOR USE WITH DRIVER PROGRAM TRESP

     SUBROUTINE D(IT,X)
       REAL X(*)
       COMMON/COMMAND/U(1)
       THRESH= 1.E-4

       SW= X(1) + 0.5*X(2)*ABS(X(2))
       IF (ABS(SW).LT.THRESH) THEN
            IF(X(1).GT.0.) U(1)=  1.
            IF(X(1).LT.0.) U(1)= -1.
       ELSE
            IF(SW.LT.0.) U(1)=  1.
            IF(SW.GT.0.) U(1)= -1.
       END IF
       IF( (X(1)**2 + X(2)**2) .LT. THRESH) U(1)= 0.

     RETURN
     END


     SUBROUTINE F(TIME,X,XP)
       REAL X(*), XP(*)
       COMMON/COMMAND/U(1)

       XP(1)= X(2)
       XP(2)= U(1)

     RETURN
     END
```

Figure 3.4-5 Software for simulation of bang-bang control

Note that *SW* is considered equal to zero if it is within a threshold of 10^{-4} on either side of zero. Note also that the control input u must be turned off (i.e., set to zero) when the state is sufficiently close to the origin, in order to bring the plant to rest there.

Subroutine F(*TIME, X, XP*) in the figure describes the plant dynamics (1), (2). These subroutines are used by the time response driver program TRESP in Appendix A, which uses a Runge-Kutta integrator.

The state trajectories resulting from the simulation when $y_0 = 10$ and $v_0 = 10$ are shown in Fig. 3.4-6. The Runge-Kutta integration step was 25 msec. It can be seen from the slope of $v(t)$ that the control input switches from $u = -1$ to $u = 1$ at $t_s \approx 18$ sec, and that the minimum time to the origin $T \approx 26$ sec.

Compare Fig. 3.4-6 to the simulation in Example 3.3-4.

e. Computation of Time to the Origin

It is not difficult to compute the switching time t_s and the minimum time to the origin T in terms of the initial state (y_0, v_0).

Let us suppose that the initial state is above the switching curve, that is

$$y_0 > -\frac{1}{2} v_0 |v_0|. \tag{23}$$

This situation is shown in Fig. 3.4-4. Then the initial control of $u = -1$ is applied to drive the state along the parabola passing through (y_0, v_0) to the switching curve, at which time t_s the control is switched to $u = 1$ to bring the state to the origin.

The switching curve for $v < 0$ is described by $y = \frac{1}{2}v^2$. We can find the switching

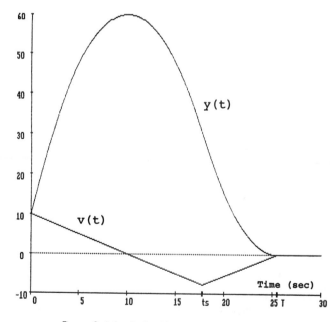

Figure 3.4-6 Optimal bang-bang state trajectories

time t_s by determining when the state is on this curve. Using (17) and (18) with $u = -1$ yields

$$y(t) = y_0 + v_0 t - t^2/2 = \frac{v^2(t)}{2} = \frac{v_0^2}{2} - v_0 t + t^2/2$$

on the switching curve, or

$$t^2 - 2v_0 t + \frac{v_0^2}{2} - y_0 = 0. \tag{24}$$

The switching time is therefore

$$t_s = v_0 + \sqrt{y_0 + v_0^2/2} \tag{25}$$

where the positive root of (24) is selected to make t_s positive for all v_0.

At the switching time, the state is on the switching curve (at point "a" in Fig. 3.4-4), and using (17)

$$v(t_s) = v_0 - t_s. \tag{26}$$

Also using (17) for the remaining time $(T - t_s)$ yields (now $u = 1$!)

$$0 = v(T) = v(t_s) + (T - t_s). \tag{27}$$

Taking (26) into account gives the minimum time to the origin of

$$T = 2t_s - v_0$$

or

$$T = v_0 + 2\sqrt{y_0 + v_0^2/2} \tag{28}$$

To check (25) and (28), let $y_0 = 10$, $v_0 = 10$. Then $t_s = 17.75$ and $T = 25.49$. These numbers agree with Fig. 3.4-6.

A similar development holds if (y_0, v_0) is below the switching curve. ∎

PROBLEMS FOR CHAPTER 3

Problems for Section 3.1

3.1-1 Derivation of Optimal Controller. According to Leibniz's rule, the differential of the functional

$$J(x) = \int_{t_0}^{T} h(x(t), t) \, dt$$

is given by

$$dJ = h(x(T), T) \, dT - h(x(t_0), t_0) \, dt_0 + \int_{t_0}^{T} [h_x^T(x(t), t) \, \delta x] \, dt,$$

with $h_x \equiv \partial h / \partial x$ and δx the variation in x. Use this formula to derive (3.1.7). Now, continue to derive (3.1.10).

3.1-2 Minimization of Functionals. It is desired to minimize

$$J = \int_0^T h(t)\, dt$$

with $h(t) = \dot{x}^2(t)$.

a. Taking $T = \pi$, formulate this as an optimal control problem. Find and sketch the optimal trajectory $x(t)$. Find the optimal cost.

b. Repeat for $T = 1$.

3.1-3 Control of Bilinear Systems. Consider the bilinear system

$$\dot{x} = ax + dxu + bu$$

where $x(t)$ and $u(t)$ are scalars. Define the associated PI

$$J = \frac{1}{2} x^T(T) s(T) x(T) + \frac{1}{2} \int_0^T (qx^2 + ru^2)\, dt.$$

a. Find the Hamiltonian system; that is, eliminate $u(t)$ in the state/costate equations. The result will be a set of coupled cubic equations.

b. Take $a = d = b = q = r = s(T) = T = 1$ and write a computer program to solve for the optimal trajectory using the shooting point method.

Problems for Section 3.2

3.2-1 Prove that (3.2.19) is the solution to (3.2.18). (Use Leibniz's rule to differentiate $P(t)$.)

3.2-2 Open-Loop Control. A plant is given by

$$\dot{x} = \begin{bmatrix} 0 & 1 \\ 2 & -1 \end{bmatrix} x + \begin{bmatrix} 0 \\ 1 \end{bmatrix} u.$$

a. Verify that the plant is unstable and reachable.

b. Compute the open-loop optimal controller over the time interval [0, 1]. It may be necessary to compute the gramian by explicit integration. This part takes several pages of work.

c. Write subroutines to compute the optimal control and apply it to the plant using a slightly modified version of program TRESP in Appendix A. Use the Lyapunov equation to compute the gramian. To invert the gramian use a subroutine such as LINV2F in [IMSL]. Very little analysis is needed for this part.

d. To see the usefulness of LQ control for multivariable systems, repeat part c using

$$B = \begin{bmatrix} 0 & 1 \\ 1 & -1 \end{bmatrix}.$$

3.2-3 Derive the Joseph stabilized formulation (3.2.48).

3.2-4 Simulation of LQR. Modify program TRESP in Appendix A so that it implements the LQR simulation procedure in Fig. 3.2-3. Before $F(T, X, XP)$ is called, you will need to add do loops to integrate the Riccati equation (3.2.51) using the Runge-Kutta integrator and then flip the solution according to (3.2.52). Using this program and the subroutines in Example 3.2-4, perform the simulation and obtain the graphs displayed there.

3.2-5 Simulate the LQR for the SISO system of Problem 3.2-2.

3.2-6 LQR for Newton's System. Consider the Newton's law system

$$\dot{x}_1 = x_2$$
$$\dot{x}_2 = u$$

with PI

$$J = \frac{1}{2} x^T(T)x(T) + \int_0^T (x^T x + ru^2)\, dt.$$

a. Write the Riccati equation as three scalar differential equations. Find the Kalman gain in terms of the entries $s_i(t)$ of $S(t)$.
b. Write subroutines to simulate the LQR on a digital computer.
c. Find analytic expressions for the steady-state (e.g., $T \to \infty$) Riccati solution and Kalman gain. To accomplish this, simply set $\dot{S} = 0$ in the Riccati equation and solve the resulting three algebraic equations for the components s_i of S.

3.2-7 LQR For Damped Harmonic Oscillator. Repeat Problem 3.2-6 for the system

$$\dot{x} = \begin{bmatrix} 0 & 1 \\ -\omega_n^2 & -2\zeta\omega_n \end{bmatrix} x + \begin{bmatrix} 0 \\ 1 \end{bmatrix} u.$$

3.2-8 LQ Regulator with State/Input Cross Weighting. Let the plant

$$\dot{x} = Ax + Bu \tag{1}$$

have the modified performance index

$$J(t_0) = \frac{1}{2} x^T(T)S(T)x(T) + \frac{1}{2} \int_{t_0}^T [x^T \ u^T] \begin{bmatrix} Q & V \\ V^T & R \end{bmatrix} \begin{bmatrix} x \\ u \end{bmatrix}, \tag{2}$$

where the block matrix is positive semidefinite and $R > 0$.

a. Define a modified Kalman gain as

$$K = R^{-1}(V^T + B^T S) \tag{3}$$

where $S(t)$ is the solution to the Riccati equation

$$-\dot{S} = A^T S + SA - K^T R K + Q. \tag{4}$$

[Compare (4) to (3.2.47)]. Show that the optimal control is

$$u(t) = -K(t)x(t). \tag{5}$$

b. Show that the optimal remaining cost on any subinterval $[t, T]$ is

$$J(t) = \frac{1}{2} x^T(t)S(t)x(t). \tag{6}$$

In summary, if the cost index contains a weighting V that picks up the state-input inner product, the only required modification to the LQ regulator is that the Kalman gain must be modified, and the Riccati-equation formulation (4) should be used.

3.2-9 Zero-Final-State Controller. The final state may be driven exactly to zero by using the fixed-final-state controller of Table 3.2-1. As an alternative, a feedback law that accomplishes this may be derived from the LQR as follows.

This exercise indicates that we may design a LQR to obtain a final state that is as small as we desire by simply increasing the final-state weighting $S(T)$ in the performance index. To achieve a final state of exactly zero, we may set $Q = 0$ and let $S(T)$ go to infinity in the performance index (3.2.25). Then, if the optimal $J(t_0)$ is bounded, the final state $x(T)$ must be equal to zero.

Using the fact that, for any nonsingular matrix M

$$\frac{d}{dt} M^{-1} = -M^{-1} \dot{M} M^{-1} \qquad (1)$$

define $P = S^{-1}$ and show that, if S satisfies the Riccati equation, then P satisfies the Lyapunov equation

$$\dot{P} = AP + PA^T - BR^{-1}B^T. \qquad (2)$$

Since we would like $S(T)$ tending to infinity, we may select $P(T) = 0$ as the final condition for (2). Complete the specification of the control law and compare with the open-loop controller in Table 3.2-1.

Problems for Section 3.3

3.3-1 Steady-State LQ Design for Harmonic Oscillator. Repeat Example 3.3-1 using the damped harmonic oscillator given in Example 2.1-7.

3.3-2 Optimal Versus Suboptimal LQR
  a. Using program TRESP in Appendix A, perform a computer simulation of the steady-state LQR for Newton's system to verify the results given in Example 3.3-4.
  b. Modify TRESP in Appendix A so that it can be used to simulate the optimal LQR as given in Fig. 3.2-3. Using this program, simulate the *optimal* LQR for Newton's system using the same state weights as in part a. Experiment with different choices for the final state weighting $S(T)$. Compare your results to part a.

3.3-3 Optimal Versus Suboptimal LQR. Compare the optimal and suboptimal performance of the LQR for the damped harmonic oscillator of Example 2.1-7. Use computer simulation as was done in Problem 3.3-2.

3.3-4 Eigenstructure LQR Design for Harmonic Oscillator. Repeat Example 3.3-5 for the harmonic oscillator given in section 2.1. For tractability, use the undamped case $\alpha = 0$.

3.3-5 Software for Eigenstructure LQR Design
  a. Write a computer program to build the Hamiltonian matrix H from the plant and PI matrices, find the eigenstructure of H, and then determine the optimal steady-state Kalman gain. Subroutines that find the eigenstructure of a matrix may be found in [IMSL], LINPACK [Dongarra et al. 1979], and elsewhere.
  b. Using this program, repeat Example 3.3-5 for the damped harmonic oscillator.

Problems for Section 3.4

3.4-1 Bang-Bang Control of a Scalar System. Let

$$\dot{x} = x - u$$

with $x(t)$ a scalar. It is desired to drive any initial state $x(0)$ to the origin in minimum time while ensuring that $|u(t)| \leq 1$. Solve the minimum-time problem along the lines of the work in Example 3.4-1.

3.4-2 Bang-Bang Control of Harmonic Oscillator. Repeat Example 3.4-1 for the undamped harmonic oscillator (i.e., take $\alpha = 0$) in section 2.1.

4

Output-Feedback Design

SUMMARY

In Chapter 3 we discussed design with full state-variable feedback. Unfortunately, in real design problems only some of the states are available as measured outputs. In this chapter we develop design techniques using reduced state, or output feedback. This also allows us to use a compensator with any desired dynamics. First, the LQR with output feedback is covered. Next, the tracking or servo problem is discussed; tracking a reference input is one of the most fundamental problems in control. Finally, we show how to design controls that make the system behave like a reference model with desirable characteristics.

INTRODUCTION

In the previous chapter we laid the foundations of modern optimal control by discussing systems that have all their states available for feedback. In fact, in most real situations only some of the states are measured as outputs. In this chapter we cover the linear quadratic regulator (LQR) using *output feedback*. As we shall see in section 4.1, the design equations will be more complicated than the ones in Table 3.2-2 for full state feedback.

On the other hand, output-feedback design allows the use of compensators with any desired structure in the control system. We shall show this in section 4.2, where we use output feedback design to attack the problem of *tracking a reference input*. The tracking, or *servo* problem is one of the most fundamental design problems in control. It arises in problems as varied as home temperature control, automobile speed control, radar tracking of aircraft, and spacecraft trajectory following. We might say that the servodesign technique of section 4.2 is our fundamental approach to controller design throughout the book.

In section 4.3 we present a simplified servodesign approach that is sometimes appropriate. It involves first designing a regulator, and then adding a feedforward term to ensure perfect tracking.

In section 4.4 we treat a Command Generator Tracker (CGT) approach to servodesign that relies on knowing the dynamics of the reference input. Finally, in section 4.5 we cover controls design when it is desired for the system to behave like a given reference model with good characteristics.

For the student, all parts of each section are important. Those primarily interested in the results and their use may skip the subsections labeled "Derivation."

4.1 LINEAR QUADRATIC REGULATOR WITH OUTPUT FEEDBACK

Our objective in this section is to regulate nonzero initial states to zero using output-feedback controls. This amounts to the problem of stabilizing the plant.

We assume the plant is given by the linear time-invariant state-variable model

$$\dot{x} = Ax + Bu \tag{4.1.1}$$

$$y = Cx, \tag{4.1.2}$$

with $x(t) \in \mathbf{R}^n$ the state, $u(t) \in \mathbf{R}^m$ the control input, and $y(t) \in \mathbf{R}^p$ the measured output. Since the input and output are both vectors, we are talking about general multi-input/multi-output (MIMO) systems.

The admissible controls are output feedbacks of the form

$$u = -Ky, \tag{4.1.3}$$

where K is an $m \times p$ matrix of constant feedback coefficients to be determined by the design procedure. Since the control input $u(t)$ is only allowed to depend on the output $y(t)$ and not on the entire state $x(t)$, this control law reflects the fact that we are constrained by the requirement that only measurable quantities may be used to determine $u(t)$. Thus, we are dealing with reduced state information, or *output feedback*.

The objective of state regulation for the system may be attained by selecting the control input $u(t)$ to minimize a performance index (PI) of the type

$$J = \frac{1}{2} \int_0^\infty (x^T Q x + u^T R u) \, dt, \tag{4.1.4}$$

where $Q \geq 0$ and $R > 0$ are symmetric weighting matrices. To see this, note that if we select the feedback gains K to make J small, then since the integration limits are infinite, the integrand must vanish as $t \to \infty$. Thus, with a suitable choice of Q, the state $x(t)$ will go to zero with time, as we desire. In fact, we should choose Q so that $(\sqrt{Q}, A)$ is observable, for then $x^T Q x = \|\sqrt{Q}\, x\|^2 \to 0$ along with the dynamical relation (4.1.1) guarantees that $x(t) \to 0$.

4.1 Linear Quadratic Regulator with Output Feedback

The linear quadratic regulator (LQR) problem with output feedback is the following: given the linear system (4.1.1)/(4.1.2), find the feedback coefficient matrix K in the control input (4.1.3) that minimizes the value of the quadratic PI (4.1.4).

By substituting the control (4.1.3) into (4.1.1), the closed-loop system equations are found to be

$$\dot{x} = (A - BKC)x \equiv A_c x. \tag{4.1.5}$$

Output Feedback LQR Design Equations

We shall now derive the design equations in Table 4.1-1. These equations may be solved for the optimal output-feedback gain K that minimizes the PI J, as illustrated in the examples.

Derivation

The PI may be expressed in terms of K as

$$J = \frac{1}{2} \int_0^\infty x^T(Q + C^T K^T RKC)x \, dt. \tag{4.1.6}$$

The design problem is now to select the gain K so that J is minimized subject to the dynamical constraint (4.1.5).

This *dynamical* optimization problem may be converted into an equivalent *static* one that is easier to solve as follows. Suppose we can find a constant, symmetric, positive-semidefinite matrix P so that

$$\frac{d}{dt}(x^T P x) = -x^T(Q + C^T K^T RKC)x. \tag{4.1.7}$$

Then, J may be written as

$$J = \frac{1}{2} x^T(0) P x(0) - \frac{1}{2} \lim_{t \to \infty} x^T(t) P x(t). \tag{4.1.8}$$

Assuming that the closed-loop system is asymptotically stable so that $x(t)$ vanishes with time, this becomes

$$J = \frac{1}{2} x^T(0) P x(0). \tag{4.1.9}$$

If P satisfies (4.1.7), then we may use (4.1.5) to see that

$$\begin{aligned} -x^T(Q + C^T K^T RKC)x &= \frac{d}{dt}(x^T P x) = \dot{x}^T P x + x^T P \dot{x} \\ &= x^T(A_c^T P + P A_c)x. \end{aligned} \tag{4.1.10}$$

Since this must hold for all initial conditions, and hence for all state trajectories $x(t)$, we may write

$$g \equiv A_c^T P + P A_c + C^T K^T R K C + Q = 0. \tag{4.1.11}$$

If K and Q are given and P is to be solved for, then this is called a *Lyapunov* equation. (A Lyapunov equation is a symmetric linear matrix equation. Note that the equation does not change if its transpose is taken.)

In summary, for any fixed feedback matrix K, if there exists a constant, symmetric, positive-semidefinite matrix P that satisfies (4.1.11), then the control cost J for the closed-loop system is given in terms of P by (4.1.9). Note that, to find this cost, only the initial condition $x(0)$ is needed.

By using the trace identity

$$\text{tr}(AB) = \text{tr}(BA) \tag{4.1.12}$$

for any compatibly dimensioned matrices A and B (with the trace of a matrix the sum of its diagonal elements) we may write (4.1.9) as

$$J = \frac{1}{2} \text{tr}(PX) \tag{4.1.13}$$

where the $n \times n$ symmetric matrix X is defined by

$$X \equiv x(0) x^T(0) \tag{4.1.14}$$

It is now clear that the problem of selecting K to minimize (4.1.6) subject to the dynamical constraint (4.1.5) on the states is equivalent to the *algebraic* problem of selecting K to minimize (4.1.13) subject to the constraint (4.1.11) on the auxiliary matrix P.

To solve this modified problem, we use the Lagrange multiplier approach [Lewis 1986] to modify the problem yet again. Thus, adjoin the constraint to the PI by defining the Hamiltonian

$$H = \text{tr}(PX) + \text{tr}(gS) \tag{4.1.15}$$

with S a symmetric $n \times n$ matrix of Lagrange multipliers which still needs to be determined. Then, our constrained optimization problem is equivalent to the simpler problem of minimizing (4.1.15) without constraints. To accomplish this, we need only set the partial derivatives of H with respect to all the independent variables P, S, and K equal to zero. Using the facts that for any compatibly dimensioned matrices A, B, and C and any scalar y

$$\frac{\partial}{\partial B} \text{tr}(ABC) = A^T C^T \tag{4.1.16}$$

4.1 Linear Quadratic Regulator with Output Feedback

and

$$\frac{\partial y}{\partial B^T} = \left[\frac{\partial y}{\partial B}\right]^T, \qquad (4.1.17)$$

the necessary conditions for the solution of the LQR problem with output feedback are easily found to be the design equations given in Table 4.1-1 (see the problems).

Discussion

To obtain the output-feedback gain K minimizing the PI (4.1.4), we need to solve the three coupled design equations in Table 4.1-1. The first two of these are Lyapunov equations and the third is an equation for the gain K. If R is positive definite and CSC^T is nonsingular, then (4.1.23) may be solved for K to obtain

$$K = R^{-1}B^T PSC^T(CSC^T)^{-1}. \qquad (4.1.18)$$

The output-feedback LQR design procedure is quite strange, for to find K we must determine along the way the values of two auxiliary and apparently unnecessary $n \times n$ matrices, P and S. These auxiliary quantities may, however, not be as unnecessary as it appears, for note that the optimal cost may be determined directly from P and the initial state by using (4.1.9).

This is an important result in that the $n \times n$ auxiliary matrix P is independent of the state. Given a feedback matrix K, the auxiliary matrix P may be computed from the Lyapunov equation (4.1.21). Then, only the initial condition $x(0)$ is required to compute the closed-loop cost under the influence of the feedback control. That is, we may compute the cost of applying the feedback control *before we actually apply it*. If it turns out to be too large, we can just reselect the weighting matrices Q and R and try another design.

Unfortunately, the dependence of X in (4.1.14) on the initial state $x(0)$ is undesirable, since it makes the optimal gain dependent on the initial state through equation (4.1.22). This dependence is typical of output-feedback design; recall that in the case of state feedback it does not occur (see Table 3.2-2).

Since, in many applications $x(0)$ may not be known, it is usual [Levine and Athans 1970] to sidestep this problem by minimizing not the PI (4.1.4) but its *expected value*, that is, $E\{J\}$. Then, (4.1.9) and (4.1.14) are replaced by

$$E\{J\} = \frac{1}{2} E\{x^T(0)Px(0)\} = \frac{1}{2} \operatorname{tr}(PX), \qquad (4.1.19)$$

where the symmetric $n \times n$ matrix

$$X \equiv E\{x(0)x^T(0)\} \qquad (4.1.20)$$

is the initial autocorrelation of the state. This is the formulation given in Table 4.1-1.

Thus, it is usual to assume that nothing is known of $x(0)$ except its initial autocorrelation X. In fact, the usual assumption is that the initial states are uniformly distributed

TABLE 4.1-1 LQR WITH OUTPUT FEEDBACK

System Model:

$$\dot{x} = Ax + Bu$$
$$y = Cx$$

Feedback Control:

$$u = -Ky$$

Performance Index:

$$J = \frac{1}{2} E \left[\int_0^\infty (x^T Q x + u^T R u) \, dt \right]$$

Design Equations for the Optimal Output Feedback Gain:

$$0 = \frac{\partial H}{\partial S} = A_c^T P + P A_c + C^T K^T R K C + Q \quad (4.1.21)$$

$$0 = \frac{\partial H}{\partial P} = A_c S + S A_c^T + X \quad (4.1.22)$$

$$0 = \frac{1}{2} \frac{\partial H}{\partial K} = RKCSC^T - B^T PSC^T. \quad (4.1.23)$$

where

$$A_c = A - BKC, \quad X = E\{x(0) x^T(0)\}$$

Optimal Cost:

$$J = \frac{1}{2} \operatorname{tr}(PX) \quad (4.1.24)$$

on the unit sphere; then $X = I$, the identity. This is a sensible assumption for the regulator problem, where we are trying to drive arbitrary nonzero initial states to zero.

Closed-Loop Stability

We saw in section 3.3 that, in the case of full state feedback, the closed-loop plant is guaranteed to be stable if (A, B) is reachable and $(\sqrt{Q}, A)$ is observable. This is one of the fundamental results of modern control, for the algebraic Riccati equation (ARE) leads to a numerically efficient way to find the Kalman gain, thus offering a way to stabilize any multivariable plant. In view of the difficulty of stabilizing multi-input/multi-output systems using classical one-loop-at-a-time procedures, this contribution is significant.

Unfortunately, in the case of output feedback, there are not such firm results. It is, however, generally the case that if:

1. (A, B, C) is stabilizable by output feedback: that is, if there does exist a K such that $(A - BKC)$ is stable,
2. $(\sqrt{Q}, A)$ is observable,

then the LQR with output feedback stabilizes the plant.

It is interesting to note that the condition of output stabilizability is stronger than either reachability or observability. Note that

$$A_c = A - BKC = A - BF$$

with $F = KC$. Thus, output feedback is a form of state feedback with, however, a state-feedback matrix F of a restricted form. Namely, F must be a linear combination of the rows of C. Clearly, placing the poles with such a restricted feedback is more difficult than placing the poles with a general F, which latter is equivalent to the reachability of (A, B).

On the other hand

$$A_c = A - BKC = A - LC$$

with $L = BK$. Thus, output feedback is a form of output injection with, however, the output-injection matrix L required to be a linear combination of the columns of B. Clearly, placing the poles with such a restricted output injection is more difficult than placing the poles with a general L, which latter is equivalent to the observability of (C, A).

To make matters worse, there is no convenient test for output stabilizability. The closest thing may be provided by Kwon and Youn [1987], where there are some results in terms of the solutions to two Lyapunov equations. It has been shown in Syrmos [1991] that an alternative test for output stabilizability is provided in terms of the solution to a *bilinear* generalized Lyapunov equation. Thus, while the state-feedback problem is linear, the output-feedback problem is bilinear.

For design purposes, things are not so grim, for it is generally found that as long as the above two conditions hold, the closed-loop system is stable using the gain provided by Table 4.1-1. What this means is that, generally speaking, for any admissible choices for the PI weighting matrices Q and R, closed-loop stability is guaranteed. Thus, Q and R may be treated as design parameters which are interactively varied until a suitable control gain K is obtained. That is, Q and R are chosen and the equations of Table 4.1-1 are solved for K. Then, the step response and robustness (see Chapter 8) are examined. If they are not suitable, Q and R are varied and the process is repeated.

We should contrast this with a naive approach based on classical techniques, which might involve directly varying the elements of K in an attempt to obtain suitable closed-loop performance. In this approach, closed-loop stability is not even guaranteed for all values of K, so that an acceptable design would be far more difficult to achieve. Hence the need for time-consuming one-loop-at-a-time design using root locus, and so on, with any classical approach.

Thus, the design problem of tuning the entries of K for good performance, where closed-loop stability may not even hold, has been replaced by the problem of tuning the entries of Q and R, where closed-loop stability is usually guaranteed and need not be worried about.

In the remainder of this chapter we shall give further insight on selecting Q and R in a reasonable and intuitive fashion. Indeed, this is where the engineering design aspect enters into modern LQ control theory, for once Q and R have been selected, the feedback gain K is uniquely determined.

Deriving the LQR Using Full State Feedback

It is not difficult to derive the steady-state design equations for full state feedback using the output-feedback design equations. Indeed, if $C = I$, and S is nonsingular, then according to (4.1.23)

$$K = R^{-1}B^T P \qquad (4.1.25)$$

is the Kalman gain, so that matrix S is not needed to determine the optimal control.

Using $C = I$ in the Lyapunov equation (4.1.21) yields

$$0 = (A - BK)^T P + P(A - BK) + Q + K^T R K$$

whence (4.1.25) results in

$$0 = A^T P + PA + Q - PBR^{-1}B^T P \qquad (4.1.26)$$

which is nothing but the ARE from section 3.3 (with a slight change in notation).

Since S is not needed to determine the Kalman gain, it is not necessary to solve (4.1.22). Therefore, the initial state autocorrelation X need not be known in the case of full state feedback.

Solution of the Design Equations

The design equations for P, S, and K in Table 4.1-1 are coupled nonlinear matrix equations in three unknowns. There are several basic approaches for solving these design equations for the optimal output-feedback gain K. They may all be implemented using available software routines on a digital computer.

Thus, the selection of suitable output-feedback gains K is an interactive computer-aided design procedure where the design parameters Q and R are selected, then K is determined using software. Once the step response and other closed-loop properties have been examined, Q and R may be varied and the design repeated to improve them if necessary.

Given good software and a reasonable technique for selecting Q and R, this procedure is a direct and quick one. Note especially that closed-loop stability is generally guaranteed for any admissible Q and R. In the next sections of this chapter we shall discuss sensible approaches for selecting Q and R.

Let us enumerate some techniques for solving for K. Then, in subsequent discus-

4.1 Linear Quadratic Regulator with Output Feedback

sions, we shall assume that some software routine for determining the LQR output-feedback gains is available. Such software is discussed in Appendix A.

The solution techniques we shall describe are iterative numerical ones that vary K based on changes in J. It should be emphasized that, in the output-feedback problem, there may be more than one *local minimum* of the PI. This means that the optimal value determined for the gain K may depend on the initial guess K_0 selected for the minimization routine.

Numerical optimization routines

There are many numerical routines for optimizing the PI J by varying K. Good success has been achieved using the Simplex routine in Press et al. [1986].

The best way to apply these routines is to use (4.1.21) to determine P for a given value of K, and then (4.1.24) to find the value of J. Given this value, the Simplex algorithm will vary K to minimize J. Thus, the remaining design equations are not needed using this approach.

If $(Q + C^T K^T RKC) > 0$, then the Lyapunov equation (4.1.21) has a unique positive definite solution P if and only if $A - BKC$ is asymptotically stable [Kailath 1980]. Thus, the software should check for this condition and disallow any K that destabilizes the plant.

To solve the Lyapunov equation, one may use standard software (e.g., subroutine $A^T XPXA$ in ORACLS [Armstrong 1980], MATLAB [Moler et al. 1987]).

A major problem is the determination of an initial gain K_0 that stabilizes the plant to start the search procedure. One technique for doing this is given in Halyo and Broussard [1981].

Gradient-based routines

Numerical routines that use the gradient $\partial J/\partial K$ as additional information require fewer iterations than the schemes just mentioned.

The gradient may be determined from Table 4.1-1 as follows. Given a value for K, solve the two Lyapunov equations for P and S. Again, stability of $A - BKC$ will be required. Then, use these values of P and S in (4.1.23). Under these conditions, we have [see (4.1.15)]

$$H = \operatorname{tr}(PX) + \operatorname{tr}(gS) = \operatorname{tr}(PX) = J,$$

since $g = 0$ [see (4.1.11)]. Then

$$\frac{\partial H}{\partial K} = \frac{\partial J}{\partial K} \tag{4.1.27}$$

so that (4.1.23) yields the gradient required by the search routine. Note that in this case all the equations of Table 4.1-1 are used.

Good results were obtained using the Davidon-Fletcher-Powell routine [Press et al. 1986].

Specialized Routines

An efficient iterative solution algorithm specifically for the output-feedback problem was presented in Moerder and Calise [1985]. It is given in Table 4.1-2. Software to implement this algorithm may be based on the ORACLS subroutine ATXPXA [Armstrong 1980] which solves the Lyapunov equation.

A few words of discussion on the algorithm in Table 4.1-2 are in order. First, it was shown in Moerder and Calise [1985] that the algorithm converges to a local minimum for J if the following conditions hold.

CONDITIONS FOR CONVERGENCE OF THE LQ SOLUTION ALGORITHM:

1. There exists a gain K such that A_c is stable; that is, (A, B, C) is output stabilizable.
2. The output matrix C has full row rank p.
3. Control weighting matrix R is positive definite. This means that all the control inputs should be weighted in the PI.
4. Q is positive semidefinite and $(\sqrt{Q}, A)$ is *detectable* (Section 2.5).

TABLE 4.1-2 OPTIMAL OUTPUT-FEEDBACK SOLUTION ALGORITHM

1. **Initialize:**
   Set $k = 0$
   Determine a gain K_0 so that $A - BK_0C$ is asymptotically stable
2. **k-th iteration:**
   Set $A_k = A - BK_kC$
   Solve for P_k and S_k in

   $$0 = A_k^T P_k + P_k A_k + C^T K_k^T R K_k C + Q$$

   $$0 = A_k S_k + S_k A_k^T + X$$

   Set $J_k = \text{tr}(P_k X)$.
   Evaluate the gain update direction

   $$\Delta K = R^{-1} B^T P S C^T (CSC^T)^{-1} - K_k$$

   Update the gain by

   $$K_{k+1} = K_k + \alpha \Delta K$$

   where α is chosen so that

   $$A - BK_{k+1}C \text{ is asymptotically stable}$$

   $$J_{k+1} \equiv \text{tr}(P_{k+1} X) \leq J_k$$

   If $J_k + 1$ and J_k are close enough to each other, go to 3.
   Otherwise, set $k = k + 1$ and go to 2.
3. **Terminate:**
   Set $K = K_k+1$, $J = J_k+1$.
   Stop.

4.1 Linear Quadratic Regulator with Output Feedback

The detectability condition basically means that Q should be chosen so that all unstable states are weighted in the PI. Then, if J is bounded so that $\sqrt{Q}x(t)$ vanishes for large t, the unstable states will be forced to zero through the action of the control. The stable states, of course, will automatically go to zero. A stronger condition than detectability is observability, which is easier to check than detectability. The observability of $(\sqrt{Q}, A)$ means basically that *all* states are weighted in the PI.

The issue of selecting an initial stabilizing output feedback gain K_0 remains in the algorithm of Table 4.1-2. At this point it is worth discussing an example to illustrate these notions.

Example 4.1-1: Multivariable Regulator Using Output Feedback

In Example 2.5-2 we designed a decoupling controller for the circuit in Fig. 4.1-1. Assuming values of 1 for all components yields the state equations

$$\dot{x} = \begin{bmatrix} -1 & 0 & 1 & 0 \\ 0 & -1 & 0 & -1 \\ -1 & 0 & -1 & -1 \\ 0 & 1 & -1 & -1 \end{bmatrix} x + \begin{bmatrix} 0 & 0 \\ 0 & 0 \\ 1 & 0 \\ 0 & 1 \end{bmatrix} u. \tag{1}$$

$$y = \begin{bmatrix} 1 & 1 & 0 & 0 \\ 0 & 1 & 0 & 0 \end{bmatrix} x. \tag{2}$$

The poles of this system are at

$$\begin{aligned} s &= -0.5 \pm j0.866, \\ & -1.5 \pm j0.866 \end{aligned} \tag{3}$$

and the open-loop output response to initial conditions of $x_1 = x_2 = x_3 = x_4 = 1$ is shown in Fig. 4.1-2. The output takes about 8 sec. to die to zero.

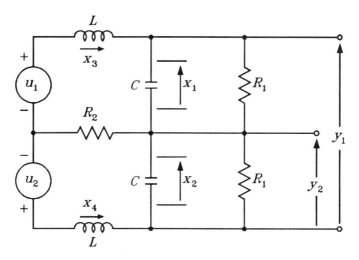

Figure 4.1-1 Two-input/two-output circuit

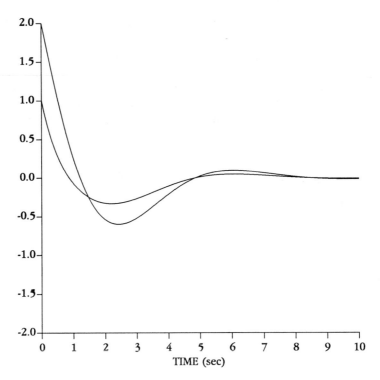

Figure 4.1-2 Open-loop output response

We should like to increase the speed of this circuit to achieve better regulation of the output about zero. For this purpose we may use the output feedback

$$u = -Ky. \tag{4}$$

This is a two-input/two-output feedback design problem with four gains to be selected. We could vary each gain separately and use classical root locus techniques. However, this is time consuming.

As a more direct approach, let us select the PI

$$J = \frac{1}{2} \int_0^\infty (y^T y + r u^T u) \, dt, \tag{5}$$

with $r > 0$. The optimal gain K that minimizes $E\{J\}$ is found by solving the design equations in Table 4.1-1 with $R = rI$ and, since $y^T y = x^T H^T H x$,

$$Q = H^T H = \begin{bmatrix} 1 & 1 & 0 & 0 \\ 1 & 2 & 0 & 0 \\ 0 & 0 & 0 & 0 \\ 0 & 0 & 0 & 0 \end{bmatrix}. \tag{6}$$

4.1 Linear Quadratic Regulator with Output Feedback

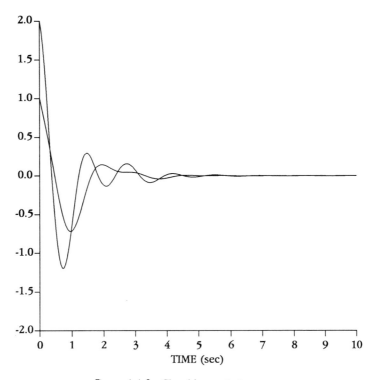

Figure 4.1-3 Closed-loop output response

Using the software described in Appendix A, we found K and plotted the outputs for various choices of r. Suitable responses were obtained with $r = 1/1000$; they are shown in Fig. 4.1-3. Note that the output has died to zero for all practical purposes by 5 sec. The closed-loop poles are at

$$s = -0.88 \pm j4.58, \\ -1.12 \pm j2.24 \tag{7}$$

and the feedback gain matrix is

$$K = \begin{bmatrix} 21.71 & -16.20 \\ 1.687 & -4.906 \end{bmatrix}. \tag{8}$$

We should like to emphasize a few points. First, the naive approach of attempting to vary all four entries of K directly does not even guarantee closed-loop stability, as well as having the disadvantage of four design parameters. In the LQ approach, there was only one design parameter, namely r. The software that solves for K is fast and easy to use, so that after a few selections of r (namely, $r = 1$, $r = 10$, $r = 100$), an appropriate value was found. The entire process took 10 minutes.

Since (1), (2) is observable and reachable, closed-loop stability can be practically assumed for all $r > 0$. Improved response in this circuit may easily be obtained using a dynamic compensator, as discussed in the next section. ∎

4.2 TRACKING A REFERENCE INPUT

We have seen how to design a linear quadratic feedback regulator that drives the state to zero. In many practical applications, however, we are interested in making an output follow or *track* a *reference input* signal. This is called the *servo* design problem. If the reference input is constant, it is called the *nonzero setpoint* problem. An example is a speed control unit on an automobile, which tries to keep the speed at a nonzero constant value.

The solution of the general LQ tracker problem is not straightforward, even using full state feedback [Athans and Falb 1966, Kwakernaak and Sivan 1972, Lewis 1986]. Moreover, the optimal solution is not causal, but contains a feedforward term that is generated by a backward differential equation.

In this section we shall consider approaches to the tracker problem that are more suitable for our purposes. To derive results that are relevant for practical problems, we shall use output feedback and not state feedback. This allows us to design controllers with any desired structure, as we shall soon see.

The design approach in this section might be said to be the basic approach used in the book. It is a general technique that applies to a wide variety of situations due to the general form of the PI, and good software to implement Table 4.2-1 (see Appendix A) makes it easy to apply in a computer-aided design environment.

Compensators of Known Structure

In many controls design applications there is some experience and knowledge that dictates what sort of compensator dynamics yield good performance. For example, it may be necessary to augment some feedforward channels with integrators to obtain a steady-state error of exactly zero. Again, low-pass filters may be required to filter measurements that are noisy.

A dynamic compensator of prescribed structure may be incorporated into the system description as follows [Davison and Ferguson 1981].

Consider the situation in Fig. 4.2-1 where the plant is described by

$$\dot{x} = Ax + Bu \quad (4.2.1)$$

$$y = Cx \quad (4.2.2)$$

with state $x(t)$, control input $u(t)$, and $y(t)$ the *measured output* available for feedback purposes. In addition

$$z = Hx \quad (4.2.3)$$

is a *performance output*, which must track the given *reference input* $r(t)$. It is not generally equal to $y(t)$.

The dynamic compensator has the form

$$\begin{aligned} \dot{w} &= Fw + Ge \\ v &= Dw + Je \end{aligned} \quad (4.2.4)$$

4.2 Tracking a Reference Input

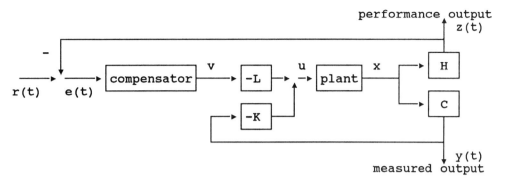

Figure 4.2-1 Plant with compensator of desired structure

with state $w(t)$, output $v(t)$, and input equal to the *tracking error*

$$e(t) = r(t) - z(t). \quad (4.2.5)$$

F, G, D, and J are known matrices chosen to include the desired structure in the compensator.

The allowed form for the plant control input is

$$u = -Ky - Lv, \quad (4.2.6)$$

where the constant gain matrices K and L are to be chosen in the controls design step to result in satisfactory tracking of $r(t)$. This formulation allows for both *feedback and feedforward* compensator dynamics.

These dynamics and output equations may be written in augmented form as

$$\frac{d}{dt}\begin{bmatrix} x \\ w \end{bmatrix} = \begin{bmatrix} A & 0 \\ -GH & F \end{bmatrix}\begin{bmatrix} x \\ w \end{bmatrix} + \begin{bmatrix} B \\ 0 \end{bmatrix}u + \begin{bmatrix} 0 \\ G \end{bmatrix}r \quad (4.2.7)$$

$$\begin{bmatrix} y \\ v \end{bmatrix} = \begin{bmatrix} C & 0 \\ -JH & D \end{bmatrix}\begin{bmatrix} x \\ w \end{bmatrix} + \begin{bmatrix} 0 \\ J \end{bmatrix}r \quad (4.2.8)$$

$$z = [H \quad 0]\begin{bmatrix} x \\ w \end{bmatrix} \quad (4.2.9)$$

and the control input may be expressed as

$$u = -[K \quad L]\begin{bmatrix} y \\ v \end{bmatrix}. \quad (4.2.10)$$

Note that, in terms of the augmented plant/compensator state description, the admissible controls are represented as a *constant* output feedback $[K \quad L]$. In the augmented description, all matrices are known except the gains K and L, which need to be selected to yield acceptable closed-loop performance.

Linear Quadratic Servo Design Using Output Feedback

By redefining the state, the measured output, and the matrix variables to streamline the notation, we see that the augmented equations (4.2.7)–(4.2.9) that contain both the plant and the compensator are of the form

$$\dot{x} = Ax + Bu + Er \qquad (4.2.11)$$

$$y = Cx + Fr \qquad (4.2.12)$$

$$z = Hx, \qquad (4.2.13)$$

with $z(t)$ the performance output which is required to track the reference input $r(t)$ and $y(t)$ additional measured outputs.

In this description, let us take the state $x(t) \in \mathbf{R}^n$, control input $u(t) \in \mathbf{R}^m$, reference input $r(t) \in \mathbf{R}^q$, $y(t) \in \mathbf{R}^p$, and $z(t) \in \mathbf{R}^q$. In terms of the new variables, the admissible controls (4.2.10) are proportional output feedbacks of the form

$$u = -Ky = -KCx - KFr \qquad (4.2.14)$$

with constant gain K to be determined. See Fig. 4.2-2.

Using these equations, the closed-loop system is found to be

$$\begin{aligned}\dot{x} &= (A - BKC)x + (E - BKF)r \\ &\equiv A_c x + B_c r.\end{aligned} \qquad (4.2.15)$$

Our formulation is taken from Stevens, Lewis, and AL-Sunni [1991] and differs sharply from the traditional formulations of the optimal tracker problem [Kwakernaak and Sivan 1972, Lewis 1986]. Note that (4.2.14) includes both feedback and *feedforward* terms, so that both the closed-loop poles *and zeros* may be affected by varying the gain K. Thus, we should expect better success in shaping the step response than by placing only the poles. Note also that, in contrast to Kwakernaak and Sivan [1972], we have specified the allowed form (4.2.6) for the control at the outset. Thus, our controller will have a structure that makes sense in terms of the classical intuition of the controls designer.

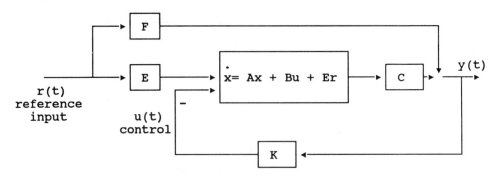

Figure 4.2-2 Augmented plant/feedback structure

4.2 Tracking a Reference Input

Since the performance specifications of many control systems are given in terms of time-domain criteria, and these criteria are closely related to the step response, we shall assume henceforth that the reference input $r(t)$ is a step command with magnitude r_0.

It should be clearly understood that the assumption of a constant $r(t)$ is for optimal design purposes only. The resulting tracker will work for any arbitrary reference command $r(t)$.

Deviation system

Let us denote steady-state values by overbars and deviations from the steady-state values by tildes. Then, the state, output, and control deviations are given by

$$\tilde{x}(t) = x(t) - \bar{x} \qquad (4.2.16)$$

$$\tilde{y}(t) = y(t) - \bar{y} = K\tilde{x} \qquad (4.2.17)$$

$$\tilde{z}(t) = z(t) - \bar{z} = H\tilde{x} \qquad (4.2.18)$$

$$\tilde{u}(t) = u(t) - \bar{u} = -KCx - KFr_0 - (-KC\bar{x} - KFr_0) = -KC\tilde{x}(t)$$

or

$$\tilde{u} = -K\tilde{y}. \qquad (4.2.19)$$

The tracking error $e(t) = r(t) - z(t)$ is given by

$$e(t) = \tilde{e}(t) + \bar{e} \qquad (4.2.20)$$

with $\bar{e}$ the steady-state error and the transient error given by

$$\tilde{e}(t) = e(t) - \bar{e} = (r_0 - Hx) - (r_0 - H\bar{x}) = -H\tilde{x}$$

or

$$\tilde{e} = -\tilde{z}. \qquad (4.2.21)$$

Since in any acceptable design the closed-loop plant will be asymptotically stable, A_c is nonsingular. According to (4.2.15), at steady state

$$0 = A_c\bar{x} + B_c r_0, \qquad (4.2.22)$$

so that the steady-state state response $\bar{x}$ is

$$\bar{x} = -A_c^{-1}B_c r_0 \qquad (4.2.23)$$

and

$$\bar{e} = r_0 - H\bar{x} = (I + HA_c^{-1}B_c)r_0. \qquad (4.2.24)$$

Using (4.2.16), (4.2.19), and (4.2.23) in (4.2.15), the dynamics of the state deviation are seen to be

$$\dot{\tilde{x}} = A_c\tilde{x}. \qquad (4.2.25)$$

$$\tilde{y} = C\tilde{x} \tag{4.2.26}$$

$$\tilde{z} = H\tilde{x} = -\tilde{e} \tag{4.2.27}$$

and the control input to this *deviation system* is

$$\tilde{u} = -K\tilde{y}. \tag{4.2.28}$$

Thus, the step-response shaping problem has been converted to a *regulator problem* for the deviation system.

Again, we emphasize the difference between our approach and traditional ones (e.g., Kwakernaak and Sivan 1972). Once the gain K in (4.2.28) has been found, the control for the plant is given by (4.2.14), which inherently has both feedback and feedforward terms. Thus, no extra feedforward term need be added.

Note that we do not assume a Type I system, which would force $\bar{e}$ to be equal to zero. This can be important in some applications, where it may not be desirable to force the system to be of Type I by augmenting all control channels with integrators. This augmentation complicates the servo structure. Moreover, it is well known from classical control theory that suitable step responses may often be obtained without resorting to inserting integrators in all the feedforward channels.

Performance index

To make the tracking error $e(t)$ small, we propose to attack two equivalent problems: the problem of regulating the error deviation $\tilde{e}(t) = -\tilde{z}(t)$ to zero, and the problem of making small the steady-state error $\bar{e}$.

To make small both the error deviation $\tilde{e}(t) = -H\tilde{x}(t)$ and the steady-state error $\bar{e}$, we propose selecting K to minimize the performance index (PI)

$$J = \frac{1}{2}\int_0^\infty (t^k \tilde{e}^T \tilde{e} + \tilde{u}^T R \tilde{u})\, dt + \frac{1}{2}\bar{e}^T V \bar{e} + \frac{1}{2}\sum_i \sum_j g_{ij} k_{ij}^2, \tag{4.2.29}$$

with $R > 0$, $V \geq 0$, $g_{ij} \geq 0$ design parameters. The integrand is the standard quadratic PI with, however, time weighting t^k of the error deviation, V is a weighting on the steady-state error, and g_{ij} is a weight on element k_{ij} of the gain matrix K.

In some applications we should like some of the elements of the gain K to be zero so that the controller has more structure (see Example 4.2-1). To make element k_{ij} small, we may select a large value (e.g., 1000) for the corresponding weighting g_{ij} [Moerder and Calise 1985].

The time-varying weighting t^k in the PI places a heavy penalty on errors that occur late in the response, and is thus very effective in suppressing the effect of a slow pole, as well as in eliminating lightly damped settling behavior. This is well known from classical control theory, where the idea was used in ITAE and ISTSE PIs [D'Azzo and Houpis 1988].

It is known that, if time-varying weighting is used in the PI, the optimal feedback gains are time-varying [Fortin and Parkins 1972]. However, here we assume that K is time-invariant and so obtain a suboptimal solution which is more suitable for our purposes. This has been the common approach in the literature [Abdel-Moneim and Sorial 1982, Kwon and Youn 1986, Subbayyan and Vaithilingam 1977].

To make $\bar{e}$ smaller, V may be selected larger. If the system is of Type I, which may be ensured by adding integrators in all feedforward paths, then V may be set to zero since $\bar{e}$ is automatically zero [D'Azzo and Houpis 1988]. However, even if this is not the case, our approach will allow selection of K to make $\bar{e}$ *small enough* (see Example 4.2-3).

If the system is of Type 0, then it is known from classical control theory that to make the steady-state error $\bar{e}$ exactly zero it is generally necessary to increase the feedback gains to infinity. Thus, often as V is increased, the elements of K will increase without bound. This effect may be counteracted by selecting larger values of g_{ij}. What this amounts to is a design technique that allows trade-offs between small steady-state errors and reasonable values of the gain K by selecting V and g_{ij}.

Making small the error deviation $\tilde{e}(t)$ improves the transient response, while making small the steady-state error $\bar{e}(t)$ improves the steady-state response. These effects involve a trade-off, so that if the system is of Type 0 there is also a design trade-off between selecting large values of k and large values of V (see Example 4.2-3).

Note that the PI weights the state and control *deviations*, and not the states and controls themselves.

We can generally select $R = rI$ and $V = vI$, with r and v scalars. This simplifies the design since now only a few parameters must be tuned during the interactive design process.

We should point out that the proposed approach is suboptimal in the sense that minimizing the PI does not necessarily minimize a quadratic function of the total error $e(t) = \bar{e} + \tilde{e}(t)$.

Using (4.2.27) and (4.2.28), the PI may be written as

$$J = \frac{1}{2} \int_0^\infty (t^k \tilde{x}^T Q \tilde{x} + \tilde{x}^T C^T K^T R K C \tilde{x})\, dt + \frac{1}{2} \bar{e}^T V \bar{e} + \frac{1}{2} \sum_i \sum_j g_{ij} k_{ij}^2, \quad (4.2.30)$$

with $Q = H^T H$.

An important advantage of the time weighting t^k should be clearly understood. Namely, if $k > 1$ it is not necessary for (H, A) to be observable to obtain good designs with closed-loop stability. This overcomes one of the prime disadvantages of design using the standard LQR. We shall mention this again subsequently.

Derivation

We shall now derive the design equations in Table 4.2-1 for the determination of the control gain K that minimizes the PI.

As is well known [MacFarlane 1963] (see the problems), for any fixed value of K, the value of the PI is given by

$$J = \frac{1}{2} \tilde{x}^T(0) P_k \tilde{x}(0) + \frac{1}{2} \bar{e}^T V \bar{e} + \frac{1}{2} \sum \sum g_{ij} k_{ij}^2, \quad (4.2.31)$$

with $P_k \geq 0$ the solution to the nested Lyapunov equation set

$$0 = f_0 \equiv A_c^T P_0 + P_0 A_c + Q$$
$$0 = f_1 \equiv A_c^T P_1 + P_1 A_c + P_0$$
$$\vdots \tag{4.2.32}$$
$$0 = f_{k-1} \equiv A_c^T P_{k-1} + P_{k-1} A_c + P_{k-2}$$
$$0 = f_k \equiv A_c^T P_k + P_k A_c + k! P_{k-1} + C^T K^T R K C.$$

and $\bar{e}$ given by (4.2.24).

We saw in section 4.1 that in the case of output feedback the optimal gains depend on the initial condition. In the case of the LQ regulator it is traditional to assume that the initial conditions are uniformly distributed on a surface with known characteristics [Levine and Athans 1970], that is, that $E\{\tilde{x}(0)\tilde{x}^T(0)\}$ is known, with $E\{\ \}$ the expected value. Then, $E\{J\}$ is minimized instead of J.

While this may be a satisfactory assumption for the regulator problem, it is unsatisfactory for the tracker problem. In the latter situation the system starts at rest and must achieve a given final state that is dependent on the reference input, namely (4.2.23). To find the correct value of $\tilde{x}(0)$, we note that, since the plant starts at rest (i.e., $x(0) = 0$), according to (4.2.16)

$$\tilde{x}(0) = -\bar{x}, \tag{4.2.33}$$

so that the optimal cost (4.2.31) becomes

$$J = \frac{1}{2}\operatorname{tr}(P_k X) + \frac{1}{2}\bar{e}^T V \bar{e} + \frac{1}{2}\sum\sum g_{ij}k_{ij}^2 \tag{4.2.34}$$

with tr( ) the trace, P_k given by (4.2.32), $\bar{e}$ given by (4.2.24), and

$$X \equiv \bar{x}\bar{x}^T = A_c^{-1} B_c r_0 r_0^T B_c^T A_c^{-T}, \tag{4.2.35}$$

with $A_c^{-T} \equiv (A_c^{-1})^T$.

The optimal solution to the unit-step tracking problem, with (4.2.11) initially at rest, may now be determined by minimizing J in (4.2.34) over the gains K, subject to the constraints (4.2.32) and the value (4.2.24) for $\bar{e}$. This may be achieved using an algorithm like SIMPLEX in Press et al. [1986]. Alternatively, K can be efficiently found using a gradient-based algorithm like Davidon-Fletcher-Powell [Press et al. 1986] if we can determine an expression for the gradient of J with respect to K.

Such an expression for the gradient may be found by converting the minimization problem to an equivalent one using Lagrange multipliers. Thus, define the Hamiltonian

$$H = \frac{1}{2}\operatorname{tr}(P_k X) + \operatorname{tr}(f_0 S_0) + \ldots + \operatorname{tr}(f_k S_k)$$
$$+ \frac{1}{2}\bar{e}^T V \bar{e} + \frac{1}{2}\sum\sum g_{ij}k_{ij}^2, \tag{4.2.36}$$

4.2 Tracking a Reference Input

TABLE 4.2-1 LQ TRACKER WITH OUTPUT FEEDBACK

Model of System Plus Controller:

$$\dot{x} = Ax + Bu + Er$$
$$y = Cx + Fr$$
$$z = Hx$$

Tracking Error:

$$e = r - z$$

Control Input:

$$u = -Ky$$

Performance Index:

$$J = \frac{1}{2}\int_0^\infty (t^k \tilde{e}^T \tilde{e} + \tilde{u}^T R \tilde{u}) \, dt + \frac{1}{2}\bar{e}^T V \bar{e} + \frac{1}{2}\sum_i \sum_j g_{ij} k_{ij}^2$$

where: $\tilde{e}(t)$ = error deviation from steady state

$\bar{e}$ = $(I + HAc^{-1}B_c)r_0$ = steady-state error

$\tilde{u}(t)$ = control deviation from steady state

k_{ij} are the elements of K

Design Equations for the LQ Tracker:
Lyapunov Equation Set:

$$0 = A_c^T P_0 + P_0 A_c + Q$$
$$0 = A_c^T P_1 + P_1 A_c + P_0$$
$$\vdots$$
$$0 = A_c^T P_{k-1} + P_{k-1} A_c + P_{k-2}$$
$$0 = A_c^T P_k + P_k A_c + k!P_{k-1} + C^T K^T RKC$$

(4.2.40)

Lyapunov Equation Set:

$$0 = A_c S_k + S_k A_c^T + X$$
$$0 = A_c S_{k-1} + S_{k-1} A_c^T + k!S_k$$
$$0 = A_c S_{k-2} + S_{k-2} A_c^T + S_{k-1}$$
$$\vdots$$
$$0 = A_c S_0 + S_0 A_c^T + S_1$$

(4.2.41)

Gain Equation:

$$0 = \frac{1}{2}\frac{\partial H}{\partial K} = RKCS_k C^T - B^T(P_0 S_0 + \ldots + P_k S_k)C^T + B^T A_c^{-T}(P_k + H^T VH)\bar{x}\,\bar{y}^T$$
$$- B^T A_c^{-T} H^T V r_0 \bar{y}^T + g^* K$$

(4.2.42)

(continued)

TABLE 4.2-1 (continued)

where: $g*K$ is the matrix with elements $g_{ij}k_{ij}$

$r(t)$ is a unit step of magnitude r_0

$\bar{x} = -A_c^{-1}B_c r_0$

$\bar{y} = C\bar{x} + Fr_0$

$X = \bar{x}\bar{x}^T = A_c^{-1}B_c r_0 r_0^T B_c^T A_c^{-T}$

$A_c = A - BKC, \qquad B_c = E - BKF$

Optimal Cost:

$$J = \frac{1}{2}\text{tr}(P_k X) + \frac{1}{2}\bar{e}^T V \bar{e} + \frac{1}{2}\sum\sum g_{ij}k_{ij}^2 \qquad (4.2.43)$$

---

with S_i Lagrange multipliers (cf. Moerder and Calise 1985, Subbayyan and Vaithilingam 1977).

Using the same approach as in section 4.1, with a bit more patience, along with the basic matrix calculus identities

$$\frac{\partial Y^{-1}}{\partial x} = -Y^{-1}\frac{\partial Y}{\partial x}Y^{-1} \qquad (4.2.37)$$

$$\frac{\partial UV}{\partial x} = \frac{\partial U}{\partial x}V + U\frac{\partial V}{\partial x} \qquad (4.2.38)$$

$$\frac{\partial y}{\partial x} = \text{tr}\left[\frac{\partial y}{\partial z}\cdot\frac{\partial z^T}{\partial x}\right] \qquad (4.2.39)$$

the necessary conditions for optimality in Table 4.2-1 are found.

Discussion

The feedback design equations for K in Table 4.2-1 are coupled nonlinear equations that appear intimidating. However, we may make a few reassuring remarks. First, those interested in deriving them may refer to the problems, where we outline the approach. The derivation will make the equations more familiar.

Second, it is straightforward to write a computer program to solve the design equations for the optimal gain K using, for instance, subroutine ATXPXA in Armstrong [1980] which solves the Lyapunov equations. In fact, (4.2.40) may be solved using the same code in a do loop; for it is only necessary to solve the same equation $k + 1$ times using as input each time the solution P_i of the previous equation. The same holds for (4.2.41). The software that solves the design equations is described in Appendix A.

A comparison with Table 4.1-1 shows that the new equations are more complicated than the LQ regulator equations with output feedback for two reasons. First, the occurrence of the time weighting t^k in the PI requires that each of the two basic Lyapunov equations be solved $k + 1$ times. Second, the more complicated expression for the initial conditions (4.2.33), with $\bar{x}$ given in the table, has resulted in some extra terms being added in the gain equation (4.2.42). This is because both X and $\bar{e}$ depend on K.

One aspect of the equations in Table 4.2-1 is quite important. In modern LQ control, the PI weights are generally the design parameters, for once k, R, V, and g_{ij} have been selected then the optimal gain K is determined. One serious drawback with modern design techniques is that all of the elements of the state and control weighting matrices must usually be chosen (e.g., Q and R in Table 3.2-2), and there are often no easily understood guidelines for so doing. Moreover, $(\sqrt{Q}, A)$ must be observable, so that the state weighting Q must generally have at least a certain number of nonzero elements.

Note, however, that there are only a few engineering design parameters in Table 4.2-1. Indeed, it is usually suitable to select $R = rI$, $V = vI$, with r and v scalars. Moreover, the selection for g_{ij} is made based simply on which elements of K we want to make equal to zero. The time weighting power k is also a design parameter, but $k = 2$ is almost always suitable.

What this amounts to then is the following. By carefully including all relevant information in the necessary conditions, they become more complicated. However, the payoff is that, given software to solve them, the design is made *easier*, since there are only a few easily understood parameters to adjust. We shall soon make our point using some examples.

In connection with the small number of design parameters in Table 4.2-1, examine the PI where $\bar{e}^T\bar{e} = \bar{x}^T H^T H \bar{x}$. Much of the simplicity of our approach derives from the fact that the state weighting in the PI is equal to $H^T H$, which is known. Even if (H, A) is not observable (cf. Lewis 1986), good designs usually result. This is due to the fact that, even if (H, A) is not observable, $[(k!P_{k-1} + C^T K^T RKC)^{1/2}, A]$ may be observable for some k in the last of equations (4.2.40). This is an important feature of the time weighting t^k in the PI.

As a final comment, we note that the equations in the table assume a constant reference input $r(t) = r_0$. Thus, they are for optimal step-response design. This is sensible, since many performance specifications are given in terms of percent overshoot, rise time, settling time, and other quantities that are intimately related to the step response. On the other hand, once the optimal tracker has been designed, it will, of course, work for any arbitrary time-varying reference command $r(t)$ if the control structure has been properly selected by the designer.

Solution of the design equations

As was the case in section 4.1, there are several basic approaches to solving the design equations in Table 4.2-1.

The formal algebraic optimization problem of minimizing J subject to (4.2.40) may be solved using any numerical optimization routine [Press et al. 1986]. Thus, for each fixed value of K in the iteration, Lyapunov equations (4.2.40) are solved and the PI evaluated using (4.2.43). A good approach for a fairly small number ($mp \leq 10$) of gain elements in K is the SIMPLEX minimization routine [Press et al. 1986].

As an alternative solution procedure, one may use gradient-based techniques (e.g., the Davidon-Fletcher-Powell algorithm, see Press et al. [1986]), which generally require fewer iterations than non-gradient-based approaches [Choi and Sirisena 1974]. To use the equations in Table 4.2-1 to find K by a gradient minimization algorithm, for a fixed value of K solve (4.2.40), (4.2.41) for P_i and S_i. Under these circumstances $\partial J/\partial K = \partial H/\partial K$, which may be found using (4.2.42).

It is worth pointing out that there is an attractive alternative to using gain weighting g_{ij} in the PI to zero out certain gain elements. If the SIMPLEX method is used, it is only necessary to hold certain gain elements fixed, allowing the SIMPLEX to vary only the other elements. Thus, an optimal solution is achieved over the set of all gains K with those elements fixed. An advantage of this approach is that some elements of the gain matrix may be fixed at any desired, possibly nonzero, values.

If a gradient technique is used to solve the design equations, to fix certain elements of K it is only necessary to set the corresponding elements of the gradient $\partial J/\partial K$ equal to zero. Then, those elements are never updated.

The software described in Appendix A incorporates all these options.

Interactive design procedure

The design procedure we now propose is an interactive one. In the PI, values of the design parameters k, $R = rI$, $V = vI$, and g_{ij} are selected and a numerical minimization routine is used to determine the feedback gain K. Then, the closed-loop poles are examined, as is the step response using a computer simulation (e.g., TRESP in Appendix A). If these are not suitable, then the design parameters may be varied and the process repeated. The availability of good software is important, but given this the design procedure is straightforward and rapid.

We emphasize that there are only a few design parameters in our approach, namely k, r, v, and sometimes g_{ij}. Thus, it is not difficult or time-consuming to come up with good designs. (In some cases, more general R and V are needed.)

Let us demonstrate the power of this approach using some examples.

Example 4.2-1: Multivariable Regulator with Gain Element Constraints

Let us reconsider the multi-input/multi-output circuit of Example 4.1-1. Note that the feedback gain found there had two elements that were smaller than the others, namely k_{21} and k_{22}. The fewer nonzero elements of K, the easier the control is to implement. Therefore, let us attempt a design that has only k_{11} and k_{12} nonzero.

To achieve this, select the PI

$$J = \frac{1}{2} \int_0^\infty (y^T y + r u^T u)\, dt + 1000 k_{21}^2 + 1000 k_{22}^2. \tag{1}$$

4.2 Tracking a Reference Input

Since we are interested in the regulator problem, and not in tracking, we may simply set $X = I$, $V = 0$, and $\bar{y} = 0$ in Table 4.2-1. This corresponds to assuming that the initial conditions are uniformly distributed on a sphere of radius 1, and deciding to minimize $E\{J\}$. See the discussion in section 4.1-1.

Since $k = 0$ in t^k, the design equations in Table 4.2-1 reduce to those in Table 4.1-1, with only the addition of the gain weighting terms in (4.2.42) and (4.2.43). These equations are easily solved using the software described in Appendix A.

With $r = 1/1000$, the closed-loop poles were at

$$
\begin{aligned}
s = &-0.72 \pm j3.82, \\
&-1.28 \pm j0.98
\end{aligned} \tag{2}
$$

and the output response with $x_1 = x_2 = x_3 = x_4 = 1$ is shown in Fig. 4.2-3. It dies out after about 5.5 sec, and is not much slower than the response in Fig. 4.1-3 where all the elements of K were nonzero.

The gain with the bottom row elements weighted turned out to be

$$
K = \begin{bmatrix} 14.38 & -6.669 \\ -2 \times 10^{-4} & -5 \times 10^{-5} \end{bmatrix}, \tag{3}
$$

so that in an implementation the bottom row may be assumed zero without having much affect on the response or poles.

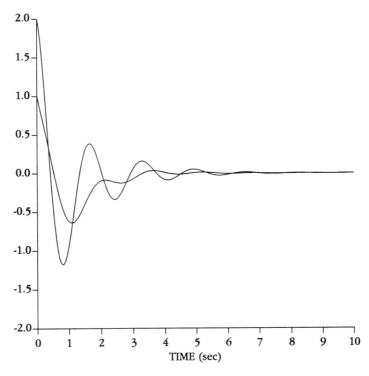

Figure 4.2-3 Output response with a single control input

What we see, then, is that for the desired performance objectives, only one control input is actually needed, namely, u_1. This will result in a considerable simplification during any implementation.

As an alternative to gain element weighting in the PI, we may simply instruct the SIMPLEX to vary only elements k_{11} and k_{12} of K, keeping the bottom row of K fixed at zero. This option requires less computation and is included in the software described in Appendix A.

∎

Example 4.2-2: Proportional-Plus-Integral Compensator For Tracking

In this example we shall see that a sensible formulation of the problem and selection of the control system structure are the keys to success using output-feedback LQ design.

a. Problem Formulation and Control System Structure

Let there be prescribed the scalar model for a DC motor

$$\dot{\omega} = -a\omega + bu \qquad (1)$$

with $1/a$ the mechanical time constant, $\omega(t)$ the motor speed, and $u(t)$ the armature control voltage. We discussed this system in Examples 3.2-3 and 3.3-3.

We should like the speed $\omega(t)$ to follow a prescribed constant reference step command of magnitude r. That is, starting from an initial speed of zero, we want $\omega(t)$ to increase to a value of r rad/s while exhibiting desirable properties in terms of speed of response, overshoot, and zero steady-state error. To obtain zero steady-state error $\bar{e}$, we may use an integrator as a precompensator so that the open-loop plant is of Type 1 [D'Azzo and Houpis 1988]. See Fig. 4.2-4.

We define the tracking error as

$$e = r - \omega \qquad (2)$$

and close an outer loop with unity feedback around the plant plus compensator. The motor control voltage is selected of the form

$$u = -k_1 e - k_2 \varepsilon = k_1 \omega - k_2 \varepsilon - k_1 r, \qquad (3)$$

with $\varepsilon(t)$ the integrator output. This corresponds to a practical proportional-plus-integral (PI) scheme designed using the intuition of classical control. We should now like to use modern

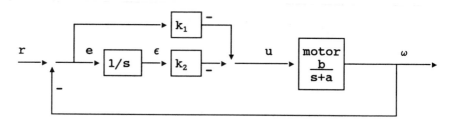

Figure 4.2-4 DC motor step-response control scheme

4.2 Tracking a Reference Input

control theory to select optimal values of the proportional and integral feedback gains k_1 and k_2.

b. LQ Output-Feedback Servodesign

To select the control gains, write the augmented state equation of the plant plus compensator as

$$\dot{x} = \begin{bmatrix} -a & 0 \\ -1 & 0 \end{bmatrix} x + \begin{bmatrix} b \\ 0 \end{bmatrix} u + \begin{bmatrix} 0 \\ 1 \end{bmatrix} r \tag{4}$$

$$y = \begin{bmatrix} -1 & 0 \\ 0 & 1 \end{bmatrix} x + \begin{bmatrix} 1 \\ 0 \end{bmatrix} r = \begin{bmatrix} e \\ \varepsilon \end{bmatrix} \tag{5}$$

$$u = -[k_1 \quad k_2] y \tag{6}$$

with the state defined as $x = [\omega \quad \varepsilon]^T$. We symbolize the augmented system as

$$\dot{x} = Ax + Bu + Er \tag{7}$$

$$y = Cx + Fr \tag{8}$$

$$u = -Ky. \tag{9}$$

The performance output is

$$z = \omega = [1 \quad 0]x \equiv Hx \tag{10}$$

so that

$$e = r - z. \tag{11}$$

c. PI and Solution

These equations are exactly of the form given in Table 4.2-1. Therefore, we may use the equations in the table to determine the control gains k_1 and k_2. Since the steady-state error $\bar{e}$ is equal to zero, we have

$$e(t) = \tilde{e}(t) = -H\tilde{x}, \tag{12}$$

and a suitable PI is

$$J = \frac{1}{2} \int_0^\infty (q\tilde{e}^2 + \tilde{u}^2) \, dt, \tag{13}$$

which reflects our concern to keep small the error and control deviations. The error weighting $q > 0$ is a design parameter whose effects we shall soon see.

Note that we could include a scalar control weighting of R; however, dividing through by R we simply scale the error weight and J. This does not change the optimal solution, for minimizing J/R is equivalent to minimizing J. There is, therefore, no advantage to using two weights in (13).

We used software like that in Press et al. [1986] (see Appendix A) to determine the optimal feedback gains using the equations of Table 4.2-1. Using these gains in the control law (3), we obtained the step responses for system (4) shown in Fig. 4.2-5. We used $a = 1$, $b = 1$, $r = 1$, and varied the value of the weighting q.

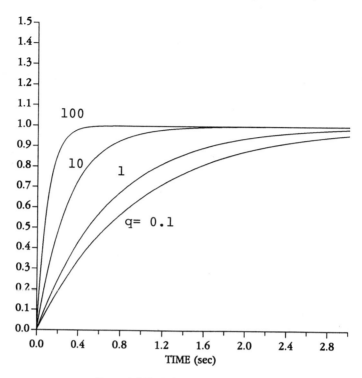

Figure 4.2-5 DC motor step responses

The values of the gain K and the closed-loop poles are shown in Table 1.

TABLE 1

q	K		Closed-loop poles	
0.1	[−1.049	−1.049]	−1,	−1.049
1	[−1.414	−1.414]	−1,	−1.414
10	[−3.318	−3.318]	−1,	−3.318
100	[−10.05	−10.05]	−1,	−10.05

We should like to emphasize that the entire design process took about 10 minutes. This is a function of having good software to implement the design equations in Table 4.2-1.

d. Discussion

We can now emphasize the real power of modern steady-state LQ design. A naive approach to controls design for this system would be to vary the feedback gains k_1 and k_2 blindly in an attempt to obtain acceptable responses. However, for some values of k_1, k_2 the closed-loop system might be unstable. Classical control provides a formal, one-look-at-a-time technique for design. However, the LQ theory has allowed us to replace k_1 and k_2 by one parameter q

4.2 Tracking a Reference Input

which may be varied to obtain acceptable behavior. The importance of this is that for all values of q the closed-loop system is stable.

One may notice that some of the responses in Fig. 4.2-5 are significantly faster than the closed-loop pole in Table 1 would indicate. Recall that, according to (4.2.15), this scheme uses optimal positioning of *both the poles and zeros* to attain step-response shaping. Indeed, a quick calculation of the closed-loop transfer function

$$H_c(s) = H(sI - (A - BKC))^{-1}(E - BKF) \tag{14}$$

from $r(t)$ to $z(t)$ shows that the zero is always at approximately $s = -1$, thereby canceling the slow pole.

e. The Observability Issue

An important final point must be made. Note that

$$q\bar{e}^2 = \tilde{x}^T Q \tilde{x} \tag{15}$$

with

$$Q = H^T H = \begin{bmatrix} 1 & 0 \\ 0 & 0 \end{bmatrix}. \tag{16}$$

Thus, H is a square root of Q; but (H, A) is not observable. Thus, a naive application of the simplified output feedback equations in Table 4.1-1 will not yield gains that stabilize the plant. Indeed, there is a fixed pole at $s = 0$, since the integrator pole is unobservable. The design equations in Table 4.1-1 are the traditional ones used for output feedback design [Levine and Athans 1970, Moerder and Calise 1985].

This deficiency is not shared by the design procedure in Table 4.2-1. Thus, by including the extra terms in $\bar{x}$ [specifically in (4.2.42)], the required form of the state weighting matrix Q is simplified. This means that Q may have fewer nonzero entries so that only a few design parameters must be selected, and these are intuitively sensible. Indeed, we had no trouble obtaining very nice step responses using the simple PI (13) with only one design parameter.

∎

Example 4.2-3: Speed Control of an Armature-Controlled DC Motor

For some applications, the simplified motor model used in the previous example may not be sufficiently accurate. Therefore, let us now discuss the armature-controlled DC motor we studied in Examples 3.1-3 and 3.2-4.

With the state $x = [i \quad \omega]^T$, we may write

$$\dot{x} = \begin{bmatrix} -a & -k' \\ k & -\alpha \end{bmatrix} x + \begin{bmatrix} b \\ 0 \end{bmatrix} u \tag{1}$$

with $i(t)$ the armature current, $\omega(t)$ the motor speed, control input $u(t)$ the armature voltage, $1/a$ the electrical time constant, $1/\alpha$ the mechanical time constant, and the remaining parameters as previously defined. Assuming values for the parameters, let us take

$$\dot{x} = \begin{bmatrix} -2.29 & -0.003 \\ 1.172 & -1.32 \end{bmatrix} x + \begin{bmatrix} 0.337 \\ 0 \end{bmatrix} u \equiv Ax + Bu. \tag{2}$$

The objective is to force the motor speed $\omega(t)$ to track a constant reference speed of r rads/sec.

a. Type I System Design

It is well known from classical control theory that we may guarantee perfect tracking of a unit step if the system is of Type I [D'Azzo and Houpis 1988]. Thus, let us include an integrator in the feedforward path as in Fig. 4.2-6. The integrator output will be represented by ε; it is the integral of the tracking error

$$e(t) = r - \omega(t). \tag{3}$$

Redefining the state as $x = [i \quad \omega \quad \varepsilon]^T$, we may write the state equations of the motor plus integral compensator as

$$\dot{x} = \begin{bmatrix} -2.29 & -0.003 & 0 \\ 1.172 & -1.32 & 0 \\ 0 & -1 & 0 \end{bmatrix} x + \begin{bmatrix} 0.337 \\ 0 \\ 0 \end{bmatrix} u + \begin{bmatrix} 0 \\ 0 \\ 1 \end{bmatrix} r \equiv Ax + Bu + Er. \tag{4}$$

The measured output corresponding to Fig. 4.2-6 is

$$y = \begin{bmatrix} \omega \\ \varepsilon \end{bmatrix} = \begin{bmatrix} 0 & 1 & 0 \\ 0 & 0 & 1 \end{bmatrix} x \equiv Cx \tag{5}$$

and the performance output which should track the reference speed r is

$$z = \omega = [0 \quad 1 \quad 0] x \equiv Hx. \tag{6}$$

Thus

$$e(t) = r - z(t). \tag{7}$$

According to the figure, the admissible armature voltage input is

$$u = -k_\omega \omega - k_\varepsilon \varepsilon = -[k_\omega \quad k_\varepsilon] y \equiv -Ky. \tag{8}$$

Note that k_ε is a feedforward gain while k_ω is a feedback gain.

This scenario is exactly the one described by the model in Table 4.2-1. Therefore, the software used to solve the design equations in that table may be used in this motor tracking design problem.

Since the integrator guarantees a zero steady-state error $\bar{e}$, we have $e(t) = \tilde{e}(t)$ the error deviation. Therefore, a suitable and entirely natural PI is

$$J = \frac{1}{2} \int_0^\infty (qt^k \tilde{e}^2 + \tilde{u}^2) \, dt, \tag{9}$$

which reflects our concern to keep small the error and control deviations. The error weighting $q > 0$ and time-weighting power k are design parameters.

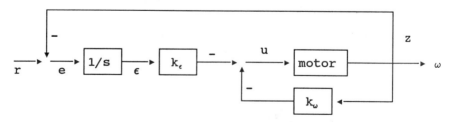

Figure 4.2-6 Armature-controlled DC motor step-response control scheme

4.2 Tracking a Reference Input

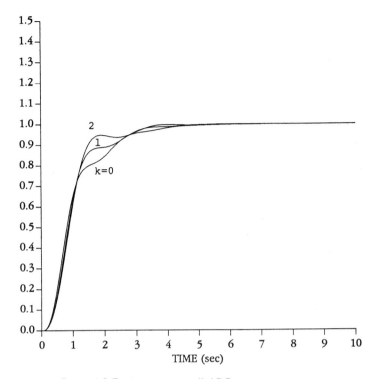

Figure 4.2-7 Armature-controlled DC motor step responses

We chose $q = 100$ and used the software described in Appendix A to solve the design equations in Table 4.2-1 for several values of k. The control gains and closed-loop poles are shown in Table 1. The step responses for $r = 1$ rad/sec starting at zero initial conditions are shown in Fig. 4.2-7.

TABLE 1

k	K	Closed-loop poles	
0	[42.54 −45.48]	−1.05,	−1.28 ± j3.94
1	[31.24 −36.92]	−1.17,	−1.22 ± j3.32
2	[24.78 −32.20]	−1.3,	−1.16 ± j2.91

The step responses show that any of the gains would result in suitable regulation of the speed about the desired value of r rad/sec. The gains corresponding to time weighting t^2 give the fastest response. See the closed-loop poles in Table 1.

b. Type 0 System Design

It is well known from classical control theory that acceptable steady-state error may often be obtained without using a forward path integrator. This can be important in view of the windup and saturation problems associated with implementing integrators (see Chapter 6). It also results in a simpler compensator.

Let us, therefore, further demonstrate the power of the design equations in Table 4.2-1 by considering the modified servo control scheme shown in Fig. 4.2-8, where no integrator is used. Thus, the open-loop system is of Type 0 [D'Azzo and Houpis 1988].

We shall discover a rather remarkable fact. Namely, using modern control with adjustment of more than one control gain, we can often achieve a steady-state error of exactly zero even if the system is of Type 0.

The state equations without the integrator and the output equations corresponding to Fig. 4.2-8 are

$$\dot{x} = \begin{bmatrix} -2.29 & -0.003 \\ 1.172 & -1.32 \end{bmatrix} x + \begin{bmatrix} 0.337 \\ 0 \end{bmatrix} u \equiv Ax + Bu \tag{10}$$

$$y = \begin{bmatrix} \omega \\ e \end{bmatrix} = \begin{bmatrix} 0 & 1 \\ 0 & -1 \end{bmatrix} x + \begin{bmatrix} 0 \\ 1 \end{bmatrix} r \equiv Cx + Fr \tag{11}$$

$$z = \omega = \begin{bmatrix} 0 & 1 \end{bmatrix} x \equiv Hx, \tag{12}$$

and the allowed control voltage is

$$u = -k_\omega \omega - k_e e = \begin{bmatrix} k_\omega & k_e \end{bmatrix} y \equiv -Ky. \tag{13}$$

The state is now $x = [i \quad \omega]^T$.

Now, the steady-state error $\bar{e}$ is not guaranteed to be equal to zero, so that

$$e(t) = \tilde{e}(t) + \bar{e}, \tag{14}$$

and we must take steps to make both $\bar{e}$ and the error deviation $\tilde{e}(t)$ small. This may easily be achieved by selecting the PI

$$J = \frac{1}{2} \int_0^\infty (t^k \tilde{e}^2 + \tilde{u}^2) \, dt + v\bar{e}^2, \tag{15}$$

which also weights the steady-state error.

Using the software described in Appendix A, we found the control gains and closed-loop poles, and plotted the step response, for several values of k and v. Suitable closed-loop performance was obtained with $k = 2$ and $v = 1000$. Then, the gains were

$$K = [-7.662 \quad -7.758] \tag{16}$$

and the closed-loop poles were at $s = -2.25, -1.36$.

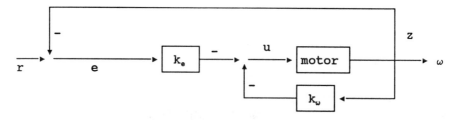

Figure 4.2-8 DC motor step-response control scheme with no forward path integrator

4.2 Tracking a Reference Input

The step response is shown in Fig. 4.2-9. The remarkable fact about this response is that the steady-state error is *exactly equal to zero*. It is worth examining this point further.

With the gains given in (16), the closed-loop system

$$\dot{x} = (A - BKC)x + (E - BKF)u \tag{17}$$

$$z = Hx \tag{18}$$

has a transfer function of

$$H_c(s) = \frac{3.06}{(s + 2.25)(s + 1.36)}, \tag{19}$$

so that the closed-loop DC gain is *exactly equal to one*. This accounts for the zero steady-state error.

To see why the DC gain is one, the open-loop transfer function breaking the loop at $e(t)$ in Fig. 4.2-8 was computed. It is

$$H_L(s) = \frac{3.06}{s(s + 3.61)}. \tag{20}$$

What this means is that the control gain of the inner loop k_ω has been selected to place one of the motor poles at $s = 0$ before closing the outer loop. Thus, our LQ design approach has converted the motor itself into a Type I system to achieve zero steady-state error.

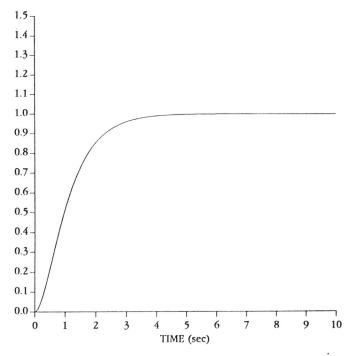

Figure 4.2-9 DC motor step response with no forward path integrator

Example 4.2-4: Control of an Inverted Pendulum

Figure 4.2-10 shows a rod attached to a cart through a pivot. A force $u(t)$ is applied to the cart through a motor attached to an axle. The control objective is to use $u(t)$ to balance the pendulum upright while simultaneously keeping the horizontal position $p(t)$ of the cart small. This is known as the *inverted pendulum* problem. In Example 2.5-1 we showed how to stabilize the system using pole-placement design with full state-variable feedback. The response obtained there was fairly good, but had a large excursion in the cart position $p(t)$. That example should be reviewed at this time.

a. Sensors and Actuators

We shall measure the rod angle $\theta(t)$ and the cart position $p(t)$. To accomplish this, we place potentiometers at the rod pivot point [for $\theta(t)$] and on one wheel [for $p(t)$]. To keep down the number of measurements and avoid unnecessary measuring instruments, we shall avoid measurements of the angular velocity $\dot\theta(t)$ and the cart velocity $\dot p(t)$. As will be seen, this will in no way detract from the quality of the controlled behavior.

A motor would be used to provide the force $u(t)$. If desired, the motor dynamics could easily be included in a design like this one.

b. Inverted Pendulum Dynamics

The state and measured outputs are

$$x = [\theta \quad \dot\theta \quad p \quad \dot p]^T, \qquad y_m = [\theta \quad p]^T. \qquad (1)$$

Assuming that $M = 5$ kg, $m = 0.5$ kg, $L = 1$ m, we obtain the dynamics

$$\dot x = \begin{bmatrix} 0 & 1 & 0 & 0 \\ 10.78 & 0 & 0 & 0 \\ 0 & 0 & 0 & 1 \\ -0.98 & 0 & 0 & 0 \end{bmatrix} x + \begin{bmatrix} 0 \\ -0.2 \\ 0 \\ 0.2 \end{bmatrix} u. \qquad (2)$$

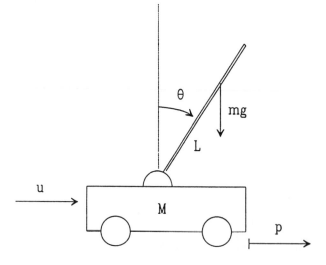

Figure 4.2-10 Inverted pendulum on a cart

4.2 Tracking a Reference Input

The output matrix is

$$y_m = \begin{bmatrix} 57.2958 & 0 & 0 & 0 \\ 0 & 0 & 1 & 0 \end{bmatrix} x. \tag{3}$$

The factor of 57.2958 has been added to convert the angle $\theta(t)$ from radians to degrees.

In practice, one would also need to include in the C matrix the constant factors involved in the potentiometers in converting from angles to volts, as well as a factor depending on the wheel radius to convert from position $p(t)$ to the potentiometer angle.

In the B matrix, we would include a factor that converts from motor voltage to the force $u(t)$ applied to the cart. It will include motor constants, gear ratios, and so on.

The open-loop poles are at

$$s = 0, 0, \pm 3.283, \tag{4}$$

so that with no control input the rod will clearly fall over due to the unstable pole at $s = 3.283$.

c. Control System Structure

We propose the simple control structure shown in Fig. 4.2-11. We have drawn it in a way to emphasize that it may be considered as a tracker system with reference inputs of zero. The motivation for this structure is as follows.

The inverted pendulum is a single-input/two-output system. Having in mind the root locus theory of classical control, to move the poles in (4) to the left-half plane we should add some compensator zeros in the left-half plane. It is well known from classical control theory that a lead compensator can often stabilize a system [D'Azzo and Houpis 1988]. Thus, let us

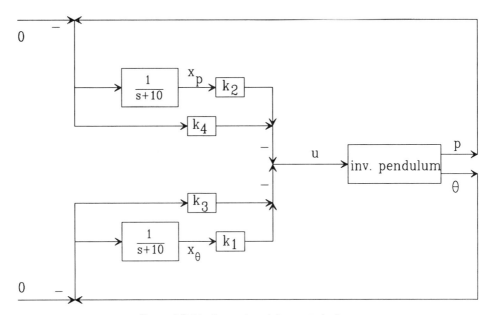

Figure 4.2-11 Inverted pendulum control scheme

place a compensator in each of the feedback loops, selecting the poles relatively far to the left. We have chosen $s = 10$ for the compensator poles here.

The important point to note is that, by varying the four control gains, the LQ solution algorithm can automatically select the zeros of the compensators. Indeed, the transfer functions of the compensators are of the form

$$\frac{v}{\theta} = \frac{k_1}{s + 10} + k_3 = k_3 \frac{[s + (10 + k_1/k_3)]}{s + 10}, \qquad (5)$$

so that by varying k_1 and k_3 both the compensator gain and its zero may be selected. Presumably, the optimal LQ gains will yield stable compensator zeros nearer the origin than $s = -10$, so that the final design will have two lead compensators.

Note that a state variable representation of (5) is

$$\dot{x}_\theta = -10 x_\theta + \theta \qquad (6)$$

$$v = k_1 x_\theta + k_3 \theta. \qquad (7)$$

Thus, we may incorporate the dynamics of the compensators into the state equations by defining the augmented state as

$$x = [\theta \ \dot{\theta} \ p \ \dot{p} \ x_\theta \ x_p]^T, \qquad (8)$$

with x_θ and x_p the compensator states. Then

$$\dot{x} = \begin{bmatrix} 0 & 1 & 0 & 0 & 0 & 0 \\ 10.78 & 0 & 0 & 0 & 0 & 0 \\ 0 & 0 & 0 & 1 & 0 & 0 \\ -0.98 & 0 & 0 & 0 & 0 & 0 \\ 1 & 0 & 0 & 0 & -10 & 0 \\ 0 & 0 & 1 & 0 & 0 & -10 \end{bmatrix} x + \begin{bmatrix} 0 \\ -0.2 \\ 0 \\ 0.2 \\ 0 \\ 0 \end{bmatrix} u \equiv Ax + Bu. \qquad (9)$$

The compensator output equations (7) may be incorporated by defining a new augmented output as

$$y = \begin{bmatrix} x_\theta \\ x_p \\ \theta \\ p \end{bmatrix} = \begin{bmatrix} 0 & 0 & 0 & 0 & 57.2958 & 0 \\ 0 & 0 & 0 & 0 & 0 & 1 \\ 57.2958 & 0 & 0 & 0 & 0 & 0 \\ 0 & 0 & 1 & 0 & 0 & 0 \end{bmatrix} x = Cx. \qquad (10)$$

Then, according to Fig. 4.2-11, the control input $u(t)$ is given by

$$u = -k_1 x_\theta - k_2 x_p - k_3 \theta - k_4 p = -[k_1 \ k_2 \ k_3 \ k_4] y \equiv -Ky. \qquad (11)$$

We are now in a position to perform the controls design to select the control gains k_i.

d. Controls Design

It is desired to select K in (11) to regulate the state x of (9) to zero. This amounts to a servo problem with reference commands of zero, or simply to a regulator problem. For the system (9) with outputs (10), let us select the PI

$$J = \frac{1}{2} \int_0^\infty (t^k x^T Q x + r u^T u) \, dt. \qquad (12)$$

4.2 Tracking a Reference Input

Then, K is determined by using the design equations of Table 4.2-1. Since, however, this is a regulator problem where the desired steady-state values are zero, we should select $\bar{y} = 0$, $X = I$ in the table. This amounts to assuming that the initial states are uniformly distributed on the unit sphere and minimizing not J but its expected value. See section 4.1.

With these simplifications, the equations in Table 4.2-1 reduce to those in Table 4.1-1, with, however, the additional iterations needed in the Lyapunov equations for the time weighting t^k. All of these special situations are easily handled using the software described in Appendix A, which solves the design equations of Table 4.2-1 for any choice of the parameters.

We chose $Q = \text{diag}\{100, 100, 1, 1, 0, 0\}$. The motivation for selecting this Q was to place heavy emphasis on keeping the angle $\theta(t)$ small; the cart position control does not matter if the rod falls over. The compensator states are of no concern and were not weighted.

A few computer-aided design iterations were performed: values of the design parameters k and r were selected, the optimal gain K was found, and the closed-loop system response was plotted. It was discovered that good behavior was obtained using $k = 2$ and $r = 0.01$.

Using this k, Q, and r the control gains were

$$K = [48.51 \quad 817.5 \quad -8.0 \quad -87.59] \tag{13}$$

and the closed-loop poles were

$$\begin{aligned} s = &-1.87 \pm j5.66 \\ &-0.35 \pm j0.68 \\ &-5.54, -10. \end{aligned} \tag{14}$$

The angle $\theta(t)$ and position $p(t)$ in response to an initial condition offset of $\theta(0) = 0.1$ rad $\approx 6°$, $p(0) = 0.1$ m are shown in Fig. 4.2.12a. The required control force $u(t)$ is shown in Fig. 4.2.12b. These plots are quite interesting and bear discussion.

e. Discussion

Let us use our imaginations to picture the behavior of the cart. Due to the initial offset of 6° in angle, a large control must be applied immediately to push the cart under the rod to catch it so it does not fall. Subsequent smaller control oscillations stabilize the rod in an upright position. Toward the end of these gyrations, a slower control motion (barely visible in the figure) begins to move the cart slowly back to the desired horizontal position of $p = 0$.

Thus, the first (fast) complex pole pair in (14) corresponds to the control motion needed to balance the rod. The second pole pair is associated with a slower control motion involved with cart position control. Note that the control in Fig. 4.2-12b has a fast component superimposed on a slower component of very low amplitude. The two real poles are associated with the compensators.

If the control magnitude in Fig. 4.2-12b is larger than the motor can apply, it is necessary to choose a larger control weighting r and repeat the design. The result will be smaller controls and poles nearer the origin, thus yielding a slower closed-loop response.

We should like to mention that a root of Q is

$$H = \begin{bmatrix} 10 & 0 & 0 & 0 & 0 & 0 \\ 0 & 10 & 0 & 0 & 0 & 0 \\ 0 & 0 & 1 & 0 & 0 & 0 \\ 0 & 0 & 0 & 1 & 0 & 0 \end{bmatrix}, \tag{15}$$

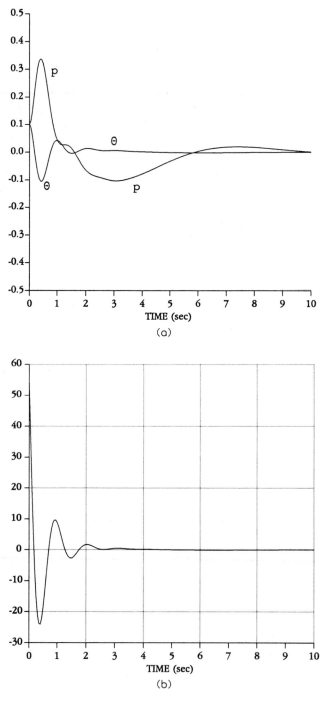

Figure 4.2-12 Inverted pendulum closed-loop response. (a) Rod angle $\theta(t)$ (rads) and cart position $p(t)$ (m). (b) Control input $u(t)$.

4.3 Tracking By Regulator Redesign

so that a test involving the observability matrix shows that the compensator states x_θ and x_p are not observable in the PI. The traditional LQ design equations (see Table 4.1-1) would require weighting these states as well. However, the time weighting t^2 has allowed us to obtain a stabilizing solution in spite of the unobservability of $(\sqrt{Q}, A)$ (see the discussion earlier in this section). Thus, we did not need to select design parameters that did not correspond to our control objectives (i.e., elements (5, 5) and (6, 6) of Q). Note that we really do not care what the compensator states do.

Finally, let us examine the compensator zeros selected by the LQ algorithm. According to (5), with the gains in (13) the compensator in the angle channel is

$$\frac{v_\theta}{\theta} = -8.0 \frac{(s + 3.9)}{s + 10} \qquad (16)$$

and the compensator in the position channel is

$$\frac{v_p}{p} = -87.59 \frac{(s + 0.7)}{s + 10}. \qquad (17)$$

These are both lead compensators as anticipated. However, due to its high zero/pole ratio, (17) is more of a filtered differentiator. Thus, we see that some information on the derivative of the position $p(t)$ is desirable to achieve good regulation of the inverted pendulum. ∎

4.3 TRACKING BY REGULATOR REDESIGN

In this section we shall discuss an alternative tracker design technique that amounts to first designing a regulator, and then adding some *feedforward terms* to guarantee tracking behavior.

This technique does not have the advantages of the direct design approach of the previous section. There, we were able to:

1. Select the form of the compensator, including a unity outer loop to allow feedforward of the error.
2. Simplify the design stage by using only a few design parameters in the PI.

However, the approach to be presented here is simple to understand and may be quite useful in some applications. It will also give us some more insight on the tracking problem. Specifically, for good tracking the number of control inputs should be at least as large as the number of reference inputs to be tracked.

Let us suppose that the plant-plus-compensator in Fig. 4.2-1 is described, using the technique described in section 4.2, as

$$\dot{x} = Ax + Bu + Er \qquad (4.3.1)$$

$$y = Cx + Fr \qquad (4.3.2)$$

$$z = Hx \qquad (4.3.3)$$

where $y(t)$ is the measurable output available for feedback and the performance output $z(t)$ is required to track the reference input $r(t)$. The tracking error is

$$e = r - z. \qquad (4.3.4)$$

Thus, this augmented description contains the dynamics of both the plant and the compensator.

It is desired to select the control input $u(t)$ so that the tracking error goes to zero.

Deviation System

For perfect tracking, there must exist an *ideal plant state* x^* and an *ideal plant input* u^* such that

$$\dot{x}^* = Ax^* + Bu^* + Er \qquad (4.3.5)$$

$$y^* = Cx^* + Fr \qquad (4.3.6)$$

$$z^* = Hx^* = r. \qquad (4.3.7)$$

If this is not so, then we cannot have tracking with zero error. See Chyung [1987] and O'Brien and Broussard [1978].

What this assumption means is that there is indeed a control input $u^*(t)$ that results in a performance output $z^*(t)$ equal to the desired $r(t)$.

Defining the state, control, and output deviations as

$$\tilde{x} = x - x^*, \quad \tilde{u} = u - u^*, \quad \tilde{y} = y - y^*, \quad \tilde{z} = z - z^*, \qquad (4.3.8)$$

we may subtract (4.3.5)–(4.3.7) from (4.3.1)–(4.3.3) to obtain the dynamics of the "deviation system" given by

$$\dot{\tilde{x}} = A\tilde{x} + B\tilde{u} \qquad (4.3.9)$$

$$\tilde{y} = C\tilde{x} \qquad (4.3.10)$$

$$\tilde{z} = H\tilde{x} = -e. \qquad (4.3.11)$$

Thus, to regulate the tracking error to zero we may simply design a regulator to control the state of the deviation system to zero. To do this, it is only necessary to select a reasonable PI that weights $\tilde{x}$ and $\tilde{u}$, and then use the design equations in Table 4.1-1, not the more complicated development relating to Table 4.2-1.

Let us note, however, that now the usual restrictions on the PI weights Q and R in Table 4.1-1 apply. That is, $(\sqrt{Q}, A)$ should be observable. What this means is that we will generally be faced with selecting too many design parameters (i.e., the elements of Q and R). Thus, an important advantage of using the approach in section 4.2 is lost.

Regulator Redesign Adding Feedforward Terms

Suppose we have obtained output feedback gains K that are optimal with respect to (4.3.9)–(4.3.11). Then

$$\tilde{u} = -K\tilde{y} \qquad (4.3.12)$$

4.3 Tracking By Regulator Redesign

so that the required control for the plant is

$$u = u^* + \bar{u} = u^* + Ky^* - Ky. \quad (4.3.13)$$

That is, the resulting control for the servo or tracker problem is the optimal regulator feedback control Ky plus some *feedforward terms* that are required to guarantee perfect tracking.

We should emphasize that, while the control (4.3.12), designed for the deviation system, is optimal, the control (4.3.13) is *not* an optimal solution to the tracker problem for the original plant. However, the closed-loop response resulting from (4.3.13) may be satisfactory for many applications; hence, this is not too severe a drawback.

It is now necessary to determine the ideal plant control u^* and output y^* in order to complete the design of the servo control law (4.3.13). To accomplish this, take the Laplace transform of the ideal plant to obtain $R = Z^* = H[(sI - A)^{-1}[BU^* + ER]$, or

$$H(sI - A)^{-1}BU^*(s) = [I - H(sI - A)^{-1}E]R(s). \quad (4.3.14)$$

Define the transfer function from $u(t)$ to $z(t)$ as

$$H(s) = H(sI - A)^{-1}B. \quad (4.3.15)$$

There exists a solution $U^*(s)$ to (4.3.14) for all $R(s)$ if and only if $H(s)$ has full row rank. Thus, *the number of control inputs should be at least equal to the number of performance outputs*. This is an important and fundamental restriction on the tracking problem.

Let us assume that the number of control inputs is equal to the number of performance outputs so that $H(s)$ is square. If in addition $H(s)$ is nonsingular, then

$$U^*(s) = H^{-1}(s)[I - H(sI - A)^{-1}E]R(s) \quad (4.3.16)$$

$$Y^* = C(sI - A)^{-1}[BU^*(s) + ER(s)] + FR(s). \quad (4.3.17)$$

Using these values of $u^*(t)$ and $y^*(t)$ in (4.3.13) yields the tracker control law.

Tracking a Unit Step

If $r(t)$ is a unit step, then the feedforward terms simplify and we can gain more intuition on the servo control problem. In this case, the ideal responses are nothing but the steady-state responses since $\dot{x}^* = 0$. See Kwakernaak and Sivan [1972]. Substituting the control (4.3.13) into (4.3.1) yields

$$\dot{x} = A_c x + Er + B(u^* + Ky^*) \quad (4.3.18)$$

$$z = Hx \quad (4.3.19)$$

where

$$A_c = A - BKC. \quad (4.3.20)$$

Noting that x^* is constant, we see that, at steady state

$$0 = A_c x^* + Er + B(u^* + Ky^*) \quad (4.3.21)$$

Thus, if $r = z^*$ as desired

$$r = Hx^* = -HA_c^{-1}[B(u^* + Ky^*) + Er], \qquad (4.3.22)$$

since A_c is stable, and hence invertible, in any useful design. Therefore

$$-HA_c^{-1}B(u^* + Ky^*) = [I + HA_c^{-1}E]r. \qquad (4.3.23)$$

Let us define the closed-loop transfer function from $u(t)$ to $z(t)$ as

$$H_c(s) = H(sI - (A - BKC))^{-1}B. \qquad (4.3.24)$$

Then, (4.3.23) may be solved for the feedforward terms $(u^* + Ky^*)$ if and only if *the closed-loop DC gain $H_c(0)$ is invertible.* (Note that $H_c(0) = -HA_c^{-1}B$.) In that event

$$(u^* + Ky^*) = H_c^{-1}(0)[I + HA_c^{-1}E]r, \qquad (4.3.25)$$

and the servo control (4.3.13) is given by

$$u = -Ky + H_c^{-1}(0)[I + HA_c^{-1}E]r. \qquad (4.3.26)$$

The second term is a feedforward term added to achieve the correct steady-state value of $z(t)$. Thus, the servo control is equal to the optimal regulator feedback control $-Ky$ plus a term involving the inverse of the closed-loop DC gain. Equation (4.3.26) is the fundamental design equation of this section.

The feedback does not change the system zeros. Consequently, the zeros of $H(s)$ and those of $H_c(s)$ are the same. Thus, $H_c(0)$ is invertible if and only if the plant has no system zeros at $s = 0$. This is the condition for perfect tracking of a unit step command, which has a pole at $s = 0$. It makes sense, since the DC behavior is related to the steady-state response.

Example 4.3-1: Tracking by Regulator Redesign

Consider the plant

$$\dot{x} = \begin{bmatrix} -\alpha & 1 \\ 0 & 0 \end{bmatrix} x + \begin{bmatrix} 0 \\ 1 \end{bmatrix} u = Ax + Bu \qquad (1)$$

$$z = [1 \quad 0]x = Hx. \qquad (2)$$

It is desired to use only x_1 for feedback purposes so that

$$y = [1 \quad 0]x = Cx. \qquad (3)$$

The control objective is to select $u(t)$ to make the tracking error

$$e = r - z \qquad (4)$$

go to zero. We assume that $r(t)$ is an arbitrary command input, not necessarily the unit step.

To achieve this goal, we can select any reasonable PI that weights $\tilde{x}$ and $\tilde{u}$, and then use the equations in Table 4.1-1 to obtain the optimal LQ regulator gain K in the deviation system control law

$$\tilde{u} = -K\tilde{y}. \qquad (5)$$

4.4 Command-Generator Tracker

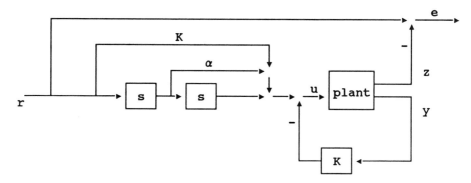

Figure 4.3-1 Tracker obtained using regulator redesign

To convert this regulator to a tracker, we can use equations (4.3.16) and (4.3.17) to write (verify!)

$$u^*(t) = \ddot{r}(t) + \alpha \dot{r}(t) \tag{6}$$

$$y^*(t) = r(t). \tag{7}$$

Therefore, the control that ensures tracking of $r(t)$ by $z(t)$ is

$$u = \ddot{r} + \alpha \dot{r} + Kr - Ky. \tag{8}$$

An implementation of this control scheme appears in Fig. 4.3-1. Note that, generally, the feedforward term that is added to the feedback control is a linear combination of $r(t)$ and its derivatives. This may not be satisfactory in some applications, since differentiation generally makes noisy signals noisier; consequently, it is usually advisable to avoid it.

In our applications, we should like to have control schemes of the form of Fig. 4.2-1 which have a unity feedback outer loop from $z(t)$ and feedforward of the tracking error $e(t)$. This is a structure that takes into account the intuition of classical control theory.

As it turns out, in this example $y = z$, so that the control law (8) may be reformulated as

$$u = \ddot{r} + \alpha \dot{r} + Ke. \tag{9}$$

However, in general $y \neq z$, so that it is not generally possible to formulate the control schemes resulting from the approach of this section using a unity feedback outer loop. ■

4.4 COMMAND-GENERATOR TRACKER

In Examples 4.2-2 and 4.2-3 we selected a PI compensator to make the loop gain of Type I in order to obtain zero steady-state tracking error in response to a step command input (i.e., a position command). If the reference input is not a unit step, then a single integrator will no longer guarantee a steady-state error of zero.

In section 4.3 we saw how to convert a regulator into a tracker by adding additional feedforward terms. However, if the reference input $r(t)$ is not constant, the feedforward terms generally contain derivatives of $r(t)$.

In this subsection we shall demonstrate *command-generator tracker* (CGT) design, which is a powerful design technique that automatically gives the precompensator required to obtain zero-steady-state error for a large class of command inputs $r(t)$. See Franklin et al. [1986]. In this approach, we shall incorporate a model of the dynamics of $r(t)$ into the control system.

As we shall see, CGT design may be used for tracking and also for disturbance rejection.

Tracking

The plant

$$\dot{x} = Ax + Bu \tag{4.4.1}$$

has measured outputs available for control purposes given by

$$y = Cx \tag{4.4.2}$$

and its performance output

$$z = Hx \tag{4.4.3}$$

is required to track the reference input $r(t)$.

Command generator system

Let us suppose that, for some initial conditions, the reference command satisfies the differential equation

$$r^{(d)} + a_1 r^{(d-1)} + \ldots + a_d r = 0 \tag{4.4.4}$$

for a given degree d and set of coefficients a_i. Most command signals of interest satisfy such an equation. For instance, the unit step of magnitude r_0 satisfies

$$\dot{r} = 0 \tag{4.4.5}$$

with $r(0) = r_0$, while the ramp (velocity command) with slope v_0 satisfies

$$\ddot{r} = 0. \tag{4.4.6}$$

with $r(0) = 0$, $\dot{r}(0) = v_0$.

We may express (4.4.4) in state-variable (observability canonical) form [Kailath 1980]. Illustrating for the case of scalar $r(t)$ and $d = 3$, this is

$$\dot{\rho} = \begin{bmatrix} 0 & 1 & 0 \\ 0 & 0 & 1 \\ -a_3 & -a_2 & -a_1 \end{bmatrix} \rho \equiv F\rho$$

$$r = [1 \quad 0 \quad 0]\rho. \tag{4.4.7}$$

4.4 Command-Generator Tracker

Note that in this form the plant matrix is zero except for a superdiagonal of 1's and the bottom row of coefficients. We call (4.4.4)/(4.4.7) the *command generator* system.

Let us define the command generator characteristic polynomial as

$$\Delta(s) = s^d + a_1 s^{d-1} + \ldots + a_d. \quad (4.4.8)$$

Then, denoting d/dt in the time domain by s, we may write (4.4.4) as

$$\Delta(s) r = 0. \quad (4.4.9)$$

To make $z(t)$ follow $r(t)$, define the tracking error

$$e = r - z = r - Hx. \quad (4.4.10)$$

We should like to convert the servo or tracking problem into a *regulator* problem where the error must be regulated to zero.

Modified system

To accomplish this, write

$$\Delta(s) e = \Delta(s) r - \Delta(s) Hx = -H\xi, \quad (4.4.11)$$

where we have used (4.4.9) and defined the modified plant state vector

$$\xi = \Delta(s) x = x^{(d)} + a_1 x^{(d-1)} + \ldots + a_d x. \quad (4.4.12)$$

Note that (4.4.11) may be written in the observability canonical form

$$\dot{\varepsilon} = F\varepsilon + \begin{bmatrix} 0 \\ -H \end{bmatrix} \xi, \quad (4.4.13)$$

where $\varepsilon(t) = [e \ \dot{e} \ \ldots \ e^{(d-1)}]^T$ is the vector of the error and its first $d-1$ derivatives.

To determine the dynamics of $\xi(t)$, operate on (4.4.1) with $\Delta(s)$ to obtain

$$\dot{\xi} = A\xi + B\mu \quad (4.4.14)$$

where the modified control input is

$$\mu = \Delta(s) u = u^{(d)} + a_1 u^{(d-1)} + \ldots + a_d u. \quad (4.4.15)$$

Now we may put all the dynamics (4.4.13), (4.4.14) into a single augmented state representation by writing

$$\frac{d}{dt} \begin{bmatrix} \varepsilon \\ \xi \end{bmatrix} = \begin{bmatrix} F & 0 \\ \hline -H & \\ 0 & A \end{bmatrix} \begin{bmatrix} \varepsilon \\ \xi \end{bmatrix} + \begin{bmatrix} 0 \\ \hline B \end{bmatrix} \mu. \quad (4.4.16)$$

Using this system, we may now perform a LQ regulator design, since if its state goes to zero, then the tracking error $e(t)$ vanishes. For this design, we shall take the outputs available for feedback as

$$v = \begin{bmatrix} I & 0 \\ 0 & C \end{bmatrix} \begin{bmatrix} \varepsilon \\ \xi \end{bmatrix}. \quad (4.4.17)$$

We can select any reasonable PI that weights $[\varepsilon^T \; \xi^T]^T$ and μ and use the design equations in Table 4.1-1 to obtain optimal feedback gains so that

$$\mu = -[K_\varepsilon \; K_y]\begin{bmatrix} \varepsilon \\ C\xi \end{bmatrix} \qquad (4.4.18)$$

or

$$\Delta(s)u = -K_\varepsilon \varepsilon - K_y C \Delta(s)x. \qquad (4.4.19)$$

Servo compensator

To determine the control input $u(t)$ for the original system, write this as

$$\Delta(s)(u + K_y y) = -K_\varepsilon \varepsilon \equiv -[K_d \ldots K_2 \; K_1]\begin{bmatrix} e \\ \dot{e} \\ \vdots \\ e^{(d-1)} \end{bmatrix}. \qquad (4.4.20)$$

Thus, we obtain the transfer function

$$\frac{u + K_y y}{e} = -\frac{K_1 s^{d-1} + \ldots + K_{d-1} s + K_d}{s^d + a_1 s^{d-1} + \ldots + a_d}, \qquad (4.4.21)$$

which may be implemented in reachability canonical form to obtain the servo control structure shown in Fig. 4.4-1.

If $d = 1$ so that $r(t)$ is a unit step, this yields exactly the PI controller shown in Fig. 4.2-4. If $d = 2$ so that $r(t)$ is a ramp, it yields two integrators in the feedforward compensator, which results in a Type 2 system and gives zero steady-state error.

Note that the CGT is a servo controller that has a sensible structure with no derivatives of $r(t)$. Note further that its dynamics reflect the dynamics of the reference input. In fact, the controller is said to contain an *internal model of the reference input generator*.

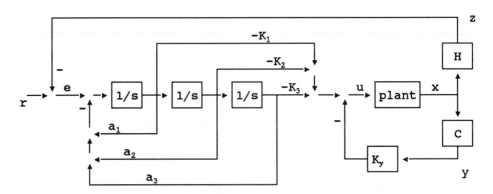

Figure 4.4-1 Command generator tracker for $d = 3$

4.4 Command-Generator Tracker

Discussion

It should be emphasized that this technique is extremely direct to apply. Indeed, given the command generator polynomial $\Delta(s)$, the system (4.4.16), (4.4.17) may be written down immediately, and the equations in Table 4.1-1 used to select the feedback gains.

It remains to say that we may place all the poles of the modified system (4.4.16) using full state feedback (i.e., $C = I$) if and only if the system is reachable. That is, the reachability matrix should have full rank. This should logically be a necessary condition if we want to accomplish the more difficult problem of pole placement using reduced-state, that is output, feedback with $C \neq I$. It is not too difficult to show by a straightforward determination of the reachability matrix of (4.4.16) that the modified system is reachable if:

1. The original system (4.4.1) is reachable, and
2. The open-loop transfer function from $u(t)$ to $z(t)$ given by

$$H(s) = H(sI - A)^{-1}B \qquad (4.4.22)$$

has no zeros at the roots of $\Delta(s) = 0$.

The second condition represents a restriction on the sorts of command inputs that a given plant may follow with zero steady-state error. For instance, to track a unit step the plant can have no system zeros at $s = 0$.

A word on the command generator assumption (4.4.4) is in order. For control systems design it is not necessary to determine the coefficients a_i that describe the actual reference command, which may be a complicated function of time (e.g., the pilot's command in an aircraft design example). Instead, the performance objectives should be taken into account to select $\Delta(s)$. For instance, if it is desired for the plant to follow a position command, then for design purposes we may select the command generator $\dot{r} = 0$. On the other hand, if the plant should follow a rate (velocity) command, we may select $\ddot{r} = 0$. Then, when the actual command input is applied (which may be neither a unit step nor a unit ramp) the system will exhibit the appropriate closed-loop behavior.

Example 4.4-1: Track Following for a Disk Drive Head-Positioning System

The head-positioning mechanism for a disk drive can be either rotary or of the linear voice coil actuator type. The latter is described by

$$\ddot{y} = \frac{k}{m} u \qquad (1)$$

where m is the mass of the coil and carriage assembly, k (nt/V) is the product of the motor force constant and power amplifier gain, $y(t)$ is the head position, and control input $u(t)$ is motor voltage [Bell et al. 1984]. By absorbing the constant k/m into the definition of the input, we may write the state-variable description

$$\dot{x} = \begin{bmatrix} 0 & 1 \\ 0 & 0 \end{bmatrix} x + \begin{bmatrix} 0 \\ 1 \end{bmatrix} u = Ax + Bu, \qquad (2)$$

with $x = [y \; v]^T$ and $v(t)$ the head velocity, where $u(t)$ is now in units of acceleration. This is just the Newton's law system we discussed in Examples 3.2-2, 3.3-1, 3.3-4, 3.4-1.

There are two control problems associated with the head-positioning mechanism. One is the *track access* or *seek* problem. Here, the head must be driven from an initial position to a certain track in minimum time. The second problem is that of *track following*, where a servo control system must be designed to hold the head above a specified track while reading from or writing to the disk. In this example we should like to deal with the track following problem [Franklin, Powell, and Emami-Naeini 1986].

Let the rotational velocity of the disk be ω_0, which is usually between 2500–4000 rpm depending on the application. Due to the fact that the disk is not quite centered, the tracks defined as circles on the disk actually trace out ellipses with respect to a stationary head position. The spindle design keeps the once-around runout component of the disk motion within 300 microinches. However, for reading and writing it is desired that the head follow the track with an error of less than 75 microinches.

To design a servo control system under these circumstances, we must account for the fact that the reference input $r(t)$ (i.e., the desired radius from the center of the spindle) satisfies the equation

$$\ddot{r} + \omega_0^2 r = 0. \tag{3}$$

Therefore, the command generator polynomial is

$$\Delta(s) = s^2 + \omega_0^2. \tag{4}$$

It is desired for the head position $y(t)$ to follow $r(t)$, so that the performance output is

$$z = [1 \; 0]x = Hx. \tag{5}$$

The augmented system (4.4.16) may now be written down as

$$\frac{d}{dt}\begin{bmatrix} e \\ \dot{e} \\ \hline \xi \end{bmatrix} = \begin{bmatrix} 0 & 1 & 0 & 0 \\ -\omega_0^2 & 0 & -1 & 0 \\ 0 & 0 & 0 & 1 \\ 0 & 0 & 0 & 0 \end{bmatrix} \begin{bmatrix} e \\ \dot{e} \\ \hline \xi \end{bmatrix} + \begin{bmatrix} 0 \\ 0 \\ \hline 0 \\ 1 \end{bmatrix} \mu. \tag{6}$$

Now, the LQ regulator design equations from Table 4.1-1 may be used to obtain the optimal feedback gains for this system so that

$$\mu = -[k_e \; k_{ep} \; k_s \; k_v] \begin{bmatrix} e \\ \dot{e} \\ \hline \xi \end{bmatrix}. \tag{7}$$

For this stage of the controls design we can select the PI in Table 4.1-1 with $Q = qI$, $R = I$, where q is a design parameter which may be chosen large for good regulation (e.g., $q = 100$).

Finally, the track following servo controller is given by (4.4.21), which has the form of Fig. 4.4-1 with $d = 2$. The command generator controller has an oscillator with a frequency of ω_0 as a precompensator. It is called the *internal model* of the reference generator [Franklin et al. 1986].

It is worth noting that in this situation (7) is a *full state-variable feedback*. Thus, if we prefer we may use the Riccati equation approach in section 3.3 to select the control gains. However, we prefer to use Table 4.1-1 with $C = I$, since the software described in Appen-

4.4 Command-Generator Tracker

dix A is convenient to use. It also allows the incorporation of time weighting t^k into the PI as shown in Table 4.2-1. This in turn allows us to select a simpler Q matrix, resulting in fewer design parameters to tune during the design phase.

∎

Tracking with Disturbance Rejection

The CGT approach may also be used in the disturbance rejection problem.

Disturbance generator system

If the system is driven by an unknown disturbance $d(t)$, then (4.4.1) must be modified to read

$$\dot{x} = Ax + Bu + Dd. \tag{4.4.23}$$

Suppose $d(t)$ satisfies the differential equation

$$d^{(q)} + p_1 d^{(q-1)} + \ldots + p_q d = 0 \tag{4.4.24}$$

for some degree q and known coefficients p_i. In illustration, d could be a constant unknown disturbance so that $\dot{d} = 0$, or a sinusoidal disturbance which satisfies a differential equation of order two.

Define

$$\Delta_d(s) = s^q + p_1 s^{q-1} + \ldots + p_q, \tag{4.4.25}$$

so that

$$\Delta_d(s)d = 0, \tag{4.4.26}$$

where s represents d/dt in the time domain. With

$$\xi \equiv \Delta_d(s)x, \quad \mu \equiv \Delta_d(s)u \tag{4.4.27}$$

it follows that (4.4.23) may be written as

$$\dot{\xi} = A\xi + B\mu, \tag{4.4.28}$$

which does not involve the disturbance.

Tracking and disturbance rejection

Now, suppose that the reference input $r(t)$ satisfies

$$\Delta_r(s)r = 0 \tag{4.4.29}$$

for some given $\Delta_r(s)$. Defining

$$\Delta(s) = \Delta_d(s)\Delta_r(s), \tag{4.4.30}$$

we may use the CGT technique to derive a controller that results in tracking of $r(t)$ by the performance output

$$z = Hx \qquad (4.4.31)$$

in the presence of the disturbance $d(t)$ (see the problems).

Indeed, if the measured output is $y = Cx$, the required controller is given exactly by (4.4.21) with the modified $\Delta(s)$ of (4.4.30).

Referring to the discussion following (4.4.21), we see that perfect disturbance rejection may be achieved only if the system has no zeros at the poles of the disturbance.

In point of fact, we need not select the polynomial $\Delta(s)$ given by (4.4.30) if $\Delta_d(s)$ and $\Delta_r(s)$ have common factors. Instead, we should select the least common multiple of $\Delta_d(s)$ and $\Delta_r(s)$.

4.5 EXPLICIT MODEL-FOLLOWING DESIGN

Here, we consider the problem of controlling a plant so that it has a response like that of a *prescribed model* with desirable behavior. The model has desirable qualities in terms of speed of response, percent overshoot, robustness, and so on. In aircraft control, for instance, a series of performance models for different situations is tabulated in [Mil. Spec. 1797]. Thus, aircraft controls design often has the objective of making the aircraft behave like the specified model.

There are two fundamentally different sorts of model-following control, "explicit" and "implicit," which result in controllers of different structure [Armstrong 1980, Kreindler and Rothschild 1976]. The latter, however, yields an inconvenient form of controller for the servo design or tracking problem; specifically, it usually requires derivatives of the performance output [Stevens and Lewis 1992]. Therefore, in this section we shall only consider explicit model following. Implicit model following is important for a different application; namely, selecting the performance index weighting matrices in the LQR problem.

Regulator with Model Following

First, we shall consider the regulator problem, where the objective is to drive the plant state to zero. Then, we shall treat the more difficult tracker or servo problem, where the plant is to follow a reference command with behavior like the prescribed model.

Let the plant be described in state-variable form by

$$\dot{x} = Ax + Bu \qquad (4.5.1)$$

$$y = Cx \qquad (4.5.2)$$

$$z = Hx \qquad (4.5.3)$$

with state $x(t) \in \mathbf{R}^n$ and control input $u(t) \in \mathbf{R}^m$. The performance output is $z(t)$, and $y(t)$ is a vector of additional measurements available for feedback purposes.

4.5 Explicit Model-Following Design

A model is prescribed with dynamics

$$\underline{\dot{x}} = \underline{A}\,\underline{x} \tag{4.5.4}$$

$$\underline{z} = \underline{H}\,\underline{x}, \tag{4.5.5}$$

where the model matrix $\underline{A}$ reflects a system with desirable handling qualities such as speed of response, overshoot, and so on. The additional model states available for feedback purposes are given by

$$\underline{y} = \underline{C}\,\underline{x}. \tag{4.5.6}$$

Model quantities shall be denoted by underbars or the subscript "m".

It is desired for the plant performance output $z(t)$ to match the model output $\underline{z}(t)$, for then the plant will exhibit the desirable time response of the model. That is, we should like to make small the *model mismatch error*

$$e = \underline{z} - z = \underline{H}\,\underline{x} - Hx. \tag{4.5.7}$$

To achieve this control objective, let us select the performance index

$$J = \frac{1}{2}\int_0^\infty (e^T Q e + u^T R u)\,dt, \tag{4.5.8}$$

with $Q \geq 0$ and $R > 0$.

We can cast this model-matching problem into the form of the regulator problem whose solution appears in Table 4.1-1 as follows.

Define the augmented state $x' = [x^T \quad \underline{x}^T]^T$ and the augmented system

$$\dot{x}' = \begin{bmatrix} A & 0 \\ 0 & \underline{A} \end{bmatrix} x' + \begin{bmatrix} B \\ 0 \end{bmatrix} u \equiv A'x' + B'u \tag{4.5.9}$$

$$y' = \begin{bmatrix} y \\ \underline{y} \end{bmatrix} = \begin{bmatrix} C & 0 \\ 0 & \underline{C} \end{bmatrix} x' \equiv C'x' \tag{4.5.10}$$

so that

$$e = [-H \quad \underline{H}]x' \equiv H'x'. \tag{4.5.11}$$

Then, the PI (4.5.8) may be written

$$J = \frac{1}{2}\int_0^\infty ((x')^T Q' x' + u^T R u)\,dt, \tag{4.5.12}$$

with

$$Q' = \begin{bmatrix} H^T Q H & -H^T Q \underline{H} \\ -\underline{H}^T Q H & \underline{H}^T Q \underline{H} \end{bmatrix}. \tag{4.5.13}$$

At this point, it is clear that the design equations of Table 4.1-1 apply if the primed quantities A', B', C', Q', are used. The conditions for convergence of the algorithm in Table 4.1-2 require that (A', B', C') be output stabilizable and $(\sqrt{Q'}, A')$ be detectable. Since the model matrix $\underline{A}$ is certainly stable, the block diagonal form of A' and C' shows that output stabilizability of the plant (A, B, C) is required. The second condition requires detectability of $(\sqrt{Q} H, A)$.

It should be noted that the detectability condition on $(\sqrt{Q} H, A)$ may be avoided by including time weighting of the form $t^k(x')^T Q' x'$ in the PI. Then, the control gains may be computed by using a simplified version of the equations in Table 4.2-1. Specifically, since the equations there deal with the tracker problem, we can take $X = I$ and $\bar{y} = 0$ to solve the regulator design problem. This corresponds to minimizing not J but its expected value. See the discussion in sections 4.1 and 4.2.

The form of the resulting control law is quite interesting. Indeed, the optimal feedback is of the form

$$u = -K'y' \equiv -[K_p \quad K_m]y' = -K_p y - K_m \underline{y}. \tag{4.5.14}$$

Thus, not only the plant output but also the *model* output is required. That is, the model acts as a *compensator* to drive the plant states to zero in such a fashion that the performance output $z(t)$ follows the model output $\underline{z}(t)$.

Tracker with Model Following

Unfortunately, while the model-following regulator problem has a direct solution that is easy to obtain, the model-following *tracker* problem is not so easy. In this situation, we should like the plant (4.5.1)–(4.5.3) to behave like the model

$$\underline{\dot{x}} = \underline{A}\,\underline{x} + \underline{B}\,r \tag{4.5.15}$$

$$\underline{z} = \underline{H}\,\underline{x}, \tag{4.5.16}$$

which is driven by the reference input $r(t)$. The approach above yields

$$\dot{x}' = \begin{bmatrix} A & 0 \\ 0 & \underline{A} \end{bmatrix} x' + \begin{bmatrix} B \\ 0 \end{bmatrix} u + \begin{bmatrix} 0 \\ \underline{B} \end{bmatrix} r \equiv A'x' + B'u + G'r \tag{4.5.17}$$

which contains a term in $r(t)$.

Let us approach this control problem by using the *command-generator* technique of section 4.4. Thus, suppose for some initial conditions the reference command satisfies the differential equation

$$r^{(d)} + a_1 r^{(d-1)} + \ldots + a_d r = 0 \tag{4.5.18}$$

for a given degree d and set of coefficients a_i. Define the command generator characteristic polynomial as

4.5 Explicit Model-Following Design

$$\Delta(s) = s^d + a_1 s^{d-1} + \ldots + a_d. \tag{4.5.19}$$

Then, denoting d/dt in the time domain by s, we may write

$$\Delta(s)r = 0. \tag{4.5.20}$$

Multiplying the augmented dynamics (4.5.17) by $\Delta(s)$ results in

$$\dot{\xi} = A'\xi + B'\mu \tag{4.5.21}$$

where the modified state and control input are

$$\xi = \Delta(s)x' = (x')^{(d)} + a_1(x')^{(d-1)} + \ldots + a_d x'. \tag{4.5.22}$$

$$\mu = \Delta(s)u = u^{(d)} + a_1 u^{(d-1)} + \ldots + a_d u. \tag{4.5.23}$$

We note the important point that $r(t)$ has vanished by virtue of (4.5.20). Let us denote

$$\xi = \begin{bmatrix} \xi_p \\ \xi_m \end{bmatrix}, \tag{4.5.24}$$

with ζ_p the modified plant state and ζ_m the modified model state.

Applying $\Delta(s)$ to the model mismatch error (4.5.7) results in

$$\Delta(s)e = [-H \quad H]\xi = H'\xi. \tag{4.5.25}$$

This may be expressed in terms of state variables using the observability canonical form [Kailath 1980], which for scalar $e(t)$ and $d = 3$ is

$$\dot{\varepsilon} = \begin{bmatrix} 0 & 1 & 0 \\ 0 & 0 & 1 \\ -a_3 & -a_2 & -a_1 \end{bmatrix} \varepsilon + \begin{bmatrix} 0 \\ H' \end{bmatrix} \xi \equiv F\varepsilon + \begin{bmatrix} 0 \\ H' \end{bmatrix} \xi \tag{4.5.26}$$

$$e = [1 \quad 0 \quad 0]\varepsilon, \tag{4.5.27}$$

where $\varepsilon(t) = [e \quad \dot{e} \quad \ldots \quad e^{(d-1)}]^T$ is the vector of the error and its first $d - 1$ derivatives.

Collecting all the dynamics (4.5.21), (4.5.26) into one system yields

$$\frac{d}{dt}\begin{bmatrix} \varepsilon \\ \xi \end{bmatrix} = \begin{bmatrix} F & 0 \\ \hline 0 & A' \end{bmatrix}\begin{bmatrix} \varepsilon \\ \xi \end{bmatrix} + \begin{bmatrix} 0 \\ \hline B' \end{bmatrix}\mu. \tag{4.5.28}$$

Using this system, we may now perform a LQ regulator design, since if its state goes to zero, then the tracking error $e(t)$ vanishes. For this design, we shall take the outputs available for feedback as

$$v = \begin{bmatrix} I & 0 & 0 \\ 0 & C & 0 \\ 0 & 0 & \underline{C} \end{bmatrix}\begin{bmatrix} \varepsilon \\ \xi_p \\ \xi_m \end{bmatrix}. \tag{4.5.29}$$

To achieve small error without using too much control energy, we may select the PI

(4.5.8) [with $u(t)$ replaced by $\mu(t)$]. According to (4.5.27), the error is given in terms of the state of (4.5.28) by

$$e = h\begin{bmatrix} \varepsilon \\ \xi \end{bmatrix}, \tag{4.5.30}$$

with $h = [1 \ 0 \ldots 0]$ the first row of the identity matrix. Therefore, in the PI we should weight the state of (4.5.28) using

$$Q' = h^T Q h. \tag{4.5.31}$$

Since the observability canonical form is always observable, the augmented system (4.5.28) is detectable if the plant (H, A) and the model $(\underline{H}, \underline{A})$ are both detectable.

Now, applying the equations of Table 4.1-1 to the system (4.5.28) with outputs (4.5.29) and PI weights Q' and R yields the optimal control law

$$\mu = -[K_\varepsilon \quad K_p \quad K_m] \begin{bmatrix} \varepsilon \\ C\xi_p \\ \underline{C}\xi_m \end{bmatrix} \tag{4.5.32}$$

or

$$\Delta(s)u = -K_\varepsilon \varepsilon - K_p C \Delta(s) x - K_m \underline{C} \Delta(s) \underline{x}. \tag{4.5.33}$$

To determine the optimal control input $u(t)$, write this as

$$\Delta(s)(u + K_p y + K_m \underline{y}) = -K_\varepsilon \varepsilon \equiv -[K_d \ldots K_2 \quad K_1] \begin{bmatrix} e \\ \dot{e} \\ \vdots \\ e^{(d-1)} \end{bmatrix}. \tag{4.5.34}$$

Thus, we obtain the transfer function

$$\frac{u + K_p y + K_m \underline{y}}{e} = -\frac{K_1 s^{d-1} + \ldots + K_{d-1} s + K_d}{s^d + a_1 s^{d-1} + \ldots + a_d}, \tag{4.5.35}$$

which may be implemented in reachability canonical form to obtain the control structure shown in Fig. 4.5-1.

The structure of this *model-following command generator tracker* (CGT) is very interesting. It consists of an output feedback K_p, a feedforward compensator which is nothing but the reference model, and an additional feedforward filter in the error channel that guarantees perfect tracking. Note that if $d = 1$ so that $r(t)$ is a unit step, the error filter is a PI controller like that shown in Fig. 4.2-4. If $d = 2$ so that $r(t)$ is a ramp, the error filter consists of two integrators, resulting in a Type 2 system that gives zero steady-state error.

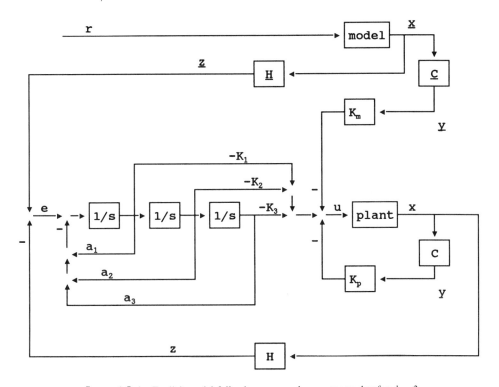

Figure 4.5-1 Explicit model-following command generator tracker for $d = 3$

It is extremely interesting to note that the augmented state description (4.5.28) is nothing but the state description of Fig. 4.5-1 (see the problems).

It should be emphasized that this technique is extremely direct to apply. Indeed, given the prescribed model and the command generator polynomial $\Delta(s)$, the system (4.5.28), (4.5.29) may be written down immediately, and the design equations in Table 4.1-1 used to select the feedback gains. See section 4.4 for some discussion on the reachability of the augmented system.

PROBLEMS FOR CHAPTER 4

Problems for Section 4.1

4.1-1 Derive the necessary conditions in Table 4.1-1.

4.1-2 Output-Feedback Design for Scalar Systems

  a. Consider the case where $x(t)$, $u(t)$, $y(t)$ are all scalars. Show that the solution S to the second Lyapunov equation in Table 4.1-1 is not needed to determine the output-feedback gain K. Find an explicit solution for P and hence for the optimal gain K. Compare to the case for full state feedback.

b. Repeat for the case where $x(t)$ and $y(t)$ are scalars, but $u(t)$ is an m-vector.

4.1-3 Use (4.1.23) to eliminate K in the Lyapunov equations of Table 4.1-1, hence deriving two coupled nonlinear equations which may be solved for the optimal auxiliary matrices S and P. Does this simplify the solution of the output-feedback design problem?

4.1-4 Software for Output-Feedback Design. Write a program that finds the gain K minimizing the PI in Table 4.1-1 using the SIMPLEX algorithm in Press et al. [1986]. Use it to verify the results of Example 4.1-1.

4.1-5 For the Newton's system

$$\dot{x} = \begin{bmatrix} 0 & 1 \\ 0 & 0 \end{bmatrix} x + \begin{bmatrix} 0 \\ 1 \end{bmatrix} u, \qquad y = \begin{bmatrix} 1 & 1 \end{bmatrix} x$$

Find the output-feedback gain that minimizes the PI in Table 4.4-1 with $Q = I$. Try various values of R to obtain a good response. You will need the software from Problem 4.1-4. The closed-loop step response may be plotted using program TRESP in Appendix A.

4.1-6 Gradient-Based Software for Output-Feedback Design. Write a program that finds the gain K minimizing the PI in Table 4.1-1 using the Davidon-Fletcher-Powell algorithm in Press et al. [1986]. Use it to verify the results of Example 4.1-1.

Problems for Section 4.2

4.2-1 Derive (4.2.31), (4.2.32). You will need to use integration by parts [MacFarlane 1963].

4.2-2 Derive the necessary conditions for optimality given in Table 4.2-1.

4.2-3 Verification of Inverted Pendulum Design on the Full Nonlinear Equations. In Example 4.2-4 we verified the LQ design by employing TRESP in Appendix A to simulate the closed-loop system $(A - BKC)$ using the linearized equations presented in the example. However, a more conscientious procedure would be to verify the closed-loop behavior of the actual nonlinear equations describing the inverted pendulum.

In the problems for section 2.1 was presented the full nonlinear model of the inverted pendulum. Use TRESP to close the loop on this nonlinear system using the gains found in Example 4.2-4.

4.2-4 Software for LQ Output-Feedback Design. Write a program to solve for the optimal gain K in Table 4.2-1 using the SIMPLEX algorithm in Press et al. [1986]. Use it to verify the examples presented.

4.2-5 Gradient-Based Software for LQ Output-Feedback Design. Write a program to solve for the optimal gain K in Table 4.2-1 using the Davidon-Fletcher-Powell algorithm in Press et al. [1986]. Use it to verify the examples presented.

4.2-6 LQ Design for Ball Balancer. In the problems for section 2.1 the equations for the ball balancer system were given.

a. Repeat Example 4.2-4 for this system. Use $R = 0.5$ m, $M = 0.2$ kg, $m = 0.5$ kg. The control structure given in the example will work for this system. The major problem will be determining an initial stabilizing gain for the optimization algorithm.

b. Verify the design using TRESP to simulate the closed-loop response by closing the loop on the nonlinear equations of the ball balancer.

Problems for Chapter 4

Problems for Section 4.3

4.3-1 Complete the design of Example 4.3-1. That is:
  a. Select a value for α, and use Table 4.1-1 to find the regulator gain K. Tune the values of Q and R until the response to nonzero initial conditions is suitable.
  b. Find the tracker control law. To verify the design, simulate the step response of the closed-loop system using TRESP in Appendix A.

4.3-2 Regulator Redesign Servo for DC Motor. Use the approach of this section to design a servo for the scalar DC motor model in Example 4.2-2. Simulate the step response of the closed-loop system.

4.3-3 Regulator Redesign Servo for DC Motor. Use the approach of this section to design a servo for the armature-controlled DC motor model in Example 4.2-3. Simulate the step response of the closed-loop system.

4.3-4 Regulator Redesign Servo for Inverted Pendulum. Use the approach of this section to design a servo for the inverted pendulum in Example 4.2-4. Simulate the step response of the closed-loop system.

Problems for Section 4.4

4.4-1 Find the reachability matrix of (4.4.16) to verify the tracking conditions for full state feedback relating to (4.4.22).

4.4-2 Derive the CGT for a system with an unknown disturbance $d(t)$ and verify that it is given by (4.4.21) with $\Delta(s)$ modified as in (4.4.30).

4.4-3 Complete the design in Example 4.4-1. That is, select $\omega_0 = 3000$ rpm and perform a regulator design on the augmented system using the equations in Table 4.1-1. Tune q to obtain suitable time responses of the augmented system to nonzero initial conditions. To verify the performance, simulate the CGT on the system using TRESP in Appendix A.

4.4-4 Tracking with Disturbance Rejection. Redo Example 4.4-1 if there is a constant bias disturbance on the head position.

4.4-5 Use the CGT approach to redo Example 4.2-2.

Problems for Section 4.5

4.5-1 Write the state variable description of Fig. 4.5-1, verifying that it is nothing but (4.5.28).

4.5-2 It is desired to make the scalar plant

$$\dot{x} = x + u, \qquad y = x, \qquad z = x$$

behave like the scalar model

$$\dot{\underline{x}} = -2\underline{x} + r, \qquad \underline{y} = \underline{x}, \qquad \underline{z} = \underline{x}$$

with reference input equal to the unit step. Use explicit model following to design a servosystem:

a. Draw the controller structure.
b. Select the control gains using LQR design on the augmented system. You will need to use the software written for the problems of section 4.1. Simulate the response.

4.5-3 A plant is described by Newton's law

$$\dot{x}_1 = x_2, \qquad \dot{x}_2 = u.$$

The velocity should follow the model output and measurements of position are taken so that

$$y = x_1, \qquad z = x_2.$$

The prescribed model with desirable characteristics is given by

$$\underline{\dot{x}} = -3\underline{x} + r, \qquad \underline{y} = \underline{x}, \qquad \underline{z} = \underline{x}$$

with $r(t)$ the unit step. Use explicit model following to design a servosystem:
a. Draw the compensator structure.
b. Select the control gains using LQR design on the augmented system. You will need to use the software written for the problems of section 4.1.

PART III

DIGITAL CONTROL

In Part II we showed how to design continuous-time controllers by a variety of means, focusing primarily on output-feedback design of servo controllers. In this part of the book we discuss the design of digital control systems. In light of the high speed and small size of today's microprocessors, most advanced control systems are implemented using digital techniques.

There are two basic approaches to the design of digital control systems. The first is to convert a given continuous-time control system into digital form using the bilinear transformation or some such technique. We call this *continuous controller redesign*, and cover it in Chapter 5. The second approach is *direct discrete-time design* of digital controllers, which we treat in Chapter 7. In building a digital control system it is important to be aware of some implementation issues like actuator saturation, quantization, overflow, and controller realizations; such topics are addressed in Chapter 6.

We give examples implementing digital controllers on an actual digital signal processor (DSP), including the code that must be compiled on the DSP.

5

Digital Control by Continuous Controller Redesign

SUMMARY

In the next three chapters we shall discuss digital control. Two design approaches will be covered: continuous controller redesign and direct discrete-time design. In this chapter we treat the former, showing how to convert existing continuous controllers to digital controllers. Any of the techniques covered in Part II can be used for the initial continuous-time design.

INTRODUCTION

In Part II of the book we showed how to design continuous-time controllers. In Chapter 3 we covered the case of full state feedback, developing some of the fundamental notions of modern control theory. In Chapter 4 we discussed design using output feedback. In that chapter we were able to design controllers with any prespecified structure, developing in section 4.2 the fundamental design approach to be used in this book.

However, with microprocessors so fast, light, accurate, versatile, and economical, control laws are often implemented in digital form. Therefore, in this part of the book we shall address the design of digital, or discrete-time, controllers.

Two fundamental approaches to digital design will be covered. In the first approach, covered in this chapter, we show how to convert an already designed continuous-time controller to a discrete-time controller. An advantage of this *continuous controller redesign* approach, which is developed in section 5.2, is that all of the techniques covered in Chapters 3 and 4 may be used to design the continuous controller, which is then simply discretized. Another advantage is that the sample period T does not have to be selected until after the continuous controller has been designed.

Unfortunately, these controller discretization schemes are approximations. We discuss this in section 5.3, where we go into the characteristics of the sampling process, hold devices, and computation delays. There, we show how to modify the continuous-time design so that a more suitable digital controller is obtained. In section 5.4 we cover the design of minimum-time digital controllers.

Regardless of any modifications in the design procedure, in continuous controller redesign the sampling period T must be small to make sure the digital controller performs like the continuous version from which it was designed. However, small sampling periods can be problematic for several reasons, as we shall see. Moreover, the continuous redesign approach gives little insight into the properties of the sampling process, such as the appearance of nonminimum-phase zeros, or the properties of discrete systems, such as deadbeat behavior.

Therefore, in Chapter 7 we cover the *direct discrete design* approach to digital controls. This is an exact technique which usually allows significantly larger sample periods than the continuous redesign approach. It also gives added insight into digital controls, as well as guaranteed performance at the sample points. Moreover, with the right software it is no more difficult than continuous controller design.

In Chapter 6 we discuss some implementation considerations, such as fixed-point arithmetic, computer quantization and roundoff error effects, and the type of realization structure of the digital controller. We provide a sample digital controller implementation on an actual digital signal processor (DSP).

There are many excellent references on digital control. Some of them are listed at the end of the book. We shall draw most heavily on Franklin and Powell [1980], Åström and Wittenmark [1984], and Phillips and Nagle [1984].

5.1 SIMULATION OF DIGITAL CONTROLLERS

A digital control scheme is shown in Fig. 5.1-1. The plant to be controlled $G(s)$ is a continuous-time system, and $K(z)$ is the dynamic digital controller, where s and z are respectively the Laplace and Z-transform variables (i.e., $1/s$ represents integration and z^{-1} represents a unit time delay). The digital controller $K(z)$ is implemented using software code in a microprocessor.

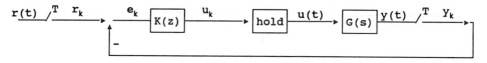

Figure 5.1-1 Digital controller

The hold device in the figure is a digital-to-analog (D/A) converter that converts the discrete control samples u_k computed by the software controller $K(z)$ into the continuous-time control $u(t)$ required by the plant. It is a *data reconstruction* device. The input u_k and output $u(t)$ for a *zero-order hold* (ZOH) are shown in Fig. 5.1-2. Note that $u(kT) = u_k$, with T the sample period, so that $u(t)$ is continuous from the right. That is, $u(t)$ is updated at times kT. We shall discuss hold devices further in section 5.3.

5.1 Simulation of Digital Controllers

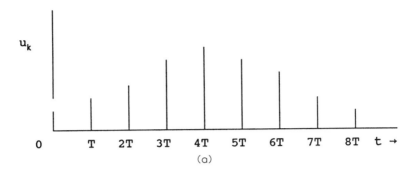

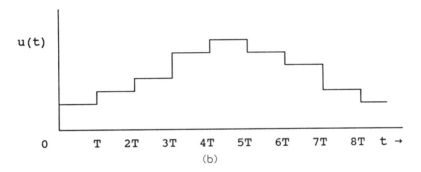

Figure 5.1-2 Data reconstruction using a ZOH. (a) Discrete control sequence u_k. (b) Reconstructed continuous signal $u(t)$.

The sampler with sample period T is an analog-to-digital (A/D) converter that takes the samples $y_k = y(kT)$ of the output $y(t)$ that are required by the software controller $K(z)$.

In this chapter we shall discuss the design of the digital controller $K(z)$. Once the controller has been designed, it is important to *simulate* it before it is implemented to determine if the closed-loop response is suitable. The simulation should provide the response at all times, including times between the samples.

To simulate a digital controller on a computer we may use the scheme shown in Fig. 5.1-3. There, the continuous dynamics $G(s)$ are contained in the subroutine $F(t, x, \dot{x})$; they are integrated using a Runge-Kutta integrator. The figure assumes a ZOH; thus, the control input $u(t)$ is updated to u_k at each time kT, and then held constant until time $(k + 1)T$. Note that two time intervals are involved: the sampling period T and the *Runge-Kutta integration period* $T_R \ll T$. T_R should be selected as an integral divisor of T.

This simulation technique provides $x(t)$ as a continuous function of time, even at values *between* the sampling instants (in fact, it provides $x(t)$ at multiples of T_R). This is essential in verifying acceptable *intersample behavior* of the closed-loop system prior to implementing the digital controller on the actual plant.

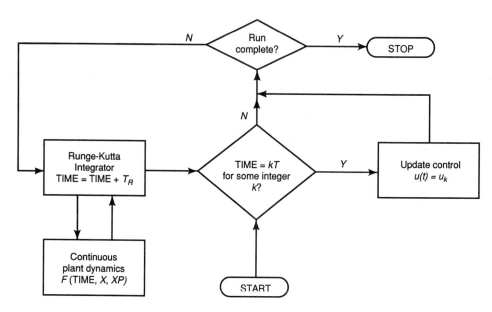

Figure 5.1-3 Digital control simulation scheme

A program that realizes Fig. 5.1-3 is given in Appendix A. In fact, it is the same program TRESP that we used for obtaining time responses in Part II of the book. It is written in a modular fashion to apply to a wide variety of situations. We shall illustrate its use in several examples.

5.2 DISCRETIZATION OF CONTINUOUS CONTROLLERS

A digital controls design approach that could directly use all of the continuous-time techniques of Chapters 3 and 4 would be extremely appealing. Therefore, in this section we shall discuss digital control by redesign of an existing continuous-time controller. In this approach, the continuous controller is first designed using any desired technique. Then, the controller is discretized to obtain the digital control law, which is finally programmed on the microprocessor.

Suppose a continuous-time controller $K^c(s)$ has been designed for the plant $G(s)$ by some means, such as LQ design. We shall discuss two approximate schemes for converting $K^c(s)$ into a discrete-time controller $K(z)$ that can be implemented on a microprocessor: the bilinear transformation and the matched pole/zero technique.

A major topic in this section will be digital proportional-integral-derivative (PID) controllers. We shall show how to convert continuous-time PID gains into digital PID gains, and discuss the Ziegler-Nichols technique for tuning the PID controller.

5.2 Discretization of Continuous Controllers

In the last subsection we discuss the control of processes with time delays, showing how to implement a Smith Predictor in discrete time.

Bilinear Transformation

Suppose there is available a continuous-time controller $K^c(s)$, and it is desired to convert it for implementation purposes into a digital controller. Let the sample period be T sec so that the *sampling frequency* is

$$f_s = \frac{1}{T}, \quad \omega_s = \frac{2\pi}{T}. \tag{5.2.1}$$

A popular way to convert a continuous transfer function to a discrete one is the *bilinear transformation* (BLT) or *Tustin's approximation*. As is well known [Oppenheim and Schafer 1975], the relation between the Laplace transform variable s and the Z-transform variable z is given by $z = e^{sT}$. Since, as may be seen by series expansion

$$z = e^{sT} \approx \frac{1 + sT/2}{1 - sT/2} \tag{5.2.2}$$

we may propose inverting this transformation and defining

$$s' = \frac{2}{T} \frac{z-1}{z+1}, \tag{5.2.3}$$

then selecting the digital controller according to

$$K(z) = K^c(s'). \tag{5.2.4}$$

The BLT corresponds to approximating integration using the trapezoid rule, since if

$$\frac{Y(z)}{U(z)} = \frac{2}{T} \frac{z-1}{z+1} = \frac{2}{T} \frac{1 - z^{-1}}{1 + z^{-1}}$$

then (recall that z^{-1} is the unit delay in the time domain so that $z^{-1} u_k = u_{k-1}$)

$$u_k = u_{k-1} + \frac{T}{2}(y_k + y_{k-1}). \tag{5.2.5}$$

Suppose the continuous transfer function is

$$K^c(s) = \frac{\prod_{i=1}^{m}(s + t_i)}{\prod_{i=1}^{n}(s + s_i)} \tag{5.2.6}$$

with poles and zeros respectively at $s = -s_i$, $s = -t_i$, and a relative degree of $r = n - m > 0$. Then the BLT yields

$$K(z) = \frac{\prod_{i=1}^{m}\left[\frac{2(z-1)}{T(z+1)} + t_i\right]}{\prod_{i=1}^{n}\left[\frac{2(z-1)}{T(z+1)} + s_i\right]}$$

$$K(z) = \left[\frac{T}{2}(z+1)\right]^r \frac{\prod_{i=1}^{m}[(z-1) + (z+1)t_i T/2]}{\prod_{i=1}^{n}[(z-1) + (z+1)s_i T/2]} \tag{5.2.7}$$

$$K(z) = \left[\frac{T}{2}(z+1)\right]^r \frac{\prod_{i=1}^{m}[(1 + t_i T/2)z - (1 - t_i T/2)]}{\prod_{i=1}^{n}[(1 + s_i T/2)z - (1 - s_i T/2)]}$$

It can be seen that the poles and finite zeros map to the z-plane according to

$$z = \frac{1 + sT/2}{1 - sT/2}. \tag{5.2.8}$$

However, the r zeros at infinity in the s-plane map into zeros at $z = -1$. This is sensible, since $z = -1$ corresponds to the *Nyquist frequency* ω_N, where $z = e^{j\omega_N T} = -1$ so that $\omega_N T = \pi$ or

$$\omega_N = \frac{\pi}{T} = \frac{\omega_s}{2}. \tag{5.2.9}$$

This is the highest frequency before folding of $|K(e^{j\omega T})|$ occurs (see Fig. 5.2-1). We shall discuss folding or aliasing in section 5.3.

Since, according to (5.2.8), the BLT maps the left-half of the s-plane into the unit circle, it maps stable continuous systems $K^c(s)$ into stable discrete systems $K(z)$.

Since the numbers of poles and zeros in (5.2.7) are both equal to $n = m + r$, regardless of the relative degree of $K^c(s)$, the BLT gives discretized transfer functions $K(z)$ that have a *relative degree of zero*; that is, the degrees of the numerator and denominator are the same. Since the relative degree is zero, then

$$K(z) = \frac{b_0 z^n + b_1 z^{n-1} + \cdots + b_n}{z^n + a_1 z^{n-1} + \cdots + a_n}. \tag{5.2.10}$$

Since $Y(z) = K(z)U(z)$, the difference equation relating y_k and u_k is

$$y_k = -a_1 y_{k-1} - \cdots - a_n y_{k-n} + b_0 u_k + b_1 u_{k-1} + \cdots + b_n u_{k-n} \tag{5.2.11}$$

and the current output y_k depends on the current input u_k. This is usually an undesirable state of affairs, since it takes some computation time for the microprocessor to compute y_k. Techniques for including the computation time will be discussed later.

5.2 Discretization of Continuous Controllers

If the continuous-time controller is given in the state-space form

$$\dot{x} = A^c x + B^c u \qquad (5.2.12)$$
$$y = Cx + Du$$

then one may use the Laplace transform and (5.2.3) to show that the discretized system using the BLT is given by [Hanselmann 1987]

$$x_{k+1} = Ax_k + B_1 u_{k+1} + B_0 u_k \qquad (5.2.13)$$
$$y_k = Cx_k + Du_k$$

with

$$A = \left[I - A^c \frac{T}{2} \right]^{-1} \left[I + A^c \frac{T}{2} \right]$$
$$B_1 = B_0 = \left[I - A^c \frac{T}{2} \right]^{-1} \frac{T}{2} B^c. \qquad (5.2.14)$$

Note that the discretized system is not a traditional state-space system since x_{k+1} depends on u_{k+1}. Aside from computation time delays, this is not a problem in our applications, since all we require of (5.2.13) is to implement it on a microprocessor. Since (5.2.13) is only a set of difference equations, this is easily accomplished.

Matched Pole-Zero

A second approximation technique for converting a continuous transfer function to a discrete one is the *matched pole-zero* (MPZ) method. Here, both the poles and finite zeros are mapped into the z-plane using the transformation e^{sT}, as follows:

1. If $K^c(s)$ has a pole (or finite zero) at $s = s_i$, then $K(z)$ will have a pole (or finite zero) at

$$z_i = e^{s_i T}. \qquad (5.2.15)$$

2. If the relative degree of $K^c(s)$ is r, so that it has r zeros at infinity, then r zeros of $K(z)$ are taken at $z = -1$ by multiplying by the factor $(1 + z)^r$.

3. The gain of $K(z)$ is selected so that the DC gains of $K^c(s)$ and $K(z)$ are the same; that is, so that

$$K(1) = K^c(0). \qquad (5.2.16)$$

An alternative to step 2 is to map only $r - 1$ of the infinite s-plane zeros into $z = -1$. This leaves the relative degree of $K(z)$ equal to one, which allows one sample period for control computation time. We shall call this the *modified MPZ* method.

Thus, if

$$K^c(s) = \frac{\prod_{i=1}^{m} (s + t_i)}{\prod_{i=1}^{n} (s + s_i)} \qquad (5.2.17)$$

and the relative degree is $r = n - m$, then the MPZ discretized transfer function is

$$K(z) = k(z + 1)^r \frac{\prod_{i=1}^{m} (z - e^{-t_i T})}{\prod_{i=1}^{n} (z - e^{-s_i T})} \qquad (5.2.18)$$

where the gain k is chosen to ensure (5.2.16).

Note that if $K^c(s)$ is stable, then so is the $K(z)$ obtained by the MPZ. This is due to the fact that poles at $s = -s_i$ are mapped into discrete poles at $z = e^{-s_i T}$, which are inside the unit circle if $s_i > 0$.

Although the MPZ requires simpler algebra than the BLT, the latter is more popular in industry.

Example 5.2-1: BLT and MPZ on a First-Order System

Let the continuous filter be

$$K^c(s) = \frac{a}{s + a}. \qquad (1)$$

a. BLT

Using the BLT

$$K_1(z) = \frac{a}{\frac{2}{T}\frac{z-1}{z+1} + a} = \frac{aT}{aT + 2} \cdot \frac{z + 1}{z - \frac{1 - aT/2}{1 + aT/2}}. \qquad (2)$$

Evaluating the DC gains, we see that $K^c(0) = 1$, and $K(1) = 1$. Note that the pole of $K_1(z)$ is at

$$z = \frac{1 - aT/2}{1 + aT/2}. \qquad (3)$$

b. BLT with Frequency Prewarping

According to (2)

$$K_1(e^{j\omega T}) = \frac{a}{\frac{2}{T}\frac{e^{j\omega T}-1}{e^{j\omega T}+1} + a} = \frac{a}{\frac{2}{T} j \tan(\omega T/2) + a}, \qquad (4)$$

so the half-power point occurs at $|K_1(e^{j\omega T})|^2 = \frac{1}{2}$, or

$$a = \frac{2}{T} \tan(\omega T/2) \qquad (5)$$

or

$$\omega = \frac{2}{T} \tan^{-1}(aT/2). \qquad (6)$$

Thus, while the cutoff frequency $\omega_{1/2}$ of the original continuous filter (1) is a rad/sec, the discretized version has a cutoff frequency of (6). That is, the BLT warps the frequency axis.

5.2 Discretization of Continuous Controllers

For small values of the sampling period T, (6) is approximately equal to a. However, for larger values of T it may be necessary to "prewarp" the continuous frequency so that, at some specified design frequency ω_1, $K^c(s)$ and $K(z)$ have the same value. This frequency ω_1 could be, for instance, the cutoff frequency $\omega_{1/2}$ or the crossover frequency ω_c. The BLT with prewarping is given by

$$s' = \frac{\omega_1}{\tan(\omega_1 T/2)} \frac{z-1}{z+1} \tag{7}$$

$$K(z) = K^c(s'). \tag{8}$$

For some purposes, such as digital filter design, prewarping is important. However, for the purposes of digital control we shall not require it.

c. Matched Pole-Zero

Using the MPZ technique to discretize $K^c(s)$ yields

$$K(z) = k\frac{a(z+1)}{z - e^{-aT}} \tag{9}$$

and the DC gain requirement $K(1) = K^c(0)$ yields $k = (1 - e^{-aT})/2a$ so that

$$K(z) = \frac{1 - e^{-aT}}{2} \cdot \frac{z+1}{z - e^{-aT}}. \tag{10}$$

Using the MPZ with the modified step 2 that leaves the relative degree equal to one yields

$$K(z) = k\frac{a}{z - e^{-aT}}, \tag{11}$$

and the DC gain requirement $K(1) = K^c(0)$ yields $k = (1 - e^{-aT})/a$ so that

$$K(z) = \frac{1 - e^{-aT}}{z - e^{-aT}}. \tag{12}$$

d. Digital Filter Frequency Response

For comparison [cf. Franklin and Powell 1980], the frequency responses of the continuous filter and the discretized filters using the BLT, BLT with prewarping, and MPZ are shown in Fig. 5.2-1 with $a = 10$ and two sampling periods. The frequency for prewarping ω_1 was selected as a rad/sec. Note that as the sampling frequency

$$\omega_s = 2\pi/T \tag{13}$$

increases, the response of the digital filters approaches that of $K^c(s)$.

The digital filter response is symmetric about the Nyquist frequency

$$\omega_N = \omega_s/2 = \pi/T \tag{14}$$

and periodic with period ω_s. Thus, for the digital filter to match the continuous frequency response fairly well, it appears that the sampling frequency ω_s and the cutoff frequency $\omega_{1/2}$ should be approximately related by

$$\omega_s > 10\omega_{1/2}. \tag{15}$$

(Note, here $\omega_{1/2} = a = 10$ rad/s.)

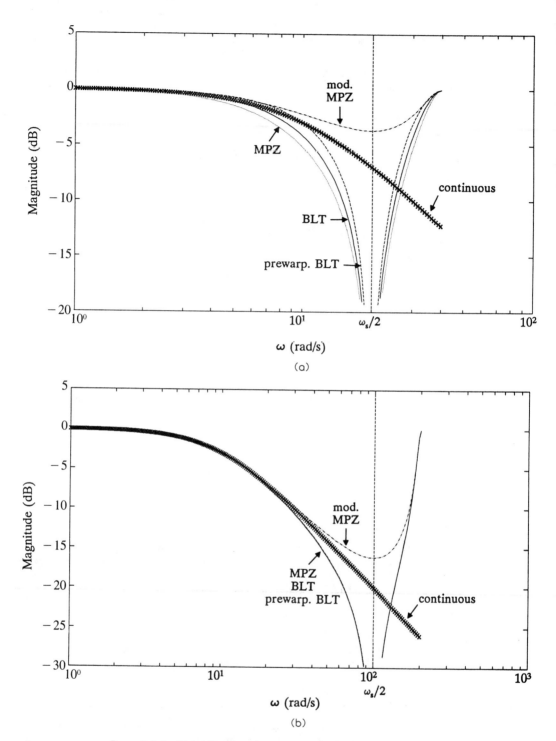

Figure 5.2-1 Digital filter frequency response. (a) Sampling frequency of $\omega_s = 30$ rad/s. (b) Sampling frequency of $\omega_s = 200$ rad/s.

5.2 Discretization of Continuous Controllers

Alternatively, there seems to be good agreement in the figure for frequencies up to about $\omega_s/10$.

∎

Example 5.2-2: Digital Inverted Pendulum Controller Via BLT

In Example 4.2-4 we designed a continuous-time controller for an inverted pendulum using LQ output-feedback techniques. Here, we shall demonstrate how to convert that continuous control system into a digital control system.

The BLT is popular in industry; therefore, we shall use it here.

The continuous controller is illustrated in Fig. 5.2-2, where

$$K_p^c(s) = \frac{k_2}{s+10} + k_4 = k_4 \frac{s+t_1}{s+10} \tag{1}$$

$$K_\theta^c(s) = \frac{k_1}{s+10} + k_3 = k_3 \frac{s+t_2}{s+10}, \tag{2}$$

with

$$t_1 = 10 + k_2/k_4, \qquad t_2 = 10 + k_1/k_3. \tag{3}$$

Suitable feedback gains in Example 4.2-4 were found to be

$$k_1 = 48.51, \qquad k_2 = 817.5, \qquad k_3 = -8.0, \qquad k_4 = -87.59. \tag{4}$$

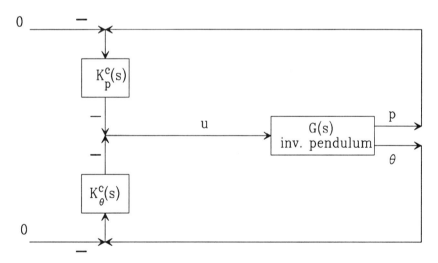

Figure 5.2-2 Continuous inverted pendulum controller

A digital control scheme with the same structure is shown in Fig. 5.2-3. We have added samplers with period T to produce the samples of cart position p and rod angle θ as well as a hold device to convert the control samples u_k computed by the digital controller back to a continuous-time control input $u(t)$ for the plant.

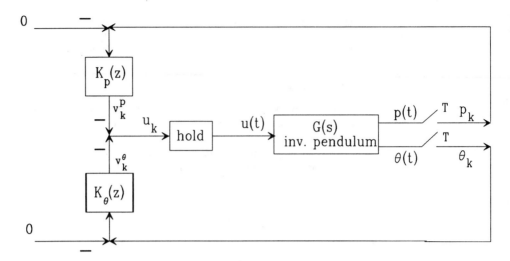

Figure 5.2-3 Digital inverted pendulum controller

Using the same numbers as in Example 4.2-4, the continuous dynamics $G(s)$ in Fig. 5.2-3 are given by

$$x = \begin{bmatrix} 0 & 1 & 0 & 0 \\ 10.78 & 0 & 0 & 0 \\ 0 & 0 & 0 & 1 \\ -0.98 & 0 & 0 & 0 \end{bmatrix} x + \begin{bmatrix} 0 \\ -0.2 \\ 0 \\ 0.2 \end{bmatrix} u \qquad (5)$$

$$y = \begin{bmatrix} 57.2958 & 0 & 0 & 0 \\ 0 & 0 & 1 & 0 \end{bmatrix} x \qquad (6)$$

where

$$x = [\theta \ \dot\theta \ p \ \dot p]^T, \qquad y = [\theta \ p]^T, \qquad (7)$$

Using the BLT, the discrete equivalents to (1) $-$ (2) are found to be

$$K_p(z) = g_1 \frac{z - z_1}{z - \pi_1}$$

$$z_1 = \frac{1 - t_1 T/2}{1 + t_1 T/2}, \qquad \pi_1 = \frac{1 - 5T}{1 + 5T}, \qquad g_1 = k_4 \frac{1 + t_1 T/2}{1 + 5T} \qquad (8)$$

and

$$K_\theta(z) = g_2 \frac{z - z_2}{z - \pi_2}$$

$$z_2 = \frac{1 - t_2 T/2}{1 + t_2 T/2}, \qquad \pi_2 = \frac{1 - 5T}{1 + 5T}, \qquad g_2 = k_3 \frac{1 + t_2 T/2}{1 + 5T}. \qquad (9)$$

Defining the intermediate signals v_k^p, v_k^θ shown in Fig. 5.2-3 and denoting the unit delay in the time domain by z^{-1}, we may express (8), (9) in terms of difference equations as follows:

5.2 Discretization of Continuous Controllers

$$v_k^p = g_1 \frac{1 - z_1 z^{-1}}{1 - \pi_1 z^{-1}} p_k \tag{10}$$

or

$$v_k^p = \pi_1 v_{k-1}^p + g_1(p_k - z_1 p_{k-1}), \tag{11}$$

and

$$v_k^\theta = g_2 \frac{1 - z_2 z^{-1}}{1 - \pi_2 z^{-1}} \theta_k \tag{12}$$

or

$$v_k^\theta = \pi_2 v_{k-1}^\theta + g_2(\theta_k - z_2 \theta_{k-1}). \tag{13}$$

The control samples u_k are thus given by

$$u_k = - v_k^p - v_k^\theta. \tag{14}$$

These difference equations describe the digital controller, and are easily implemented on a microprocessor, as we shall see later. First, however, the controller should be simulated. The FORTRAN subroutine in Fig. 5.2-4 may be used with the driver program TRESP in

```
C
C     DIGITAL INVERTED PENDULUM CONTROLLER
C
      SUBROUTINE DIG(IK,T,X)
      REAL X(*), K(4)
      COMMON/CONTROL/u
      COMMON/OUTPUT/ THETA, POS, UPLOT
C  UPLOT is used for plotting purposes only
      DATA (K(I),I= 1,4) / 48.51, 817.5, -8.0, -87.59 /
C
      t1= 10 + K(2)/K(4)
      t2= 10 + K(1)/K(3)
      z1= (1 - t1*T/2) / (1 + t1*T/2)
      p1= (1 - 5*T)   /  (1 + 5*T)
      z2= (1 - t2*T/2) / (1 + t2*T/2)
      p2=  p1
      g1= K(4) * (1 + t1*T/2) / (1 + 5*T)
      g2= K(3) * (1 + t2*T/2) / (1 + 5*T)
C
      vP= p1*vP + g1*(POS    - z1*POSM1  )
      vT= p2*vT + g2*(THETA  - z2*THETAM1)
      u = - vP - vT
      UPLOT= u
C
      POSM1= POS
      THETAM1= THETA
C
      RETURN
      END
```

Figure 5.2-4 Digital simulation software. FORTRAN subroutine to simulate digital inverted pendulum controller for use with TRESP in Appendix A.

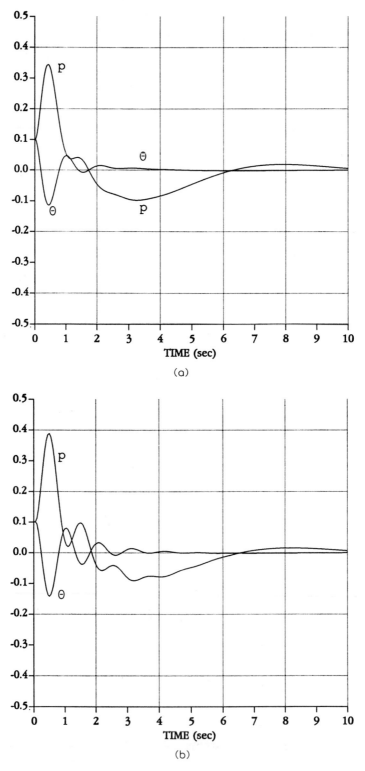

Figure 5.2-5 Effect of sampling period. (a) Response using $T = 0.01$ sec. (b) Response using $T = 0.05$ sec. (c) Response using $T = 0.1$ sec.

5.2 Discretization of Continuous Controllers

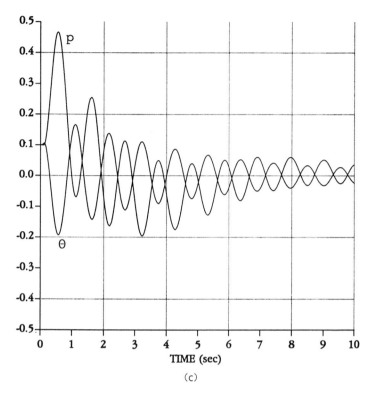

Figure 5.2-5 (*continued*)

Appendix A to simulate the digital control law. A file must also be constructed containing the continuous plant dynamics (5).

The response to initial conditions of $\theta(0) = 0.1$ rad $\approx 6°$, $p(0) = 0.1$ m using this digital controller was plotted for several sampling periods T in Fig. 5.2-5. A Runge-Kutta step size of 0.001 sec was selected. A zero-order hold was used to reconstruct $u(t)$ from u_k. Note that the response improves as T becomes small. Indeed, the response for $T = 0.01$ sec is indistinguishable from the response using a continuous controller in Example 4.2-4.

The motivation for selecting $T = 0.01$ sec was as follows. The closed-loop poles found in Example 4.2-4 were

$$\begin{aligned} s = &-1.87 \pm j5.66 \\ &-0.35 \pm j0.68 \\ &-5.54, \; -10. \end{aligned} \qquad (15)$$

For good response, the sampling period T should be selected less than about one-fifth of the fastest closed-loop time constant, or less than about $0.1/5 = 0.02$ sec. Indeed, the response in Fig. 5.2-5 using $T = 0.05$ sec is beginning to deteriorate.

The response using $T = 0.1$ sec is unsatisfactory, displaying oscillations of large amplitude.

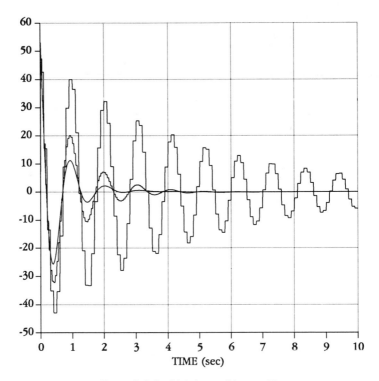

Figure 5.2-6 Digital control inputs $u(t)$

The control inputs $u(t)$ associated with Fig. 5.2-5 appear in Fig. 5.2-6. As the sampling period increases, the control magnitudes become larger.

∎

Digital PID Controller

A fundamental controller that has wide-ranging usefulness is the proportional-integral-derivative (PID) controller.

A standard continuous-time PID (proportional-integral-derivative) controller has the transfer function [Åström and Wittenmark 1984]

$$K^c(s) = k\left[1 + \frac{1}{T_I s} + \frac{T_D s}{1 + T_D s/N}\right] \qquad (5.2.19)$$

where k is the proportional gain, T_I is the integration time constant or "reset" time, T_D is the derivative time constant. Rather than use pure differentiation, a "filtered derivative" is used which has a pole far left in the s-plane at $s = -N/T_D$. The value for N is usually in the range 3–10; it is usually fixed by the manufacturer of the controller.

Let us consider a few methods of discretizing (5.2.19) with sample period T sec.

5.2 Discretization of Continuous Controllers

BLT

Using the BLT, the discretized version of (5.2.19) is found to be

$$K(z) = k\left[1 + \frac{1}{T_I \frac{2(z-1)}{T(z+1)}} + \frac{T_D \frac{2(z-1)}{T(z+1)}}{1 + \frac{T_D}{N}\frac{2(z-1)}{T(z+1)}}\right] \quad (5.2.20)$$

or, on simplifying

$$K(z) = k\left[1 + \frac{T}{T_{Id}}\frac{z+1}{z-1} + \frac{T_{Dd}}{T}\frac{z-1}{z-\nu}\right] \quad (5.2.21)$$

with the discrete integral and derivative time constants

$$T_{Id} = 2T_I \quad (5.2.22)$$

$$T_{Dd} = \frac{NT}{1 + NT/2T_D} \quad (5.2.23)$$

and the derivative-filtering pole at

$$\nu = \frac{1 - NT/2T_D}{1 + NT/2T_D}. \quad (5.2.24)$$

MPZ

Using the MPZ approach to discretize the PID controller yields

$$K(z) = k\left[1 + \frac{k_1(z+1)}{T_I(z-1)} + \frac{k_2 N(z-1)}{z - e^{-NT/T_D}}\right], \quad (5.2.25)$$

where k_1 and k_2 must be selected to match the DC gains. At DC, the D terms in (5.2.19) and (5.2.25) are both zero, so we may select $k_2 = 1$. The DC values of the I terms in (5.2.19) and (5.2.25) are unbounded. Therefore, to select k_1 let us match the low-frequency gains. At low frequencies, $e^{j\omega T} \approx 1 + j\omega T$. Therefore, for small ω, the I terms of (5.2.19) and (5.2.25) become

$$K^c(j\omega) = \frac{1}{j\omega T_I}$$

$$K(e^{j\omega T}) \approx \frac{2k_1}{T_I(j\omega T)},$$

and to match them, we require $k_1 = T/2$.

Thus, using the MPZ the discretized PID controller again has the form (5.2.21), but now with

$$T_{Id} = 2T_I \tag{5.2.26}$$

$$T_{Dd} = NT \tag{5.2.27}$$

$$\nu = e^{-NT/T_D}. \tag{5.2.28}$$

Modified MPZ

If we use the modified MPZ method, then in the I term in (5.2.25) the factor $(z + 1)$ does not appear. Then, the normalizing gain k_1 is computed to be T. In this case, the discretized PID controller takes on the form

$$K(z) = k\left[1 + \frac{T}{T_{Id}}\frac{1}{z-1} + \frac{T_{Dd}}{T}\frac{z-1}{z-\nu}\right] \tag{5.2.29}$$

with

$$T_{Id} = T_I \tag{5.2.30}$$

$$T_{Dd} = NT \tag{5.2.31}$$

$$\nu = e^{-NT/T_D}. \tag{5.2.32}$$

Now, there is a control delay of 1 sample period (T sec) in the integral term, which could be advantageous if there is a computation delay.

Difference equation implementation

Let us illustrate how to implement the modified MPZ PID controller (5.2.29) using difference equations, which are easily placed into a software computer program.

As we shall see in Chapter 6, it is best from the point of view of numerical accuracy in the face of computer roundoff error to implement digital controllers as several first- or second-order systems in parallel. Such a parallel implementation may be achieved as follows.

First, write $K(z)$ in terms of z^{-1}, which is the unit delay in the time domain (i.e., a delay of T sec), as

$$K(z^{-1}) = k\left[1 + \frac{T}{T_{Id}}\frac{z^{-1}}{1-z^{-1}} + \frac{T_{Dd}}{T}\frac{1-z^{-1}}{1-\nu z^{-1}}\right]. \tag{5.2.33}$$

(Note: there is some abuse in notation in denoting (5.2.33) as $K(z^{-1})$; this we shall accept.)

Now, suppose that the control input u_k is related to the tracking error as

$$u_k = K(z^{-1})e_k. \tag{5.2.34}$$

Then, u_k may be computed from past and present values of e_k using auxiliary variables as follows:

$$v_k^I = v_{k-1}^I + (T/T_{Id})e_{k-1} \tag{5.2.35}$$

5.2 Discretization of Continuous Controllers

$$v_k^D = \nu v_{k-1}^D + (T_{Dd}/T)(e_k - e_{k-1}) \qquad (5.2.36)$$

$$u_k = k(e_k + v_k^I + v_k^D). \qquad (5.2.37)$$

The variables v_k^I and v_k^D represent the integral and derivative portion of the PID controller respectively.

These difference equations are easily implemented in software, much like the code for the inverted pendulum controller in Fig. 5.2-4.

Ziegler-Nichols tuning of PID controller

Ziegler and Nichols [1942] gave two techniques for selecting the PID parameters. The advantage of their approach is that no knowledge of the system model is needed; all the information required to choose the parameters is obtained from simple experiments on the open-loop system [Åström and Wittenmark 1984; Franklin, Powell, and Workman 1990].

The tuning rules are based on continuous-time systems, and will apply to digital PID controllers if the sampling period T is small.

In the *transient-response method*, the step response of the open-loop system is measured. See Fig. 5.2-7. Denote the steepest slope by $R = M/\tau$ and the delay time by L. In terms of these quantities, the parameters for P, PI, and PD controllers that give a damping ration of about $\zeta = 0.2$ are given in Table 5.2-1.

In the *stability-limit tuning method*, a proportional feedback controller is closed around the system. Then, the proportional gain is slowly increased until continuous oscillations result. At this point the gain and oscillation period are recorded and called k_o and

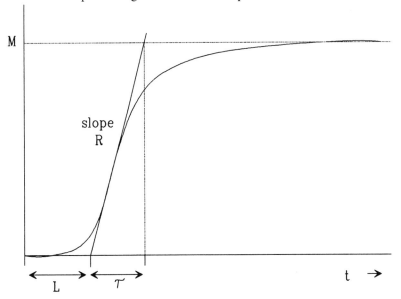

Figure 5.2-7 Process open-loop step response

TABLE 5.2-1 PID PARAMETERS USING TRANSIENT-RESPONSE METHOD

Controller Type	k	T_{Id}	T_{Dd}
P	$1/RL$		
PI	$0.9/RL$	$3L$	
PID	$1.2/RL$	$2L$	$0.5L$

TABLE 5.2-2 PID PARAMETERS USING STABILITY-LIMIT METHOD

Controller Type	k	T_{Id}	T_{Dd}
P	$0.5k_o$		
PI	$0.45k_o$	$T_o/1.2$	
PID	$0.6k_o$	$T_o/2$	$T_o/8$

T_o. In terms of these quantities, the P, PI, and PID controller parameters are given in Table 5.2-2.

These rules are used for an initial choice for the PID parameters. Final values are selected by on-line tuning of the closed-loop system.

Once the PID parameters have been selected, the sampling period may be chosen. It is generally suggested that T/T_{Dd} be in the range $0.1-0.5$. For Ziegler-Nichols tuning, this requires that T/L be in the range $0.2-1.0$ or T/T_o be in the range $0.01-0.05$.

Digital Control of Processes with Time Delays

Time delays often occur in industrial processes. If the transfer function of the basic system is $G(s)$ and the time delay is τ, then the process with delay is described by $G(s)e^{-s\tau}$. Continuous controller design is inconvenient for such systems.

Digital control of time delay processes, on the other hand, is not difficult. In this book we mention several possible approaches. Here, we suppose that a continuous controller has been designed for the basic system with no delay $G(s)$. Then, we show a straightforward way to design a digital controller for $G(s)e^{-s\tau}$ [Åström and Wittenmark 1984; Franklin, Powell, and Emami-Naeini 1986].

Examine the control structure shown in Fig. 5.2-8. According to Mason's theorem, the overall transfer function is

$$\frac{Z(s)}{R(s)} = \frac{K(s)G(s)e^{-s\tau}}{1 + K(s)G(s)e^{-s\tau} + (1 - e^{-s\tau})K(s)G(s)} \quad (5.2.38)$$

$$= \frac{K(s)G(s)}{1 + K(s)G(s)} e^{-s\tau}.$$

5.2 Discretization of Continuous Controllers

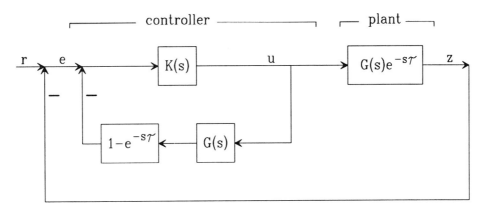

Figure 5.2-8 Continuous Smith predictor controller for time delay process

What this means is the following. Suppose a continuous controller $K(s)$ has been designed in the usual way for the plant with no delay $G(s)$. Then, the control structure shown in the figure yields the closed-loop system corresponding to the controller-plus-undelayed plant, but with the response delayed by τ. The controller shown in the figure is known as a *Smith predictor*.

Unfortunately, it is awkward to implement the Smith predictor in continuous time due to the term $e^{-s\tau}$. However, a digital controller can easily include such delay terms. Therefore, let us use the BLT, MPZ, or modified MPZ to discretize the controller. Each portion of the controller may be discretized separately. Assuming that the sample period T has been selected to be an integral divisor of the process delay τ, and recalling that $z^{-1} = e^{-sT}$, the result is shown in Fig. 5.2-9. Superscript "d" denotes the discretized versions of the continuous blocks.

This is an easily implementable digital controller that gives the same response as $K(s)$ used on the process $G(s)$ (if T is small), but with a delay of τ sec.

It is worth discussing the structure of the digital Smith predictor. Note that the controller contains a discretized model $G^d(z)$ of the process. Thus, the signal y_k in the figure corresponds to a prediction of the undelayed process output $z(t)$.

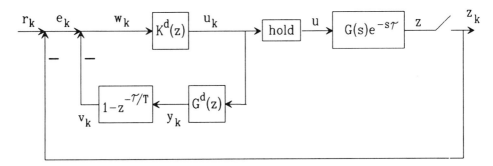

Figure 5.2-9 Digital Smith predictor controller for time delay process

An example will serve to demonstrate how easily the digital Smith predictor may be simulated using the software in Appendix A, and consequently how easily software may be generated for its implementation on a digital signal processor.

Example 5.2-3: Digital Control of Process with Time Delay

A model that can describe several sorts of manufacturing processes is

$$G(s)e^{-s\tau} = \frac{e^{-s\tau}}{s+1}. \tag{1}$$

This is a simplified description of a heat exchanger, a paper machine, and so on [Åström and Wittenmark 1984; Franklin, Powell, and Emami-Naeini 1986]. Let us use the digital Smith predictor to control it. Suppose the process delay is $\tau = 0.3$ sec.

a. Design of Continuous Controller for Undelayed System

To make the process without delay follow a desired constant set point of $r = 1$, we propose the continuous PI controller shown in Fig. 5.2-10. The dynamics of the process without delay $G(s)$ and the integrator are given by

$$\begin{bmatrix} \dot{z} \\ \dot{\varepsilon} \end{bmatrix} = \begin{bmatrix} -1 & 0 \\ -1 & 0 \end{bmatrix} \begin{bmatrix} z \\ \varepsilon \end{bmatrix} + \begin{bmatrix} 1 \\ 0 \end{bmatrix} u + \begin{bmatrix} 0 \\ 1 \end{bmatrix} r \equiv Ax + Bu + Er, \tag{2}$$

with performance output

$$z = [1 \quad 0]x \equiv Hx, \tag{3}$$

tracking error

$$e = r - z \tag{4}$$

and ε the integrator output.

The output corresponding to the desired control structure should be selected as

$$y = \begin{bmatrix} e \\ \varepsilon \end{bmatrix} = \begin{bmatrix} -1 & 0 \\ 0 & 1 \end{bmatrix} \begin{bmatrix} z \\ \varepsilon \end{bmatrix} + \begin{bmatrix} 1 \\ 0 \end{bmatrix} r \equiv Cx + Fr, \tag{5}$$

for then the control input is given by

$$u = -k_p e - k_I \varepsilon = -[k_p \quad k_I]y \equiv -Ky. \tag{6}$$

This is the formulation needed for the design approach of section 4.2. (Do not confuse y in (5) with the predicted plant output y_k in Fig. 5.2-9. We have unfortunately selected the same symbol for both.)

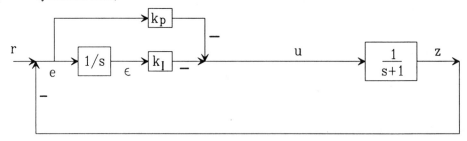

Figure 5.2-10 Continuous-time PI tracking system

5.2 Discretization of Continuous Controllers

Select the PI

$$J = \frac{1}{2}\int_0^\infty (100 t^2 e^2 + u^2)\, dt \tag{7}$$

and use the software described in Appendix A to solve the design equations in Table 4.2-1. The result is the control gain

$$K = [k_p \quad k_I] = [-3.415 \quad -3.766] \tag{8}$$

the closed-loop poles

$$s = -1.16, \; -3.26 \tag{9}$$

and the response shown in Fig. 5.2-11.

b. Digital Control of Undelayed Plant

In the form given in the previous section, the PI controller in Fig. 5.2-10 may be written

$$K(s) = -k[1 + 1/T_I s], \tag{10}$$

so that, according to (8)

$$k = -k_p = 3.415$$
$$T_I = k_p/k_I = 0.907. \tag{11}$$

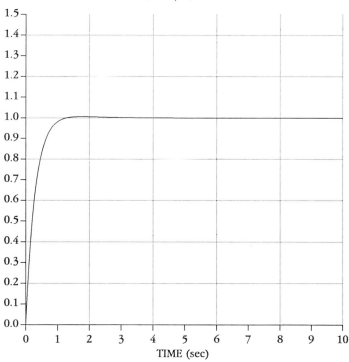

Figure 5.2-11 Simulation response $z(t)$ using continuous PI controller on undelayed process

Using the modified MPZ to discretize the PI controller [see (5.2.29)] yields

$$K^d(z) = k\left[1 + \frac{T}{T_I}\frac{1}{z-1}\right] = k\left[1 + \frac{T}{T_I}\frac{z^{-1}}{1-z^{-1}}\right]. \qquad (12)$$

The digital PI controller is shown in Fig. 5.2-12. Defining the auxiliary signals as shown there, we may write difference equations to implement the digital PI controller as

$$e_k = r - z_k = 1 - z_k \qquad (13)$$

$$w_k^I = w_{k-1}^I + \frac{T}{T_I}e_{k-1} \qquad (14)$$

$$u_k = k(e_k + w_k^I). \qquad (15)$$

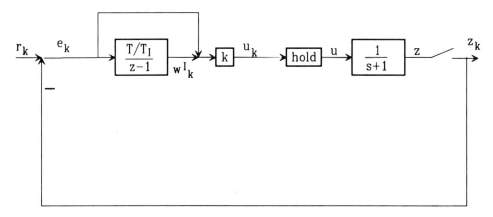

Figure 5.2-12 Digital PI controller for undelayed plant

This digital controller may be implemented using a subroutine similar to that in Fig. 5.2-4 (see also Fig. 5.2-15). It can be simulated using this subroutine and program TRESP in Appendix A. Using this digital PI controller with $T = 0.1$ sec on the undelayed plant, a simulation yields an output $z(t)$ exactly like that shown in Fig. 5.2-11 and the control $u(t)$ shown in Fig. 5.2-13.

c. Digital Smith Predictor for Delayed Process

If we now add the delay of $\tau = 0.3$ sec to the process, the digital PI controller just designed yields the output shown in Fig. 5.2-14. This figure clearly shows the deleterious effect of a delay in the controlled system that is not accounted for in the controller design.

To regain performance like that shown in 5.2-11 is easy using a digital Smith predictor, as we now see.

Using the modified MPZ to discretize the process $G(s)$ yields (see Example 5.2-1 part c)

$$G^d(z) = \frac{b}{z-p} = \frac{bz^{-1}}{1-pz^{-1}}, \qquad (16)$$

with

$$b = 1 - e^{-T}, \qquad p = e^{-T}. \qquad (17)$$

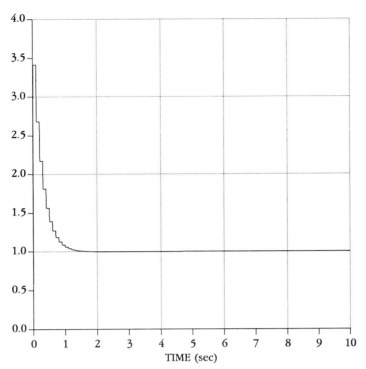

Figure 5.2-13 Control input $u(t)$ using digital PI controller on undelayed process

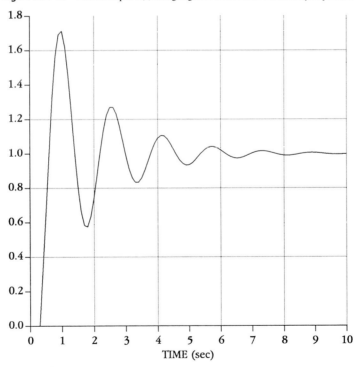

Figure 5.2-14 Simulation response $z(t)$ using digital PI controller on delayed process. Sampling period $T = 0.1$ sec, delay $\tau = 0.3$ sec.

Selecting the sampling period as $T = 0.1$ sec yields

$$b = 0.0952, \qquad p = 0.90484. \tag{18}$$

The digital control system using a Smith predictor is shown in Fig. 5.2-9, where $\tau/T = 3$ is the delay in terms of sample periods and $K^d(z)$ is shown in detail in Fig. 5.2-12. Defining the auxiliary variables shown in these figures, the difference equations implementing the digital controller are

$$y_k = py_{k-1} + bu_{k-1} \tag{19}$$

$$v_k = y_k - y_{k-3} \tag{20}$$

$$e_k = r - z_k = 1 - z_k \tag{21}$$

$$w_k = e_k - v_k \tag{22}$$

$$w'_k = w'_{k-1} + \frac{T}{T_I} w_{k-1} \tag{23}$$

$$u_k = k(w_k + w'_k). \tag{24}$$

Signal y_k is a prediction of the output of the undelayed process.

A subroutine implementing this digital controller is shown in Fig. 5.2-15. It is for use with program TRESP in Appendix A. Using this software, the simulation results in Fig. 5.2-16 were obtained.

```
C   DIGITAL CONTROLLER FOR PROCESS WITH DELAY

    SUBROUTINE DIG(IK,T,x)
    REAL x(*),k
    COMMON/CONTROL/u
    DATA k,Ti/3.415,0.907/

    p= exp(-T)
    b= 1 - p
    e= 1 - x(1)
    y= p*yM1 + b*uM1
    v= y - yM3
    w= e - v
    w1= w1M1 + (T/Ti)*wM1
    u= k*(w + w1)

C   Store signal values for next iteration:

    vM1= v
    wM1= w
    w1M1= w1
    yM3= yM2
    yM2= yM1
    yM1= y
    uM1= u

    RETURN
    END
```

Figure 5.2-15 Smith predictor digital control subroutine

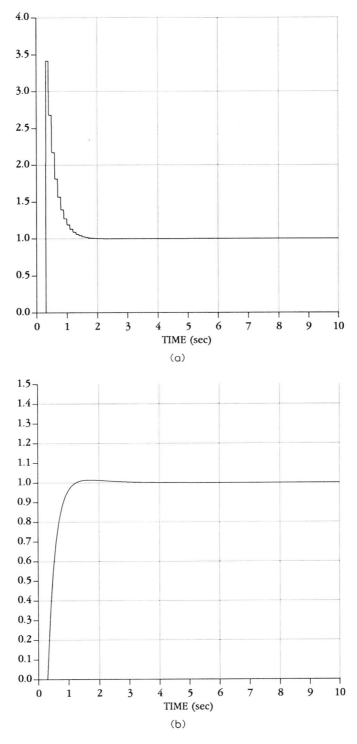

Figure 5.2-16 Simulation responses using Smith predictor on delayed process. (a) Output $z(t)$. (b) Control input $u(t)$.

```
C     PROCESS WITH TIME DELAY Tau

      SUBROUTINE F(time,x,xP)
      REAL x(*), xP(*), B(0:999)
      COMMON/CONTROL/u
      COMMON/DELAY/IT,Tr
C     Tr= Runge-Kutta step size
      DATA ITM1,Tau/-1,.3/

C     RING BUFFER FOR CONTROL DELAY

      IF(IT.EQ.0) NBUF= NINT(Tau/Tr) - 1
      IF(IT.GT.ITM1) THEN
        B(IR)= u
        IR= MOD(IR+1,NBUF+1)
        udel= B(IR)
      END IF
      ITM1= IT

C     SYSTEM

      xP(1)= -x(1) + udel

      RETURN
      END
```

Figure 5.2-17 Subroutine containing continuous dynamics with process delay τ

Note that the output $z(t)$ is the same as the output using the continuous controller on the undelayed plant (see Fig. 5.2-11), but with a delay of $\tau = 0.3$ sec. Note further that the control $u(t)$ is exactly the same as the control in part b, but delayed by $\tau = 0.3$ sec.

It is interesting to show the subroutine $F(\text{time},X,XP)$ used with TRESP that contains the continuous process dynamics with control delay. It is given in Fig. 5.2-17. The process delay has been injected using a *ring buffer*. This is a very useful piece of software with a wide variety of applications.

The code required to implement the digital Smith predictor on a digital signal processor (DSP) is exactly the subroutine DIG shown in Fig. 5.2-15. We shall demonstrate DSP implementations in subsequent chapters.

∎

5.3 SAMPLING, HOLD DEVICES, AND COMPUTATION DELAYS

In this section we should like to examine some additional aspects of controller discretization and implementation. The motivation is to provide some insight which can be used to obtain improved performance of a digital controller.

Continuous controller redesign ignores the dynamics of the sampling and hold processes, so that it is only an approximate scheme for designing digital controllers. As we have seen in Example 5.2-2, the performance of these digital controllers deteriorates with

5.3 Sampling, Hold Devices, and Computation Delays

increasing T, so that sample periods are required which may be very small. We note that smaller values of T require faster computation to determine u_k; thus, a faster, and more expensive, microprocessor may be required for small T.

The digital controller resulting from continuous redesign may be improved by two methods:

1. Modifying the digital controller $K(z)$ once it has been obtained from the continuous controller $K^c(s)$;
2. Modifying the continuous controller design to result in a more suitable digital controller $K(z)$.

In Åström and Wittenmark [1984, p. 189] a technique is given for the first method. There, after it is designed the digital controller is modified by changing the gains slightly to include the effects of the delay introduced by the ZOH. This allows the use of larger sample periods. Since the method in Åström and Wittenmark [1984] only works for full state variable feedback, we shall not discuss it here.

Instead, we shall examine method number 2—modifying the continuous controller before it is discretized. To see how to modify the design of the continuous controller so that, on discretization, it results in an improved digital controller, we will examine more closely sampling and aliasing, hold devices, and computation delays.

Sampling and Aliasing

In this subsection let us study the sampler in Fig. 5.1-1.

As illustrated in Example 5.2-1, the digital frequency response is symmetric with respect to the Nyquist frequency $\omega_N = \omega_s/2 = \pi/T$, and periodic with respect to the sampling frequency $\omega_s = 2\pi/T$. Thus, the highest frequency of any continuous signal that appears in the system should be less than ω_N; this is known as the *sampling theorem* of Shannon and Nyquist. See Oppenheim and Schafer [1975], Franklin and Powell [1980], and Åström and Wittenmark [1984].

To obtain some additional insight on the sampling theorem, let us picture the output $y^*(t)$ of the sampler with input $y(t)$ as the string of impulses

$$y^*(t) = \sum_{k=-\infty}^{\infty} y(t)\delta(t - kT) \qquad (5.3.1)$$

where $\delta(t)$ is the unit impulse. Since the impulse train is periodic, it has a Fourier series which may be computed to be

$$\sum_{k=-\infty}^{\infty} \delta(t - kT) = \frac{1}{T} \sum_{n=-\infty}^{\infty} e^{jn\omega_s t}. \qquad (5.3.2)$$

Using this in (5.3.1) and taking the Laplace transform yields

$$Y^*(s) = \frac{1}{T} \int_{-\infty}^{\infty} y(t) \left[\sum_{n=-\infty}^{\infty} e^{jn\omega_s t} \right] e^{-st} \, dt$$

$$Y^*(s) = \frac{1}{T} \sum_{n=-\infty}^{\infty} \int_{-\infty}^{\infty} y(t) e^{-(s-jn\omega_s)t} \, dt \quad (5.3.3)$$

$$Y^*(s) = \frac{1}{T} \sum_{n=-\infty}^{\infty} Y(s - jn\omega_s),$$

where $Y(s)$ is the Laplace transform of $y(t)$ and $Y^*(s)$ is the Laplace transform of the sampled signal $y^*(t)$.

Due to the factor $1/T$ appearing in (5.3.3), the sampler is said to have a gain of $1/T$.

Sketches of a typical $Y(j\omega)$ and $Y^*(j\omega)$ are shown in Fig. 5.3-1, where ω_H is the highest frequency contained in $y(t)$. Notice that at frequencies less than ω_N, the spectrum

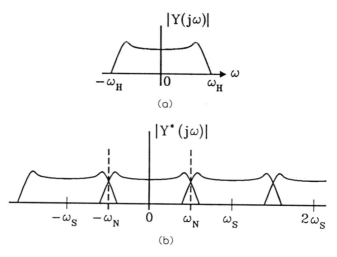

Figure 5.3-1 Sampling in the frequency domain. (a) Spectrum of $y(t)$. (b) Spectrum of sampled signal $y^*(t)$.

of $Y^*(j\omega)$ has two parts: one part comes from $Y(j\omega)$ and is the portion that should appear. However, there is an additional portion from $Y(j(\omega - \omega_s))$; the "tail" of $Y(j(\omega - \omega_s))$, which contains high-frequency information about $y(t)$, is "folded" back or "aliased" into the lower frequencies of $Y^*(j\omega)$. Thus, the high frequency content of $y(t)$ appears at low frequencies. If $\omega_H < \omega_N$, then the tail of $Y^*(j(\omega - \omega_s))$ does not appear to the left of $\omega = \omega_N$. This condition is equivalent to

$$\omega_s > 2\omega_H, \quad (5.3.4)$$

which is the sampling theorem.

It is interesting to see what the sampling theorem means in the time domain. Examine Fig. 5.3-2, where we show two continuous signals that have the same samples. If the original signal was the higher-frequency signal, then the D/A reconstruction process will produce the lower-frequency signal. Thus, aliasing can result in high-frequency sig-

5.3 Sampling, Hold Devices, and Computation Delays

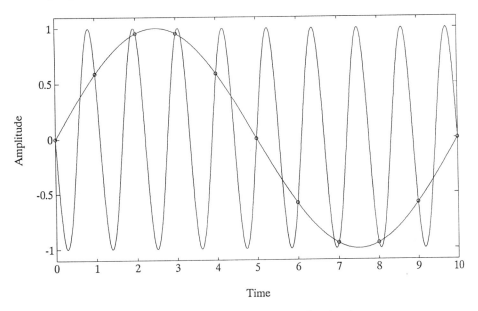

Figure 5.3-2 Example of aliasing in the time domain

nals being misinterpreted as low-frequency signals. If the sampling frequency ω_s is greater than twice the highest frequency ω_H appearing in the continuous signal, then the problem depicted in the figure does not occur and the signal can be accurately reconstructed from its samples.

For controls design, the sampling frequency ω_s must generally be significantly greater than twice the highest frequency of any signal appearing in the system. That is, the sampling theorem does not usually provide much insight in selecting ω_s. Some guides for selecting the sampling period T were given in Examples 5.2-1 and 5.2-2; a few additional guidelines are now discussed.

If the continuous-time system has a single dominant complex pole pair with natural frequency of ω, the rise time is given approximately by

$$t_r = 1.8/\omega. \tag{5.3.5}$$

It is reasonable to have at least 2–4 samples per rise time so that the error induced by ZOH reconstruction is not too great during the fastest variations of the continuous-time signal [Åström and Wittenmark 1984]. Then we have $t_r = 1.8/\omega \geq 4T$, or approximately

$$T \leq 1/2\omega. \tag{5.3.6}$$

However, if high-frequency components are present up to a frequency of ω_H rads and it is desired to retain them in the sampled system, a rule of thumb is to select

$$T \leq 1/4\omega_H. \tag{5.3.7}$$

These formulae should be used with care, and to select a suitable T it may be necessary to perform digital control designs for several values of T, for each case carrying out a computer simulation of the behavior of the plant under the influence of the proposed controller. Note particularly that when using continuous design of digital controllers even smaller sample periods may be required, since the technique is only an approximate one.

Hold Devices

The D/A hold device in Fig. 5.1-1 is needed to reconstruct the plant control input $u(t)$ from the samples u_k provided by the digital control scheme. A common method for data reconstruction is *polynomial extrapolation* [Phillips and Nagle 1984]. The Taylor series expansion for $u(t)$ about $t = kT$ is

$$u(t) = u(kT) + u'(kT)(t - kT) + \frac{u''(kT)}{2!}(t - kT)^2 + \ldots \quad (5.3.8)$$

where the derivatives may be approximated by backward differences as

$$u'(kT) = \frac{u(kT) - u((k-1)T)}{T} \quad (5.3.9)$$

$$u''(kT) = \frac{u'(kT) - u'((k-1)T)}{T} \quad (5.3.10)$$

and so on.

The zero-order hold (ZOH) results if $u(t)$ is approximated retaining only the first term of the Taylor series. Then we take

$$u(t) = u(kT) = u_k, \quad kT \leq t < (k+1)T, \quad (5.3.11)$$

with u_k the k-th sample of $u(t)$. The ZOH yields the sort of behavior in Fig. 5.1-2 and has the impulse response shown in Fig. 5.3-3. This impulse response may be written as

$$h(t) = u_{-1}(t) - u_{-1}(t - T),$$

with $u_{-1}(t)$ the unit step. Thus, the transfer function of the ZOH is

$$G_0(s) = \frac{1 - e^{-sT}}{s}. \quad (5.3.12)$$

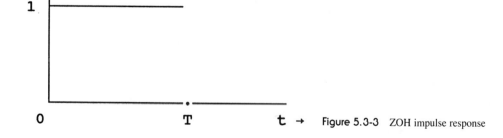

Figure 5.3-3 ZOH impulse response

5.3 Sampling, Hold Devices, and Computation Delays

To determine the Bode magnitude and phase of $G_0(s)$, write

$$G_0(j\omega) = \frac{1 - e^{-j\omega T}}{j\omega} = e^{-j\omega T/2}\frac{[e^{j\omega T/2} - e^{-j\omega T/2}]}{j\omega} \quad (5.3.13)$$

$$G_0(j\omega) = Te^{-j\omega T/2}\frac{\sin \omega T/2}{\omega T/2} = Te^{-j\omega T/2} \operatorname{sinc}(\omega/\omega_s),$$

where $\operatorname{sinc} x \equiv (\sin \pi x)/\pi x$. The magnitude and phase of the ZOH are shown in Fig. 5.3-4. Note that the ZOH is a low-pass filter of magnitude $T|\operatorname{sinc}(\omega/\omega_s)|$ with a phase of

$$\angle \text{ZOH} = \frac{-\omega T}{2} + \theta = \frac{-\pi\omega}{\omega_s} + \theta, \quad \theta = \begin{cases} 0, & \sin \omega T/2 > 0 \\ \pi, & \sin \omega T/2 < 0 \end{cases} \quad (5.3.14)$$

According to (5.3.13), for frequencies ω much smaller than ω_s, the ZOH may be approximated by

$$G_0(s) \approx Te^{-sT/2}, \quad (5.3.15)$$

that is, by a pure delay of half the sampling period and a scale factor of T.

The first-order hold (FOH) results if $u(t)$ is approximated by the first two terms of (5.3.8). Then, we take

$$u(t) = u_k + \frac{(t - kT)}{T}(u_k - u_{k-1}), \quad kT \leq t < (k + 1)T. \quad (5.3.16)$$

Data reconstruction using this sort of hold device is shown in Fig. 5.3-5. Note that the FOH needs the previous sample u_{k-1} and so requires memory for its implementation. Since it also requires some computation, the ZOH is generally used instead. Thus, commercially available digital microprocessors often have a ZOH built in by the manufacturer.

The FOH has the impulse response shown in Fig. 5.3-6. To understand this figure, first consider the interval $0 \leq t < T$, where the response $u(t)$ is given by (5.3.16) with $k = 0$, $u_{-1} = 0$, $u_0 = 1$. (Note: If $u(kT)$ has a value of $d\delta(t)$, with d the impulse magnitude, we say that $u_k = d$.) The response in the interval $T \leq t < 2T$ is given by (5.3.16) with $k = 1$, $u_0 = 1$, $u_1 = 0$.

The impulse response in Fig. 5.3-6 is described by

$$h(t) = u_{-1}(t) + \frac{1}{T}tu_{-1}(t) - 2u_{-1}(t - T) - \frac{2}{T}(t - T)u_{-1}(t - T)$$

$$+ u_{-1}(t - 2T) + \frac{1}{T}(t - 2T)u_{-1}(t - 2T),$$

so that the transfer function of the FOH (see the problems) is

$$G_1(s) = \frac{1 + Ts}{T}\left[\frac{1 - e^{-sT}}{s}\right]^2. \quad (5.3.17)$$

284   Chap. 5   Digital Control by Continuous Controller Redesign

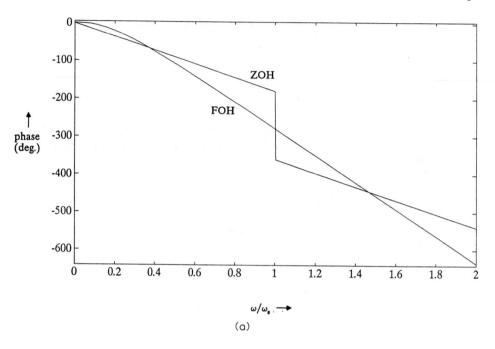
(a)

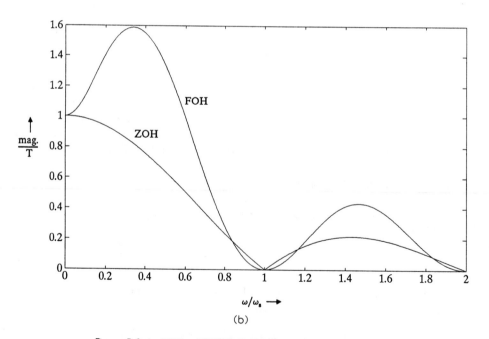
(b)

Figure 5.3-4 ZOH and FOH Bode plots. (a) Magnitude. (b) Phase.

5.3 Sampling, Hold Devices, and Computation Delays

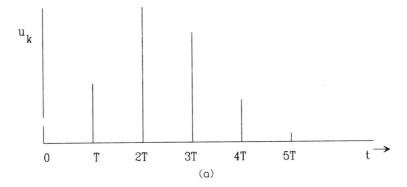

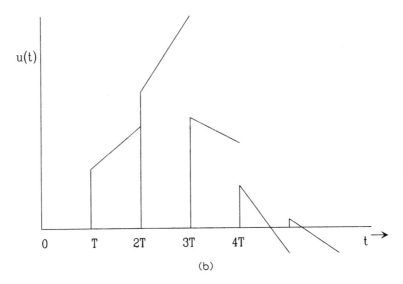

Figure 5.3-5 Data reconstruction using a FOH. (a) Discrete control sequence u_k. (b) Reconstructed continuous signal $u(t)$.

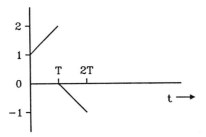

Figure 5.3-6 FOH impulse response

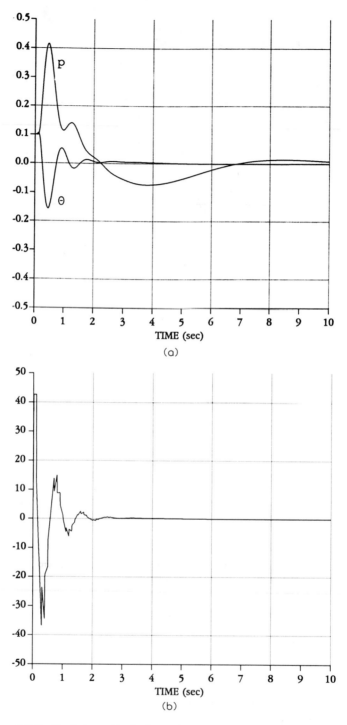

Figure 5.3-7 Simulation results for inverted pendulum with digital controller using a FOH. Sampling period $T = 0.1$ sec. (a) Rod angle $\theta(t)$ and cart position $p(t)$. (b) Control input $u(t)$.

The magnitude and phase of the FOH are given by (see the problems)

$$|G_1(j\omega)| = T\sqrt{1 + \omega^2 T^2} \, \text{sinc}^2 \, \omega/\omega_s \quad (5.3.18)$$

$$\angle G_1(j\omega) = -(\omega T - \tan^{-1} \omega T), \quad (5.3.19)$$

which are plotted in Fig. 5.3-4.

Notice that for ω small compared to ω_s the phase lag introduced by the FOH is much less than that of the ZOH. Thus, it may be advantageous to use the FOH when implementing a digital controller that has been designed from a continuous controller using the BLT or the MPZ method. In fact, it is quite instructive to simulate the digital inverted pendulum controller designed in Example 5.2-2 using a FOH instead of a ZOH.

Example 5.3-1: Inverted Pendulum Digital Control Using a FOH

In Example 4.2-4 we designed a continuous-time controller for an inverted pendulum. In Example 5.2-2 we showed how to convert this to a digital controller using the BLT. The simulation results for these two examples should now be examined.

Note that, in Example 5.2-2, the response was good for $T = 0.01$ sec, and unacceptable for $T = 0.1$ sec. Indeed, using $T = 0.1$ sec resulted in oscillatory behavior and large excursions in the control signal. In effect, the inverted pendulum is operating in open loop with a constant control signal between the sampling instants. In that example a ZOH was used to reconstruct $u(t)$.

In this example we used the same digital controller that was designed in Example 5.2-2, but modified the software to use a FOH instead of a ZOH, according to (5.3.16). The simulation results, with $T = 0.1$ sec, are shown in Fig. 5.3-7. They are quite good. Indeed, the response using a FOH and $T = 0.1$ sec is comparable to the response using a ZOH and $T = 0.01$ sec. Thus, the FOH has allowed us to increase the sampling period tenfold with no appreciable degradation in performance!

In spite of the improved performance offered by the FOH in some situations, in industry the ZOH is commonly used. In fact, most industrial digital microprocessors have a built-in ZOH.

As we shall see, if the controller is designed using the discrete design techniques in Chapter 8, the ZOH is quite adequate in digital controls applications. The poor performance in Example 5.2-2 with $T = 0.1$ sec is actually a consequence of the fact that discretization of continuous controllers is only an approximate technique for the design of digital controllers, requiring small sample periods to be effective. Direct discrete design allows the use of far larger sample periods, and only requires the implementation of a ZOH, not a FOH.

It is also possible to improve the performance of a digital controller obtained by the continuous controller redesign technique covered in this chapter so that the ZOH can be used. At the end of this section we will show a technique for so doing. ■

Computation Delay

If the microprocessor is fast so that the time Δ required to compute the digital control law is negligible, then Δ will have little effect if a digital controller is designed from the continuous controller. However, if the computation delay Δ is appreciable, then it may be necessary to account for it.

If Δ is an integral multiple of T, then the Smith predictor at the end of section 5.2 may be used to account for it.

If $\Delta \leq T$, then the computation delay may be accounted for by ensuring that u_k depends only on *previous* values of the outputs. This may often be achieved by using the modified MPZ approach for digital controller design. However, the BLT is more popular and it always yields a u_k that depends on *current* values of the outputs. Moreover, the digital PID controller (see section 5.2) always has a dependence on the current outputs through the derivative term, even if the modified MPZ is used.

If there is noise present in the system, then using outputs delayed by an entire sample period to compute u_k can, for large sample periods, lead to significant deterioration over using more recent outputs to compute u_k. Thus, if the computation delay is not negligible, but is only a fraction of T, it seems inefficient to allow it to cause a delay of a full T seconds in applying the control to the plant.

Thus, it may be necessary to account for the computation delay Δ during the design of the continuous controller. Let us now discuss such notions.

Modified Design of Continuous Controllers for Discretization

It is clear at this point that there are some considerations involved in discretization that are not apparent when a digital controller is computed using the BLT or MPZ from a continuous controller. Here, we shall discuss a few techniques for taking these into account during the design of the continuous controller $K^c(s)$.

A disadvantage of these modified continuous design techniques is that the sample period T must be selected prior to the continuous controller design. However, good software makes it easy to redesign the continuous controller with a different value of T. The advantage of the approach is that the effects of the sampling and hold operations, computation delay, and aliasing are apparent while the continuous design is being performed. Thus, they may be to some extent compensated for.

Modified continuous design can often allow significantly larger sample periods than direct application of the BLT or MPZ to a continuous controller designed with no consideration that the next step will be conversion to a digital control law.

Let us discuss continuous design modifications to compensate for aliasing, computation delays, and then the ZOH.

Anti-aliasing filters

The plant $G(s)$ is generally a low-pass filter. We have seen in Fig. 5.3-1 that, as long as the sampling frequency ω_s is selected at least twice as large as the plant cutoff frequency ω_H, the effects of aliasing will be small.

However, one type of signal appearing in the closed-loop system that may not be band-limited is *measurement noise*. Thus, high-frequency measurement noise may be aliased down to lower frequencies that are within the plant bandwidth and thus have a detrimental effect on system performance. To avoid this, low-pass *anti-aliasing filters* of the form

5.3 Sampling, Hold Devices, and Computation Delays

$$H_a(s) = \frac{a}{s + a} \quad (5.3.20)$$

may be inserted after the measuring devices and before the samplers. The cutoff frequency a should be selected less than $\omega_N = \omega_s/2$ so that there is good attenuation beyond ω_N rad/sec.

If the cutoff frequency of the anti-aliasing filter is not much higher than the plant cutoff frequency, then the filter will affect the closed-loop performance, and it should be appended to the plant *at the design stage* so that the continuous controller is designed taking it into account. That is, the controller should be designed not for the plant $G(s)$, but for $G(s)H_a(s)$.

See Fig. 5.3-8, which represents the actual plant $G(s)$ augmented by various filters, some still to be discussed, that should be taken into account while the continuous controller is being designed.

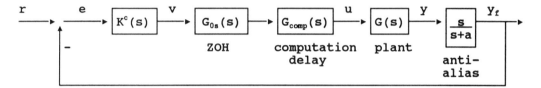

Figure 5.3-8 Plant with anti-aliasing filter and compensation to model hold device and computation delays

Computation delay dynamics

Our next topic is the computation delay. The delay associated with a computation time of Δ has a transfer function of

$$G_{\text{delay}}(s) = e^{-s\Delta}, \quad (5.3.21)$$

which has a magnitude of one and a phase of $-\omega\Delta$ rads. To account for this delay, we may perform the continuous controller design not on the plant $G(s)$, but on $G(s)e^{-s\Delta}$. However, it is awkward to design a controller for a plant whose transfer function is not rational. It is more convenient to approximate the delay with a rational transfer function.

For this purpose, we may use Padé approximants to $G_{\text{delay}}(s) = e^{-s\Delta}$, which match the first few terms of the Taylor's series expansion [Su 1971; Franklin, Powell, and Emami-Naeini 1986]. In Table 5.3-1 are given several Padé approximants to $e^{-s\Delta}$. These approximants match the first $n + m + 1$ terms of the Taylor series expansion, where n is the denominator degree and m the numerator degree.

To perform a modified continuous design which takes into account the computation delay Δ, it is only necessary to incorporate a Padé approximant $G_{\text{comp}}(s)$ to $e^{-s\Delta}$ of suitable order into the plant as shown in Fig. 5.3-8. The continuous controller $K^c(s)$ designed for this modified plant $G_{\text{comp}}(s)G(s)$ is then discretized using the BLT or MPZ to produce a digital controller $K(z)$.

TABLE 5.3-1 PADÉ APPROXIMANTS TO $e^{-s\Delta}$

$\dfrac{1}{1 + s\Delta}$	$\dfrac{1 - s\Delta/2}{1 + s\Delta/2}$	
$\dfrac{1}{1 + s\Delta + (s\Delta)^2/2}$	$\dfrac{1 - s\Delta/3}{1 + 2s\Delta/3 + (s\Delta)^2/6}$	$\dfrac{1 - s\Delta/2 + (s\Delta)^2/12}{1 + s\Delta/2 + (s\Delta)^2/12}$
$\dfrac{1}{1 + s\Delta + (s\Delta)^2/2 + (s\Delta)^3/6}$	$\dfrac{1 - s\Delta/4}{1 + 3s\Delta/4 + (s\Delta)^2/4 + (s\Delta)^3/24}$	$\dfrac{1 - 2s\Delta/5 + (s\Delta)^2/20}{1 + 3s\Delta/5 + 3(s\Delta)^2/20 + (s\Delta)^3/60}$

Notice that the Padé approximants in Table 5.3-1 that have finite zeros are *nonminimum phase*; that is, there are zeros in the right-half plane. This is a property of a pure time delay. The advantage of the modified continuous design approach is that the nonminimum-phase nature of the delayed plant manifests itself at the controller design stage, where it can be taken into account so that the resulting digital controller compensates for this problem automatically. As we see in Chapters 2 and 10, there are fundamental limitations in what may be achieved in controlling nonminimum phase plants.

ZOH and sampler dynamics

Finally, let us discuss modified continuous design taking into account the ZOH and sampler. Since the sampler has a gain of $1/T$, the sampler plus ZOH has a transfer function of

$$G_{Zs}(s) = \frac{1 - e^{-sT}}{sT}. \qquad (5.3.22)$$

Some useful approximants to $G_{Zs}(s)$ are given in Table 5.3-2. These have been computed using Padé approximants of e^{-sT} and so they are not strictly speaking Padé approximants, since they only match the first $n + m$ terms of the Taylor series. They are, however, sufficiently accurate for our purposes. Note that the approximants to $G_{Zs}(s)$ have unstable zeros.

TABLE 5.3-2 APPROXIMANTS TO $(1 - e^{-sT})/sT$

$$\frac{1}{1 + sT/2}$$

$$\frac{1 - sT/6}{1 + sT/3}$$

$$\frac{1 - sT/10 + (sT)^2/60}{1 + 2sT/5 + (sT)^2/20}$$

$$\frac{1 - sT/14 + 23(sT)^2/840 - (sT)^3/840}{1 + 3sT/7 + (sT)^2/14 + (sT)^3/120}$$

5.3 Sampling, Hold Devices, and Computation Delays

Modified continuous controller design taking into account $G_{Zs}(s)$ involves designing a controller for $G(s)G_{0s}(s)$, where $G_{0s}(s)$ is a suitable approximant to $G_{Zs}(s)$. See Fig. 5.3-8.

It is important to realize that the anti-aliasing filter should be implemented using analog circuitry as part of the plant $G(s)$. It should immediately precede the sampler. $G_{\text{comp}}(s)$, on the other hand, is not implemented since it is a model of the computation delay. $G_{0s}(s)$ is implemented by the ZOH and the sampler. $K^c(s)$ is discretized using the BLT or MPZ and becomes the digital controller $K(z)$.

The next example illustrates modified continuous design for discretization.

Example 5.3-2: Digital Inverted Pendulum Controller via Modified Continuous Design

In Example 4.2-4 we designed a continuous-time controller for an inverted pendulum. In Example 5.2-2 we showed how to use the BLT to convert that controller into digital form. It was seen that the response was good for $T = 0.01$ sec, marginal for $T = 0.05$ sec, and unacceptable for $T = 0.1$ sec.

In Example 5.3-1 we used a FOH instead of a ZOH to implement the digital controller, and it was discovered that the response for $T = 0.1$ sec was then very good.

We should like to avoid using a FOH for two reasons. First, it requires memory and is harder to implement than the ZOH. Second, a ZOH is generally built into most commercially available digital microprocessors. We also desire to use larger sample periods to allow for more control computation time, as well as to avoid some of the problems associated with small T that will be discussed in subsequent chapters.

Therefore, in this example, let us design a modified continuous controller which, on discretization, will yield a better digital controller using large sample periods than the one of Example 5.2-2. We will select the sampling period in this example to be $T = 0.1$ sec.

a. Modified Continuous-Time Plant

To accomplish this, we shall incorporate a model of the sampling and hold processes into the continuous-time dynamical model of the inverted pendulum, as shown in Fig. 5.3-8. Let us use a Padé approximant to (5.3.22). Specifically, examining Table 5.3-2, select

$$G_{0s}(s) = \frac{1 - sT/6}{1 + sT/3} = \frac{-1}{2} + \frac{9/2T}{s + 3/T}. \quad (1)$$

According to Fig. 5.3-8, the ZOH/sampler approximant should act as a filter on the plant control input $u(t)$. Thus, a state-variable representation of $G_{0s}(s)$ is given by

$$\dot{x}_z = \frac{-3}{T}x_z + \frac{9}{2T}v$$
$$u = x_z - \frac{1}{2}v, \quad (2)$$

where $v(t)$ is the new input shown in Fig. 5.3-8. With $T = 0.1$ sec this becomes

$$\dot{x}_z = -30x_z + 45v$$
$$u = x_z - 0.5v. \quad (3)$$

We should like to propose the same control structure used in Example 4.2-4. There, compensators with states x_θ and x_p were used in the angle and position feedforward channels.

The ZOH/sampler dynamics (3) may be augmented into the system-plus-compensator state equations by defining the augmented state

$$x = [\theta \quad \dot\theta \quad p \quad \dot p \quad x_\theta \quad x_p \quad x_z]^T, \tag{4}$$

Then

$$\dot x = \begin{bmatrix} 0 & 1 & 0 & 0 & 0 & 0 & 0 \\ 10.78 & 0 & 0 & 0 & 0 & 0 & -.2 \\ 0 & 0 & 0 & 1 & 0 & 0 & 0 \\ -0.98 & 0 & 0 & 0 & 0 & 0 & .2 \\ 1 & 0 & 0 & 0 & -10 & 0 & 0 \\ 0 & 0 & 1 & 0 & 0 & -10 & 0 \\ 0 & 0 & 0 & 0 & 0 & 0 & -30 \end{bmatrix} x + \begin{bmatrix} 0 \\ .1 \\ 0 \\ -.1 \\ 0 \\ 0 \\ 45 \end{bmatrix} v \equiv Ax + Bv. \tag{5}$$

The output is the same as in Example 4.2-4, namely

$$y = \begin{bmatrix} x_\theta \\ x_p \\ \theta \\ p \end{bmatrix} = \begin{bmatrix} 0 & 0 & 0 & 0 & 57.2958 & 0 & 0 \\ 0 & 0 & 0 & 0 & 0 & 1 & 0 \\ 57.2958 & 0 & 0 & 0 & 0 & 0 & 0 \\ 0 & 0 & 1 & 0 & 0 & 0 & 0 \end{bmatrix} x \equiv Cx. \tag{6}$$

Then, according to Fig. 4.2-11 the control input $v(t)$ is given by

$$v = -k_1 x_\theta - k_2 x_p - k_3 \theta - k_4 p = -[k_1 \quad k_2 \quad k_3 \quad k_4] y \equiv -Ky. \tag{7}$$

We are now in a position to perform the controls design to select the control gains k_i.

b. PI and Continuous Controls Design

To design the continuous-time controller, let us select the same PI as in Example 4.2-4, namely

$$J = \frac{1}{2} \int_0^\infty (t^k x^T Q x + r u^T u) \, dt. \tag{8}$$

with $Q = \text{diag}\{100, 100, 1, 1, 0, 0, 0\}$.

Using $k = 2$, $r = 0.01$ and the software described in Appendix A, we implemented the uniform initial condition design equations in Table 4.1-1, including the extra Lyapunov equations from Table 4.2-1 to account for the time weighting t^2. The resulting gain matrix was

$$K = [49.30 \quad 539.5 \quad -6.933 \quad -56.52], \tag{9}$$

which gave the closed-loop poles

$$\begin{aligned} s = &-1.63 \pm j5.07 \\ &-0.66 \pm j0.68 \\ &-1.75, \ -10, \ -33.66. \end{aligned} \tag{10}$$

The closed-loop response to initial conditions of $\theta = 0.1$ rad and $p = 0.1$ m is shown in Fig. 5.3-9. Note that it is slightly worse than the response achieved in Example 4.2-4, though quite satisfactory.

Let us note that the transfer function from $v(t)$ to either $\theta(t)$ or $p(t)$ contains the approximate ZOH/sampler dynamics described by (1), (3). These include a pole far to the left at $s = -30$ which has no significant effect. However, they also include a nonminimum

5.3 Sampling, Hold Devices, and Computation Delays

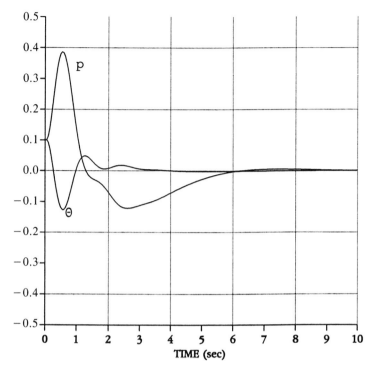

Figure 5.3-9 Inverted pendulum angle $\theta(t)$ (rads) and position $p(t)$ (m) using modified continuous-time controller

phase zero at $s = 60$. This zero significantly changes the root locus, and the control gains (9) selected automatically by the LQ approach take this nonminimum phase zero into account.

It should also be realized that, in contrast to the situation in Example 5.2-2, which relied on the continuous design from Example 4.2-4, the sampling period is now needed to write the continuous dynamics (5), and hence to design the continuous-time controller.

c. Digital Controller

The modified continuous controller just designed is described by exactly the same equations as in Example 5.2-2, with, however, the modified gains k_i given in (9). Thus, the new digital controller is exactly the same as the one described in that example, though using the modified gains.

To examine the performance of the modified digital controller, we may use program DIGCTL in Appendix A along with the continuous-time pendulum dynamics and the subroutine DIG(IT,T,X) from Example 5.2-2 with the gains in (9). The response for $T = 0.1$ sec is shown in Fig. 5.3-10a. Note that, at this design sample period of $T = 0.1$ sec, the digital control response is much like the response using the continuous-time controller shown in Fig. 5.3-9.

We have also shown in Fig. 5.3-10b the digital control response with a sample period of $T = 0.05$ sec. A sample period of $T = 0.01$ sec yields the same result.

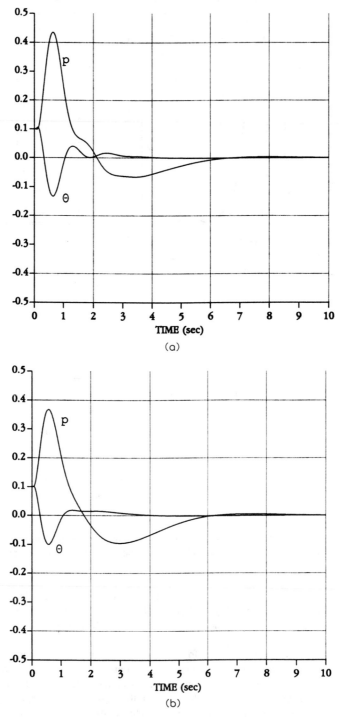

Figure 5.3-10 Inverted pendulum angle $\theta(t)$ and position $p(t)$ using digital controller from modified continuous-time redesign. (a) Sample period $T = 0.1$ sec. (b) Sample period $T = 0.05$ sec.

5.4 Minimum-Time Control

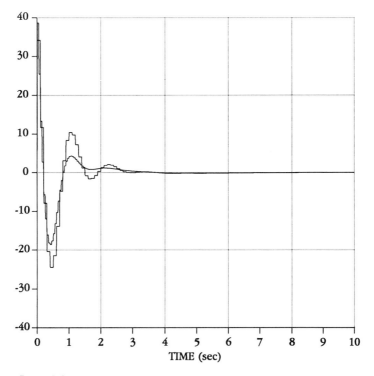

Figure 5.3-11 Digital control inputs $u(t)$ for $T = 0.05$ sec and $T = 0.1$ sec

The control inputs $u(t)$ required for $T = 0.05$ sec and $T = 0.1$ sec are shown in Fig. 5.3-11, with the larger-magnitude response corresponding to $T = 0.1$ sec.

Clearly, the response shown in the figures is excellent. It far surpasses the digital control response in Example 5.2-2 for $T = 0.1$ sec. Thus, we have demonstrated that a sensible technique for taking into account some of the properties of the sample-and-hold process in the design stage of the continuous controller results in improved digital controllers that may be used with larger sample periods T.

∎

5.4 MINIMUM-TIME CONTROL

In section 3.4 we considered the problem of determining the control input $u(t)$ for the system

$$\dot{x} = Ax + Bu, \qquad (5.4.1)$$

$x \in \mathbf{R}^n$, that drives the state from a given initial value $x(0)$ to a desired final value $x(T)$ in minimum time T, while satisfying the control magnitude constraint

$$|u(t)| \leq 1 \qquad (5.4.2)$$

for all $t \in [t_0, T]$.

We saw that, if the time-optimal problem is normal, then the components of the optimal control are always at either their maximum or minimum values. The switchings between these values occur instantaneously, and each control component is allowed to switch at most $n - 1$ times. This sort of control is called *bang-bang control*.

As we saw in Example 3.4-1, the control switching times may be determined, and indeed the optimal bang-bang control may be expressed as a state-variable feedback.

A little thought will reveal that the bang-bang control strategy is automatically a digital control law. That is, the control changes its values only at discrete times, namely, at the switching times. Thus, to implement the minimum-time controller as a digital control, it is only necessary to select a sample period T that is small enough so that the control switchings occur for all practical purposes at integral sampling times.

PROBLEMS FOR CHAPTER 5

Problems for Section 5.2

5.2-1 Prove (5.2.13), (5.2.14) by taking the Laplace transform of $\dot{x} = Ax + Bu$ and then using the BLT.

5.2-2 **Discretization of Harmonic Oscillator.** Redo Example 5.2-1 for the harmonic oscillator

$$K^c(s) = \frac{\omega_n^2}{s^2 + 2\alpha\omega_n s + \omega_n^2}.$$

That is:
  a. Discretize using the BLT.
  b. Discretize using BLT with frequency prewarping. Use ω_n for the prewarping frequency.
  c. Discretize using the MPZ.
  d. Discretize using the modified MPZ.
  e. Using software like MATLAB [Moler et al. 1987] or MATRIX$_x$ [1989], plot the frequency responses of the various digital filters for several sample periods T.

5.2-3 Simulate the digital inverted pendulum controller designed in Example 5.2-2 using a FOH instead of a ZOH. For comparison, use the same sample periods that were used there.

5.2-4 Design a digital inverted pendulum controller (see Example 5.2-2) using the MPZ technique. Simulate the step response for a few sample periods T and compare to the digital controller designed using the BLT.

5.2-5 **Digital Controller for Ball Balancer.** In the problems for section 2.1 the equations for a ball balancer were given. In the problems for section 4.2 a continuous controller for this system was designed.
  a. Discretize that continuous control system using the BLT, repeating Example 5.2-2 for the balancer.
  b. Simulate the digital controller using a ZOH with program TRESP in Appendix A.
  c. Simulate the digital controller using a FOH.

5.2-6 **MPZ Digital Controller for Ball Balancer.** Repeat Problem 5.2-5 using the MPZ.

Problems for Chapter 5

5.2-7 Digital Control of a DC Motor. A continuous-time PI controller for a DC motor using a scalar model was designed in Example 4.2-2. Design and simulate a digital version of this control system for several sample periods T. Use the BLT. Repeat using the modified MPZ.

5.2-8 Digital Control of Armature-Controlled DC Motor. Repeat Problem 5.2-7 for the continuous PI controller designed in Example 4.2-3. Repeat for the Type 0 control system designed in that example.

5.2-9 Ziegler-Nichols Tuning. Consider the armature-controlled DC motor of Example 4.2-3.
  a. Simulate the step response of the system using TRESP. Hence, use the Ziegler-Nichols method to select the PI controller parameters. Simulate the continuous closed-loop system with these parameters.
  b. Convert the continuous controller of part a into a digital controller using the BLT and simulate the closed-loop system.

5.2-10 Armature-Controlled DC Motor with Delay. Consider the armature-controlled DC motor of Example 4.2-3. Suppose there is an additional control delay of $\tau = 0.3$ sec.
  a. Repeat Problem 5.2-9. Is the response suitable?
  b. Now, design a Smith predictor digital control system. Discretize the controller for the undelayed plant using the control gains of Example 4.2-3. Try several sample periods to recover the quality of the response in Example 4.2-3.

Problems for Section 5.3

5.3-1 Prove (5.3.17), beginning with Fig. 5.3-6. Thus, derive (5.3.18), (5.3.19).

5.3-2 A Padé approximant for e^{-sT} is

$$G(s) = \frac{1 - 2sT/3 + (sT)^2/6}{1 + sT/3}.$$

  a. Use long division to determine how many terms of the Taylor series of e^{-sT} are matched by $G(s)$.
  b. Use $G(s)$ to derive one of the approximants for the ZOH plus sampler shown in Table 5.3-2. How many terms of the Taylor series are matched by this approximant?

5.3-3 Modified Design for DC Motor Digital Controller. In Example 4.2-2 a continuous PI controller was designed for a DC motor assuming an approximate scalar model. In Problem 5.2-7 a digital controller for this plant was designed.
  a. For the digital controller of Problem 5.2-7, select a large sample period T that yields bad responses.
  b. For this T, perform a modified continuous controller design that incorporates the effect of the sampler and ZOH hold device, proceeding as in Example 5.3-2. Find the digital controller using the BLT, and simulate the closed-loop system using TRESP in Appendix A. Verify that the performance using this large T is quite suitable using modified continuous design.

5.3-4 Modified Design for Ball Balancer Digital Controller. Repeat Problem 5.3-3 for the ball balancer of Problem 5.2-5.

6

Implementation of Digital Controllers

SUMMARY

In Chapter 5 we showed how to design digital controllers by redesign of an already existing continuous-time controller. In Chapter 7 we shall show how to design digital controllers by direct discrete design.

Once a digital controller has been designed, it must be simulated on a computer and then implemented in hardware on the actual plant. We showed how to simulate digital controllers in section 5.1, giving several examples of the procedure throughout Chapter 5. In this chapter, we discuss the actual implementation of digital controllers on a digital signal processor.

INTRODUCTION

In Chapter 4 we showed how to design continuous-time controllers of desired structure using output feedback. The fundamental design approach was given in section 4.2, where we showed that the selection of a sensible performance index (PI) is the key to successful design. We derived the design equations for the output-feedback control gain K given in Table 4.2-1. The software described in Appendix A conveniently solves the design equations, and the design approach was demonstrated in several examples.

In Chapter 5 we showed how to convert a continuous-time controller into a digital controller using approximation techniques. The next step is to simulate the digital controller. We showed how to accomplish this using simple computer programs in design examples throughout Chapter 5. The final step is to implement the digital controller in hardware on the actual plant. In this chapter we show how this may be accomplished.

We discuss some important issues in the implementation of digital controllers. First, realizing that practical electrical and hydraulic control actuators have physical limitations on their ranges of motion, we discuss actuator saturation and integrator windup. Integrator windup is not a problem peculiar to digital controllers, but must also be taken into account in implementing continuous controllers. However, this is an appropriate place to discuss the issue since it is easily corrected using digital techniques.

Next, we mention some problems that are peculiar to implementing digital controllers on microprocessors, which have a finite wordlength. In section 6.2 we cover signal quantization and round-off errors, and in section 6.3 numerical overflow and scaling.

There are many ways to implement a given transfer function in hardware, and some are better than others. We cover this issue in section 6.4.

Finally, to complete the picture, in section 6.5 we give some subroutines that are useful in implementing digital controllers on an actual commercial digital signal processor (DSP). To illustrate, we have chosen to use the Texas Instruments TMS320C25 DSP. In section 6.6 we give a design example showing how to implement digital controllers using a DSP.

6.1 ACTUATOR SATURATION AND WINDUP

Actuator saturation leading to integrator windup is a problem that occurs in both continuous-time and digital control systems. Correction techniques are similar in both cases, and are easily implemented digitally.

A digital controller may be represented in the dynamic state-space form

$$x_{k+1} = Fx_k + Gw_k \quad (6.1.1)$$

$$v_k = Cx_k + Dw_k, \quad (6.1.2)$$

where $x_k \in \mathbf{R}^n$ is the controller state and w_k the controller input, composed generally of the tracking error and the plant measured output. The discrete control sequence v_k is passed through a hold device to generate the continuous plant control input $u(t)$.

We have assumed thus far that the plant control input $v_k \in \mathbf{R}^m$ which is computed by the controller can actually be applied to the plant. However, in practical systems the plant inputs (such as motor voltages, etc.) are limited by *maximum* and *minimum* allowable values. Thus, the relation between the *desired plant input* v_k and the *actual plant input* u_k is given by the sort of behavior shown in Fig. 6.1-1, where u_H and u_L represent respectively the maximum and minimum control effort allowed by the mechanical actuator. If there are no control limits, then we may set $u_k = v_k$.

Thus, to describe the actual case in a practical control system with actuator saturation, we are forced to include *nonlinear saturation functions* in the control channels as shown in Fig. 6.1-2.

Consider the simple case where the controller is an integrator with input w_k and output v_k. Then, all is well as long as v_k is between u_L and u_H, for in this region the plant input u_k equals v_k. However, if v_k exceeds u_H then u_k is limited to its maximum value u_H. This in itself may not be a problem. The problem arises if w_k remains positive, for then the integrator continues to integrate and v_k may increase well beyond u_H. Then, when w_k becomes negative, it may take considerable time for v_k to decrease below u_H. In the meantime, u_k is held at u_H, giving an incorrect control input to the plant. This effect of

300 Chap. 6 Implementation of Digital Controllers

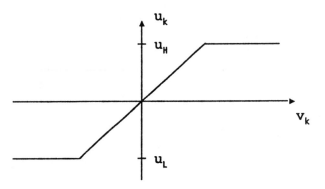

Figure 6.1-1 Actuator saturation function

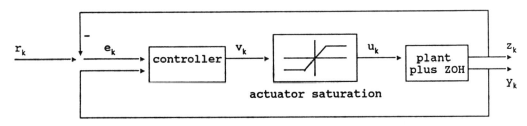

Figure 6.1-2 Control system including actuator saturation

integrator saturation is called *windup*. It arises because the controllers we design are generally dynamical in nature, which means that they store information or energy.

To correct integrator windup, it is necessary to limit the state of the controller so that it is consistent with the saturation effects being experienced by the plant input u_k. This is not difficult to achieve [Åström and Wittenmark 1984]. Indeed, set $u_k = v_k$ and write (6.1.2) in the form

$$0 = u_k - Cx_k - Dw_k.$$

Now, multiply it by L, which shall soon be selected, and add it to (6.1.1) to obtain

$$x_{k+1} = (F - LC)x_k + (G - LD)w_k + Lu_k. \qquad (6.1.3)$$

A little thought shows that actuator windup occurs when F is not asymptotically stable. For then, as long as w_k in (6.1.1) is nonzero, x_k will continue to increase. However, by using the formulation (6.1.3) for the digital controller and selecting L so that

$$F_o = F - LC \qquad (6.1.4)$$

is asymptotically stable, this problem is averted.

A special case occurs when L is selected so that F_o has all poles at the origin. Then, x_k displays deadbeat behavior; after n time steps it remains limited to an easily computed value dependent on the values of w_k and u_k (see the problems).

The *antiwindup gain* L may be selected to place the poles of F_o arbitrarily if (L, F)

6.1 Actuator Saturation and Windup

is observable. However, as long as (L, F) is *detectable* (i.e., has all its unstable poles observable), then windup may be eliminated using this technique.

To complete the antiwindup design, the plant control input may be selected according to

$$u_k = \text{sat}(Cx_k + Dw_k), \quad (6.1.5)$$

where the *saturation function* is defined for scalars as

$$\text{sat}(v) = \begin{cases} u_H, & v \geq u_H \\ v, & u_L < v < u_H, \\ u_L, & v \leq u_L \end{cases} \quad (6.1.6)$$

with u_H and u_L the maximum and minimum allowable values respectively. For vectors, the saturation function is defined as

$$\text{sat}(v) = \begin{bmatrix} \text{sat}(v_1) \\ \text{sat}(v_2) \\ \vdots \\ \text{sat}(v_m) \end{bmatrix}. \quad (6.1.7)$$

The values of u_H and u_L for each component v_i should be selected to correspond to the actual limits on the components of the plant input u_k.

Note that the limited signal u_k is used in (6.1.3), providing a feedback arrangement. What we have in effect done is include an *observer* with dynamics F_o in the digital controller. Since F_o is asymptotically stable, the observer will provide reasonable "estimates" even in the event of saturation.

Where u_k is not saturated, the controller with antiwindup compensation (6.1.3), (6.1.5) is identical to (6.1.1), (6.1.2).

If the controller is given in transfer function form

$$R(z^{-1})u_k = T(z^{-1})r_k - S(z^{-1})w_k, \quad (6.1.8)$$

where r_k is the reference command and z^{-1} is interpreted in the time domain as a unit delay of T sec, then antiwindup compensation may be incorporated as follows.

Select a desired stable observer polynomial $A_o(z^{-1})$ and add $A_o(z^{-1})u_k$ to both sides to obtain

$$A_o u_k = T r_k - S w_k + (A_o - R) u_k. \quad (6.1.9)$$

A regulator with antiwindup compensation is then given by

$$A_o v_k = T r_k - S w_k + (A_o - R) u_k. \quad (6.1.10)$$

$$u_k = \text{sat}(v_k). \quad (6.1.11)$$

Example 6.1-1: Digital PI Controller with Antiwindup Compensation

From section 5.2 a general digital PI controller with sampling period T sec is given by

$$u_k = k\left[1 + \frac{T}{T_I}\frac{1}{z-1}\right]e_k, \quad (1)$$

where we have used design by the modified MPZ to obtain a delay of T sec in the integrator to allow for computation time. The proportional gain is k and the reset time is T_I; both are fixed in the design stage. The tracking error is $e_k = r_k - z_k$, as usual.

Multiply by $(1 - z^{-1})$ to write

$$(1 - z^{-1})u_k = k[(1 - z^{-1}) + Tz^{-1}/T_I]e_k, \qquad (2)$$

which is in the transfer function form (6.1.8). The corresponding difference equation form for implementation is

$$u_k = u_{k-1} + ke_k + k(-1 + T/T_I)e_{k-1}. \qquad (3)$$

This controller will experience windup problems since the autoregressive polynomial $R = 1 - z^{-1}$ has a root at $z = 1$, making it marginally stable. Thus, when u_k is limited, the integrator will continue to integrate, "winding up" beyond the saturation level.

a. Antiwindup Compensation

To correct this problem, select an observer polynomial of

$$A_o(z^{-1}) = 1 - \alpha z^{-1}, \qquad (4)$$

which has a pole at some desirable location $|\alpha| < 1$. The design parameter α may be selected by simulation studies. Then, the controller with antiwindup protection (6.1.10), (6.1.11) is given by

$$(1 - \alpha z^{-1})v_k = k\left[1 + \left(-1 + \frac{T}{T_I}\right)z^{-1}\right]e_k + (1 - \alpha)z^{-1}u_k \qquad (5)$$

$$u_k = \text{sat}(v_k). \qquad (6)$$

The corresponding difference equations for implementation are

$$v_k = ke_k + \alpha v_{k-1} + k\left(-1 + \frac{T}{T_I}\right)e_{k-1} + (1 - \alpha)u_{k-1} \qquad (7)$$

$$u_k = \text{sat}(v_k). \qquad (8)$$

A few lines of FORTRAN code implementing this digital controller are given in Fig. 6.1-3. This subroutine may be used as the control update routine DIG for the digital simulation driver program TRESP in Appendix A.

If $\alpha = 1$ we obtain the special case (2), which is called the *position form* and has no antiwindup compensation. If $\alpha = 0$, we obtain the *deadbeat antiwindup compensation*

$$v_k = k\left[1 + \left(-1 + \frac{T}{T_I}\right)z^{-1}\right]e_k + u_{k-1}, \qquad (9)$$

with corresponding difference equation implementation

$$v_k = u_{k-1} + ke_k + k\left(-1 + \frac{T}{T_I}\right)e_{k-1}. \qquad (10)$$

If u_k is not in saturation, then this amounts to updating the plant control by adding the second and third terms on the right-hand side of (10) to u_{k-1}. These terms are therefore nothing but $u_k - u_{k-1}$. The compensator with $\alpha = 0$ is thus called the *velocity form* of the PI controller.

6.1 Actuator Saturation and Windup

```
DIGITAL PI CONTROLLER WITH ANTIWINDUP COMPENSATION

    SUBROUTINE DIG(IK,T,x)
    REAL x(*), k
    COMMON/CONTROL/ u
    COMMON/OUTPUT/ z
    DATA k,AL,TI,ULOW,UHIGH/ 3.318, 0.9, 1., -1.5 , 1.5/
    DATA r/1./

    v= AL*v + k*(-1 + T/TI)*e + (1-AL)*u
    e= r - z
    v= k*e + v
    u= AMAX1(ULOW,v)
    u= AMIN1(UHIGH,u)

    RETURN
    END
```

Figure 6.1-3 FORTRAN code implementing PI controller with antiwindup compensation.

b. Digital Control of DC Motor

In Example 4.2-2 we designed a continuous-time PI controller for speed control for a DC motor described by

$$\dot{\omega} = -a\omega + bu, \tag{11}$$

with ω the angular velocity. The controller was of the form

$$u = -[k_1 + k_2/s]e, \tag{12}$$

where $e = r - \omega$ is the tracking error, with r the desired command angular velocity. Taking $a = 1$ and $b = 1$, with $k_1 = k_2 = -3.318$, we obtained poles at $s = -1, -3.318$. The slower pole was canceled by a zero, so that the step response had only a mode like $e^{-3.318t}$.

Writing the PI controller as

$$u = k[1 + 1/T_I s]e, \tag{13}$$

we see that

$$\begin{aligned} k &= -k_1 = 3.318 \\ T_I &= k_1/k_2 = 1 \text{ sec.} \end{aligned} \tag{14}$$

The digital controller obtained using the modified MPZ is given by (1).

The time constant of the closed-loop system is $\tau = 1/3.318 = 0.3$ sec, so that a sampling period of $T = 0.05$ sec is reasonable. The sampling period should be about one tenth the time constant.

Program TRESP in Appendix A was used to obtain the response shown in Fig. 6.1-4a. No saturation limits were imposed on u_k. The corresponding digital control $u(t)$ is shown in Fig. 6.1-5a.

Next, a saturation limit of $u_H = 1.5$ v was imposed on the control u_k. No antiwindup compensation was used (i.e., $\alpha = 1$ in Fig. 6.1-3). The resulting behavior is shown in

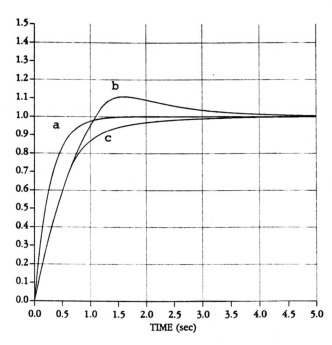

Figure 6.1-4 Angular velocity step responses using digital controller. (a) Digital PI controller with no saturation limits. (b) Digital PI controller with saturation limit $u_H = 1.5\ v$ and no antiwindup compensation. (c) Digital PI controller with saturation limit $u_H = 1.5\ v$ and antiwindup compensation with $\alpha = 0.9$.

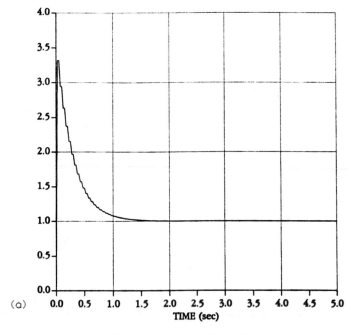

Figure 6.1-5 Digital control inputs. (a) u_k with no saturation limit.

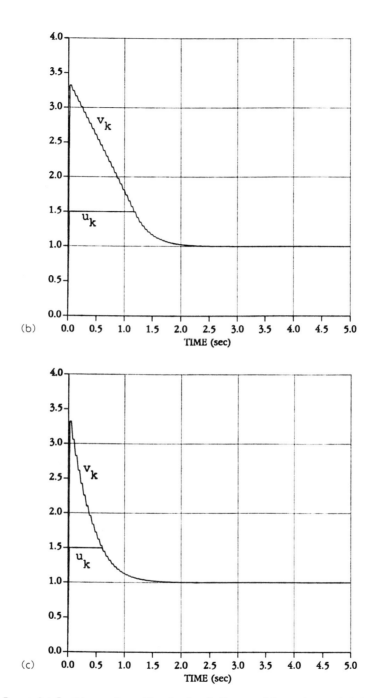

Figure 6.1-5 (b) u_k and v_k with saturation limit $u_H = 1.5$ v and no antiwindup compensation. (c) u_k and v_k with saturation limit $u_H = 1.5$ v and antiwindup compensation with $\alpha = 0.9$.

Figs. 6.1-4b and 6.1-5b and $\omega(t)$ displays an unacceptable overshoot. Note that u_k remains at its upper limit of $u_H = 1.5\ v$ until 1.2 sec due to the integrator windup.

Finally, Figs. 6.1-4c and 6.1-5c show that the overshoot problem is easily corrected using $\alpha = 0.9$ in the digital PI controller with antiwindup protection in Fig. 6.1-3.

In this example, as α decreases the step response slows down. The value of $\alpha = 0.9$ was selected after several simulation runs with different values of α.

■

6.2 QUANTIZATION AND ROUNDOFF

We have assumed throughout that infinite precision is available in representing numbers in the digital controller. In fact, however, microprocessors are limited in their accuracy. Commercially available digital signal processors (DSPs) use either fixed-point arithmetic or floating-point arithmetic. Fixed-point DSPs are generally cheaper and faster, but they are limited in their dynamic range available for representing numbers. Since fixed-point DSPs are adequate for many controls applications and require greater care in implementing the controller, we shall focus on them. Understanding controls implementation on a fixed-point DSP will be more than enough to allow one to deal with a floating-point unit.

Deterministic Analysis of Quantization Effects

Let us consider a fixed-point DSP with a 16-bit word. One way to represent numbers on this machine is to scale them so that their magnitude is less than one. Then, one bit is reserved for the sign and 15 bits are devoted to representing the magnitude: this is called *fractional representation* as a Q15 number. In this case, the *quantization error* is given by

$$q = 2^{-15} = 0.0000305, \tag{6.2.1}$$

corresponding to a relative error of 0.00305%. That is, two numbers may differ by this amount but still have the same finite wordlength representation. An alias for q is the *quantization step size*.

There are several ways to quantize numbers. In *truncation*, the least significant bits are simply ignored. In *roundoff*, the last bit is set to zero if the first bit lost is a "0," and to one if the first bit lost is a "1." We shall represent the quantized value of a real number x as $Q[x]$.

Many DSPs use the 2's-complement number system [Phillips and Nagle 1984]. In this number system, the average error using roundoff is zero, but the average error using truncation is $q/2$. Plots showing quantization for the 2's-complement using truncation and roundoff are shown in Fig. 6.2-1.

Errors arise due to the quantization of the *control signals*, and due to the quantization of the *controller coefficients*. The control signals are quantized during the sampling or A/D process.

6.2 Quantization and Roundoff

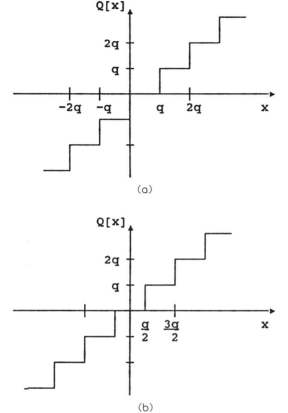

Figure 6.2-1 Quantizer characteristic for 2's-complement. (a) Truncation. (b) Roundoff.

In view of the fact that the entire left half of the s-plane is mapped by discretization schemes into the interior of the unit circle in the z-plane, great accuracy in the digital controller coefficients is needed to place the poles as desired. That is, small changes in coefficient values can yield large changes in the z-plane closed-loop pole locations. We shall mention this again in the subsection on "Controller Realization Structures."

Once the signals and coefficients have been represented by the finite computer wordlength, there arise additional errors whenever a multiplication is performed. To minimize these, many digital signal processors accumulate products in a double-length word. After multiplication, the result must be reduced to standard word size, for instance by truncation or roundoff.

In addition to improper closed-loop pole locations, and hence potentially unsatisfactory time responses, it is important to be aware of two other effects of quantization. A linear system with a quantizer in the loop is a *nonlinear system*. Therefore, it can exhibit some of the features of nonlinear systems. We shall mention the occurrence of *multiple equilibrium points* and *limit cycles*. The next example serves for illustration.

Example 6.2-1: Multiple Equilibria and Limit Cycles Under Quantization

This example is from Åström and Wittenmark [1984] and Franklin and Powell [1980].
Consider the scalar system

$$y_{k+1} = Q[\alpha y_k], \tag{1}$$

where $Q[\ ]$ represents quantization using truncation. Due to the presence of quantization, this is a nonlinear system. Let us assume that $|\alpha| < 1$ so that the linear system without $Q[\ ]$ is stable with a single equilibrium point at the origin.

Systems with quantization may be analyzed using several techniques. Here, we shall show a graphical approach.

Figure 6.2-2 shows the graphs of $Q[\alpha y]$ and of the line y, which has a slope of $1/\alpha$ in the figure. This figure may be used to investigate the dynamical development of (1). Specifically, given y_k, we may use the figure to determine y_{k+1} in the presence of quantization.

Suppose that the value of y_k is given by the point a on the y-axis in Fig. 6.2-2a. Move horizontally from a to the line y and then down to the point b on the αy axis to find αa. Then, include the quantization effect by moving up from b to the function $Q[\alpha y]$ and then horizontally to the y-axis; this yields $Q[\alpha a]$, denoted as the point c in the figure, which is equal to y_{k+1}.

To determine the trajectory y_k, a shortcut for this procedure is simply to follow the dashed "staircase" between the graphs of y and $Q[\alpha y]$ shown in the figure. Note that, in the case we are discussing, the staircase eventually approaches the origin, which is thus the equilibrium point.

a. Multiple Equilibria

Now suppose that the initial condition for (1) is negative. In this event we may perform an exercise like the one just described to obtain the dashed staircase shown in Fig. 6.2-2b. Note that the equilibrium point is now $y = -2q$, with q the quantization step size.

In fact, any point of intersection of the line y and $Q[\alpha y]$ is an equilibrium point of the nonlinear system (1) for the appropriate initial condition!

It would be desirable to find a value for $|kq|$, the largest possible value of an equilibrium point. From Fig. 6.2-2b it may be seen that the most negative intersection of y and $Q[\alpha y]$ occurs at a value of $-kq$ such that

$$-kq\alpha < -kq + q \tag{2}$$

or

$$k < \frac{1}{1 - \alpha}. \tag{3}$$

Therefore, the largest magnitude of $y = -kq$ which is an equilibrium point is given by

$$|y| < \frac{q}{1 - \alpha}. \tag{4}$$

b. Limit Cycles

Suppose now that $-1 < \alpha < 0$. Then, the plots of y and $Q[\alpha y]$ appear as in Fig. 6.2-3. A staircase procedure like the one described above shows that there is now no equilibrium

6.2 Quantization and Roundoff

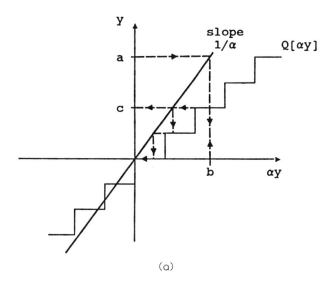

(a)

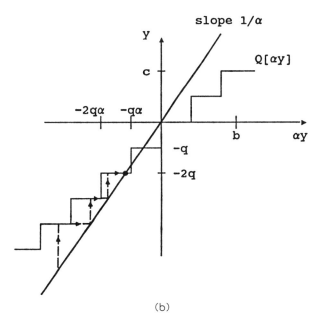

(b)

Figure 6.2-2 Graphical determination of equilibrium points. (a) Equilibrium at the origin. (b) Nonzero equilibrium.

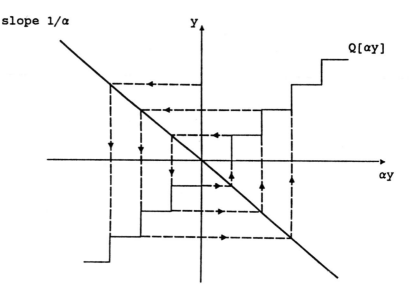

Figure 6.2-3 Graphical determination of limit cycles.

point, but that there is a limit cycle. That is, following the staircase, one proceeds around the origin in the same pattern after an initial transient phase.

A bound like (4) on the magnitude of the limit cycle may be found (see the problems).

∎

One way to eliminate multiple equilibria is to add a high-frequency noise of low amplitude called *dither* to the control input. From Fig. 6.2-2b, it appears that adding a noise component n_k of amplitude $3q$ to the system to obtain

$$y_{k+1} = Q[\alpha y_k] + n_k \qquad (6.2.2)$$

would result in an average drift toward the origin. This occurs because, if y_k approaches the equilibrium point of $-2q$, it will be forced to take on positive values, for which the origin is the only equilibrium.

If the dither occurs at frequencies outside the system bandwidth, it should have minimal deleterious effects on system performance.

Conditions for the absence of limit cycles are given in Phillips and Nagle [1984] and Åström and Wittenmark [1984], where some more techniques for analyzing them are given.

Example 6.2-2: Quantization Effects in a Digital Inverted Pendulum Controller

To illustrate the sort of quantization effects to be expected in a more complicated digital control system, let us consider the inverted pendulum. In Example 5.2-2 we designed a digital controller for this system which gave very good results for $T = 0.01$ sec.

6.2 Quantization and Roundoff

It is easy to write a 5-line function to simulate quantization of signals in a digital controller. In FORTRAN V, such a function, assuming roundoff is used, is

```
REAL FUNCTION Q(y)
DATA WL/7/
z= NINT(y*2**WL)
Q= z/(2**WL)
RETURN
END
```
(1)

where WL is the binary wordlength. Before implementing any controller on a DSP, it is important to simulate it including quantization effects corresponding to the accuracy of the DSP.

Thus, suppose we take a seven-bit wordlength, corresponding to $2^{-7} = .0078125$ or about 0.78% relative accuracy, and redo our simulation in Example 5.2-2 including quantization effects. Let us use $T = 0.01$ sec and the same control gains used in that example. To simulate the effects of quantization, it is only necessary to insert the function $Q(\ )$ at appropriate places in the digital control subroutine DIG in Example 5.2-2.

The results are shown in Fig. 6.2-4, where we used initial conditions of $\theta = 0.1$ rad,

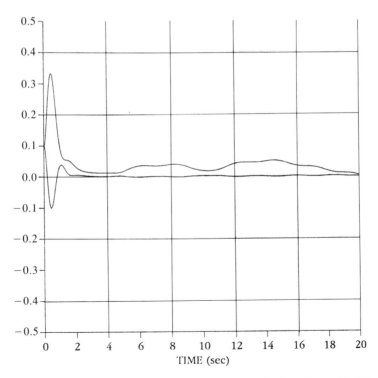

Figure 6.2-4 Inverted pendulum response including quantization effects with 7-bit word.

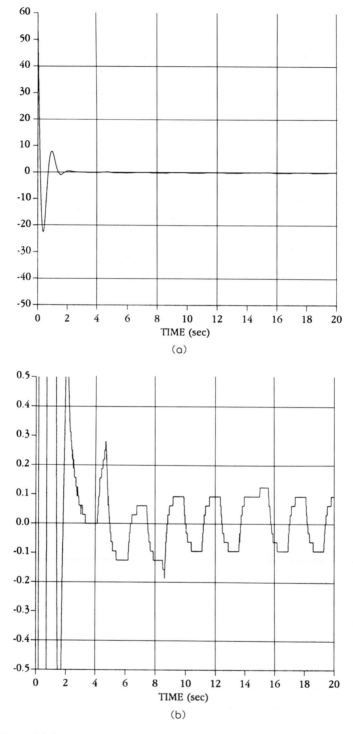

Figure 6.2-5 Control effort with quantization effects. (a) Normal scale. (b) Expanded scale showing limit cycle.

$p = 0.1\ m$. Comparing this response to that in Example 5.2-2, it may be seen that the initial portion of the plot is not too much changed. However, after about 4 sec, the responses begin to diverge. It is not clear that they ever reach zero.

The control effort in the face of quantization is shown in Fig. 6.2-5. In part a of the figure it is difficult to notice any problems. However, the expanded vertical scale in part b clearly shows the presence of a limit cycle, so that the digital controller with 7 bits of accuracy is unsatisfactory. ∎

Stochastic Analysis of Quantization Effects

An alternative to deterministic analyses of quantization is a stochastic approach. This is a linear analysis approach where the quantization error is treated as a *noise process* that is added to the quantized signals.

For stochastic analysis, the mean and variance of the noise process must be known. These may easily be determined for quantization. For 2's-complement arithmetic (see Fig. 6.2-1), the mean error using truncation is $q/2$. Thus, truncation may cause *biased* signals. On the other hand, the mean error using roundoff is zero.

Assuming that the value x giving rise to the quantized value $Q[x]$ has an equal probability of occurring anywhere in the interval of values that is mapped to $Q[x]$ (see Fig. 6.2-1), the quantization noise is uniformly distributed. Then, for both truncation and roundoff, the quantization noise has a variance of $q^2/12$.

It may be necessary to perform extensive simulation that includes the effects of quantization to test a digital controller prior to its installation.

6.3 OVERFLOW AND SCALING

Primary references for this section are Åström and Wittenmark [1984], Phillips and Nagle [1984], Slivinsky and Borninski [1987], and Hanselmann [1987].

Digital signal processors that use fixed-point arithmetic are generally cheaper and faster than those that use floating-point. Moreover, digital controllers are usually easy to implement using fixed-point. Therefore, in this section we discuss some problems unique to fixed-point arithmetic.

The DSP wordlength is finite, with a popular wordlength being 16 bits. Let us use a *fractional representation* for numbers, so that all numbers have been scaled so that their magnitudes are less than or equal to one; the first bit of a word represents the sign and b bits represent the magnitude of the number. Assume that fixed-point arithmetic is used. Then, the numbers that may be represented using the 2's-complement number system are bounded by

$$-1 \leq Q[x] \leq 1 - 2^{-b}, \qquad (6.3.1)$$

with $Q[x]$ the quantized value of a number x.

If two fractions are multiplied, the result is a fraction. Thus, no overflow will occur on multiplication.

However, even if two numbers are less than one, their *sum* may not be. If it is not, overflow is said to occur on addition. Severe problems may result in digital control if overflow is not properly handled.

Overflow Handling

The overflow characteristic of the 2's-complement number system is shown in Fig. 6.3-1, where it is assumed that in the addition any carry bit is ignored. This characteristic exhibits *wraparound*, which is not unlike aliasing (see section 5.3); the sum of two large numbers may be erroneously interpreted as a small number.

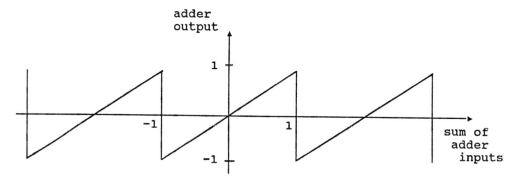

Figure 6.3-1 Two's-complement number system overflow characteristic

If overflow is ignored, limit cycles may appear in the closed-loop system which are called *overflow oscillations*. Many commercial DSPs have an automatic overflow handling capability. Otherwise, it may be necessary to write a software *overflow protection routine*.

Such a routine may automatically detect overflow, then scale the product by shifting it right. Note that a right shift of one bit corresponds to a division by 2. Alternatively, the overflow protection routine may convert the 2's-complement wraparound characteristic to a *saturation* characteristic like that in Fig. 6.3-2. The saturation overflow characteristic precludes overflow oscillations.

An important property of the 2's-complement system is that a sequence of additions or subtractions always produces the correct final result as long as the final result is in the range (6.3.1). Intermediate overflows of partial sums do not matter and can be ignored. Thus, the problem lies in guaranteeing that the result is in the allowed range. If this can indeed be guaranteed, then intermediate overflows should be allowed, since eliminating them can result in decreased accuracy in the final result.

6.3 Overflow and Scaling

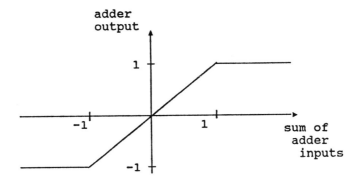

Figure 6.3-2 Saturation overflow characteristic

A test for the absence of overflow oscillations is provided as follows [Phillips and Nagle 1984]. Represent the digital controller in state-space form

$$x_{k+1} = Fx_k + Gw_k \qquad (6.3.2)$$
$$u_k = Cx_k + Dw_k.$$

If there exists a diagonal matrix S with positive diagonal elements and $S - F^TSF$ is positive definite, overflow oscillations are impossible.

For any stable controller there is always a 2's-complement implementation such that overflow oscillations are absent. Indeed, if F is stable, there is a positive definite solution P to

$$P - F^TPF = I. \qquad (6.3.3)$$

Let T be a symmetric square root of P (i.e., $P = T^2$). Then

$$P^{-1} = I - (T^{-1}FT)^T I (T^{-1}FT)$$

is positive definite and the identity I is diagonal. Thus, T is a state-space transformation to a new realization in which overflow oscillations are absent. That is, there are no overflow oscillations in the filter

$$x'_{k+1} = T^{-1}FTx'_k + T^{-1}Bu_k. \qquad (6.3.4)$$
$$u_k = CTx'_k + Dw_k.$$

A disadvantage with this is that the new realization may have a structure that is not suitable for implementation (see the next section).

Example 6.3-1: Overflow Oscillations

This example is from Parks and Burrus [1987].
The transfer function

$$H(z) = \frac{1}{z^2 - z + .5} \qquad (1)$$

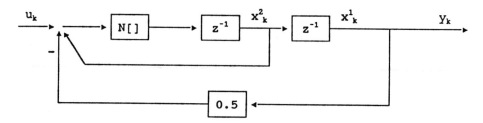

Figure 6.3-3 Reachable canonical form realization of system

can be realized in reachable canonical form (Chapter 2) as shown in Fig. 6.3-3. The nonlinear term $N[\ ]$ represents the overflow characteristics of the adder.

The difference equations associated with the figure are

$$x^1_{k+1} = x^2_k$$
$$x^2_{k+1} = N[-.5x^1_k + x^2_k + u_k] \qquad (2)$$
$$y_k = x^1_k.$$

Suppose the input u_k is zero and the initial conditions are $x^1_0 = 0.8$, $x^2_0 = -0.8$.

a. 2's-Complement Overflow Characteristic

If $N[\ ]$ represents the 2's-complement overflow characteristic shown in Fig. 6.3-1, then we may use (2) to compute

$$y_0 = 0.8$$
$$x^1_1 = -0.8$$
$$x^2_1 = N[-1.2] = 0.8$$
$$y_1 = -0.8$$
$$x^1_2 = 0.8$$
$$x^2_2 = N[1.2] = -0.8.$$
$$y_2 = 0.8$$

Therefore, the output will oscillate between 0.8 and -0.8.

b. Saturation Overflow Characteristic

Using overflow protection in the DSP coding to convert $N[\ ]$ to the saturation characteristic in Fig. 6.3-2 yields quite a different behavior. In fact, it is a good exercise to compute y_k to see that the system is stable, so that all signals go to zero.

Clearly, the location in the block diagram of the nonlinear term $N[\ ]$ makes a difference. That is, a different realization of $H(z)$ will have different overflow behavior. ∎

Signal Scaling

In 2's-complement arithmetic using a fractional representation for numbers, as long as we can guarantee that the end result of a sequence of additions and subtractions is less than

6.3 Overflow and Scaling

one, there will be no overflow problem. It does not matter whether the intermediate partial sums are less than one. We can use scaling to guarantee that the output u_k of the digital controller (6.3.2) has magnitude less than one as follows.

Let the maximum magnitude over all k of component i of w_k be w_{Mi}, and the maximum magnitude of component i of u_k be u_{Mi}. Suppose that the w_{Mi} are known, either from simulation studies or some other means. Then, the values of u_{Mi} may be determined using the impulse response as follows.

Let the transfer function of (6.3.2) be the $p \times m$ matrix given by

$$H(z) = C(zI - F)^{-1}G + D, \tag{6.3.5}$$

and h_k be the $p \times m$ impulse response matrix. That is, $H(z)$ is the z-transform of h_k. Then

$$u_k = \sum_{\ell=0}^{\infty} h_\ell w_{k-\ell}. \tag{6.3.6}$$

Denoting the i-th component of u_k as $u_i(k)$ and the (i, j)-th element of h_k as $h_{ij}(k)$, we may write

$$u_i(k) = \sum_{j=1}^{m} \sum_{\ell=0}^{\infty} h_{ij}(\ell) w_j(k - \ell), \tag{6.3.7}$$

whence

$$|u_i(k)| \leq \sum_{j=1}^{m} \sum_{\ell=0}^{\infty} |h_{ij}(\ell) w_j(k - \ell)|$$

$$\leq \sum_{j=1}^{m} \sum_{\ell=0}^{\infty} |h_{ij}(\ell)| \cdot |w_j(k - \ell)|$$

$$\leq \sum_{j=1}^{m} \sum_{\ell=0}^{\infty} |h_{ij}(\ell)| \cdot w_{Mj}.$$

Therefore, we may take

$$u_{Mi} = \sum_{j=1}^{m} \sum_{\ell=0}^{\infty} |h_{ij}(\ell)| \cdot w_{Mj}. \tag{6.3.8}$$

Set

$$u_M = \max_{i} \{u_{Mi}\}.$$

If all components of the controller input w_k are scaled by $1/u_M$, then each component of the controller output u_k is guaranteed to have a magnitude less than one. Therefore, overflow will not be a problem regardless of the magnitudes of any intermediate signals or partial sums.

We call

$$\|H\|_\infty \equiv \max_i \sum_{j=1}^{m} \sum_{\ell=0}^{\infty} |h_{ij}(\ell)| \quad (6.3.9)$$

the *system gain*. It is the L_∞ operator norm of the system.

There is a simple procedure for determining a suitable scale factor that does not involve finding the system gain. All that is required is to simulate the continuous-time controller prior to discretization. Then, the scale factor for dividing the digital controller inputs w_k is simply the maximum value of the control signal $u(t)$ in the continuous controller. See Example 6.3-2.

An alternative to scaling the input signal is to scale each value of the impulse response h_k by $1/u_M$. However, due to the structure of commercial DSPs such as the Texas Instruments TMS320C25, this is not recommended. Indeed, due to the fact that multiply accumulates are performed with 32 bits of accuracy, if the input is scaled instead of the filter coefficients, no accuracy is lost in the implementation.

To see this, note that a 16-bit data word must be shifted into the highest 16 bits of the 32-bit accumulator to begin a multiply. This amounts to a left shift of 16 and then a transfer to the accumulator. However, if the data is shifted only 15 bits, for instance, and then transferred to the accumulator, it has been divided by two and no information has been lost. Thus, signal scaling may easily be performed as the data is loaded into the accumulator.

The scaling must be inverted after all the computations are completed. Thus, the scaled digital controller output u_k should be multiplied by u_M before it is applied to the plant.

Filter Coefficient Scaling

Suppose the controller transfer function is the scalar

$$H(z) = \frac{b_0 + b_1 z^{-1} + \cdots + b_m z^{-m}}{a_0 + a_1 z^{-1} + \cdots + a_n z^{-n}} \quad (6.3.10)$$

Then, u_k is given by the autoregressive moving-average (ARMA) equation

$$a_0 u_k = -\sum_{i=1}^{n} a_i u_{k-i} + \sum_{i=0}^{m} b_i w_{k-i}. \quad (6.3.11)$$

Even if signal scaling has been performed to guarantee that u_k has all components of magnitude less than one, there will be a problem with implementation in a fractional representation if any of the filter coefficients has a magnitude larger than one.

One way to deal with this problem is to scale all the filter coefficients b_i and a_i so that all their magnitudes are less than one. Note that a_0 should also be scaled.

An alternative follows. Consider the example

$$u_k = 1.7 u_{k-1} - 0.72 u_{k-2} + w_{k-1}.$$

6.3 Overflow and Scaling

By simply dividing 1.7 by 2 and writing this as

$$u_k = 0.85 u_{k-1} + 0.85 u_{k-1} - 0.72 u_{k-2} + w_{k-1}$$

we may implement this filter without scaling.

Measurement Noise

Important topics in digital control are robustness and noise rejection. In Chapter 8 we shall cover such notions in detail, showing how to design digital controllers that are robust to noise and disturbances. Let us merely mention here that feedback control systems are inherently robust. However, overflow in the controller can destroy robustness. In fact, measurement noise can induce overflow oscillations in an improperly designed digital controller, as we show in the next example.

Example 6.3-2: Overflow Oscillations and Measurement Noise in a Digital Motor Speed Controller

In Example 4.2-3 we designed a continuous-time speed controller for the armature-controlled DC motor

$$\dot{x} = \begin{bmatrix} -2.29 & -0.003 \\ 1.172 & -1.32 \end{bmatrix} x + \begin{bmatrix} 0.337 \\ 0 \end{bmatrix} u \equiv Ax + Bu \tag{1}$$

$$z = [0 \quad 1]x \equiv Hx = \omega \tag{2}$$

with state $x = [i \quad \omega]^T$, where $i(t)$ is the armature current and $\omega(t)$ is the angular velocity. It would be useful to review that example at this point.

We used the continuous-time PI controller

$$u = -k_\omega \omega + k_I e \tag{3}$$

where the tracking error is

$$e = r - \omega, \tag{4}$$

with r the constant reference command velocity. Suitable gains were determined to be

$$k_\omega = 24.78, \qquad k_I = 32.20. \tag{5}$$

The continuous closed-loop system had poles at

$$s = -1.3, \qquad -1.16 \pm j2.91. \tag{6}$$

A simulation of this continuous-time controller is given in Fig. 6.3-4, where the step response $\omega(t)$ and the control voltage $u(t)$ are shown.

a. Digital Speed Controller

A digital version of the speed controller is shown in Fig. 6.3-5. Using the modified MPZ to digitize the continuous PI controller (see section 5.2) yields the digital PI controller

$$u_k = -k_\omega \omega_k + k_I T \frac{1}{z-1} e_k. \tag{7}$$

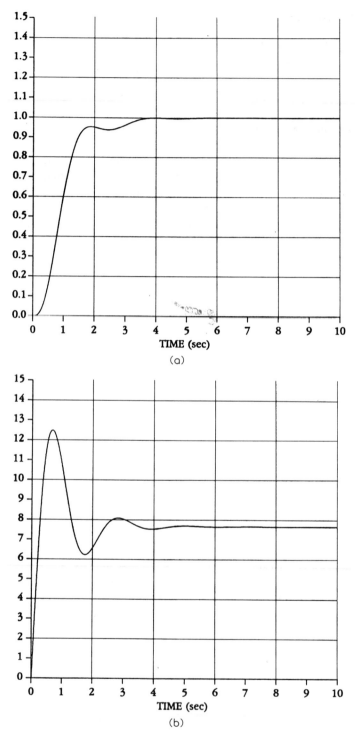

Figure 6.3-4 Simulation of continuous motor speed controller. (a) Motor speed $\omega(t)$. (b) Motor armature voltage $u(t)$.

6.3 Overflow and Scaling

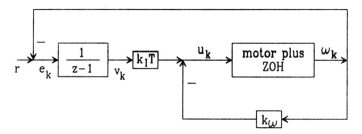

Figure 6.3-5 Digital motor speed controller

Let us select $T = 0.01$ sec. Then

$$u_k = -24.78\omega_k + 0.322 \frac{1}{z-1} e_k \qquad (8)$$

which may be implemented in the form

$$v_k = v_{k-1} + e_{k-1}$$
$$u_k = -24.78\omega_k + 0.322 v_k. \qquad (9)$$

b. Digital Controller Simulation

This digital controller may easily be simulated using program TRESP in Appendix A (see section 5.1). The FORTRAN V subroutine that simulates the controller is given in Fig. 6.3-6. It is worth closely examining this subroutine, for it is written in a modular fashion that clearly reveals the scaling needed for the DSP inputs, the simulated internal functions of the DSP, and the scaling of the DSP outputs. It also includes a function that simulates the effects of 2's-complement overflow in the DSP, as well as tachometer measurement noise injection, which we shall discuss later.

The scale factor "sca" may be determined by examining the control input in the continuous controller, which is shown in Fig. 6.3-4b. The maximum value of $u(t)$ there is 12.42 v, so that the scale factor should be larger than this value to keep the internal DSP signals less than 1. Therefore, let us be safe and select a scaling factor of 20.

To simulate the digital controller, we commented out the noise injection portion of the code to obtain a noise-free simulation. The scaling is appropriate, so that the scaled version of u_k appearing in the DSP is never greater than 1, and the sampling period T is small. Thus, the results of the simulation are indistinguishable from the results shown in Fig. 6.3-4 that use the continuous controller.

c. Effects of Improper Signal Scaling

To see the effects of improper scaling, suppose we select a scale factor of sca = 10. Then, since the scaled version of u_k in the DSP may become greater than one, overflow in the DSP will be a problem. Indeed, the results, still for the noise-free case, are shown in Fig. 6.3-7. The motor speed now responds sluggishly, and the effects of overflow are clearly seen in the control voltage.

DIGITAL MOTOR SPEED CONTROLLER

```
    SUBROUTINE DIG(IK,T,X)
    REAL X(8), Kw, Ki
    COMMON/CONTROL/ u
    COMMON/OUTPUT/ omega, uplot
uplot IS USED TO PLOT u ONLY
omega IS PROVIDED BY DIGCTL FROM MOTOR DYNAMICS
    DATA Kw, Ki, sca/ 24.78, 32.20, 20./
    DATA ref/ 1./

Add tachometer measurement noise
    CALL RANDOM(anoise)
    omegam= omega + 0.1*anoise

Scale DSP inputs by 1/sca
    omegas= omegam/sca
    es= e/sca

Simulate functions of DSP (including overflow)
    v= v + es
    us= OV(-Kw*omegas + Ki*T*v)
    e= ref - omega

Scale DSP output by sca
    u= sca*us
    uplot= u

    RETURN
    END



FUNCTION TO SIMULATE 2'S COMPLEMENT OVERFLOW

    REAL FUNCTION OV(u)
    DATA uH, uL/ 1.,-.1/

    IF (u.GT.0.) OV= MOD(u,uH)
    IF (u.LT.0.) OV= mod(u,uL)

    RETURN
    END
```

Figure 6.3-6 Digital motor speed control subroutine for use with TRESP

d. Measurement Noise Effects

The response in Fig. 6.3-7 may be suitable for some applications. Let us now show, however, that if there is tachometer measurement noise, then improper scaling results in unacceptable behavior.

Thus, let us perform a simulation including the measurement noise injection portion of the code in Fig. 6.3-6. The results for the case when proper scaling of sca = 20 are shown in Fig. 6.3-8. The motor speed response $\omega(t)$ is very suitable and looks quite like that in

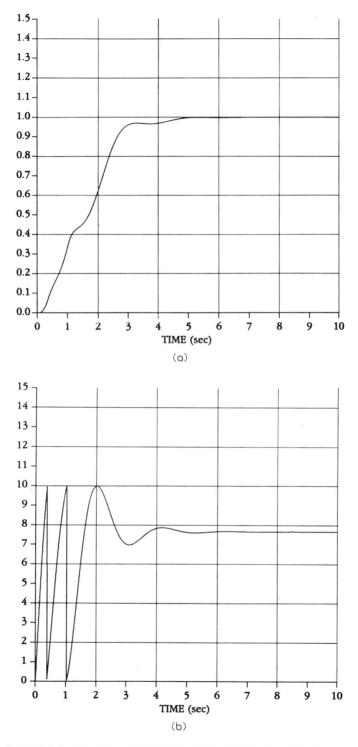

Figure 6.3-7 Simulation of digital motor speed controller with improper scaling. (a) Motor speed $\omega(t)$. (b) Motor armature voltage $u(t)$.

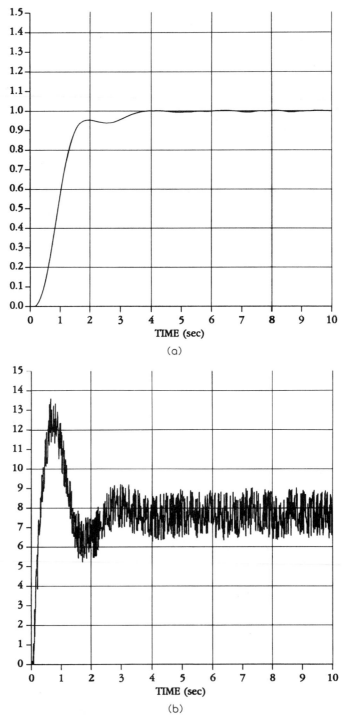

Figure 6.3-8 Simulation of digital motor speed controller with proper scaling and measurement noise. (a) Motor speed $\omega(t)$. (b) Motor armature voltage $u(t)$.

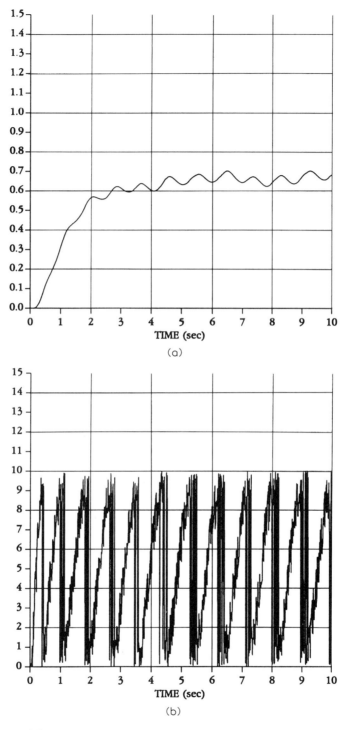

Figure 6.3-9 Simulation of digital motor speed controller with improper scaling and measurement noise. (a) Motor speed $\omega(t)$. (b) Motor armature voltage $u(t)$.

Fig. 6.3-4. This is because, due to the robustness of the closed-loop digital control system, most of the effects of the noise are relegated to the control signal $u(t)$ shown in Fig. 6.3-8b.

Next, we included tachometer measurement noise and also used an *improper scale factor* of sca = 10 which causes overflow in the DSP. The resulting response is shown in Fig. 6.3-9. Note that overflow has caused the robustness property of the feedback system to vanish. The motor speed $\omega(t)$ has ripples and does not even reach the correct steady-state value. The overflow oscillations are clearly visible in the control voltage $u(t)$.

■

6.4 CONTROLLER REALIZATION STRUCTURES

Suppose a digital controller has been designed using a technique like the approximate one in Chapter 5 or the exact one in Chapter 7. It can always be put into the state-space form

$$x_{k+1} = Fx_k + Gw_k$$
$$u_k = Cx_k + Dw_k, \quad (6.4.1)$$

with $x \in \mathbf{R}^n$, which has the transfer function

$$H(z) = C(zI - F)^{-1}G + D \quad (6.4.2)$$

Given the finite wordlength of the DSP using fixed-point arithmetic, it makes a great deal of difference how the digital filter is *implemented*.

Roundoff errors can occur every time an arithmetic operation is performed. Moreover, since all the stable behavior of a discrete system is described by the location of the poles within the unit circle, great accuracy is required in the filter coefficients to obtain desired closed-loop pole locations.

A direct implementation of the digital filter would involve simply writing n difference equations describing (6.4.1), and would be virtually guaranteed to have severe numerical problems if n is large. Specifically, the controllable and observable canonical forms (Chapter 2) are notoriously unstable numerically. That is, their poles are very sensitive to small variations in their coefficients. See Example 6.4-2.

Moreover, it can be shown that the sensitivity to coefficient variations of the impulse response and frequency response is also high in direct implementations [Hanselmann 1987].

Parallel and Cascade Implementations

For good numerical performance with fixed-point arithmetic, digital filters should be implemented as cascade or parallel combinations of first- and second-order filters.

A state-space transformation may be used to place the digital filter into an appropriate form for implementation. To obtain real coefficients, the *real Jordan form* is suitable [Phillips and Nagle 1984, Hanselmann 1987].

The real Jordan form (see Example 2.4-3) is obtained by selecting as a basis for the

6.4 Controller Realization Structures

state space the modal matrix M, which contains the eigenvectors of A. However, if $\lambda = \lambda_R + j\lambda_I$ is a complex eigenvalue with complex eigenvector $v = v_R + jv_I$, we should select as the columns of M not v and its conjugate v^*, but rather the real vectors v_R and v_I. Then, the state-space transformed system

$$x_{k+1} = M^{-1}FMx_k + M^{-1}Gw_k \qquad (6.4.3)$$
$$u_k = CMx_k + Dw_k$$

will have a system matrix in the real Jordan form

$$J = M^{-1}FM = \text{diag}(J_i), \qquad (6.4.4)$$

with Jordan blocks of the form

$$J_i = \begin{bmatrix} \Lambda & I & & & \\ & \Lambda & I & & \\ & & \ddots & \ddots & \\ & & & \Lambda & I \\ & & & & \Lambda \end{bmatrix} \qquad (6.4.5)$$

where, for complex eigenvalues $\lambda_R + j\lambda_I$, I is the 2×2 identity matrix and

$$\Lambda = \begin{bmatrix} \lambda_R & \lambda_I \\ -\lambda_I & \lambda_R \end{bmatrix} \qquad (6.4.6)$$

and for real eigenvalues λ, $I = 1$ and $\Lambda = \lambda$.

Clearly, the transformed system may now be implemented using first- and second-order blocks in cascade and parallel. Corresponding to each real pole, there will be first-order blocks, and corresponding to each complex pole there will be second-order blocks. The parallelism arises due to (6.4.4).

Some cascade structure will be needed to implement Jordan blocks of the form (6.4.5). If F is *simple* so that all the Jordan blocks have length one, then a pure parallel implementation is possible.

The form suitable for implementation may also be found by performing a partial fraction expansion (PFE) on the transfer function. A technique for doing this in terms of the eigenstructure is given in Chapter 9. However, a *real PFE* should be found that will have the form (in the simple case)

$$H(z) = D + \sum_{i=1}^{r} H_i(z), \qquad (6.4.7)$$

where $H_i(z)$ is first-order for real poles and second-order for complex poles. The direct-feed terms are all in D.

We are therefore concerned at this point with implementing first-order filters of the form

$$H_1(z^{-1}) = \frac{b_1 z^{-1}}{1 + a_1 z^{-1}} \qquad (6.4.8)$$

and second-order filters of the form

$$H_2(z^{-1}) = \frac{b_1 z^{-1} + b_2 z^{-2}}{1 + a_1 z^{-1} + a_2 z^{-2}}. \tag{6.4.9}$$

To implement $H_1(z^{-1})$, we may write

$$y_k = H_1(z^{-1})u_k$$

$$(1 + a_1 z^{-1})y_k = b_1 z^{-1} u_k,$$

or

$$y_k = -a_1 y_{k-1} + b_1 u_{k-1}, \tag{6.4.10}$$

which is a difference equation that may be programmed on the DSP.

Second-Order Modules

There are many ways to implement the second-order transfer function [Phillips and Nagle 1984]. Among these are *direct forms 1 through 4* (denoted D1, D2, D3, D4), and *cross-coupled forms 1 and 2* (denoted X1, X2).

In Fig. 6.4-1 we show D1, D3, and X1. Modules D2, D4, and X2 respectively are their duals (that is, all arrows are reversed and the roles of the input and the output are interchanged). In Table 6.4-1 we give a comparison of the number of time-delay elements, multipliers, and summing junctions for each form. Note that D1 and X1 conserve time-delay elements, while D3 conserves summing junctions.

The difference equation implementations of these second-order modules are given in Table 6.4-2, with $y_k = H_2(z^{-1})u_k$. It is interesting that the difference equations for the X1 module may be written from the complex PFE. (And hence, with a little manipulation, from the usual complex Jordan form.)

When implementing these modules, it is important to incorporate overflow protection [Slivinsky and Borninski 1987]. When interconnecting them to produce $H(z)$, scaling may be introduced between the modules [Phillips and Nagle 1984].

The next example shows some considerations that are relevant in selecting between the structures of the second-order modules.

Example 6.4-1: Pole Sensitivity to Coefficient Quantization of Second-Order Modules

This example is from Phillips and Nagle [1984].

A second-order transfer function is given by

$$H_2(z^{-1}) = \frac{b_1 z^{-1} + b_2 z^{-2}}{1 + a_1 z^{-1} + a_2 z^{-2}} \tag{1}$$

$$= \frac{N z^{-1}}{1 + p z^{-1}} + \frac{N^* z^{-1}}{1 + p^* z^{-1}}. \tag{2}$$

Since stability is not dependent on the zeros, let us examine the effect of quantization on the poles.

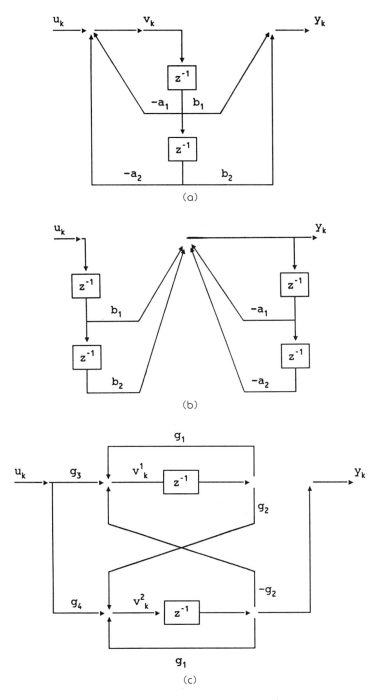

Figure 6.4-1 Implementations of second-order digital filters. (a) Direct form 1, D1. (b) Direct form 3, D3. (c) Cross-coupled form 1, X1.

TABLE 6.4-1 ELEMENTS OF SECOND-ORDER MODULES

	Structure		
	D1	D3	X1
Time-delay elements	2	4	2
Multipliers	4	4	6
Summing junctions	2	1	2

TABLE 6.4-2 DIFFERENCE EQUATION IMPLEMENTATION OF SECOND-ORDER MODULES

D1:
$$v_k = -a_1 v_{k-1} - a_2 v_{k-2} + u_k$$
$$y_k = b_1 v_{k-1} + b_2 v_{k-2}$$

D3:
$$y_k = -a_1 y_{k-1} - a_2 y_{k-2} + b_1 u_{k-1} + b_2 u_{k-2}$$

X1:
$$v_k^1 = g_1 v_{k-1}^1 - g_2 v_{k-1}^2 + g_3 u_k$$
$$v_k^2 = g_1 v_{k-1}^2 + g_2 v_{k-1}^1 + g_4 u_k$$
$$y_k = v_{k-1}^2$$

where g_i are defined by:

$$H_2(z^{-1}) = \frac{Nz^{-1}}{1+pz^{-1}} + \frac{N^* z^{-1}}{1+p^* z^{-1}}$$

$$g_1 = -\text{Re}(p)$$
$$g_2 = -\text{Im}(p)$$
$$g_3 = 2\,\text{Im}(N)$$
$$g_4 = 2\,\text{Re}(N)$$

If the poles are at $z = re^{\pm j\theta}$, then the characteristic equation is

$$\Delta(z) = (z - re^{j\theta})(z - re^{-j\theta}) = z^2 - 2r\cos\theta z + r^2. \qquad (3)$$

a. Coefficient Quantization in the D1 Module

Suppose we decide to implement $H(z^{-1})$ using the D1 module. Then, from (1) and (3)

$$a_1 = -2r\cos\theta, \qquad a_2 = r^2. \qquad (4)$$

6.4 Controller Realization Structures

If the filter coefficients a_1 and a_2 are quantized, then

$$r \cos \theta = -\frac{1}{2} Q[a_1] \tag{5}$$

$$r = (Q[a_2])^{1/2}. \tag{6}$$

According to Fig. 6.4-2a, (5) restricts the poles to a finite number of vertical lines. Requirement (6) further restricts the poles to lie on a finite number of circles. Thus, coeffi-

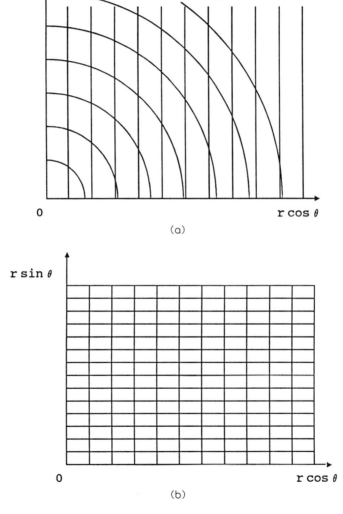

Figure 6.4-2 Effect of coefficient quantization of pole location. (a) Using the D1 module. (b) Using the X1 module.

cient quantization in the D1 implementation restricts the poles to a finite number of locations in the z-plane in the pattern shown in Fig. 6.4-2a.

b. Coefficient Quantization in the X1 Module

If we decide to implement $H(z^{-1})$ using the X1 module, then

$$-p = re^{j\theta} = r\cos\theta + jr\sin\theta \qquad (7)$$

so that from Table 6.4-2

$$g_1 = r\cos\theta \qquad (8)$$

$$g_2 = r\sin\theta. \qquad (9)$$

If the filter coefficients g_1 and g_2 are quantized, then

$$r\cos\theta = Q[g_1] \qquad (10)$$

$$r\sin\theta = Q[g_2]. \qquad (11)$$

Examining Fig. 6.4-2b, (10) restricts the poles to a finite number of vertical lines, while (11) restricts them to a finite number of horizontal lines.

Therefore, coefficient quantization in the X1 implementation restricts the poles to a finite number of z-plane locations in the pattern shown in Fig. 6.4-2b.

Consequently, the D1 structure is more suited for filters with poles near the unit circle, while the X1 structure gives a more uniform pattern of realizable pole locations throughout the unit circle.

■

Example 6.4-2: Direct Versus Parallel Implementation

This example is meant to show that higher-order digital filters should not be implemented directly, but should be implemented as combinations of first- and second-order filters in cascade or parallel.

Consider the digital filter

$$H(z) = \frac{0.05342551z^3 - 0.1386638z^2 + 0.1193337z - 0.03400936}{z^4 - 3.700715z^3 + 5.144919z^2 - 3.184936z + 0.7408184} \qquad (1)$$

a. Parallel Implementation

A PFE was performed to obtain

$$H(z) = H^1(z) + H^2(z) \qquad (2)$$

with

$$H^1(z) = \frac{0.04875051z - 0.04637224}{z^2 - 1.900081z + 0.9048375} \qquad (3)$$

$$H^2(z) = \frac{0.004674996z + 0.004373379}{z^2 - 1.800634z + 0.8187308} \qquad (4)$$

Each second-order filter was implemented using the D3 form and the responses were added to obtain the output. The step response with a 15-bit word size is shown in Fig. 6.4-3. It is virtually indistinguishable from the response assuming an infinite wordlength.

6.4 Controller Realization Structures

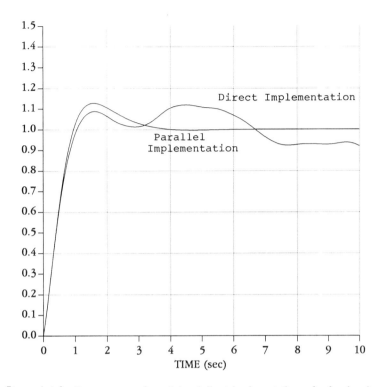

Figure 6.4-3 Step response of parallel and direct implementations of a fourth-order digital filter

b. Direct Implementation

The fourth-order filter (1) was now implemented directly using a fourth-order version of the D3 form. The step response with a 15-bit word size is shown in Fig. 6.4-3. It is unacceptable.

Therefore, even with a comparatively long wordlength, direct implementations of higher-order digital filters are to be avoided.

∎

Delta Form

We have thus far discussed digital implementations in terms of the shift operator z. We could call these "shift form" implementations. A disadvantage of these forms is that stable discrete-time systems have their poles within the unit circle, a small region, so that great accuracy in the coefficients is required to obtain accurate poles. This makes the filter performance very sensitive to coefficient quantization.

The sensitivity to quantization errors can be greatly decreased by using the *delta form* implementation [Middleton and Goodwin 1990]. The *delta transformation* is

$$\delta = \frac{z - 1}{T} \qquad (6.4.11)$$

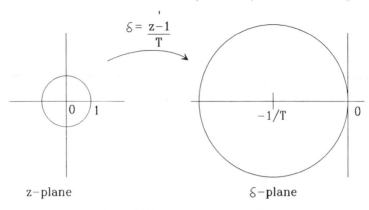

Figure 6.4-4 The Delta transformation

$$z = 1 + \delta T, \quad (6.4.12)$$

with T the sampling period. To understand its usefulness, examine Fig. 6.4-4 to see that it maps the unit circle in the z-plane into a larger circle of radius $1/T$ in the δ-plane. Thus, less coefficient accuracy in the characteristic equation is required to specify pole locations in this larger region. Moreover, as T goes to zero, the large circle in the δ-plane becomes the left-half plane. This means that the properties of delta form systems approach the properties of continuous-time systems as $T \to 0$.

It is interesting to compare the δ transformation to the w-plane transformation in section 7.4.

According to (6.4.11), the delta operator in the time domain corresponds to

$$\delta x_k = \frac{1}{T}(x_{k+1} - x_k), \quad (6.4.13)$$

which is a differencing operation. Thus, δx is quite similar to the continuous-time operation of differentiation.

If the digital controller is given in the state-space form (6.4.1), then, according to (6.4.12), the delta form implementation is

$$\delta x_k = F^\delta x_k + G^\delta w_k$$
$$u_k = C x_k + D w_k \quad (6.4.14)$$

where

$$F^\delta = \frac{F - I}{T}, \quad G^\delta = \frac{G}{T}. \quad (6.4.15)$$

Note that F^δ focuses on the difference between F and the identity matrix, amplifying it by a factor of $1/T$. An important effect of this is that pole locations near $z = 1$ are "spread

6.4 Controller Realization Structures

out" in the δ-plane for greater accuracy. Techniques for accurately computing F^δ and G^δ are given in Middleton and Goodwin [1990].

To implement (6.4.14) recursively we must use (6.4.13) to write

$$x_{k+1} = x_k + T(F^\delta x_k + G^\delta w_k). \tag{6.4.16}$$

If the digital controller is given in transfer function form, we may proceed as follows to obtain the delta form implementation.

Consider first the first-order filter (6.4.8). We may multiply by z and then use (6.4.12) to compute

$$H_1(\delta) = \frac{b_1}{T\delta + (1 + a_1)} \tag{6.4.17}$$

so that $u_k = H_1(\delta) w_k$, or

$$(T\delta + (1 + a_1))u_k = b_1 w_k$$

$$\delta u_k = -\frac{(1 + a_1)}{T} u_k + \frac{b_1}{T} w_k.$$

To implement this recursively, we use (6.4.11) to write

$$u_{k+1} = u_k - T\left[\frac{(1 + a_1)}{T} u_k + \frac{b_1}{T} w_k\right]$$

or

$$u_k = u_{k-1} - T\left[\frac{(1 + a_1)}{T} u_{k-1} + \frac{b_1}{T} w_{k-1}\right]. \tag{6.4.18}$$

Since the second two terms on the right-hand side are added to u_{k-1} to produce u_k, this could be called a *velocity form implementation*. Compare to Example 6.1-1. Although it appears that T simply cancels out, recall that the quantizer due to finite wordlength effectively appears after the quantity in the brackets has been computed. Division by T inside the bracket makes the delta form implementation more robust to quantization.

For the second-order filter (6.4.9), we may multiply by z^2 and then substitute for z to compute

$$H_2(\delta) = \frac{b_1 T\delta + (b_1 + b_2)}{T^2\delta^2 + (2 + a_1)T\delta + (1 + a_1 + a_2)} \tag{6.4.19}$$

so that $u_k = H_2(\delta) w_k$, or

$$T^2\delta^2 u_k = -(2 + a_1)T\delta u_k - (1 + a_1 + a_2)u_k + b_1 T\delta w_k + (b_1 + b_2)w_k.$$

Using (6.4.11) we now obtain the recursion

$$u_k = [2u_{k-1} - u_{k-2}] - \frac{(2 + a_1)}{T}\delta u_{k-2} - \frac{(1 + a_1 + a_2)}{T^2}u_{k-2} \qquad (6.4.20)$$
$$+ \frac{b_1}{T}\delta w_{k-2} + \frac{(b_1 + b_2)}{T^2}w_{k-2},$$

which could be called an "acceleration form implementation."

To implement this filter we may use

$$v^1_{k-1} \equiv \delta u_{k-2} = \frac{1}{T}[u_{k-1} - u_{k-2}]$$

$$v^2_{k-1} \equiv \delta w_{k-2} = \frac{1}{T}[w_{k-1} - w_{k-2}]$$

$$u_k = [2u_{k-1} - u_{k-2}] - \frac{(2 + a_1)}{T}v^1_{k-1} \qquad (6.4.21)$$
$$- \frac{(1 + a_1 + a_2)}{T^2}u_{k-2} + \frac{b_1}{T}v^2_{k-1}$$
$$+ \frac{(b_1 + b_2)}{T^2}w_{k-2}.$$

Alternatively, we could use (6.4.16) with

$$F^\delta = \begin{bmatrix} 0 & 1 \\ -\frac{(1 + a_1 + a_2)}{T^2} & -\frac{(2 + a_1)}{T} \end{bmatrix}, \quad G^\delta = \begin{bmatrix} 0 \\ 1 \end{bmatrix} \qquad (6.4.22)$$

and

$$u_k = \begin{bmatrix} \frac{b_1 + b_2}{T^2} & \frac{b_1}{T} \end{bmatrix} w_k. \qquad (6.4.23)$$

6.5 DIGITAL SIGNAL PROCESSOR SUBROUTINES

In this section we shall give some subroutines that are important for implementing digital controllers on a commercial DSP. To make the discussion concrete, we will illustrate using the Texas Instruments TMS320C25 DSP. A brief discussion of some of the operational characteristics and instructions of the C25 is given in Appendix D. Full details on

6.5 Digital Signal Processor Subroutines

the operation and instruction set of the C25 may be found in the *Second-Generation TMS320 User's Guide*, available from Texas Instruments.

In the next section we give an example of implementing a digital controller on the C25.

We have selected the C25, which is a fixed-point DSP, due to the fact that it is inexpensive and specially tailored to digital control and signal processing applications. As we shall see, it is simple to implement digital controllers on a fixed-point machine, so that the added expense of a floating-point DSP is often not warranted.

FIR Filter

A finite-impulse-response filter (FIR) is one of the form

$$y_k = \sum_{i=0}^{m} b_i u_{k-i} \tag{6.5.1}$$

that does not have any poles, but only zeros.

One useful FIR filter is

$$y_k = \frac{1}{T^2}(u_k - 2u_{k-1} + u_{k-2}), \tag{6.5.2}$$

which computes acceleration y_k from position measurements u_k given sample period T. C25 programs implementing this FIR with $T = 0.1$ sec are shown in Fig. 6.5-1; two implementations are given, one using direct addressing and one using indirect addressing. Note that we are *decrementing* through the coefficient array b_i (see Appendix D), and that a branch command (e.g., the B command) is used to repeat the filtering for incrementing values of k.

With $T = 0.1$ sec, (6.5.2) becomes

$$y_k = 100u_k - 200u_{k-1} + 100u_{k-2}.$$

Scaling may be accomplished by noting that, since the impulse response of a FIR filter is b_k, the system gain defined in section 6.3 is $|100| + |-200| + |100| = 400$. Dividing all coefficients by 400 yields

$$y_k = 0.25u_k - 0.5u_{k-1} + 0.25u_{k-2}.$$

This is the recursion implemented in Fig. 6.5-1. It has a gain of 1 and so is guaranteed to have a fractional output if the input is fractional. Thus, it may be implemented using fixed-point arithmetic with a fractional representation for numbers. The sequence y_k should be multiplied by 400 on output, or alternatively by hardware after output by the DSP.

Also shown in the figure is the technique for data input from port 1 and output to port 0. Presumably port 1 has data from a sampler (A/D converter) and port 0 returns data to a D/A hold device. The details of the hardware interfacing are not discussed here.

```
               .title  "Second-Order Direct-Form FIR Filter Using Direct Addressing"
*
*              Implements equation y(k)=b0*u(k)+b1*u(k-1)+b2*u(k-2)
*
*              Define filter coefficients (Q15 numbers)
coef0          .set    2000h   ;b0 =  0.25 (Q15)
coef1          .set    0c000h  ;b1 = -0.5  (Q15)
coef2          .set    2000h   ;b2 =  0.25 (Q15)
*
*              Reserve space in data memory
               .bss    b0,1    ;coefficient of u0
               .bss    b1,1    ;coefficient of u1
               .bss    b2,1    ;coefficient of u2
               .bss    u0,1    ;most recent input data
               .bss    u1,1
               .bss    u2,1    ;oldest input data
               .bss    y,1     ;output
*
*              Program code
               .text
fir            rovm            ;reset overflow mode
               ssxm            ;set sign extension mode
               spm     1       ;set product mode for 1 left shift between product
                               ; register and accumulator
               ldpk    b0      ;initialize data page register
               lalk    coef0   ;load coefficient in low accumulator
               sacl    b0      ;store in data memory
               lalk    coef1   ;continue bucket brigade
               sacl    b1
               lalk    coef2
               sacl    b2
loop           zac             ;zero accumulator
               lt      u2      ;load T register with oldest input data (Q15)
               mpy     b2      ;P = b2*u2 (Q30)
               ltd     u1      ;load T with u1 (Q15), move u1 to u2, add shifted
                               ; product register to accumulator (Q31)
               mpy     b1      ;P = b1*u1 (Q30)
               ltd     u0      ;load T with u0 (Q15), move u0 to u1, add shifted
                               ; product register to accumulator (Q31)
               mpy     b0      ;P = b0*u0 (Q30)
               apac            ;accumulate final product (Q31)
               sach    y       ;store high word (Q15) of accumulator
               out     y,pa0   ;send output to I/O port 0
               in      u0,pa1  ;get input from I/O port 1
               b       loop    ;continue filtering indefinitely
               .end
```

(a)

Figure 6.5-1 TMS320C25 code to implement a FIR filter. (a) Using direct addressing.

6.5 Digital Signal Processor Subroutines

```
                .title  "Second-Order Direct-Form FIR Filter Using Indirect Addressing"
*
*               Implements equation y(k)=b0*u(k)+b1*u(k-1)+b2*u(k-2)
*
*               Reserve space in data memory
                .bss    u,4         ;input data table, most recent input first
                                    ;must reserve one extra word for data move to spill
                                    ; off end of table
                .bss    y,1         ;output (Q15)
*
*               Put coefficients in program memory
*               Coefficients are in reverse order for use with MACD instruction
b2              .word   2000h       ;b2 =  0.25 (Q15)
b1              .word   0c000h      ;b1 = -0.5  (Q15)
b0              .word   2000h       ;b0 =  0.25 (Q15)
*
*               Program code
                .text
fir             rovm                ;reset overflow mode
                ssxm                ;set sign extension mode
                spm     1           ;set product mode for 1 left shift between product
                                    ; register and accumulator
                ldpk    u           ;initialize data page register
                larp    ar1         ;select auxiliary register 1
loop            zac                 ;zero accumulator
                lrlk    ar1,u+2     ;point to end of input data table (oldest data)
                mpyk    0           ;zero P register
                rptk    2           ;execute the following instruction 3 times
                macd    b2,*-       ;accumulate product, load T register with data pointed
                                    ; to by AR1, multiply by coefficient in program memory
                                    ; (product is a Q30 in P register), move input data in
                                    ; data memory up one word, decrement AR1, and point to
                                    ; next coefficient in program memory for next
                                    ; repetition
                apac                ;accumulate final product (Q31)
                sach    y           ;store high word (Q15) of accumulator
                out     y,pa0       ;send output to I/O port 0
                in      u,pa1       ;get input from I/O port 1
                b       loop        ;continue filtering indefinitely
                .end
```

(b)

Figure 6.5-1   (b) Using indirect addressing and MACD.

First- and Second-Order IIR Filters

We saw in section 6.4 that digital controllers should be implemented as a combination of first- and second-order filters in series and parallel.

Figure 6.5-2 shows the C25 code that implements the first-order filter

$$H_1(z^{-1}) = \frac{b_1 z^{-1}}{1 + a_1 z^{-1}} \tag{6.5.3}$$

```
        .title   "First-Order IIR Filter"
*
*       The filter is implemented using the following equation:
*                y(k) = a1*y(k-1)+b1*u(k-1)
*
*       Define filter coefficients (Q15 numbers)
coefb1  .set     6000h       ;b1 = 0.75  (Q15)
coefa1  .set     4000h       ;a1 = 0.5   (Q15)
*
*       Reserve space in data memory
        .bss     b1,1        ;coefficient of u1
        .bss     a1,1        ;coefficient of y1
        .bss     u0,1        ;most recent input data
        .bss     u1,1        ;oldest input data
        .bss     y0,1        ;most recent output data
        .bss     y1,1        ;oldest output data
*
*       Program code
        .text
iir     rovm                 ;reset overflow mode
        ssxm                 ;set sign extension mode
        spm      1           ;set product mode for 1 left shift between product
                             ; register and accumulator
        ldpk     b1          ;initialize data page register
        lalk     coefb1      ;load coefficient in low accumulator
        sacl     b1          ;store in data memory
        lalk     coefa1      ;continue bucket brigade
        sacl     a1
loop    zac                  ;zero accumulator
        lt       u1          ;load T register with oldest input data (Q15)
        mpy      b1          ;P = b1*u1 (Q30)
        dmov     u0          ;move u0 to u1
        lta      y1          ;load T with y1 (Q15), add shifted product register
                             ; to accumulator (Q31)
        mpy      a1          ;P = a1*y1 (Q30)
        apac                 ;accumulate final product (Q31)
        dmov     y0          ;move y0 to y1
        sach     y0          ;store high word (Q15) of accumulator
        out      y0,pa0      ;send output to I/O port 0
        in       u0,pa1      ;get input from I/O port 1
        b        loop        ;continue filtering indefinitely
        .end
```

Figure 6.5-2 TMS320C25 code to implement a first-order digital filter

using the difference equation

$$y_k = -a_1 y_{k-1} + b_1 u_{k-1}. \quad (6.5.4)$$

(Actually, we have implemented $y_k = a_1 y_{k-1} + b_1 u_{k-1}$, which is more convenient than subtracting. Thus, $-a_1$ should be set as the value of the AR coefficient in the code.)

Figure 6.5-3 shows the C25 code that implements the second-order filter

$$H_2(z^{-1}) = \frac{b_1 z^{-1} + b_2 z^{-2}}{1 + a_1 z^{-1} + a_2 z^{-2}} \quad (6.5.5)$$

6.5 Digital Signal Processor Subroutines

```
        .title  "Second-Order Direct-Form 1 IIR Filter"
*
*       The filter is implemented using the following equations:
*           v(k)=a1*v(k-1)+a2*v(k-2)+u(k)
*           y(k)=b1*v(k-1)+b2*v(k-2)
*
*       Store filter coefficients (Q15 numbers) in program memory
        .data
coefb1  .word   4000h       ;b1 =  0.5  (Q15)
coefb2  .word   2000h       ;b2 =  0.25 (Q15)
coefa1  .word   4000h       ;a1 =  0.5  (Q15)
coefa2  .word   0e000h      ;a2 = -0.25 (Q15)
*
*       Reserve space in data memory for coefficients
        .bss    b1,1
        .bss    b2,1
        .bss    a1,1
        .bss    a2,1
        .bss    u,1         ;input data
        .bss    v0,1        ;v(k)
        .bss    v1,1        ;v(k-1)
        .bss    v2,1        ;v(k-2)
        .bss    y,1         ;output data
*
*       Program code
        .text
iir     rovm                ;reset overflow mode
        ssxm                ;set sign extension mode
        spm     1           ;set product mode for 1 left shift between product
                            ; register and accumulator
        larp    ar0         ;select auxiliary register 0
        lrlk    ar0,b1      ;point to start of coefficient table in data memory
        rptk    3           ;4 iterations
        blkp    coefb1,*+   ;move coefficient from program memory to data
                            ; memory pointed to by AR0, increment AR0, and
                            ; point to next word in program memory for next
                            ; iteration
        ldpk    u           ;initialize data page register
loop    in      u,pa1       ;get input data from port 1
        zalh    u           ;zero accumulator and load high accumulator with input
                            ; data as a Q31
        lt      v1          ;load T register with v(k-1) (Q15)
        mpy     a1          ;P = a1*v1 (Q30)
        lta     v2          ;load T with v(k-2) (Q15), add shifted product register
                            ; to accumulator (Q31)
        mpy     a2          ;P = a2*v2 (Q30)
        apac                ;accumulator=u+a1*v1+a2*v2 (Q31)
        sach    v0          ;store v(0) (Q15)
        zac                 ;zero accumulator
        mpy     b2          ;P = b2*v2 (Q30)
        ltd     v1          ;load T with v1 (Q15), move v1 to v2, add shifted
                            ; product register to accumulator (Q31)
        mpy     b1          ;P = b1*v1 (Q30)
```

Figure 6.5-3 TMS320C25 code to implement a second-order digital filter in D1 form

```
apac              ;accumulator=b1*v1+b2*v2 (Q31)
dmov    v0        ;move v0 to v1
sach    y         ;store high word (Q15) of accumulator
out     y,pa0     ;send output to I/O port 0
b       loop      ;continue filtering indefinitely
.end
```

Figure 6.5-3 (*continued*)

using the D1 form in Table 6.4-2. Recall that the difference equations for the D1 form are

$$v_k = -a_1 v_{k-1} - a_2 v_{k-2} + u_k \qquad (6.5.6)$$
$$y_k = b_1 v_{k-1} + b_2 v_{k-2}.$$

The code segments given in this section may be implemented as subroutines in a larger program. An alternative to subroutines in the C25 DSP is to use macros. The advantage of this is that the six instructions required to invoke and return from a subroutine are saved. This is because a macro assembles in line in the program.

PID Controller

Figure 6.5-4 illustrates a subroutine that implements the digital PID controller

$$u_k = k\left[1 + \frac{T}{T_{Id}}\frac{1}{z-1} + \frac{T_{Dd}}{T}\frac{z-1}{z-\nu}\right] e_k \qquad (6.5.7)$$

As we mentioned in section 5.2, this controller may be implemented using the difference equations

$$v_k^I = v_{k-1}^I + (T/T_{Id})e_{k-1} \qquad (6.5.8)$$

$$v_k^D = \nu v_{k-1}^D + (T_{Dd}/T)(e_k - e_{k-1}) \qquad (6.5.9)$$

$$u_k = k(e_k + v_k^I + v_k^D). \qquad (6.5.10)$$

In the figure, it is assumed that all required scaling of coefficients and signals has been accomplished.

The integrated error signal is stored internally as a 32-bit number to provide sufficient accuracy when small errors are integrated with small integrator gains.

6.6 DIGITAL SIGNAL PROCESSOR CONTROL IMPLEMENTATION EXAMPLE

In this section we present an example showing how to implement digital controllers on a DSP. We shall illustrate using the Texas Instruments TMS320C25.

6.5 Digital Signal Processor Subroutines

```
                .title  "PID Controller"
*
*       The PID controller is implemented using the following equations:
*               vi(k)=vi(k-1)+Ki*e(k-1)
*               vd(k)=a*vd(k-1)+Kd*[e(k)-e(k-1)]
*               u(k) =Kp*[e(k)+vi(k)+vd(k)]
*       The resulting control law is:
*               u=Kp*[1+Ki/(z-1)+Kd*(z-1)/(z-a)]*e
*       It is assumed that scaling has been performed such that all signals
*       and constants can be represented by Q15 numbers.  The integrated error
*       signal vi(k) is stored internally as a Q31 so that small error signals
*       can be integrated.
*
*       Store constants in program memory
                .data
const           .word   0052h   ;Ki = 0.0025 (Q15)
                .word   4000h   ;Kd = 0.5    (Q15)
                .word   6666h   ;Kp = 0.8    (Q15)
                .word   00a4h   ;a  = 0.005  (Q15)
*
*       Reserve space in data memory
                .bss    Ki,1
                .bss    Kd,1
                .bss    Kp,1
                .bss    a,1
                .bss    e0,1    ;e(k):   most recent error signal
                .bss    e1,1    ;e(k-1): previous error signal
                .bss    vil,1   ;low word of integrated error signal
                .bss    vih,1   ;high word of integrated error signal
                .bss    vd,1    ;differentiated error signal
                .bss    u,1     ;u(k):   controller output
*
*       Program code
                .text
pid             rovm            ;reset overflow mode
                ssxm            ;set sign extension mode
                spm     1       ;set product mode for 1 left shift between product
                                ; register and accumulator
                larp    ar0     ;select auxiliary register 0
                lrlk    ar0,Ki  ;point to start of coefficient table in data memory
                rptk    3       ;4 iterations
                blkp    const,*+        ;move coefficient from program memory to data
                                ; memory pointed to by AR0, increment AR0, and
                                ; point to next word in program memory for next
                                ; iteration
                ldpk    Ki      ;initialize data page register
loop            in      e0,pa1  ;get current error signal from port 1
                zals    vil     ;zero accumulator and load low accumulator with
                                ; low word of vi(k-1)
                addh    vih     ;add high word of vi(k-1) to high accumulator so that
                                ; accumulator now contains vi(k-1) as a Q31
                lt      Ki      ;load T register with Ki (Q15)
```

Figure 6.5-4 TMS320C25 code to implement a digital PID controller

```
mpy     e1          ;P = Ki*e(k-1) (Q30)
apac                ;accumulator = vi(k-1)+Ki*e(k-1) = vi(k) (Q31)
sacl    vil         ;store low word of vi(k)
sach    vih         ;store high word of vi(k)
zac                 ;zero accumulator
lt      Kd          ;load T register with Kd (Q15)
mpy     e0          ;P = Kd*e(k) (Q30)
apac                ;add shifted product register to accumulator (Q31)
mpy     e1          ;P = Kd*e(k-1) (Q30)
lts     a           ;load T with a (Q15) and subtract shifted product
                    ; register from accumulator (Q31)
mpy     vd          ;P = a*vd(k-1) (Q30)
apac                ;accumulator = a*vd(k-1)+Kd*[e(k)-e(k-1)] = vd(k) (Q31)
sach    vd          ;store vd(k) (Q15)
dmov    e0          ;move e(k) to e(k-1)
zac                 ;zero accumulator
lt      Kp          ;load T register with Kp (Q15)
mpy     e0          ;P = Kp*e(k) (Q30)
apac                ;add shifted product register to accumulator (Q31)
mpy     vih         ;multiply the high word (Q15) of vi(k) by Kp
                    ; P = Kp*vi(k) (Q30)
apac                ;add shifted product register to accumulator (Q31)
mpy     vd          ;P =Kp*vd(k) (Q30)
apac                ;accumulator = Kp*[e(k)+vi(k)+vd(k)] = u(k) (Q31)
sach    u           ;store high word (Q15) of accumulator
out     u,pa0       ;send output to I/O port 0
b       loop        ;continue filtering indefinitely
.end
```

Figure 6.5-4 (*continued*)

Example 6.6-1: DSP Implementation of Digital Controller for an Inverted Pendulum

In Example 4.2-4 we designed a continuous-time controller for an inverted pendulum. In Example 5.2-2 we used the BLT to discretize the controller, and in Example 5.3-2 we used modified continuous-time design to obtain an improved digital controller. In this example we show how to implement the digital controller of Example 5.2-2 on an actual inverted pendulum using the Texas Instruments TMS320 C25 DSP chip. The DSP hardware used was the Chimera board of Atlanta Signal Processors, Inc. (see Appendix D).

a. Digital Control Structure

A block diagram of the digital controller appears in Fig. 6.6-1, where

$$K_p(z) = g_1 \frac{z - z_1}{z - \pi_1}$$ (1)

$$z_1 = \frac{1 - t_1 T/2}{1 + t_1 T/2}, \qquad \pi_1 = \frac{1 - 5T}{1 + 5T}, \qquad g_1 = k_4 \frac{1 + t_1 T/2}{1 + 5T}$$

6.5 Digital Signal Processor Subroutines

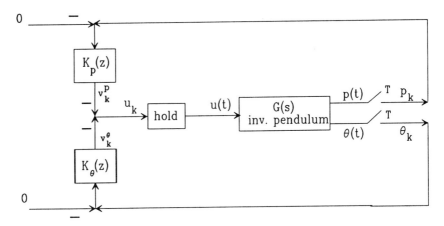

Figure 6.6-1 Digital controller for inverted pendulum

and

$$K_\theta(z) = g_2 \frac{z - z_2}{z - \pi_2} \tag{2}$$

$$z_2 = \frac{1 - t_2 T/2}{1 + t_2 T/2}, \qquad \pi_2 = \frac{1 - 5T}{1 + 5T}, \qquad g_2 = k_3 \frac{1 + t_2 T/2}{1 + 5T}.$$

The k_i are the continuous controller gains, and the parameters t_i ($-t_i$ are the continuous-time compensator zeros) are given by

$$t_1 = 10 + k_2/k_4, \qquad t_2 = 10 + k_1/k_3. \tag{3}$$

Suitable continuous-time control gains were found in Example 4.2-4 to be

$$k_1 = 48.51, \qquad k_2 = 817.5, \qquad k_3 = -8.0, \qquad k_4 = -87.59. \tag{4}$$

In Example 5.2-2 the sample period was taken as $T = 0.01$ sec. Using these values, the poles, zeros, and gains of the digital compensators were

$$\begin{aligned}
\pi_1 &= 0.9048, & z_1 &= 0.9934, & g_1 &= -83.7 \\
\pi_2 &= 0.9048, & z_2 &= 0.9614, & g_2 &= -7.8.
\end{aligned} \tag{5}$$

Defining the intermediate signals v_k^p, v_k^θ shown in Fig. 6.6-1 and denoting the unit delay in the time domain by z^{-1}, we may express (1), (2) in terms of difference equations as follows:

$$v_k^p = g_1 \frac{1 - z_1 z^{-1}}{1 - \pi_1 z^{-1}} p_k \tag{6}$$

or

$$v_k^p = \pi_1 v_{k-1}^p + g_1(p_k - z_1 p_{k-1}), \tag{7}$$

and

$$v_k^\theta = g_2 \frac{1 - z_2 z^{-1}}{1 - \pi_2 z^{-1}} \theta_k \tag{8}$$

or

$$v_k^\theta = \pi_2 v_{k-1}^\theta + g_2(\theta_k - z_2 \theta_{k-1}). \tag{9}$$

The control samples u_k are then given by

$$u_k = -v_k^p - v_k^\theta. \tag{10}$$

Therefore, to implement the digital controller requires only equations (7), (9), and (10), with θ_k the measured angle of the rod and p_k the measured position of the cart.

b. The Hardware

The experimental inverted pendulum was not mounted on a cart, but rather on a sliding metal plate mounted on rails 0.4 m long. The plate was attached to a conveyor-type belt that was stretched across two large-toothed wheels and moved back and forth using a permanent magnet DC servomotor, which turned one of the wheels. There was significant friction that was not modeled in the design equations. The dynamics of the motor were also not included in the design equations.

The rod angle was measured using a potentiometer attached to the pivot point of the rod. Thus, pivoting of the rod was the same as turning of the potentiometer. The position was measured using a potentiometer at the center of one of the gears supporting the conveyor belt. As the sliding metal plate moves back and forth the gear, and hence the potentiometer, rotates. Both potentiometers, especially the angle measurement, were noisy. In a more demanding application, it might be necessary to filter the potentiometer measurements (see Chapter 9).

The C25 chip was mounted on the Chimera DSP board from Atlanta Signal Processors, Inc. (Appendix D). The Chimera provides quite a convenient controller for processes with one control input and at most two outputs.

c. Digital Controller Implementation

The motor was driven using a servo amplifier whose input was the voltage output of the DSP board, computed according to (10). The servo amp provided some gain. The relations of the potentiometer outputs to the rod angle and cart position also involved gains.

Therefore, the poles and zeros of the digital controller given in (5) were used in the DSP program, but the gains were manually tuned for good performance using the gain adjustment on the servo amp. This allowed compensation for the many gains throughout the control system.

The difference equations (7), (9), (10) are easily implemented on the C25 DSP. Figure 6.6-2 shows the assembly language source code for implementation of the digital controller. Note that the t_1, t_2, m_1, m_2 defined there are the negatives of the definitions in the text. The control shell which interfaces with the A/D and D/A converters and determines the sample period is not shown, since it is device specific.

The source code has two primary subroutines. Subroutine CON_INIT is called only once, during initialization; it stores the controller coefficients in memory. Subroutine CNTRL

```
;       ****************************************************************
;                           PENDULUM.ASM
;       Controller for inverted pendulum.
;       This module contains only the controller initialization and
;       control algorithm subroutines.  The control shell which
;       actually sends and receives samples from the DAC/ADC and
;       determines the sample period is omitted.
;       ****************************************************************

;       define labels used to link with the control shell
dp          .set    300h>>7         ;control shell data page

            .def    con_init        ;make controller initialization
                                    ; subroutine global
            .def    cntrl           ;make control algorithm subroutine
                                    ; global

            .ref    in              ;the control shell stores the angle
                                    ; sample at address 'in' on page 'dp'
            .ref    auxin           ;the control shell stores the position
                                    ; sample at address 'auxin' on page
                                    ; 'dp'
            .ref    out             ;the output of the controller is
                                    ; returned to the control shell at
                                    ; address 'out' on page 'dp'
            .ref    setpnt          ;the control shell stores the position
                                    ; setpoint at address 'setpnt' on page
                                    ; 'dp'


;       ****************************************************************
;       The controller structure consists of two feedback paths which
;       which are summed to produce the control input.  The compensator
;       transfer function in the position feedback path is
;               Kp(z) = g1*(z-z1)/(z-p1)
;       and in the angle feedback path is
;               Ka(z) = g2*(z-z2)/(z-p2).
;       These transfer functions are obtained by discretizing (BLT) the
;       continuous compensator transfer functions
;               Kpc(s) = k2/(s-m1) + k4 = k4*(s-t1)/(s-m1) and
;               Kac(s) = k1/(s-m2) + k3 = k3*(s-t2)/(s-m2) where
;               t1 := m1 - k2/k4 and t2 := m2 - k1/k3.
;       The parameters in the discrete equivalent transfer functions
;       are given by
;               z1 := (1 + t1*T/2)/(1 - t1*T/2),
;               p1 := (1 + m1*T/2)/(1 - m1*T/2),
;               g1 :=  k4*(1 - t1*T/2)/(1 - m1*T/2),
;               z2 := (1 + t2*T/2)/(1 - t2*T/2),
;               p2 := (1 + m2*T/2)/(1 - m2*T/2), and
;               g2 := k3*(1 - t2*T/2)/(1 - m2*T/2),
;       where T is the sample period.
;       The design parameters are the pole locations m1 and m2 in the
;       continuous compensator, the feedback gains [k1 k2 k3 k4], and
;       the sample period T.
;       The controller is implemented using the following equations:
;               vp[k] = p1*vp[k-1] + g1*(ep[k] - z1*ep[k-1])
;               va[k] = p2*va[k-1] + g2*(ea[k] - z2*ea[k-1])
;               u[k] = vp[k] + va[k]
;       where ep is the position error and ea is the angle error (error
;       is defined as reference minus measurement.)
```

Figure 6.6-2   Implementation of the inverted pendulum digital controller on the TMS320C25 DSP

```
;       ****************************************************************
;       ****************************************************************
;       Controller initialization
;       ****************************************************************

        ;allocate data memory for compensator coefficients
z1              .set    200h            ;zero of discrete position compensator
p1              .set    201h            ;pole of discrete position compensator
g1              .set    202h            ;gain of discrete position compensator
z2              .set    203h            ;zero of discrete angle compensator
p2              .set    204h            ;pole of discrete angle compensator
g2              .set    205h            ;gain of discrete angle compensator
        ;allocate data memory for controller inputs and outputs
pos             .set    206h            ;position measurement k
ep              .set    207h            ;position error k
ep1             .set    208h            ;position error k-1
ang             .set    209h            ;angle measurement k
ea              .set    20ah            ;angle error k
ea1             .set    20bh            ;angle error k-1
u               .set    20ch            ;control output k
        ;allocate data memory for compensator state variables
vp              .set    20dh            ;position compensator state k-1 on
                                        ; entry, state k on exit
va              .set    20eh            ;angle compensator state k-1 on entry,
                                        ; state k on exit
        ;allocate additional data memory needed to implement controller
temp            .set    20fh            ;temporary storage

                .text
con_init
        ;initialize coefficients
                ldpk    200h>>7         ;select controller data page
                lalk    07f28h          ;z1 = 0.9934
                sacl    z1              ;position compensator zero    (Q15)
                lalk    073d0h          ;p1 = 0.9048
                sacl    p1              ;position compensator pole    (Q15)
                lalk    0f700h          ;g1 = -9.00
                sacl    g1              ;position compensator gain
                                        ; (sxxxxxxx.xxxxxxxx)
                lalk    07b0fh          ;z2 = 0.9614
                sacl    z2              ;angle compensator zero    (Q15)
                lalk    073d0h          ;p2 = 0.9048
                sacl    p2              ;angle compensator pole    (Q15)
                lalk    0f500h          ;g2 = -11.00
                sacl    g2              ;angle compensator gain
                                        ; (sxxxxxxx.xxxxxxxx)
        ;zero inputs, outputs, and states
                larp    ar2
                lrlk    ar2,pos
                zac
                rptk    8
                sacl    *+
                ret

;       ****************************************************************
;       Control algorithm
;       ****************************************************************

                .text
```

Figure 6.6-2 (*continued*)

```
cntrl
                ssxm                    ;set sign extension mode
                sovm                    ;enable overflow mode
                spm     1               ;1 left shift between product register
                                        ; and acc
                larp    ar2             ;ar2 is used as a bit shift counter
        ;get measurements
                ldpk    dp              ;select control shell data page
                lac     in              ;get angle measurement k
                ldpk    200h>>7         ;select controller data page
                sacl    ang             ;store angle    (Q15)
                ldpk    dp              ;select control shell data page
                lac     auxin           ;get position measurement k
                ldpk    200h>>7         ;select controller data page
                sacl    pos             ;store position (Q15)
        ;use control shell setpoint as position reference
                ldpk    dp              ;select control shell data page
                lac     setpnt          ;get setpoint   (Q15)
                ldpk    200h>>7         ;select controller data page
                sub     pos             ;calculate position error
                sfr                     ;shift right once to avoid overflow
                sacl    ep              ;store position error
                                        ; (sx.xxxxxxxxxxxxxxx)
        ;use zero as angle reference
                lalk    0               ;reference angle is zero
                sub     ang             ;calculate angle error
                sfr                     ;shift right once for symmetry with
                                        ; position feedback loop
                sacl    ea              ;store angle error
                                        ; (sx.xxxxxxxxxxxxxxx)
        ;calculate vp[k]
                lt      z1              ;get z1 in T register
                mpy     ep1             ;P = z1*ep[k-1]
                pac                     ;acc = P<<1
                sach    temp            ;save product   (sx.xxxxxxxxxxxxxxx)
                lac     ep              ;acc = ep[k]
                sub     temp            ;acc = ep[k] - z1*ep[k-1]
                sfr                     ;shift right once to avoid overflow
                sacl    temp            ;save difference (sxx.xxxxxxxxxxxxxx)
                lt      g1              ;get g1 in T register
                                        ; (sxxxxxxx.xxxxxxxx)
                mpy     temp            ;P = g1*(ep[k] - z1*ep[k-1])
                pac                     ;acc = P<<1
                lark    ar2,9           ;need to shift acc left 9 bits to
                                        ; make it Q31
                call    cntrl100        ;shift acc (with saturation)

                lt      p1              ;get p1 in T register  (Q15)
                mpy     vp              ;P = p1*vp[k-1]   (Q30)
                apac                    ;acc = acc + P<<1 (Q31)
                sach    vp              ;store vp[k]      (Q15)
        ;calculate va[k]
                lt      z2              ;get z2 in T register
                mpy     ea1             ;P = z2*ea[k-1]
                pac                     ;acc = P<<1
                sach    temp            ;save product   (sx.xxxxxxxxxxxxxxx)
                lac     ea              ;acc = ea[k]
                sub     temp            ;acc = ea[k] - z2*ea[k-1]
                sfr                     ;shift right once to avoid overflow
                sacl    temp            ;save difference (sxx.xxxxxxxxxxxxxx)
```

Figure 6.6-2 (*continued*)

```
                lt      g2              ;get g2 in T register
                                        ; (sxxxxxxx.xxxxxxxx)
                mpy     temp            ;P = g2*(ea[k] - z2*ea[k-1])
                pac                     ;acc = P<<1
                lark    ar2,9           ;need to shift acc left 9 bits to
                                        ; make it Q31
                call    cntrl100        ;shift acc (with saturation)

                lt      p2              ;get p2 in T register   (Q15)
                mpy     va              ;P = p2*va[k-1]   (Q30)
                apac                    ;acc = acc + P<<1   (Q31)
                sach    va              ;store va[k]   (Q15)
;calculate u[k]
                addh    vp              ;acc = va[k] + vp[k]   (Q31)
                sach    u               ;store u[k]   (Q15)
                ldpk    dp              ;select control shell data page
                sach    out             ;send u[k] to D/A
;age errors
                ldpk    200h>>7         ;select controller data page
                dmov    ep              ;ep[k]-->ep[k-1]
                dmov    ea              ;ea[k]-->ea[k-1]
                ret

;       ****************************************************************
;       This subroutine shifts the accumulator left the number of times
;       specified by the current auxiliary register and saturates the acc
;       on overflow
;       ****************************************************************

cntrl100        banz    cntrl101,*-     ;branch if shift counter is greater
                                        ; than zero and decrement counter
                b       cntrl200        ;finished

cntrl101        sfl                     ;get sign bit in carry flag
                bc      cntrl110        ;branch if acc negative

cntrl102        banz    cntrl104,*-     ;branch if more bits to shift and
                                        ; decrement counter

                bgez    cntrl200        ;branch if no overflow

cntrl103        rsxm
                lalk    7fffh,15        ;saturate positive
                sfl
                adlk    0ffffh
                ssxm
                b       cntrl200        ;done

cntrl104        sfl
                bc      cntrl103        ;branch if positive overflow

                b       cntrl102        ;loop until overflow or done shifting

cntrl110        banz    cntrl112,*-     ;branch if no more bits to shift and
                                        ; decrement counter

                blz     cntrl200        ;branch if no overflow
```

Figure 6.6-2 (*continued*)

```
cntrl111         rsxm
                 lalk    8000h,15         ;saturate negative
                 sfl
                 ssxm
                 b       cntrl200         ;done

cntrl112         sfl
                 bnc     cntrl111         ;branch if negative overflow

                 b       cntrl110         ;loop until overflow or done shifting

cntrl200         ret

                 .end
```

Figure 6.6-2 (*continued*)

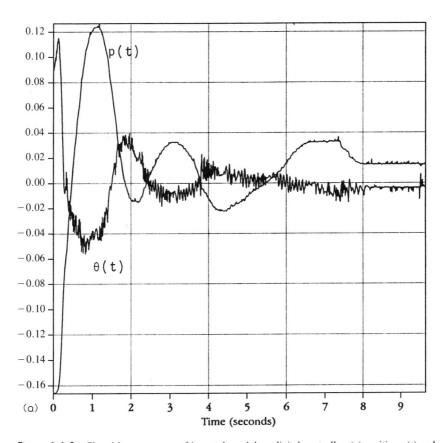

Figure 6.6-3 Closed-loop response of inverted pendulum digital controller. (a) position $p(t)$ and rod angle $\theta(t)$. (b) Motor control input. (*continues*)

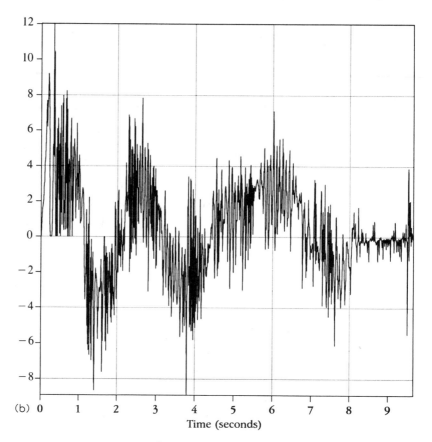

Figure 6.6-3 (*continued*)

implements the controller using the same steps that appear in Fig. 5.2-4. Note also that CNTRL calls a subroutine to shift the accumulator with saturation. It is necessary to perform this function since the C25 overflow mode, which was set during this implementation, does not affect shift operations.

The resulting closed-loop response for nonzero initial conditions $\theta(0)$ and $p(0)$ is shown in Fig. 6.6-3. It may be compared with the response shown in Example 5.2-2. Although the noisy angle measurements cause a very noisy control input, much of the noise is filtered out by the motor dynamics, and the controller actually worked very well. The overall shapes of the signals look much like the signals in the simulation in Example 5.2-2. ■

PROBLEMS FOR CHAPTER 6

Problems for Section 6.1

6.1-1 Write down the value of the state x_k in the antiwindup controller (6.1.3) for the deadbeat case where F_o has all poles at the origin. Assume that $w_k = e_k$ and u_k are constant and that $k > n$.

6.1-2 Repeat 6.1-1 for the case where the controller is just an integrator so that $x_{k+1} = x_k + (T/T_I)e_k$, $u_k = \text{sat}(x_k)$. Simplify as far as possible.

6.1-3 Redo Example 6.1-1, designing the digital PI controller using the BLT. Design the antiwindup compensator and perform the simulations for the motor speed control system.

6.1-4 In Example 4.2-3 a speed control system was designed for an armature-controlled DC motor. Design a digital control system with antiwindup protection. Perform computer simulations to test its performance.

Problems for Section 6.2

6.2-1 Repeat Example 6.2-1 if $Q[\ ]$ is quantization using roundoff.

6.2-2 In Example 6.2-1, determine conditions for the existence of a limit cycle. Determine also a bound on the magnitude of any limit cycle in terms of the quantization step size q and the time constant α.

6.2-3 In Example 4.2-3 a speed control system was designed for an armature-controlled DC motor. Find a digital controller using the BLT. Simulate the closed-loop system for several sample periods. Now, using the subroutine in Example 6.2-2, simulate the digital controller with quantization error. Try several wordlengths.

6.2-4 In Example 5.2-3 a digital control system was designed for a system with time delay. Using the subroutine in Example 6.2-2, simulate the digital controller with quantization error. Try several wordlengths.

Problems for Section 6.3

6.3-1 Find y_k in Example 6.3-1b to verify that, with saturation, the system is stable.

6.3-2 Repeat Example 6.3-1 using:
  a. the observable canonical form implementation for $H(z)$
  b. a diagonal (Jordan form) implementation
   (The nonlinear function $N[\ ]$ should be placed after each adder in the circuit.)

Problems for Section 6.4

6.4-1 Show how to determine the X1 difference equations in Table 6.4-2 directly from the complex Jordan form blocks corresponding to a complex pair of poles.

6.4-2 Fill in the details in the derivations of the delta-form implementations (6.4.18) and (6.4.20).

6.4-3 In Example 6.4-2, select a small wordlength so that even the parallel implementation of the filter is unsuitable. Correct the problem using delta-form realizations for each of the two parallel second-order filters.

6.4-4 In Example 5.2-2, a digital controller was designed for the inverted pendulum. In Example 6.2-2 it was demonstrated that the controller did not work well with a wordlength of 7 bits. Implement the digital controller using delta form. Compare the performance of this implementation to the shift operator implementation.

7
Digital Control by Direct Discrete-Time Design

SUMMARY

In Chapter 5 we showed how to design a digital controller by redesign of an existing continuous-time controller using approximation methods like the bilinear transformation. Using that approach, the sample period must be small for good performance. In this chapter we shall show how to design digital controllers directly in the discrete-time domain. This allows more exact control using larger sample periods. It also allows us to investigate some properties of the sampling process such as the appearance of nonminimum-phase zeros and the loss of observability.

INTRODUCTION

In Chapter 5 we showed how to convert continuous-time controllers into digital controllers by discretization using the BLT or the MPZ. We discovered that the sample period T should be small for good performance. However, as we shall see in section 7.3, there are problems with using sample periods that are too small. Thus, the old adage "if it doesn't work, decrease the sample period," should certainly not be used in any conscientious design.

In this chapter we shall show how to design digital controllers directly in the discrete-time domain that give more accurate performance than those found by continuous controller redesign, while also allowing larger sample periods T. The important point is that direct discrete design guarantees exact performance at the sample points, regardless of the size of T.

In direct discrete design of digital controllers, the continuous-time plant is first discretized, or converted into a discrete-time system, using a simple computer program (DISC, given in Appendix A). Then, a controller is designed for the discrete-time system.

Direct discrete design of digital controllers exactly includes the effects of the ZOH, in contrast to the continuous redesign approach using the BLT or MPZ, where its effects are usually ignored (although they may be *approximately* taken into account). This means that with discrete design the closed-loop performance is guaranteed at the sample points.

Another advantage of discrete design is that greater insight is obtained into digital design and the properties of discrete systems. Specifically, in section 7.3 we shall mention the introduction of *nonminimum-phase zeros* and the *loss of observability and controllability* through sampling. These notions give us a *lower bound* on the sample period. That is, it will become clear that *T should not be selected too small*.

The design technique in Chapter 5 required application of the BLT or MPZ to a continuous controller. This is usually done by hand for the various components of the controller. On the other hand, to perform digital design using the direct discrete approach, no hand calculations are necessary. Instead, a few computer programs described in Appendix A are used.

In section 7.1 we discuss discretization of continuous-time systems, and in section 7.2 discretization of the performance index (PI). These techniques allow us to convert a continuous-time controls design problem to a *discrete-time controls design problem*, whose solution directly yields the digital control input u_k. In section 7.4 we show how to perform discrete controls design using techniques like those in Chapter 4. Some design examples are given.

7.1 DISCRETIZATION OF CONTINUOUS SYSTEMS

In this section we shall discuss converting a continuous-time system into a discrete-time system. We shall explicitly include the ZOH required to convert the discrete control samples u_k into the plant control input $u(t)$. In section 7.2 we show how to discretize a continuous PI. Using the techniques of these two sections, a continuous controls design problem may be converted into a discrete-time design problem, whose solution directly yields the digital control input u_k. In section 7.4 we show how linear quadratic (LQ) techniques like those in Chapter 4 may be used to solve such discrete-time design problems, yielding a digital controller for the continuous-time plant.

Sampling the Plant

In this subsection we shall show how to convert a continuous-time state-variable description of a plant into a discrete-time state-variable description.

Suppose a continuous time-invariant plant is given by

$$\dot{x} = Ax + Bu \tag{7.1.1}$$

$$y = Cx + Du \tag{7.1.2}$$

with $x(t) \in \mathbf{R}^n$, measured output $y(t) \in \mathbf{R}^p$, and control input $u(t) \in \mathbf{R}^m$. For digital control purposes it is desired to define a discrete time index k such that

$$t = kT \tag{7.1.3}$$

7.1 Discretization of Continuous Systems

with T the sampling period. Then, the discrete control input u_k is to be switched at times kT, $k = 0, 1, \ldots, N-1$ by the microprocessor.

The usual procedure for controlling the plant is to hold the control input $u(t)$ constant between control switchings. This may be achieved by adding a ZOH before the plant as shown in Fig. 7.1-1. Then, the continuous plant input $u(t)$ is given in terms of the discrete control u_k by

$$u(t) = u_k, \qquad kT \le t < (k+1)T. \tag{7.1.4}$$

Note that $u(t)$ is switched at the times kT so that it is continuous from the right.

Also shown in Fig. 7.1-1 is a sampler with period T added to the output channel of the plant. This A/D device generates the samples

$$y_k = y(kT) \tag{7.1.5}$$

of the output. Let us also define the samples

$$x_k = x(kT) \tag{7.1.6}$$

of the state vector.

It is now required to determine a dynamical relation between u_k and x_k such that

$$x_{k+1} = A^s x_k + B^s u_k. \tag{7.1.7}$$

That is, we need to determine the sampled equivalents A^s, B^s of A and B.

To achieve this, first write the solution of (7.1.1), which is

$$x(t) = e^{A(t-t_0)}x(t_0) + \int_{t_0}^{t} e^{A(t-\tau)}Bu(\tau)\,d\tau. \tag{7.1.8}$$

Setting $t_0 = kT$ and $t = (k+1)T$ yields

$$x((k+1)T) = e^{AT}x(kT) + \int_{kT}^{(k+1)T} e^{A[(k+1)T-\tau]}Bu(\tau)\,d\tau.$$

Since $u(t)$ has the constant value of u_k over the sample period due to the ZOH, we may extract it from the integrand. Then, changing variables to $\lambda = \tau - kT$ yields

$$x_{k+1} = e^{AT}x_k + \int_0^T e^{A[T-\lambda]}B\,d\lambda \cdot u_k.$$

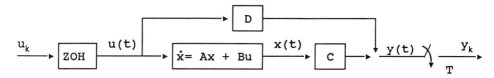

Figure 7.1-1 Plant with ZOH

Changing variables again to $\tau = T - \lambda$ yields finally

$$x_{k+1} = e^{AT}x_k + \int_0^T e^{A\tau}B\, d\tau \cdot u_k.$$

By comparison with (7.1.7) we may now identify the discretized plant matrices as

$$A^s = e^{AT} \tag{7.1.9}$$

$$B^s = \int_0^T e^{A\tau}B\, d\tau. \tag{7.1.10}$$

It is important to notice that the discretized plant matrix A^s is always nonsingular.

Since (7.1.2) is a nondynamical relation, we may simply write

$$y_k = Cx_k + Du_k. \tag{7.1.11}$$

That is, the C and D matrices are unchanged on discretization.

The design of digital controls proceeds as follows. First, a continuous-time state-variable model of the plant is derived. Then, A^s and B^s are determined using (7.1.9) and (7.1.10). Finally, the techniques to be presented in section 7.4 are used to design a discrete control sequence u_k for (7.1.7)/(7.1.11). During the implementation phase when the control is actually applied to the plant, $u(t)$ is manufactured by passing u_k through a ZOH. Commercial digital signal processors (DSPs) will usually have a ZOH built in.

It is important to clearly realize that this approach to digital controls design explicitly includes the effects of the ZOH, since $u(t)$ was assumed constant over the sample period in deriving B^s. Thus, the resulting controller takes into account the properties of the hold and sampling processes, guaranteeing exact behavior at the sample instants. This is in contrast to the situation with the continuous controller redesign approach covered in Chapter 5, where the ZOH is generally ignored, or at best only approximately included.

In order to evaluate (7.1.9) and (7.1.10) we may use

$$A^s = e^{AT} = I + AT + \frac{1}{2!}A^2T^2 + \frac{1}{3!}A^3T^3 + \ldots \tag{7.1.12}$$

and, by integrating this term-by-term and multiplying by B

$$B^s = BT + \frac{1}{2!}ABT^2 + \frac{1}{3!}A^2BT^3 + \ldots. \tag{7.1.13}$$

Based on these equations, it is easy to write a computer program to discretize a continuous plant (A, B) (see program DISC in Appendix A). The series may be terminated when the terms become smaller than some desired threshold. For large n, this technique could suffer numerical problems due to the raising of matrix A to large powers.

The poles of the continuous system are the eigenvalues of A. The poles of the discretized system are the eigenvalues of $A^s = e^{AT}$. However, if s_i are the eigenvalues of A, then for any polynomial function $f(.)$, the eigenvalues of $f(A)$ are given by $f(s_i)$

[Kailath 1980]. Therefore, if s_i are the poles of the continuous system (7.1.1), then the poles z_i of the discretized system (7.1.7) are given by

$$z_i = e^{s_i T}. \tag{7.1.14}$$

However

$$z_i = e^{Re(s_i)T} e^{jIm(s_i)T} = e^{Re(s_i)T} (\cos[Im(s_i)T] + j \sin[Im(s_i)T]),$$

where the second factor has magnitude of one. Therefore, if the continuous system is stable (i.e., has all poles with $Re(s_i) < 0$), then the discretized system has all poles inside the unit circle. That is, *stability is preserved by discretization*. In fact, the $j\omega$ axis in the s-plane is mapped into the unit circle in the z-plane by (7.1.14), with the left-half plane mapping into the interior of the unit circle.

If $u(t)$ is given by (7.1.4), then the samples $x_k = x(kT)$ are given by (7.1.7). To determine the values of $x(t)$ between the sample points, we could use

$$x(t) = e^{A(t-kT)} x_k + \int_{kT}^{t} e^{A(t-\tau)} B \, d\tau \cdot u_k, \qquad kT \leq t < (k+1)T. \tag{7.1.15}$$

However, for our purposes we shall generally design the control sequence u_k and then, to verify its performance, carry out a digital simulation as discussed in section 5.1, where u_k is applied to the *continuous plant* (7.1.1) using a Runge-Kutta integrator. Thus, the intersample values of $x(t)$ will be obtained automatically.

If the continuous system (7.1.1) has time-varying matrices $A(t)$, $B(t)$, then a derivation like the one just presented shows that the discretized system is also time-varying with

$$A_k^s = \phi((k+1)T, kT) \tag{7.1.16}$$

$$B_k^s = \int_{kT}^{(k+1)T} \phi((k+1)T, \tau) B(\tau) \, d\tau, \tag{7.1.17}$$

where $\phi(t, t_0)$ is the state transition matrix [Kailath 1980] of $A(t)$.

For some applications it may not be desirable for the control input $u(t)$ to switch between discrete values. However, if the plant contains any low-pass characteristics, as is the usual situation, then it will smooth out the jumps in the control. If additional smoothing is required, a low-pass filter may be placed in the plant input channel, and the resulting series configuration can then be sampled.

An alternative is to use higher-order hold circuits that convert the discrete control u_k into continuous plant input functions $u(t)$. Exercise 7.1-4 gives the discretized system when a *first-order hold* is used instead of the ZOH.

Although it is possible to use D/A devices other than the ZOH for converting the discrete control sequence u_k into the continuous plant control $u(t)$, it is usually not worth the trouble. The zero-order hold sampling technique just presented provides an *exact discretized equivalent* of the continuous-time plant for which discrete controls can be designed. By using the technique of ZOH sampling, we are at least guaranteed that the *samples* of $x(t)$ behave like the discrete state x_k, which has desirable behavior because of our selection of u_k. The intersample behavior depends on the choice of the sampling period T, which we shall soon discuss.

An additional argument against using higher-order hold devices is that most commercial DSPs contain a ZOH.

Let us examine ZOH discretization in some examples that frequently appear.

Example 7.1-1: Discretization of Scalar System

Consider the scalar continuous-time system

$$\dot{x} = ax + bu. \tag{1}$$

According to (7.1.9) and (7.1.10), the discretized system is given by

$$x_{k+1} = e^{aT}x_k + \int_0^T e^{a\tau}b\, d\tau \cdot u_k$$

or

$$x_{k+1} = e^{aT}x_k + \frac{(e^{aT}-1)}{a}bu_k. \tag{2}$$

The pole is at e^{aT}, which is inside the unit circle if $a < 0$.

∎

Example 7.1-2: Discretization of Newton's System

A particle obeying Newton's law satisfies

$$\dot{x} = \begin{bmatrix} 0 & 1 \\ 0 & 0 \end{bmatrix} x + \begin{bmatrix} 0 \\ 1 \end{bmatrix} u = Ax + Bu, \tag{1}$$

with state $x = [s \;\; v]^T$, where s = position and v = velocity, and $u(t)$ an acceleration input. See Example 2.1-4.

Since A is nilpotent (i.e., $A^j = 0$ for some j), in particular $A^2 = 0$, it is extremely convenient to use (7.1.12) and (7.1.13) to discretize the system. Indeed

$$A^s = I + AT = \begin{bmatrix} 1 & T \\ 0 & 1 \end{bmatrix} \tag{2}$$

$$B^s = BT + \frac{1}{2!}ABT^2 = \begin{bmatrix} T^2/2 \\ T \end{bmatrix}. \tag{3}$$

The poles of the discretized system are at the roots of

$$0 = \Delta^s(z) \equiv |zI - A^s| = \begin{vmatrix} z-1 & -T \\ 0 & z-1 \end{vmatrix} = (z-1)^2. \tag{4}$$

That is, they are both at $z = 1$. This agrees with (7.1.14).

∎

7.1 Discretization of Continuous Systems

Example 7.1-3: Discretization of Damped Harmonic Oscillator

A damped harmonic oscillator is described by

$$\dot{x} = \begin{bmatrix} 0 & 1 \\ -\omega^2 & -2\alpha \end{bmatrix} x + \begin{bmatrix} 0 \\ 1 \end{bmatrix} u, \tag{1}$$

with ω the natural frequency and α the real part of the poles. See Example 2.1-5. Defining the imaginary part of the poles as β, we have $\omega^2 = \alpha^2 + \beta^2$, and the characteristic polynomial is

$$\Delta(s) = s^2 + 2\alpha s + \omega^2 = (s + \alpha)^2 + \beta^2. \tag{2}$$

The poles are at $s = -\alpha \pm j\beta$.

The resolvent matrix is

$$(sI - A)^{-1} = \frac{1}{(s + \alpha)^2 + \beta^2} \begin{bmatrix} s + 2\alpha & 1 \\ -\omega^2 & s \end{bmatrix}, \tag{3}$$

whence the inverse Laplace transform delivers

$$A^s = e^{AT} = e^{-\alpha T} \begin{bmatrix} \cos \beta T + \frac{\alpha}{\beta} \sin \beta T & \frac{1}{\beta} \sin \beta T \\ \frac{-\omega^2}{\beta} \sin \beta T & \cos \beta T - \frac{\alpha}{\beta} \sin \beta T \end{bmatrix}. \tag{4}$$

Evaluating (7.1.10) yields

$$B^s = \begin{bmatrix} \frac{1}{\omega^2} \left[1 - e^{-\alpha T} \left(\cos \beta T + \frac{\alpha}{\beta} \sin \beta T \right) \right] \\ \frac{1}{\beta} e^{-\alpha T} \sin \beta T \end{bmatrix}. \tag{5}$$

The poles of the discretized system are located at the roots of the discrete characteristic polynomial $\Delta^s(z) = |zI - A^s|$, which is

$$\Delta^s(z) = z^2 - 2e^{-\alpha T} \cos \beta T\, z + e^{-2\alpha T} \tag{6}$$

$$= (z - e^{-\alpha T} \cos \beta T)^2 + e^{-2\alpha T} \sin^2 \beta T. \tag{7}$$

Thus, the sampled poles are at

$$z = e^{-\alpha T}(\cos \beta T + j \sin \beta T) = e^{-\alpha T} e^{-j\beta T}, \tag{8}$$

agreeing with (7.1.14). Note that, if the continuous system is stable so that $\alpha > 0$, then the sampled poles have magnitude less than one. ∎

Exercise 7.1-4: Sampling Using a First-Order Hold

As far as the controlled performance of the system goes, we shall see that the ZOH gives excellent results even using sample periods that are not especially small. This is because we are basing our design of digital controls on a sampled system that is an exact discretized equivalent of the continuous-time plant including the ZOH. Thus, the samples of the continuous state vector are exactly equal to the values of the discrete state vector sequence.

This means that it is rarely worth the trouble to use higher-order hold devices for the digital-to-analog conversion in Fig. 7.1-1.

It is interesting, however, to derive the sampled system when a first-order hold is used in Fig. 7.1-1 instead of the ZOH. As we have seen in section 5.3, the first-order hold reconstructs $u(t)$ from u_k by linear extrapolation. Thus

$$u(t) = u_k + (u_k - u_{k-1})\frac{(t - kT)}{T}, \qquad kT \leq t < (k + 1)T \tag{1}$$

or

$$u(t) = \frac{t - (k - 1)T}{T} u_k - \frac{t - kT}{T} u_{k-1}, \qquad kT \leq t < (k + 1)T. \tag{2}$$

Let the continuous-time system be given by

$$\dot{x} = Ax + Bu. \tag{3}$$

a. Show that the discretized equivalent to (3) preceded by a first-order hold is given by

$$x_{k+1} = A^s x_k + B_0 u_k + B_1 u_{k-1}, \tag{4}$$

where

$$A^s = e^{AT}, \tag{5}$$

$$B_0 = \frac{1}{T} \int_0^T e^{A\tau} B(2T - \tau) \, d\tau, \tag{6}$$

$$B_1 = -\frac{1}{T} \int_0^T e^{A\tau} B(T - \tau) \, d\tau, \tag{7}$$

Note that (4) may be brought to state-space form by writing

$$\begin{bmatrix} x_{k+1} \\ u_k \end{bmatrix} = \begin{bmatrix} A^s & B_1 \\ 0 & 0 \end{bmatrix} \begin{bmatrix} x_k \\ u_{k-1} \end{bmatrix} + \begin{bmatrix} B_0 \\ I \end{bmatrix} u_k. \tag{8}$$

b. This may be compared to Hanselmann [1987], where the FOH is implemented using linear *interpolation* so that

$$u(t) = u_k + (u_{k+1} - u_k)\frac{(t - kT)}{T}, \qquad kT \leq t < (k + 1)T. \tag{9}$$

Show that this results in a discretized system of the form

$$x_{k+1} = A^s x_k + B_0 u_k + B_1 u_{k+1}, \tag{10}$$

so that x_{k+1} depends on u_{k+1}. Find A^s, B_0, B_1. Note that (10) is not a state-space system.

Sampling Systems with Time Delays

The control of continuous-time systems with time delays can present some serious problems [Åström and Wittenmark 1984; Franklin, Powell, and Emami-Naeini 1986]. Such systems are also difficult to deal with analytically since they are infinite dimensional. This is unfortunate since delays are common in many practical systems. (One sort of delay is the computation delay of the microprocessor.)

On the other hand, it is easy to design *digital controllers* for systems with delays, since a finite-dimensional state model is easy to find for such systems. In section 5.2 we showed how to design a Smith predictor for controlling these systems. Let us now demonstrate how to discretize systems with delays, so that direct discrete design may be used to find a digital controller.

Thus, let the time-invariant plant be described by

$$\dot{x}(t) = Ax(t) + Bu(t - \tau) \tag{7.1.18}$$

with $\tau > 0$ a control delay. We may write the control delay as

$$\tau = (d - 1)T + \delta, \qquad 0 \leq \delta < T, \tag{7.1.19}$$

where the integer d is the *integral delay* and δ is the *fractional delay*. That is, the delay τ is always equal to an integral number of sample periods plus a portion of the sample period.

First, let us assume $d = 1$ so that $\tau = \delta$ is less than T. Then, according to (7.1.8), we have

$$x((k + 1)T) = e^{AT}x(kT) + \int_{kT}^{(k+1)T} e^{A[(k+1)T - \lambda]} Bu(\lambda - \delta) \, d\lambda.$$

In contrast with the situation in the previous section, the shifted $u(\lambda - \delta)$ is no longer constant over the period of integration since according to (7.1.4) $u(t - \delta)$ changes its value from u_{k-1} to u_k at $t - \delta = kT$. We may deal with this by splitting the integral into two portions. Thus

$$x((k + 1)T) = e^{AT}x(kT) + \int_{kT}^{kT+\delta} e^{A[(k+1)T - \lambda]} B \, d\lambda \cdot u_{k-1}$$

$$+ \int_{kT+\delta}^{(k+1)T} e^{A[(k+1)T - \lambda]} B \, d\lambda \cdot u_k.$$

Now, by changing variables we discover that

$$x_{k+1} = A_0 x_k + B_1 u_{k-1} + B_0 u_k, \tag{7.1.20}$$

with

$$A_0 = e^{AT} \tag{7.1.21}$$

$$B_0 = \int_0^{T-\delta} e^{A\lambda} B \, d\lambda \tag{7.1.22}$$

$$B_1 = e^{A(T-\delta)} \int_0^{\delta} e^{A\lambda} B \, d\lambda. \tag{7.1.23}$$

A state-space model for (7.1.20) is

$$\begin{bmatrix} x_{k+1} \\ u_k \end{bmatrix} = \begin{bmatrix} A_0 & B_1 \\ 0 & 0 \end{bmatrix} \begin{bmatrix} x_k \\ u_{k-1} \end{bmatrix} + \begin{bmatrix} B_0 \\ I \end{bmatrix} u_k. \tag{7.1.24}$$

The delay has added m states to the discretized plant that are required in order to store information about the previous input. Note, however, that (7.1.24) is a perfectly well-behaved discrete-time system, for which controllers may easily be designed using the techniques to be presented in section 7.4.

If the control delay τ is greater than the sampling period T so that the integral delay d is not equal to one, then we may show that (7.1.20) is replaced by

$$x_{k+1} = A_0 x_k + B_1 u_{k-d} + B_0 u_{k-d+1}, \tag{7.1.25}$$

with B_0 and B_1 the same as before. Now, a state-space model is given by

$$\begin{bmatrix} x_{k+1} \\ u_{k-d+1} \\ \vdots \\ u_{k-1} \\ u_k \end{bmatrix} = \begin{bmatrix} A_0 & B_1 & B_0 & 0 & \cdots & 0 \\ 0 & 0 & I & 0 & \cdots & 0 \\ \vdots & & & & & \vdots \\ 0 & \cdots & & & 0 & I \\ 0 & \cdots & & & & 0 \end{bmatrix} \begin{bmatrix} x_k \\ u_{k-d} \\ \vdots \\ u_{k-2} \\ u_{k-1} \end{bmatrix} + \begin{bmatrix} 0 \\ 0 \\ \vdots \\ 0 \\ I \end{bmatrix} u_k. \tag{7.1.26}$$

The characteristic polynomial of this system is

$$\Delta^s(z) = z^{md} |zI - A_0|, \tag{7.1.27}$$

with m the number of inputs. That is, there are n poles given by (7.1.14) along with md extra poles at $z = 0$ due to the pure time delays introduced into the system.

Example 7.1-5: Discretization of Newton's System with Time Delay

This example is from Åström and Wittenmark [1984]. In Example 7.1-2 we considered Newton's system. Suppose that we now introduce a control delay $0 < \tau < T$ so that

$$\dot{x} = \begin{bmatrix} 0 & 1 \\ 0 & 0 \end{bmatrix} x(t) + \begin{bmatrix} 0 \\ 1 \end{bmatrix} u(t - \tau) = Ax(t) + Bu(t - \tau). \tag{1}$$

According to (7.1.21)–(7.1.23) we obtain

$$A_0 = \begin{bmatrix} 1 & T \\ 0 & 1 \end{bmatrix} \tag{2}$$

7.1 Discretization of Continuous Systems

$$B_0 = \begin{bmatrix} (T - \tau)^2/2 \\ T - \tau \end{bmatrix}. \quad (3)$$

$$B_1 = \begin{bmatrix} \tau(T - \tau/2) \\ \tau \end{bmatrix}. \quad (4)$$

∎

Sampling the Transfer Function

It is a straightforward matter to determine the relation between the transfer functions of the plant and its discretized version including the ZOH. If there is an output

$$y = Cx \quad (7.1.28)$$

associated with (7.1.1), then, by sampling both sides we obtain

$$y_k = Cx_k. \quad (7.1.29)$$

The continuous and discretized transfer functions are defined respectively by

$$H(s) = C(sI - A)^{-1}B \quad (7.1.30)$$

$$H^s(z) = C(zI - A^s)^{-1}B^s, \quad (7.1.31)$$

where argument s is the Laplace variable and z the Z-transform variable.

The impulse response of the ZOH is

$$\mathbf{L}(\text{ZOH}) = \frac{(1 - e^{-sT})}{s}, \quad (7.1.32)$$

with $\mathbf{L}(\cdot)$ the Laplace transform. Recall that the delay e^{-sT} is represented in terms of Z-transforms as z^{-1}. Therefore, with some abuse in notation we can say

$$\mathbf{L}(\text{ZOH}) = \frac{(1 - e^{-sT})}{s} = \frac{(1 - z^{-1})}{s}. \quad (7.1.33)$$

Let us consider Fig. 7.1-1. The transfer function from u_k to y_k including the ZOH is given by

$$H^s(z) = (1 - z^{-1})\, \mathbf{Z}\!\left[\frac{H(s)}{s}\right], \quad (7.1.34)$$

with $\mathbf{Z}(.)$ the Z-transform.

This equation says that $H^s(z)/(1 - z^{-1}) = \mathbf{Z}(H(s)/s)$. Since the Z-transform of the discrete unit step is $1/(1 - z^{-1})$, this means that the discrete step response should equal the samples of the continuous step response. Thus, ZOH sampling is also called discretization by *step invariance*.

According to (7.1.34), the key to finding the discretized transfer function $H^s(z)$ is determining the Z-transform of $H(s)/s$. This may be achieved using the partial fraction expansion. If the $n + 1$ poles of $H(s)/s$ are distinct, then its PFE is

$$\frac{H(s)}{s} = \sum_{i=1}^{n+1} \frac{K_i}{s - s_i}, \qquad (7.1.35)$$

with K_i the residue of the pole at $s = s_i$. A term-by-term Z-transform of this is easy to perform, for recall that each term has an inverse Laplace transform of $K_i e^{s_i t}$. However, this is equal to $K_i (e^{s_i T})^k = K_i z_i^k$, which has a Z-transform of $K_i/(1 - z_i z^{-1})$. Therefore, according to (7.1.34) the sampled transfer function is given by

$$H^s(z) = (1 - z^{-1}) \sum_{i=1}^{n+1} \frac{K_i}{1 - e^{s_i T} z^{-1}} = (1 - z^{-1}) \sum_{i=1}^{n+1} \frac{K_i}{1 - z_i z^{-1}}, \qquad (7.1.36)$$

where

$$z_i = e^{s_i T}. \qquad (7.1.37)$$

Evaluation of $H^s(z)$ by using this formula, or by discretization of (A, B) and then application of (7.1.31), will yield the same result.

If $H(s)/s$ has repeated poles, the partial-fraction expansion will contain terms like $K/(s - s_i)^j$. If some of the poles of $H(s)/s$ are complex, then the associated residues K_i in (7.1.36) will be complex. In this case it is better to keep the complex pole pair in a single term. In either case we should refer to Table 7.1-1 to obtain the Z-transforms of the terms in the partial-fraction expansion of $H(s)/s$.

The *relative degree* of a scalar transfer function is the degree of the denominator minus the degree of the numerator. It is equivalent to the number of zeros at infinity. It is an interesting fact that, no matter what the relative degree of the continuous-time transfer function, the relative degree of the discretized equivalent $H^s(z)$ *is always equal to one*, except at some isolated values of the sampling period T. This may be seen by examining (7.1.36), for on combining the sum over a common denominator, the constant term in the numerator vanishes because $H(s)/s$ has relative degree of at least two. The numerator term in z^{-1} is the highest-order term, and its coefficient is a transcendental equation in T. That is, it vanishes only for isolated values of T.

An example would be useful at this point.

Example 7.1-6: Discretization of a Transfer Function

Suppose a continuous-time system has the transfer function

$$H(s) = \frac{-6(s - 1)}{(s + 2)(s + 3)}. \qquad (1)$$

Then, performing the partial-fraction expansion

$$\frac{H(s)}{s} = \frac{-9}{s + 2} + \frac{8}{s + 3} + \frac{1}{s}, \qquad (2)$$

7.1 Discretization of Continuous Systems

TABLE 7.1-1 TABLE OF Z-TRANSFORMS

Time Function	Laplace Transform	Z-Transform
$u_{-1}(t)$	$\dfrac{1}{s}$	$\dfrac{z}{z-1}$
t	$\dfrac{1}{s^2}$	$\dfrac{Tz}{(z-1)^2}$
$\dfrac{t^2}{2!}$	$\dfrac{1}{s^3}$	$\dfrac{T^2 z(z+1)}{2(z-1)^3}$
e^{-at}	$\dfrac{1}{s+a}$	$\dfrac{z}{z-e^{-aT}}$
te^{-at}	$\dfrac{1}{(s+a)^2}$	$\dfrac{Tze^{-aT}}{(z-e^{-aT})^2}$
$(1-at)e^{-aT}$	$\dfrac{s}{(s+a)^2}$	$\dfrac{z(z-e^{-aT}(1+aT))}{(z-e^{-aT})^2}$
$e^{-\alpha T}\sin \beta t$	$\dfrac{\beta}{(s+\alpha)^2+\beta^2}$	$\dfrac{ze^{-\alpha T}\sin \beta T}{z^2 - 2e^{-\alpha T}\cos \beta T z + e^{-2\alpha T}}$
$e^{-\alpha T}\cos \beta t$	$\dfrac{s+\alpha}{(s+\alpha)^2+\beta^2}$	$\dfrac{z(z-e^{-\alpha T}\cos \beta T)}{z^2 - 2e^{-\alpha T}\cos \beta T z + e^{-2\alpha T}}$

so that (7.1.36) yields the sampled transfer function

$$H^s(z) = (1 - z^{-1})\left[\frac{-9}{1 - e^{-2T}z^{-1}} + \frac{8}{1 - e^{-3T}z^{-1}} + \frac{1}{1 - z^{-1}}\right]. \quad (3)$$

Combining over a common factor results in

$$H^s(z) = \frac{z^{-1}(1 - 9e^{-2T} + 8e^{-3T}) + z^{-2}(8e^{-2T} - 9e^{-3T} + e^{-5T})}{(1 - e^{-2T}z^{-1})(1 - e^{-3T}z^{-1})} \quad (4)$$

Several things are worthy of note. First, there is no constant numerator term for precisely the reason that $H(s)/s$ has relative degree of 2. Second, since the Z-transform equivalent of $1/s$ is $1/(1 - z^{-1})$, the multiplier $(1 - z^{-1})$ cancels. The reader should become convinced of these facts by working through the example in detail.

Multiplying (4) by z^2 yields

$$H^s(z) = \frac{z(1 - 9e^{-2T} + 8e^{-3T}) + (8e^{-2T} - 9e^{-3T} + e^{-5T})}{(z - e^{-2T})(z - e^{-3T})}. \quad (5)$$

This has relative degree of one unless the coefficient of z vanishes. Representing the numerator as $b_1 z + b_2$, the coefficients b_1 and b_2 are plotted versus the sampling period T in Fig. 7.1-2. Note that (only) for $T = 0.86$ is the first coefficient zero and the relative degree of $H^s(z)$ two.

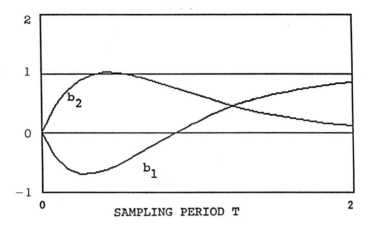

Figure 7.1-2 Numerator coefficients of discretized transfer function vs. T

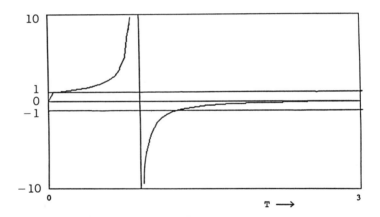

Figure 7.1-3 Zero of discretized transfer function vs. T

Shown in Fig. 7.1-3 is the location of the zero of $H^s(z)$ as a function of the sampling period T. Note that if T is larger than a certain value, the zero is inside the unit circle and the discretized system is minimum phase. We shall soon have more to say about this behavior. ∎

Systems with delays

It is not difficult to deal with systems with time delays from the point of view of the transfer function [Clarke 1981].

Thus, suppose there is a delay $\tau > 0$ so that

$$y(t) = H(s)u(t - \tau). \tag{7.1.38}$$

We are abusing notation somewhat by using $H(s)$ in the time domain, but let us consider s as a representation for the operator d/dt in this equation.

If the delay satisfies (7.1.19), then

$$Y(s) = H(s)e^{-s\tau}U(s) = H(s)e^{s(T-\delta)}U(s) \cdot e^{-sdT}.$$

However, $e^{-sdT} = z^{-d}$ represents a delay of an integral number of sampling periods. Therefore, the discretized transfer function is obtained as follows. First, perform the partial-fraction expansion

$$\frac{H(s)e^{s(T-\delta)}}{s} = \sum_{i=1}^{n+1} \frac{K_i}{s - s_i}, \tag{7.1.39}$$

which incorporates the fractional delay δ. This may be performed using standard techniques, since $e^{s_i(T-\delta)}$ is easy to evaluate. The discretized transfer function is now given by

$$H^s(z) = z^{-d}(1 - z^{-1})\sum_{i=1}^{n+1} \frac{K_i}{1 - e^{s_i T}z^{-1}} = z^{-d}(1 - z^{-1})\sum_{i=1}^{n+1} \frac{K_i}{1 - z_i z^{-1}}, \tag{7.1.40}$$

where the integral delay d has been incorporated.

7.2 DISCRETIZATION OF THE PERFORMANCE INDEX

In the continuous-time linear quadratic optimal control problem we are interested in selecting the control input $u(t)$ for the plant

$$\dot{x} = Ax + Bu \tag{7.2.1}$$

that minimizes the performance index

$$J = \frac{1}{2}x^T(t_f)S(t_f)x(t_f) + \frac{1}{2}\int_0^{t_f}(x^TQx + u^TRu)\,dt \tag{7.2.2}$$

on the time interval $[0, t_f]$. We discussed this problem in Chapters 3 and 4, with t_f generally equal to infinity.

To take advantage of today's digital signal microprocessors in the implementation of our control schemes, we must provide a discrete control sequence u_k for input into the plant. The most direct and accurate way to do this is to convert the continuous-time LQ problem into a discrete-time LQ problem, whose solution yields u_k.

In the discrete-time LQ optimal control problem we are interested in selecting the control input u_k for the plant

$$x_{k+1} = A^s x_k + B^s u_k \tag{7.2.3}$$

that minimizes the PI

$$J^s = \frac{1}{2} x_N^T S_N x_N + \frac{1}{2} \sum_{k=0}^{N-1} (x_k^T Q^s x_k + u_k^T R^s u_k) \tag{7.2.4}$$

on the time interval $[0, N]$. We shall show how to solve such discrete control problems in section 7.4.

The relation between the continuous LQ problem and the discrete LQ problem is provided by selecting a sampling period T and setting

$$t = kT. \tag{7.2.5}$$

Then $x_k \equiv x(kT)$ and $u_k \equiv u(kT)$. We have shown how to discretize the plant matrices A and B to obtain their sampled equivalents A^s and B^s. We must now discuss how the discrete PI is obtained from the continuous PI so that our performance objectives may be met by the discrete design techniques to be described in section 7.4.

Sampling the PI Weighting Matrices

To discretize the PI, we may write

$$J = \frac{1}{2} x^T(t_f) S(t_f) x(t_f) + \frac{1}{2} \sum_{k=0}^{N-1} \int_{kT}^{(k+1)T} (x^T Q x + u^T R u) \, dt, \tag{7.2.6}$$

where

$$N = t_f/T. \tag{7.2.7}$$

Now, a first-order approximation to each integral yields

$$J = \frac{1}{2} x_N^T S(t_f) x_N + \frac{1}{2} \sum_{k=0}^{N-1} (x_k^T Q x_k + u_k^T R u_k) T.$$

This is equivalent to (7.2.4) if we define

$$J^s = J \tag{7.2.8}$$

$$Q^s = QT \tag{7.2.9}$$

$$R^s = RT \tag{7.2.10}$$

$$S_N = S(t_f). \tag{7.2.11}$$

Thus, once we have selected weighting matrices $S(t_f)$, Q, and R suitable to our performance objectives, they must be converted using these relations before the discrete design is performed.

It is possible to obtain exact expressions for Q^s and R^s. See for example Åström and Wittenmark [1984]. However, except in rare cases it is not worth the trouble to use formulae more complicated than the ones provided here. This is because Q and R are not in any event selected by a hard-and-fast rule, but rather they are selected to have a rough form and range of values using engineering judgment, and, to be honest, trial and error. Finding exact equivalents to quantities whose exact values are of no concern is not often useful.

One item, however, is interesting and may be useful to know. Even if the continuous PI has no weighting on the state/input cross term $x^T u$, if we rigorously discretize the PI there does appear in J^s a cross term of the form $x_k^T W u_k$. The magnitude of the matrix W, though, is only on the order of T^2. Another point is that, even if Q is diagonal, Q^s generally has off-diagonal terms of order T^2.

Discretization of Time-Domain Performance Specifications

In connection with discretization of the performance index, we may as well discuss the discretization of performance specifications of other sorts.

A complex pole pair at $s = -\alpha \pm j\beta$ in the s-plane gives rise to a term in the transfer function denominator which can be written in the various forms

$$\Delta(s) = s^2 + 2\alpha s + \omega^2 = (s + \alpha)^2 + \beta^2 \\ = s^2 + 2\zeta\omega s + \omega^2, \tag{7.2.12}$$

with ω the natural frequency, β the oscillation frequency, ζ the damping ratio, and the various quantities related by

$$\omega^2 = \alpha^2 + \beta^2 \tag{7.2.13}$$

$$\zeta = \alpha/\omega. \tag{7.2.14}$$

The corresponding time functions depend on the residues of the poles but have the general form

$$f(t) = e^{-\alpha t}(k_1 \cos \beta t + k_2 \sin \beta t). \tag{7.2.15}$$

The parameters α, β, ζ, and ω are *design parameters* that can be used to specify the closed-loop poles required to obtain desirable closed-loop properties. To get a feel for this, let us discuss the step response of the system with transfer function

$$H(s) = \frac{\omega^2}{s^2 + 2\alpha s + \omega^2}. \tag{7.2.16}$$

The meaning of β is clear. The *rise time* is the time required for the step response to rise from 0.1 to 0.9 of its final value. It is approximately given by

$$t_r = 1.8/\omega. \tag{7.2.17}$$

The *settling time* is the time it takes for the step response to settle to its steady-state value, and it may be taken as

$$t_s = 5/\alpha, \tag{7.2.18}$$

with $1/\alpha$ the time constant of the pole pair. The *percent overshoot* (POV) is defined as

$$POV = \frac{(r_M - r_{ss})}{r_{ss}} \times 100\%, \tag{7.2.19}$$

with r_{ss} the steady-state value of the step response and r_M its maximum value. It is given in terms of the damping ratio by

$$POV = e^{-\pi\zeta/\sqrt{1-\zeta^2}}. \tag{7.2.20}$$

For $0 < \zeta < 0.6$, this may be approximated by $POV = 1 - \zeta/0.6$.

Given desired performance specifications in terms of the rise time, settling time, percent overshoot, and cutoff frequency, it is a simple matter using these relations to determine suitable values of α, β, and ω. According to Example 7.1-3, the discretized equivalent to $\Delta(s)$ is

$$\begin{aligned}\Delta^s(z) &= z^2 - 2e^{-\alpha T}\cos \beta T \, z + e^{-2\alpha T} \\ &= (z - e^{-\alpha T}\cos \beta T)^2 + e^{-2\alpha T}\sin^2 \beta T,\end{aligned} \tag{7.2.21}$$

so that knowing α, β, and ω allows one to compute the z-plane locations of the discretized poles. Thus, we may select the desired locations in the z-plane of the poles of the closed-loop discretized system for good time-response characteristics.

We should mention that if $H(s)$ has a zero in the finite plane, then its influence on the step response should be taken into account.

Example 7.2-1: Digital Controls Design Using Ackermann's Formula

Consider Newton's system

$$\dot{s} = v, \qquad \dot{v} = u, \tag{1}$$

with position $s(t)$, velocity $v(t)$, and an acceleration input $u(t)$. The state is $x = [s \quad v]^T$. It is desired to design a digital state feedback controller

$$u_k = -Kx_k = -[k_p \quad k_v]x_k = -k_p s_k - k_v v_k \tag{2}$$

that yields a percent overshoot of 4% and a settling time of 5 sec in the closed-loop system. The sampling time is $T = 0.1$ sec.

a. Discretization of System

According to Example 7.1-2, the sampled plant is

$$x_{k+1} = \begin{bmatrix} 1 & 0.1 \\ 0 & 1 \end{bmatrix} x_k + \begin{bmatrix} 0.005 \\ 0.1 \end{bmatrix} u_k \equiv Ax_k + Bu_k. \tag{3}$$

b. Discretization of Performance Specifications

Given that POV = 4%, $t_s = 5$ sec, one may compute

$$\alpha = 1, \qquad \zeta = 1/\sqrt{2}, \qquad \omega = \sqrt{2}, \qquad \beta = 1. \tag{4}$$

(We computed the damping ratio ζ by using (7.2.20) iteratively, selecting ζ and evaluating POV until 4% was obtained.) Therefore, the desired discrete-time characteristic polynomial is

$$\Delta^D(z) = z^2 - 2e^{-T}\cos T\, z + e^{-2T}$$
$$= z^2 - 1.8006z + 0.8187. \tag{5}$$

c. Ackermann's Formula Design

Using Ackermann's formula, the feedback K for the discretized system is computed by evaluating

$$U_2 = [B \quad AB] = \begin{bmatrix} 0.005 & 0.015 \\ 0.1 & 0.1 \end{bmatrix} \tag{6}$$

$$\Delta^D(A) = A^2 - 1.8006A + 0.8187I \tag{7}$$

$$K = [0 \quad 1]U_2^{-1}\Delta^D(A) = [1.81 \quad 1.9035]. \tag{8}$$

The digital control law is therefore given by

$$u_k = -1.81 s_k - 1.9035 v_k. \tag{9}$$

∎

7.3 PRACTICAL CONSIDERATIONS IN SAMPLING

It is important to be aware of some additional considerations in sampling a continuous system. Some good supplementary references on this topic are Franklin and Powell [1980] and Åström and Wittenmark [1984]. Here, we shall discuss the issues of the zeros of the discretized system, controllability and observability of the sampled system, and *lower* bounds on the sampling period.

Nonminimum-Phase Zeros

Let us discuss an issue that can cause problems in digital control if it is not understood. The poles of the system are well behaved under discretization, mapping according to

$$z_i = e^{s_i T}. \tag{7.3.1}$$

This formula shows that if the continuous-time system has poles in the left-half s-plane, then the discretized system has poles inside the unit circle in the z-plane. However, the situation is not so simple for the system zeros.

Let us examine (7.1.36). The residues K_i depend on the poles and zeros of the continuous-time system. The zeros of $H^s(z)$ in turn depend on the residues K_i and the discrete poles $e^{s_i T}$. Thus, the zeros of the discretized system depend on the poles and zeros of the continuous system as well as the sampling period T. It can be shown that, if the continuous system has a finite zero at $s = p_i$, then the discretized system has a zero at *approximately*

$$q_i \approx e^{p_i T}. \tag{7.3.2}$$

We have seen in section 2.3 that nonminimum-phase systems have some fundamental limitations in how well they may be controlled in closed-loop. Unfortunately, even if a continuous-time system has minimum phase, its discretized equivalent can have nonminimum-phase zeros, as we now see.

If the continuous system is stable, then the poles of the discretized system are stable. According to (7.3.2), the stable zeros of $H(s)$ should also be mapped into stable zeros of $H^s(z)$. This should be checked in each case, however, since (7.3.2) is only approximate.

The problem enters when the continuous system $H(s)$ has a relative degree r greater than one, and hence more than one system zero at infinity. Since the discretized system $H^s(z)$ generally has a relative degree of 1, it follows that some of the infinite zeros in the s-plane will become finite zeros in the z-plane on sampling. Indeed, $r - 1$ extra zeros will map into the finite plane. It is the locations of these zeros that can be a cause for concern.

Due to the relation $z = e^{sT}$, there is a correspondence between small values of the sampling period T and large frequencies s. If $H(s)$ has r infinite zeros, then for large s, we have $H(s) \approx b/s^r$, with b the coefficient of the highest power of s in the numerator of $H(s)$. Thus, according to (7.1.34), $H^s(z)$ tends to

$$H^s(z) = (1 - z^{-1})\mathbf{Z}\left[\frac{b}{s^{r+1}}\right]. \qquad (7.3.3)$$

Using the formula for conversion of the Laplace transform to the Laplace transform to the Z-transform, there results [Clarke 1981]

$$H^s(z) = \frac{b}{r!}(1 - z^{-1})D^r\left[\frac{1}{1 - z^{-1}}\right], \qquad (7.3.4)$$

where the operator D is defined by

$$D = -Tz\frac{d}{dz}. \qquad (7.3.5)$$

The numerators for the first few values of r are tabulated in Table 7.3-1. For $r = 1$ no extra finite zeros are introduced by sampling. If $r = 2$ and T is small, however, there is a zero at approximately $z = -1$. If the relative degree of $H(s)$ is $r = 3$ and T is small, there are zeros at approximately $z = -3.73, -1/3.73$.

The point is that for r greater than 2 there are always unstable zeros in the z-plane for small sampling periods. This should clearly demonstrate the fallacy of believing that all problems in digital control may be solved by decreasing the sampling period. Indeed, when $r = 3$, the zeros of $H^s(z)$ are stable for *large* values of the sampling period T, with one zero moving toward $z = -3.73$ as T is decreased.

In Fu and Dumont [1989], a Nyquist approach is given for selecting T to ensure minimum-phase behavior in the sampled system. It is shown that if the continuous system has minimum phase and T is *larger* than a certain value, the sampled system will be of minimum phase.

7.3 Practical Considerations in Sampling

TABLE 7.3-1 NUMERATORS OF $H^s(z)$ FOR SMALL T VERSUS RELATIVE DEGREE

r	Numerator of $H^s(z)$
1	1
2	$z + 1$
3	$z^2 + 4z + 1$
4	$z^3 + 11z^2 + 11z + 1$
5	$z^4 + 26z^3 + 66z^2 + 26z + 1$

Loss of Observability

The next example shows another problem that can occur in sampling of systems.

Example 7.3-1: Loss of Reachability and Observability Through Sampling

Consider the special case of Example 7.1-3 where $\alpha = 0$, $\omega = \beta$. This is the undamped harmonic oscillator. Then

$$\dot{x} = \begin{bmatrix} 0 & 1 \\ -\omega^2 & 0 \end{bmatrix} x + \begin{bmatrix} 0 \\ 1 \end{bmatrix} u. \tag{1}$$

Let us take position measurements so that the output is

$$y = [1 \quad 0] x. \tag{2}$$

Since the reachability matrix

$$U = [B \quad AB] = \begin{bmatrix} 0 & 1 \\ 1 & 0 \end{bmatrix} \tag{3}$$

and observability matrix

$$V = \begin{bmatrix} C \\ CA \end{bmatrix} = \begin{bmatrix} 0 & 1 \\ 1 & 0 \end{bmatrix} \tag{4}$$

are nonsingular, the system is reachable and observable.

From Example 7.1-3 the sampled system matrices are

$$A^s = e^{AT} = \begin{bmatrix} \cos \omega T & \frac{1}{\omega} \sin \omega T \\ -\omega \sin \omega T & \cos \omega T \end{bmatrix} \tag{5}$$

$$B^s = \begin{bmatrix} \frac{1}{\omega^2}(1 - \cos \omega T) \\ \frac{1}{\omega} \sin \omega T \end{bmatrix}. \tag{6}$$

The determinants of the reachability and observability matrices of the sampled system are

$$|U^s| = -\frac{2}{\omega^3}\sin \omega T(1 - \cos \omega T) \tag{7}$$

$$|V^s| = \frac{1}{\omega}\sin \omega T. \tag{8}$$

Thus, the sampled system is both unreachable and unobservable at

$$T = \frac{i\pi}{\omega}, \tag{9}$$

for any integer i.

■

We have seen in section 3.3 that if the continuous-time system (A, B) is reachable and $(\sqrt{Q}, A)$ is observable then the steady-state Kalman gain stabilizes the system. A similar result for discrete systems will be discovered in section 7.4. Therefore, the loss of reachability and observability through sampling can be a serious affair.

In section 7.4 we shall see that the loss of observability can also result in unsatisfactory *intersample behavior*. Thus, even though the samples $x(kT)$ behave as desired, the system performance between the sample periods may be very bad.

Fortunately, the situation is not as inauspicious as Example 7.3-1 seems to indicate. Indeed, if the continuous system is reachable and observable and we are careful in our selection of T, then all will be well, as we may now show.

A first-order approximation to (7.1.12) and (7.1.13) yields the *Euler approximation*

$$A^s = I + AT \tag{7.3.6}$$

$$B^s = TB \tag{7.3.7}$$

to the sampled system. Although this approximation is not accurate enough to use in computing A^s and B^s for design purposes, it can give us some insight here. Indeed, we see that the sampled reachability matrix is given by

$$U^s = [B^s \quad A^s B \quad \ldots \quad (A^s)^{n-1} B^s]$$
$$\approx [TB \quad (I + AT)TB \quad (I + AT)^2 TB \quad \ldots]$$

However, by multiplying out the terms and performing column operations on the result it is evident that

$$\text{rank}(U^s) = \text{rank}[TB \quad T^2 AB \quad T^3 A^2 B \quad \ldots], \tag{7.3.8}$$

so that the sampled reachability matrix should have full rank if the continuous-time system is reachable.

Indeed, U^s does have full rank for *almost* all values of T; it can lose rank only at isolated values of T. Thus, reachability may be lost at these isolated values of T. The same holds true for observability.

7.4 Discrete Design Techniques

Selecting the Sample Period

Our final topic in this section is the selection of the sampling period T. In Examples 5.2-1 and 5.2-2 and in section 5.3 we discussed *upper bounds* on T, seeing that T should be less than about one-fifth the fastest time constant in the closed-loop system, or one-fifth the period of the highest closed-loop frequencies present. Let us now discuss some other concerns, including a *lower bound* on T.

From what we have seen, at isolated values of T reachability and observability may be lost and the relative degree of the discretized transfer function may be greater than one. Although the latter may not be detrimental in a given situation, we should at least be aware of what is going on. Certainly, we should avoid those values of T that result in a loss of desirable system properties.

In Example 7.1-6 we saw that the zero of the discretized system is stable for values of T *larger* than a certain value. We have also seen from Table 7.3-1 that simply decreasing the sampling period is certainly a bad idea if the relative degree of $H(s)$ is greater than 1. In fact, it is shown in Fu and Dumont [1989] that if the continuous system is minimum phase and T is *larger* than a certain value, the sampled system is minimum phase.

This can be important when it comes to the ease with which a digital controller may be designed, since plants with unstable zeroes are notoriously difficult to control (section 2.3).

Another concern should be considered when deciding on the smallest allowed T. According to (7.3.7), as T decreases the magnitude of the terms in the control matrix B^s decreases as well. Note also from (7.3.8) that U^s may become ill-conditioned (i.e., nearly singular) if T becomes too small.

7.4 DISCRETE DESIGN TECHNIQUES

In the first part of this chapter we showed how to discretize continuous-time systems using the ZOH or step-invariance approach. The simple computer program DISC given in Appendix A performs this discretization automatically. The resulting discrete-time system exactly contains the ZOH and sampler, in contrast to the approximation approach to digital controls design given in Chapter 5. We saw in addition that sampling of continuous-time systems gives some insight into the sampling process.

In this section we give some techniques for designing controllers for discrete-time systems. Our contention is that discrete design techniques are difficult neither to understand nor to apply. First, we discuss classical control, then we extend the modern techniques of Chapter 4 to discrete systems, discussing both the tracker and regulator problems. Thus, the process of sampling the continuous-time system and then designing a controller for the resulting discrete-time system will give us an approach to the design of digital controllers. We call the process "direct discrete-time design."

The direct discrete design approach for digital controllers has some important advantages over the approximate design technique given in Chapter 5. It allows the use of larger sampling periods since it guarantees exact closed-loop performance at the sample points. Since it is straightforward to find sampled equivalents to continuous systems with time delays, it affords an easy way to control systems with delays. Moreover, the entire design procedure may be accomplished using computer software. No hand calculations such as those usually used to perform the BLT in Chapter 5 are needed.

Discrete Classical Design

Let us first discuss some classical controls design techniques for discrete-time systems. Then, in the next subsections we present some modern discrete design techniques.

Digital compensator of desired structure

Exactly as in the continuous-time case (see section 4.2), desired compensator dynamics may be included in the digital controller. Thus, suppose the discretized system equations have been computed by ZOH sampling/step invariance to be

$$x_{k+1} = Ax_k + Bu_k \qquad (7.4.1)$$

$$y_k = Cx_k \qquad (7.4.2)$$

with state x_k, control input u_k, and y_k the measured output available for feedback purposes. In addition, let

$$z_k = Hx_k \qquad (7.4.3)$$

be a performance output, which must track the given reference command input r_k.

Having in mind the discrete version of Fig. 4.2-1, the discrete dynamic compensator has the form

$$w_{k+1} = Fw_k + Ge_k$$
$$v_k = Dw_k + Je_k \qquad (7.4.4)$$

with state w_k, output v_k, and input equal to the tracking error

$$e_k = r_k - z_k. \qquad (7.4.5)$$

F, G, D, and J are known matrices chosen to include the desired structure in the compensator.

The allowed form for the plant control input is

$$u_k = -Ky_k - Lv_k, \qquad (7.4.6)$$

where the constant gain matrices K and L are to be chosen in the controls design step to result in satisfactory tracking of r_k.

7.4 Discrete Design Techniques

These dynamics and output equations may be written in augmented form as

$$\begin{bmatrix} x_{k+1} \\ w_{k+1} \end{bmatrix} = \begin{bmatrix} A & 0 \\ -GH & F \end{bmatrix} \begin{bmatrix} x_k \\ w_k \end{bmatrix} + \begin{bmatrix} B \\ 0 \end{bmatrix} u_k + \begin{bmatrix} 0 \\ G \end{bmatrix} r_k \quad (7.4.7)$$

$$\begin{bmatrix} y_k \\ v_k \end{bmatrix} = \begin{bmatrix} C & 0 \\ -JH & D \end{bmatrix} \begin{bmatrix} x_k \\ w_k \end{bmatrix} + \begin{bmatrix} 0 \\ J \end{bmatrix} r_k \quad (7.4.8)$$

$$z_k = [H \quad 0] \begin{bmatrix} x_k \\ w_k \end{bmatrix} \quad (7.4.9)$$

and the control input may be expressed as

$$u_k = -[K \quad L] \begin{bmatrix} y_k \\ v_k \end{bmatrix}. \quad (7.4.10)$$

By redefining the state, the output, and the matrix variables to streamline the notation, we see that the augmented equations (7.4.7)–(7.4.9) are of the form

$$x_{k+1} = Ax_k + Bu_k + Er_k \quad (7.4.11)$$

$$y_k = Cx_k + Fr_k \quad (7.4.12)$$

$$z_k = Hx_k. \quad (7.4.13)$$

In this description, let us take the state $x_k \in \mathbf{R}^n$, control input $u_k \in \mathbf{R}^m$, reference input $r_k \in \mathbf{R}^q$, performance output $z_k \in \mathbf{R}^q$, and measured output $y_k \in \mathbf{R}^p$.

In terms of the redefined variables, the admissible controls (7.4.10) are proportional output feedbacks of the form

$$u_k = -Ky_k = -KCx_k - KFr_k \quad (7.4.14)$$

with constant gain K to be determined. This situation corresponds to the block diagram in Fig. 4.2-2 and has the same structure as in the continuous-time case.

Now, we present some notions of controls design for discrete-time systems taken from classical control theory.

Digital compensator for zero steady-state error

Here, we present the discrete-time analog of the integrator, which could be included in the feedforward path to make the system of Type I, thus guaranteeing perfect tracking of a step command.

Using the continuous final-value theorem

$$\lim_{t \to \infty} e(t) = \lim_{s \to 0} sE(s) \quad (7.4.15)$$

it is easy to derive the well-known result that to exactly follow a reference step, ramp, and so on, the forward path in a continuous closed-loop system must contain a replica of the reference command generator. For example, to follow a unit step with zero steady-state error $\bar{e}$, it is required that the feedforward path contain an integrator (i.e., must be "of Type I"). This notion is formalized in the derivation of the Command Generator Tracker in section 4.4.

Exactly the same results hold for discrete-time systems. They may be shown using the discrete final-value theorem

$$\lim_{k \to \infty} e_k = \lim_{z \to 1} (1 - z^{-1})E(z). \tag{7.4.16}$$

In connection with this theorem, note that DC for discrete systems occurs at $z = e^{j\omega T} = e^0 = 1$.

The appropriate digital compensator needed for tracking is easily obtained by discretizing the corresponding continuous-time compensator. For instance, using the BLT, we may discretize the integrator $1/s$ to obtain

$$G_I(z) = \frac{T}{2} \frac{z + 1}{z - 1}. \tag{7.4.17}$$

For zero $\bar{e}$ in response to a discrete unit step, it is only necessary to include $G_I(z)$ as a part of the discrete feedforward compensator (see the problems).

To discover the feedforward compensator needed for exact following of a unit ramp at steady-state, we may discretize $1/s^2$.

Discrete root locus design

The root locus is drawn for discrete-time systems using exactly the same rules as for the continuous-time case. However, the interpretation of the pole locations is different [Franklin, Powell, and Emami-Naeini 1986].

Using the ZOH sampling/step invariance discretization technique, we have seen that the continuous plant poles s_i are mapped into the z-plane according to

$$z = e^{sT}. \tag{7.4.18}$$

Therefore, the $j\omega$-axis in the s-plane maps into the unit circle $e^{j\omega T}$ in the z-plane. Thus, for discrete systems, the stable region is the interior of the unit circle. Moreover, DC for discrete systems corresponds to $z = 1$. Thus, the small region near $s = 0$ is essentially equivalent to the small region around $z = 1$.

There is no location in the z-plane corresponding to frequencies higher than the Nyquist frequency $\omega_N = \omega_s/2$.

Let us consider a complex pole pair at

$$s = -\alpha \pm j\beta = -\zeta\omega_n + j\omega_n\sqrt{1 - \zeta^2}$$

7.4 Discrete Design Techniques

in the s-plane, with ζ the damping ratio and ω_n the natural frequency. Since

$$z = e^{sT} = e^{-\alpha T}e^{\pm j\beta T},$$

lines of constant α in the s-plane are mapped into circles in the z-plane of radius $e^{-\alpha T}$. Since

$$z = e^{-\zeta\omega_n T}e^{\pm j\omega_n T\sqrt{1-\zeta^2}}$$

lines of constant damping ratio ζ are mapped into logarithmic spirals, while circles of constant ω_n are mapped into curves at right angles to these spirals. The lines of constant ζ and ω_n in the z-plane are shown in Fig. 7.4-1 [Dorsey 1987].

In the next example we illustrate discrete root locus design, as well as introducing the important notion of *deadbeat control*, which has no counterpart in continuous-time systems. We also show that the *intersample behavior* can be extremely bad in digital control systems if care is not taken in the design.

Example 7.4-1: Deadbeat Control and Intersample Behavior of a System Obeying Newton's Laws

In this example we shall consider the digital control of the system

$$\dot{x} = \begin{bmatrix} 0 & 1 \\ 0 & 0 \end{bmatrix} x + \begin{bmatrix} 0 \\ 1 \end{bmatrix} u = Ax + Bu \quad (1)$$

with $x = [y \ v]^T$, where $y(t)$ is position, $v(t)$ the velocity, and $u(t)$ the input acceleration (cf. Åström and Wittenmark 1984). This is a state description of any system obeying Newton's laws $\ddot{y} = u$ (see Example 2.1-4).

To make the position $y(t)$ follow a desired reference input $r(t)$, we could use the digital control structure shown in Fig. 7.4-2, where unity feedback of the position is used to manufacture an error signal

$$e_k = r_k - y_k. \quad (2)$$

The discrete input sequence is then given by

$$u_k = k_e e_k - k_v v_k \quad (3)$$

$$= -[k_e \ k_v]x_k + k_e r_k \equiv -Kx_k + k_e r_k. \quad (4)$$

a. Discrete Root Locus Design

According to Example 7.1-2 the continuous plant, ZOH, and sampler have the discrete state description

$$x_{k+1} = \begin{bmatrix} 1 & T \\ 0 & 1 \end{bmatrix} x_k + \begin{bmatrix} T^2/2 \\ T \end{bmatrix} u_k = A^s x_k + B^s u_k \quad (5)$$

with T the sample period. Therefore, the closed-loop system is

$$x_{k+1} = (A^s - B^s K)x_k + Bk_e r_k \equiv A_c x_k + Bk_e r_k \quad (6)$$

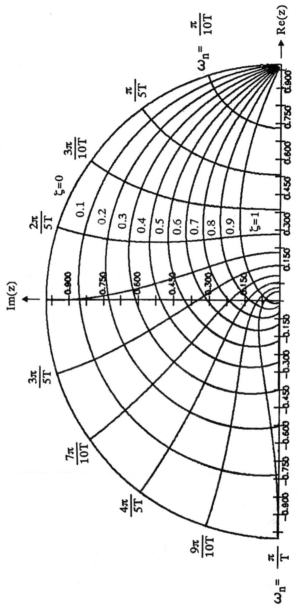

Figure 7.4-1 Lines of constant ζ and ω_n in the z-plane

7.4 Discrete Design Techniques

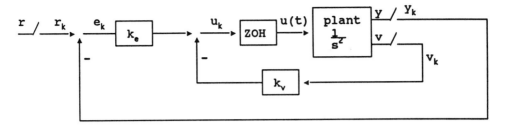

Figure 7.4-2 Digital tracker for Newton's System

or

$$x_{k+1} = \begin{bmatrix} 1 - k_e T^2/2 & T - k_v T^2/2 \\ -k_e T & 1 - k_v T \end{bmatrix} x_k + k_e \begin{bmatrix} T^2/2 \\ T \end{bmatrix} r_k, \tag{7}$$

whence the closed-loop characteristic equation is

$$\Delta_c(z) = |zI - (A^s - B^s K)| \tag{8}$$
$$= z^2 - (2 - k_e T^2/2 - k_v T)z + (1 + k_e T^2/2 - k_v T) = 0.$$

We may write this as

$$\Delta_c(z) = (z - 1)^2 + k_e[(\rho T + T^2/2)z - (\rho T - T^2/2)] \tag{9}$$

with

$$\rho \equiv k_v/k_e \tag{10}$$

the ratio of derivative to proportional feedback. If $\rho = 3T/2$ then

$$\Delta_c(z) = (z - 1)^2 + k_e 2T^2 \left(z - \frac{1}{2} \right), \tag{11}$$

and standard root locus theory may be used to obtain the root locus versus k_e that is shown in Fig. 7.4-3.

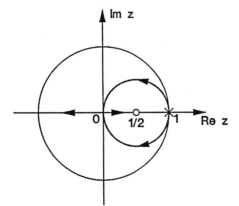

Figure 7.4-3 Discrete root locus for Newton's system

Using this figure in conjunction with Fig. 7.4-1, a value of k_e that yields a suitable closed-loop damping ratio and natural frequency may be selected.

Let us now mention several possible designs.

b. Deadbeat Control

If we select

$$k_e = 1/T^2, \qquad k_v = 3/2T \qquad (12)$$

then (8) indicates that $\Delta_c(z) = z^2$ so that both closed-loop poles are at the origin. See Fig. 7.4-3. We claim that, using these control gains, the output is brought exactly to any desired constant reference value after *two sample periods*.

To see this, note that according to the Cayley-Hamilton Theorem [Kailath 1980], since the closed-loop characteristic polynomial is $\Delta_c(z) = z^2$, we must have $\Delta_c(A) = (A_c)^2 = 0$. This may be verified by squaring A_c, which may be computed to be

$$A_c = \begin{bmatrix} 1/2 & T/4 \\ -1/T & -1/2 \end{bmatrix}. \qquad (13)$$

Therefore, if $r_k = 0$, then the system response at $k = 2$ is

$$x_2 = (A_c)^2 x_0 = 0 \qquad (14)$$

for any x_0. That is, the state vanishes after two sample periods for any initial state.

If all closed-loop poles are at the origin so that $\Delta_c(z) = z^n$ (where $x_k \varepsilon \mathbf{R}^n$), the controller is said to be *deadbeat*. Then $(A_c)^n = 0$. A matrix which yields zero when raised to a finite power is said to be *nilpotent*. The required power is always less than or equal to n. A deadbeat controller has the property that, if the input is zero, the state is driven exactly to zero in at most n sample periods no matter what the initial condition.

If $r_k = y_f u_{-1}(k)$, with $u_{-1}(k)$ the unit step and y_f the desired final value of the position y_k, then since A_c is nilpotent

$$x_k = A_c^k x_0 + Bk_e r_{k-1} + A_c Bk_e r_{k-2}, \qquad (15)$$

which the reader should verify yields

$$x_k = \begin{bmatrix} 1 \\ 0 \end{bmatrix} y_f, \qquad k \geq 2. \qquad (16)$$

The deadbeat control gains (12) thus guarantee that after 2 sampling periods $y_k = y_f$ and $v_k = 0$. This perfect tracking of a unit step results because the open-loop system in Fig. 7.4-2 is of Type 2.

Figure 7.4-4 shows simulations of the deadbeat controller for $T = 0.5, 1, 2$. We used software like that in section 5.1 and $y(0) = 0$, $v(0) = -10$, and $r(t) = y_f = 10$. The response is indeed exactly zero after two sample periods. Note that, as T decreases and the system settles to y_f more quickly, the required control magnitude increases. See the dependence on T in (12).

We should like to point out that the sampling periods used in Fig. 7.4-4 are much larger than those allowed in the continuous redesign approach in Chapter 5. This is because, whatever the value of T, the discrete design approach used in this example guarantees exact behavior at the sample points, and no approximation is involved.

Deadbeat control is a remarkable and useful phenomenon which has no counterpart in

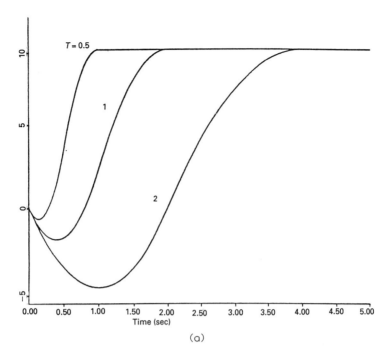

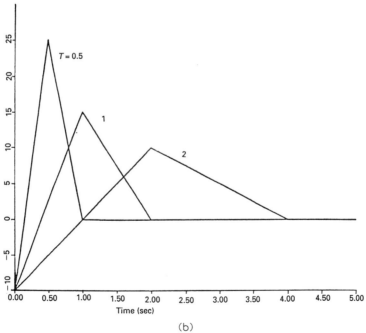

Figure 7.4-4 Deadbeat responses for Newton's system. (a) Position $y(t)$. (b) Velocity $v(t)$. (c) Control input $u(t)$.

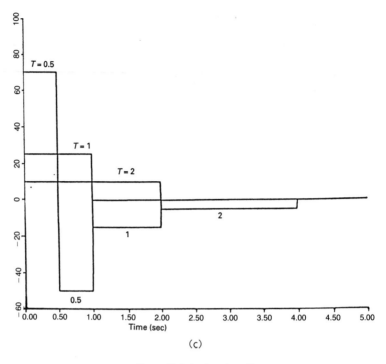

Figure 7.4-4 (*continued*)

continuous-time. That is, in continuous design we may place the closed-loop poles arbitrarily far to the left in the s-plane and obtain a response that decays to the desired value arbitrarily quickly, but in theory it never reaches the final value exactly in a finite time. (Note that $z = e^{sT}$ maps the point $z = 0$ into $s = -\infty$.) Thus, digital control can result in control performance that cannot be matched using continuous-time design techniques.

Note further that moving the desired poles further left in the s-plane requires increasingly high feedback gains, while deadbeat control in the z-plane is achieved using the finite gains (12).

In deadbeat design, there is only one design parameter, namely the sampling period. For then, the control gains are given by (12). The sampling period may be selected based on simulation so that settling is achieved quickly, but not so quickly that the plant is stressed by forcing it to respond too fast.

c. Pole-Placement Design

Suppose we want to place the closed-loop poles so that the system has the desired closed-loop polynomial

$$\Delta^d(z) = z^2 - d_1 z + d_2. \tag{17}$$

Then, we may equate the coefficients of (8) and (17) and solve to obtain the required control gains

$$k_c = \frac{1}{T^2}(1 - d_1 + d_2) \tag{18}$$

7.4 Discrete Design Techniques

$$k_v = \frac{1}{2T}(3 - d_1 - d_2). \tag{19}$$

The desired coefficients d_1 and d_2 may be chosen for suitable time-domain performance.

d. Intersample Behavior

The intersample behavior in Fig. 7.4-4 is very good. However, this is not always the case and must be checked using simulation before the proposed digital controller is implemented on the actual plant.

To illustrate the sort of problem that can arise, define the output

$$y_k = [1 \ \ 0]x_k \equiv Hx_k \tag{20}$$

so that the closed-loop transfer function from r_k to y_k is

$$H_c(z) = H(zI - A_c)^{-1}Bk_e \tag{21}$$

$$= \frac{k_e(z + 1)T^2/2}{\Delta_c(z)}. \tag{22}$$

If we select $k_v = 2/T$, then this becomes (verify!)

$$H_c(z) = \frac{k_e(z + 1)T^2/2}{(z + 1)(z - (1 - k_eT^2/2))} = \frac{k_eT^2/2}{(z - (1 - k_eT^2/2))}. \tag{23}$$

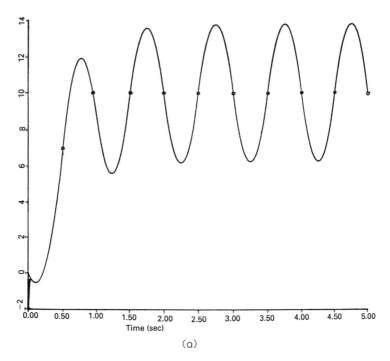

(a)

Figure 7.4-5 System response showing unobservable intersample oscillations. (a) Position $y(t)$. (b) Velocity $v(t)$. (c) Control input $u(t)$.

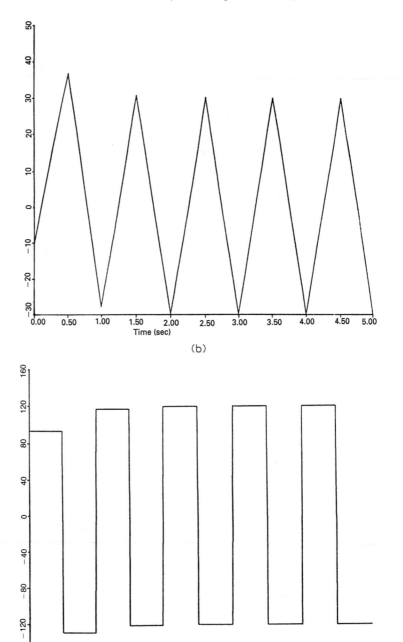

Figure 7.4-5 (continued)

7.4 Discrete Design Techniques

Note that a pole/zero pair has canceled. This choice for k_v corresponds to the point on the root locus in Fig. 7.4-3 where one closed-loop pole is at $z = -1$.

To see why the pole/zero cancellation has occurred, note that the observability matrix is

$$V = \begin{bmatrix} H \\ HA \end{bmatrix} = \begin{bmatrix} 1 & 0 \\ 1 - k_e T^2/2 & T(1 - k_v T/2) \end{bmatrix} \quad (24)$$

The choice of $k_v = 2/T$ renders the closed-loop system *unobservable*, hence the cancellation.

A simulation of the system step-response for $k_v = 2/T$ and $k_e = 4/3T^2$ is shown in Fig. 7.4-5 for $T = 0.5$ sec. Note that at the samples, $y(t) = y_f$ as desired after $2T$ sec. However, the intersample behavior is unsatisfactory. This is because a marginally stable pole and zero cancel, so that an oscillatory mode is unobservable.

■

Discrete Bode design

Given a discrete-time transfer function $G(z)$, the frequency response may be studied by plotting $|G(e^{j\omega T})|$ and $\angle G(e^{j\omega T})$. Unlike the continuous-time case, it is awkward to sketch these magnitude and phase plots using simple rules of thumb. This presents no problem for today's designer, however, since it is simple to write a computer program that plots $|G(e^{j\omega T})|$ and $\angle G(e^{j\omega T})$.

Once these plots are available, the gain and phase margin are determined the same as in the continuous case. A polar plot of $G(e^{j\omega T})$ allows the Nyquist design technique to be applied as well.

W-Plane design

There is, however, another drawback to Bode design for discrete systems [Franklin and Powell 1980]. According to the Bode gain-phase relation, for minimum-phase systems the phase is uniquely determined by an integral of the slope of the magnitude curve on a log-log plot. For systems with *rational* transfer functions, these slopes are approximated by constants. This gives the continuous-time design criterion that the magnitude curve must cross unity gain at a slope of approximately -20 dB/decade for the phase to remain above the stability boundary of $180°$. The point is that continuous-time design for minimum-phase systems may be accomplished using only the magnitude Bode curve.

Unfortunately, the discrete transfer function $G(e^{j\omega T})$ is not rational in ω and the simplicity of Bode's design technique is lost in the z-plane. It should, however, be clearly understood that if plots of both $|G(e^{j\omega T})|$ and $\angle G(e^{j\omega T})$ are available, design may still be performed.

A remedy for the loss of simplicity in Bode design for discrete systems is to transform $G(z)$ to the *w-plane* [Franklin and Powell 1980, Phillips and Nagle 1984] using the transformation

$$z = \frac{1 + wT/2}{1 - wT/2}. \quad (7.4.19)$$

The familiar Bode design techniques may be used for the resulting function, denoted

$G(w)$, to obtain a compensator $K(w)$. This compensator is then transformed back to the z-plane using

$$w = \frac{2}{T} \frac{z-1}{z+1} \tag{7.4.20}$$

to obtain the digital controller $K(z)$.

It is clear what is going on here. Note that $G(z)$ was presumably obtained from the continuous plant $G^c(s)$ using the ZOH or step invariance sampling technique of section 7.1. This results in the poles being mapped from the s-plant into the z-plane according to

$$z = e^{sT}. \tag{7.4.21}$$

However, (7.4.19) is an approximation to e^{sT}. Therefore, $G(w)$ should be an approximation to the original continuous-time transfer function $G^c(s)$, and the s-plane Bode techniques should approximately apply.

Indeed, w-plane design is nothing but the exact reverse of the continuous-time design procedure based on the BLT that was discussed in section 5.2. In fact, it has two additional steps: namely, conversion of $G^c(s)$ to $G(z)$, and reconversion of the controller $K(w)$ to $K(z)$. Note, however, that $G(w)$ includes the ZOH and sampler dynamics. Thus, it may be more akin to the "modified" continuous design approach at the end of section 5.3.

Tracking by Regulator Redesign

We should now like to select the output feedback gain K for the system-plus-compensator (7.4.11) with measured output (7.4.12) so that the performance output z_k tracks the reference input r_k. In this section we design a tracker or servocontrol system by first solving the regulator problem and then adding a feedforward term to guarantee tracking [Kwakernaak and Sivan 1972]. This will give us some insight on the structure of tracking control systems. In the next section we attack the tracker problem directly using discrete LQ output-feedback design.

Closed-loop specifications are often given in terms of time-domain requirements like rise time, percent overshoot, and so on. These qualities are closely related to the step response; therefore, to find the optimal gain K we shall assume in the design phase that the reference command r_k is a unit step of magnitude r. Note that the resulting servo system will work for any arbitrary command signal r_k; our approach simply guarantees good time-domain performance specifications. In the following, the reader should be alerted to the fact that we shall be using z_k as a signal and z as the Z-transform variable.

Define the deviations

$$\begin{aligned} \tilde{x}_k &= x_k - \bar{x}, & \tilde{u}_k &= u_k - \bar{u}, \\ \tilde{y}_k &= y_k - \bar{y}, & \tilde{z}_k &= z_k - \bar{z}, \end{aligned} \tag{7.4.22}$$

with overbar denoting the steady-state values. Note that we want $\bar{z} = r$, and then $\tilde{z}_k = -e_k$. According to (7.4.11), (7.4.12), (7.4.13), the steady-state values satisfy the equations

7.4 Discrete Design Techniques

$$\bar{x} = A\bar{x} + B\bar{u} + Er \tag{7.4.23}$$

$$\bar{y} = C\bar{x} + Fr \tag{7.4.24}$$

$$\bar{z} = H\bar{x} = r. \tag{7.4.25}$$

By subtracting the steady-state equations from the plant dynamics, we obtain the shifted system dynamics

$$\tilde{x}_{k+1} = A\tilde{x}_k + B\tilde{u}_k \tag{7.4.26}$$

$$\tilde{y}_k = C\tilde{x}_k \tag{7.4.27}$$

$$\tilde{z}_k = H\tilde{x}_k = -e_k. \tag{7.4.28}$$

Now, suppose we are able to solve the regulator problem for this *deviation system*. That is, suppose we have selected a feedback gain K in

$$\tilde{u}_k = -K\tilde{y}_k \tag{7.4.29}$$

so that the deviation system is stable and hence e_k goes to zero. We shall soon show how to achieve this using LQ output-feedback design (see Table 7.4-2). (Alternatively, classical design techniques may be used to choose K.) Then, the regulator control law (7.4.29) may be converted into a servocontrol input that makes the tracking error go to zero in the plant (7.4.11) by simply using (7.4.22) to write

$$u_k = -Ky_k + (\bar{u} + K\bar{y}) \equiv -Ky_k + u_0. \tag{7.4.30}$$

This tracking control law is just the original regulator term $-Ky_k$ plus an extra constant feedforward term u_0 that is added to guarantee a nonzero steady-state performance output of $\bar{z} = r$.

To determine u_0, substitute (7.4.30) into (7.4.11) to obtain the closed-loop dynamics

$$x_{k+1} = (A - BKC)x_k + Bu_0 + Er \equiv A_c x_k + Bu_0 + Er, \tag{7.4.31}$$

whence the closed-loop transfer function from u_0 to z_k is

$$H_c(z) = H(zI - A_c)^{-1}B. \tag{7.4.32}$$

At steady state, $x_{k+1} = x_k \equiv \bar{x}$, so that

$$\bar{x} = A_c \bar{x} + Bu_0 + Er,$$

or

$$\bar{x} = (I - A_c)^{-1}(Bu_0 + Er). \tag{7.4.33}$$

If there is perfect steady-state tracking, the steady-state error is zero and

$$r = H\bar{x} = H(I - A_c)^{-1}(Bu_0 + Er)$$

or

$$H(I - A_c)^{-1}Bu_0 = [I - H(I - A_c)^{-1}E]r. \tag{7.4.34}$$

However, the coefficient of u_0 is nothing but the closed-loop DC gain $H_c(1)$. For this equation to have a solution u_0 for all r, it is necessary that $H_c(1)$ have full row rank, so that it must have at least as many columns as rows. That is, *for perfect tracking there must be at least as many components in the plant input u_k as there are components in the performance output z_k.*

If $H_c(1)$ is square and nonsingular, then the constant offset component of the control is given by

$$u_0 = H_c^{-1}(1)[I - H(I - A_c)^{-1}E]r, \tag{7.4.35}$$

so that the servocontrol law that makes z_k follow a unit step is

$$u_k = -Ky_k + H_c^{-1}(1)[I - H(I - A_c)^{-1}E]r. \tag{7.4.36}$$

If $E = 0$ this becomes simply

$$u_k = -Ky_k + H_c^{-1}(1)r. \tag{7.4.37}$$

We have thus discovered that the servo control may be obtained from the regulator control by adding a constant feedforward term that depends on the magnitude of the *DC* gain.

Define the open-loop transfer function in (7.4.11), (7.4.13) from u_k to z_k as

$$H(z) = H(zI - A)^{-1}B. \tag{7.4.38}$$

The transmission zeros are the values of z where

$$|H(z)| = 0. \tag{7.4.39}$$

Since the zeros are not changed by output feedback, the zeros of $H(z)$ and $H_c(z)$ are the same. Therefore, $H_c(1)$ is nonsingular if and only if the system has no transmission zeros at $z = 1$. This is the condition for perfect tracking of a unit step command. Notice that the unit step has a pole at $z = 1$.

Example 7.4-2: Digital Tracker by Regulator Redesign for Motor Speed Control

In Example 4.2-3 we designed a continuous-time tracker for speed control in an armature-controlled DC motor. Here we shall illustrate digital tracker design using the regulator redesign approach.

a. Discretization of the Plant

The continuous-time description from Example 4.2-3 is

$$\dot{x} = \begin{bmatrix} -2.29 & -0.003 \\ 1.172 & -1.32 \end{bmatrix} x + \begin{bmatrix} 0.337 \\ 0 \end{bmatrix} u \equiv Ax + Bu \tag{1}$$

$$z = [0 \quad 1]x$$

where $x = [i \quad \omega]^T$, with $i(t)$ the armature current, $\omega(t)$ the angular velocity, and $u(t)$ the armature voltage. Selecting a sampling period of $T = 0.05$ sec and sampling by the ZOH technique in section 7.1 (using program DISC in Appendix A) yields

$$x_{k+1} = \begin{bmatrix} 0.89181 & -0.00014 \\ 0.05355 & 0.93613 \end{bmatrix} x_k + \begin{bmatrix} 0.01592 \\ 0.00047 \end{bmatrix} u_k \equiv Ax_k + Bu_k \tag{2}$$

$$z_k = [0 \quad 1]x_k \equiv Hx_k.$$

7.4 Discrete Design Techniques

This discrete model contains the discretized continuous dynamics, the ZOH, and the sampler. The continuous plant (1) has the transfer function

$$\frac{0.395}{(s + 1.324)(s + 2.286)} \qquad (3)$$

and the discretized system has the transfer function

$$H(z) = \frac{0.00047(z + 0.9416)}{(z - 0.892)(z - 0.936)}. \qquad (4)$$

Note that the poles have mapped according to $z = e^{sT}$.

b. Design of Regulator

Selecting the output feedback control law

$$u_k = -k_\omega z_k \qquad (5)$$

yields the closed-loop transfer function

$$H_c(z) = \frac{H(z)}{1 + k_\omega H(z)}, \qquad (6)$$

and standard techniques yield the root locus shown in Fig. 7.4-6. Selecting $k_\omega = 10$ results in the closed-loop poles

$$z = 0.9116 \pm j0.0902 \qquad (7)$$

and the closed-loop transfer function

$$H_c(z) = \frac{0.00047(z + 0.9416)}{(z - 0.9116)^2 + 0.0902^2}. \qquad (8)$$

The regulation behavior of the closed-loop system is demonstrated in Fig. 7.4-7, which shows the response for nonzero initial motor speed.

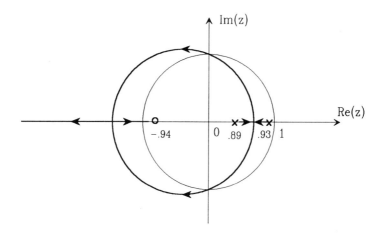

Figure 7.4-6 Root locus for motor regulator design

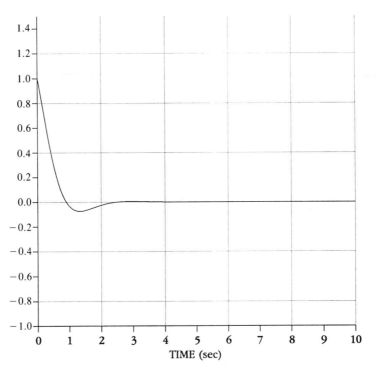

Figure 7.4-7 Closed-loop response of regulator

c. Tracker Design

To convert this regulator to a tracker that makes the motor follow a commanded angular velocity of r rad/sec, we may use the control law

$$u_k = -k_\omega z_k + k_f r, \qquad (9)$$

where a feedforward term $k_f r$ has been added. See Fig. 7.4-8.

According to the development in this section, the feedforward gain k_f should be equal to the reciprocal of the *DC* gain of the closed-loop regulator system. Thus, according to (8), $H_c(1) = 0.05663$ and

$$k_f = 1/H_c(1) = 17.658. \qquad (10)$$

Using these gains in the digital control system in Fig. 7.4-8 yields the step response shown in Fig. 7.4-9a. To obtain this plot, the digital control subroutine in Fig. 7.4-10 was used with

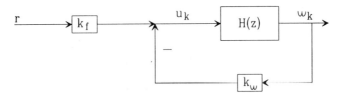

Figure 7.4-8 Digital servo system for motor speed control

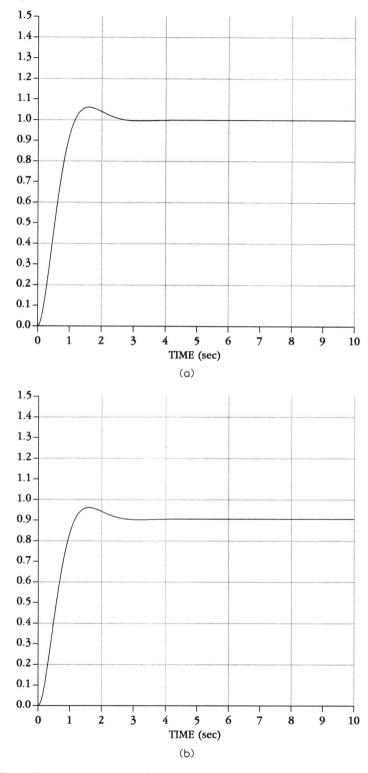

Figure 7.4-9 Step response of digital speed controller. (a) Using the correct value of the DC gain. (b) Using an incorrect value for the DC gain.

```
C    DIGITAL CONTROL SUBROUTINE FOR USE WITH DIGCTL
     SUBROUTINE DIG(IK,T,x)
     REAL x(*), kf, kw
     COMMON/CONTROL/u
     DATA kf,kw/17.658,10./
     DATA r/1.0/

     u= -kw*x(2) + kf*r

     RETURN
     END
```

Figure 7.4-10 Digital controller subroutine for use with program DIGCTL

subroutine DIGCTL as discussed in section 5.1. We emphasize that the digital control simulation was performed on the continuous dynamics (1) to obtain the response at *all* times, including the intersample behavior.

d. Discussion

Although the step response shown in Fig. 7.4-9a is quite good, the design is not robust; if the DC gain is not exactly known due to modeling inaccuracies, the steady-state error will be nonzero. Figure 7.4-9b shows the step response when k_f was selected as 16. Note that the final value of $\omega(t)$ is 0.9, and not the desired value of 1.

It is worth noting that the tracker in Fig. 7.4-8 is not of a convenient form since the tracking error is not used for feedback. See Example 4.3-1. Thus, the design technique in section 4.2, where a controller structure is chosen and the control gains then selected in an optimal fashion, is usually more suitable. We shall extend this approach to discrete systems in the next subsection.

■

Linear Quadratic Tracker with Output Feedback

We have just shown a straightforward technique for tracker design that amounts to finding stabilizing gains for the system (i.e., solving the regulator problem) and then adding a feedforward term related to the DC gain to guarantee perfect tracking. Unfortunately, this technique yields a tracker whose structure can be inconvenient. In particular, we usually like to see a unity-gain outer loop so that the tracking error is apparent, explicitly contributing to the control input. The servosystem obtained by regulator redesign often does not have such a structure. Moreover, if we do not know the plant parameters exactly and so compute the DC gain incorrectly, the steady-state tracking error will be nonzero.

In this subsection we shall present an alternative tracker design technique that overcomes these problems. It is the discrete-time equivalent to section 4.2. First, we formulate the tracker problem. Then, we shall derive the design equations for finding the optimal control gain in Table 7.4-1. These equations may be solved using the software described in Appendix A. We show how to use the equations using a design example.

7.4 Discrete Design Techniques

TABLE 7.4-1 DISCRETE LQ TRACKER WITH OUTPUT FEEDBACK

System Model:

$$x_{k+1} = Ax_k + Bu_k + Er_k$$

$$y_k = Cx_k + Fr_k$$

Control:

$$u_k = -Ky_k$$

Performance Index:

$$J = \frac{1}{2} \sum_{k=0}^{\infty} (\tilde{x}_k^T Q \tilde{x}_k + \tilde{u}_k^T R \tilde{u}_k) + \frac{1}{2} \bar{e}^T V \bar{e} + \frac{1}{2} \sum_i \sum_j g_{ij} k_{ij}^2,$$

with k_{ij} the elements of K.

Optimal Output Feedback Gain:

$$0 = \frac{\partial H}{\partial S} = A_c^T P A_c - P + Q + C^T K^T R K C \quad (7.4.62)$$

$$0 = \frac{\partial H}{\partial P} = A_c S A_c^T - S + X \quad (7.4.63)$$

$$0 = \frac{1}{2}\frac{\partial H}{\partial K} = RKCSC^T - B^T P A_c S C^T + B^T (A_c - I)^{-T}(P + H^T V H)\bar{x}\bar{y}^T$$

$$- B^T (A_c - I)^{-T} H^T V r \bar{y}^T + g*K \quad (7.4.64)$$

with r_k a unit step of magnitude r and:
$g*K$ a matrix with elements $g_{ij}k_{ij}$

$$\bar{x} = -(A_c - I)^{-1} B_c r$$

$$\bar{y} = C\bar{x} + Fr$$

$$X = \bar{x}\bar{x}^T$$

$$A_c = A - BKC, \qquad B_c = E - BKF$$

Optimal Cost:

$$J = \frac{1}{2} tr(PX) + \frac{1}{2}\bar{e}^T V \bar{e} + \frac{1}{2} \sum \sum g_{ij} k_{ij}^2$$

where:

$$\bar{e} = [I + H(A_c - I)^{-1} B_c] r$$

Formulation of the tracker problem

Let us assume that a compensator of desired structure has been selected. For instance, to achieve perfect steady-state tracking of a unit step command, we may incorporate the discrete integrator (7.4.17) into the compensator. Then, as we have seen, the discrete-time equations of the plant-plus-compensator may be written as

$$x_{k+1} = Ax_k + Bu_k + Er_k \qquad (7.4.40)$$

$$y_k = Cx_k + Fr_k \qquad (7.4.41)$$

$$z_k = Hx_k, \qquad (7.4.42)$$

with $x_k \in \mathbf{R}^n$ the state, $u_k \in \mathbf{R}^m$ the plant input, r_k the reference input, $y_k \in \mathbf{R}^p$ the measured output, and z_k the performance output.

The control objective is to make small the tracking error given by

$$e_k = r_k - z_k. \qquad (7.4.43)$$

For design purposes, we shall assume that r_k is a unit step of magnitude r. This is motivated by the fact that many system performance parameters, such as speed of response, percent overshoot, and so on, are given in terms of the step response. The optimal design determined using this assumption will then result in tracking of any arbitrary command r_k with good time-response behavior.

The plant control input is of the form

$$u_k = -Ky_k = -KCx_k - KFr_k, \qquad (7.4.44)$$

which has a feedback portion and a feedforward portion. Thus, the design amounts to selecting the gain K for suitable performance. One should appreciate the fact that the power of using *output feedback* design as opposed to state feedback design is that the structure of the control system is retained. Moreover, since K appears in the feedback and feedforward portions of (7.4.44), this approach amounts to determining K so that both the closed-loop poles and zeros are placed in optimal locations.

Using (7.4.44) in (7.4.40) yields the closed-loop system

$$x_{k+1} = (A - BKC)x_k + (E - BKF)r_k \equiv A_c x_k + B_c r_k.$$

Denoting the steady-state values by overbars, the steady-state response is (note that at steady-state $x_{k+1} = x_k = \bar{x}$)

$$\bar{x} = -(A_c - I)^{-1} B_c r. \qquad (7.4.45)$$

LQ performance index

To influence the step response, let us define the deviations (7.4.22) and note that

$$\begin{aligned} \tilde{u}_k &= u_k - \bar{u} = -KCx_k - KFr - (-KC\bar{x} - KFr) \\ &= -KC\tilde{x}_k. \end{aligned} \qquad (7.4.46)$$

The steady-state error in response to a unit step of magnitude r is

$$\bar{e} = r - \bar{z} = r - H\bar{x} = [I + H(A_c - I)^{-1}B_c]r. \qquad (7.4.47)$$

The error deviation is

$$\begin{aligned} \tilde{e}_k &= e_k - \bar{e} = (r - Hx_k) - (r - H\bar{x}) = -H\tilde{x}_k \\ &= -\tilde{z}_k, \end{aligned} \qquad (7.4.48)$$

7.4 Discrete Design Techniques

where the performance output deviation is

$$\tilde{z}_k = z_k - \bar{z} = Hx_k - H\bar{x} = H\tilde{x}_k. \tag{7.4.49}$$

To make small the tracking error $e_k = \tilde{e}_k + \bar{e}$, we may weight both $\tilde{e}_k$ and $\bar{e}$ in a quadratic performance index (PI). Thus, we propose the PI

$$J = \frac{1}{2}\sum_{k=0}^{\infty}(\tilde{e}_k^T\tilde{e}_k + \tilde{u}_k^T R\tilde{u}_k) + \frac{1}{2}\bar{e}^T V\bar{e} + \frac{1}{2}\sum_i\sum_j g_{ij}k_{ij}^2. \tag{7.4.50}$$

$$J = \frac{1}{2}\sum_{k=0}^{\infty}(\tilde{x}_k^T Q\tilde{x}_k + \tilde{u}_k^T R\tilde{u}_k) + \frac{1}{2}\bar{e}^T V\bar{e} + \frac{1}{2}\sum_i\sum_j g_{ij}k_{ij}^2. \tag{7.4.51}$$

where $Q = H^T H$, $R > 0$, and $V \geq 0$ is the steady-state error weighting matrix. It is generally acceptable to select $R = \rho I$, $V = vI$, with ρ and v scalar design parameters.

If each channel of the compensator contains a discrete integrator (or, more generally, a pole at $z = 1$), then the system is of Type I and the steady-state error will automatically be equal to zero. Then, we may select $V = 0$ in the PI. Otherwise, to ensure small steady-state error we should select nonzero V. The usefulness of this approach is that small steady-state errors can be guaranteed even when some channels are of Type 0.

The last term in the PI weights the elements k_{ij} of the control gain matrix K. The motivation is to allow for more structure in the control system. Thus, if error component number one should not influence control input number two, the appropriate entry of K may be made as small as desired by selecting its weight g_{ij} large in the PI. Then, in an implementation, this gain element can be set to zero.

It is easy to show that the closed-loop dynamics of the deviation are

$$\tilde{x}_{k+1} = A_c\tilde{x}_k. \tag{7.4.52}$$

Using (7.4.46) in (7.4.51) yields

$$J = \frac{1}{2}\sum_{k=0}^{\infty}\tilde{x}_k^T(Q + C^T K^T RKC)\tilde{x}_k + \frac{1}{2}\bar{e}^T V\bar{e} + \frac{1}{2}\sum\sum g_{ij}k_{ij}^2. \tag{7.4.53}$$

The design problem is now to determine the output-feedback gain K to minimize the cost (7.4.53) subject to (7.4.52).

Note that weighting $\tilde{e}_k$ and $\bar{e}$ separately in the PI is suboptimal in the sense that minimizing J does not minimize a quadratic function of the total error e_k. It does, though, generally result in excellent tracker designs.

An expression for the optimal cost

Suppose we can find a constant matrix $P > 0$ such that

$$\tilde{x}_{k+1}^T P\tilde{x}_{k+1} - \tilde{x}_k^T P\tilde{x}_k = -\tilde{x}_k^T(Q + C^T K^T RKC)\tilde{x}_k^T. \tag{7.4.54}$$

Then

$$J = -\frac{1}{2}\sum_{k=0}^{\infty}(\tilde{x}_{k+1}^T P\tilde{x}_{k+1} - \tilde{x}_k^T P\tilde{x}_k) + \frac{1}{2}\bar{e}^T V\bar{e} + \frac{1}{2}\sum\sum g_{ij}k_{ij}^2.$$

However, the sum telescopes, and if the closed-loop system is an asymptotically stable servo so that $\tilde{x}_k$ goes to zero with k, then

$$J = \frac{1}{2}\tilde{x}_0^T P \tilde{x}_0 + \frac{1}{2}\bar{e}^T V \bar{e} + \frac{1}{2}\sum\sum g_{ij}k_{ij}^2. \tag{7.4.55}$$

To find such a P, use (7.4.52) in conjunction with (7.4.54) to write

$$0 = \tilde{x}_k^T (A_c^T P A_c - P + Q + C^T K^T R K C)\tilde{x}_k. \tag{7.4.56}$$

Since this must hold for all initial conditions and hence all state trajectories, it is equivalent to

$$0 = g \equiv A_c^T P A_c - P + Q + C^T K^T R K C. \tag{7.4.57}$$

Thus, for any fixed value of K, if we determine the solution P to this equation, then the value of J is given by (7.4.55).

To determine the correct value of the initial deviation $\tilde{x}_0$ to use in (7.4.55), note that the plant starts at rest so that $x_0 = 0$. Therefore, according to (7.4.22)

$$\tilde{x}_0 = -\bar{x} \tag{7.4.58}$$

so that (7.4.55) becomes

$$\begin{aligned}J &= \frac{1}{2}\bar{x}^T P \bar{x} + \frac{1}{2}\bar{e}^T V \bar{e} + \frac{1}{2}\sum\sum g_{ij}k_{ij}^2 \\ &= \frac{1}{2}\operatorname{tr}(PX) + \frac{1}{2}\bar{e}^T V \bar{e} + \frac{1}{2}\sum\sum g_{ij}k_{ij}^2\end{aligned} \tag{7.4.59}$$

where tr( ) is the matrix trace and

$$X = \bar{x}\,\bar{x}^T \tag{7.4.60}$$

with the steady-state value $\bar{x}$ given by (7.4.45).

The upshot of these manipulations is that we may determine the optimal feedback gain K by minimizing (7.4.59) subject to the algebraic constraint (7.4.57). This minimization problem may be solved using one of the numerical techniques in, for instance Press et al. [1986]. Good results have been obtained for small values of mp using a SIMPLEX routine (note that K has mp entries). Equation (7.4.57) may be solved for P using a subroutine in ORACLS [Armstrong 1980] or using MATLAB. See Appendix A for a description of the software used for design in this book.

Finding the optimal gains using gradient algorithms

To find the optimal control gain K using faster gradient-based routines such as the Davidon-Fletcher-Powell algorithm [Press et al. 1986], we need to know $\partial J/\partial K$. This gradient may be determined as follows.

7.4 Discrete Design Techniques

Define the Hamiltonian

$$H = tr(PX) + tr(gS) + \frac{1}{2}\bar{e}^T V \bar{e} + \frac{1}{2}\sum\sum g_{ij}k_{ij}^2. \quad (7.4.61)$$

with g the function in (7.4.57) and S an undetermined Lagrange multiplier. Now, by setting all partial derivatives equal to zero, we may select K to minimize H. Necessary conditions for a minimum are given in Table 7.4-1 (see the problems). They should be compared to the continuous-time design equations in Table 4.2-1.

To use these equations in a gradient algorithm to find K, for each choice of K solve (7.4.62) and (7.4.63) for P and S, then use these intermediate quantities to evaluate J and the gradient $\partial J/\partial K$, which is equal to $\partial H/\partial K$ for this choice of P and S.

To find the optimal gain K using a non-gradient-based algorithm such as SIMPLX, equations (7.4.63) and (7.4.64) are not needed.

A design consideration

Any minimization algorithm will require the selection of an initial stabilizing gain. Such a gain may or may not be easy to find. One reliable, though tedious, way to find a stabilizing gain is to use discrete root locus techniques by closing one loop at a time in the control system. The gain can then be optimized using the design equations in Table 7.4-1. Another technique for finding a stabilizing K is given in Halyo and Broussard [1981].

As in the continuous-time case (see section 4.2), to find suitable gains K it will generally be necessary for $(\sqrt{Q}, A)$ to be detectable and (A, B) to be controllable. The controllability condition is generally satisfied in well-designed problems. The detectability of $(\sqrt{Q}, A)$ is guaranteed by its observability, which is easier to check that detectability, since it is equivalent to the full rank of the observability matrix. However, we would often like to select $Q = H^T H$ to weight $\bar{e}_k^T \bar{e}_k$ in the PI, and it is common for (H, A) not to be detectable, especially if there is a discrete integrator in the feedforward path. Thus, we may be forced to add extra state weighting terms in the PI purely due to the detectability requirement. Specifically, it is found that good results are usually obtained if we weight $\bar{e}_k$ and the integrator states in the PI.

A discrete-time version of time-dependent weighting, presented in section 4.2 for continuous-time systems, does exist [Fukuta and Tamura 1984]. This allows one to find a suitable gain K without needing to consider the observability issue. However, the equations are complicated and are omitted due to space limitations.

Example 7.4-3: Digital Inverted Pendulum Controller via Discrete LQ Design

In Example 4.2-4 we designed a continuous-time controller for an inverted pendulum on a cart. In Examples 5.2-2, 5.3-1, and 5.3-2 we converted this continuous-time controller into a digital controller using approximation methods. Here, we should like to design a digital controller for the inverted pendulum using the exact approach of direct discrete-time design.

Using BLT redesign of the continuous controller, we were able to obtain good responses using a digital controller in Example 5.2-2 with a sample period of $T = 0.01$ sec.

Using redesign of modified continuous-time controller in Example 5.3-2 we were able to obtain good responses for $T = 0.1$ sec. In this example we shall demonstrate that using direct discrete design also allows one to obtain good responses for $T = 0.1$ sec.

a. Discrete Inverted Pendulum Dynamics

The continuous-time pendulum dynamics are given in Example 4.2-4. Using the simple program DISC in Appendix A to compute the ZOH/step invariant sampled dynamics (see section 7.1) with a sampling period of $T = 0.1$ sec we obtain

$$x_{k+1} = \begin{bmatrix} 1.054386 & 0.101806 & 0 & 0 \\ 1.097473 & 1.054386 & 0 & 0 \\ -0.004944 & -0.000164 & 1 & 0.1 \\ -0.099770 & -0.004944 & 0 & 1 \end{bmatrix} x_k + \begin{bmatrix} -0.001009 \\ -0.020361 \\ 0.001001 \\ 0.020033 \end{bmatrix} u_k \quad (1)$$

$$z_k = \begin{bmatrix} 57.2958 & 0 & 0 & 0 \\ 0 & 0 & 1 & 0 \end{bmatrix} x_k, \quad (2)$$

The state is $x_k = [\theta_k \; \dot{\theta}_k \; p_k \; \dot{p}_k]^T$ and the performance output is $z_k = [\theta_k \; p_k]^T$, with the factor 57.2958 yielding θ_k in degrees.

The continuous system has poles at $s = 0, 0, 3.28, -3.28$. The discrete system has poles at

$$z = 1, 1, 1.3886, 0.7201, \quad (3)$$

which corresponds to the transformation $z = e^{sT}$.

b. Digital Controller Structure

To obtain the structure of the digital controller we simply map the poles of the continuous compensator in Example 4.2-4 using $z = e^{sT}$, obtaining the digital controller shown in Fig. 7.4-11. The digital compensator poles are at

$$z = e^{-10T} = 0.368. \quad (4)$$

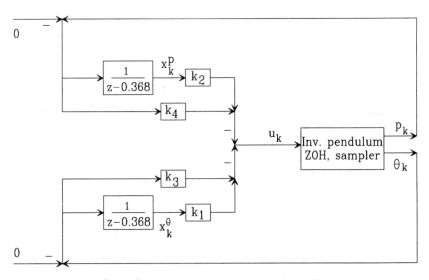

Figure 7.4-11 Inverted pendulum digital controller

7.4 Discrete Design Techniques

Note that the compensator zeros depend on the gains k_i. Thus, as in Example 4.2-4, the compensator zeros will be determined in an optimal fashion by selecting the control gains k_i using discrete LQ techniques.

The discrete dynamics of the plant-plus-ZOH-plus-sampler and the compensators as shown in Fig. 7.4-11 may be written as

$$x_{k+1} = \begin{bmatrix} 1.054386 & 0.101806 & 0 & 0 & 0 & 0 \\ 1.097473 & 1.054386 & 0 & 0 & 0 & 0 \\ -0.004944 & -0.000164 & 1 & 0.1 & 0 & 0 \\ -0.099770 & -0.004944 & 0 & 1 & 0 & 0 \\ 1 & 0 & 0 & 0 & 0.368 & 0 \\ 0 & 0 & 1 & 0 & 0 & 0.368 \end{bmatrix} x_k \quad (5)$$

$$+ \begin{bmatrix} -0.001009 \\ -0.020361 \\ 0.001001 \\ 0.020033 \\ 0 \\ 0 \end{bmatrix} u_k + \begin{bmatrix} 0 & 0 \\ 0 & 0 \\ 0 & 0 \\ 0 & 0 \\ 0 & -1 \\ -1 & 0 \end{bmatrix} r_k$$

$$y_k = \begin{bmatrix} 0 & 0 & 0 & 0 & 57.2958 & 0 \\ 0 & 0 & 0 & 0 & 0 & 1 \\ 57.2958 & 0 & 0 & 0 & 0 & 0 \\ 0 & 0 & 1 & 0 & 0 & 0 \end{bmatrix} x_k + \begin{bmatrix} 0 & 0 \\ 0 & 0 \\ 0 & -1 \\ -1 & 0 \end{bmatrix} r_k \quad (6)$$

where the state and output have been redefined as

$$x_k = [\theta_k \; \dot{\theta}_k \; p_k \; \dot{p}_k \; x_k^\theta \; x_k^p]^T \quad (7)$$

$$y_k = [x_k^\theta \; x_k^p \; e_k^\theta \; e_k^p]^T$$

and the reference input is $r_k = [r_k^p \; r_k^p]^T$, which is zero in this case. The tracking error is

$$e_k = z_k - r_k \equiv \begin{bmatrix} e_k^\theta \\ e_k^p \end{bmatrix}. \quad (8)$$

We denote this augmented system as

$$x_{k+1} = Ax_k + Bu_k + Er_k \quad (9)$$

$$y_k = Cx_k + Fr_k. \quad (10)$$

The control input is now expressed as

$$u_k = -Ky_k = -k_1 x_k^\theta - k_2 x_k^p - k_3 \theta_k - k_4 p_k, \quad (11)$$

as in Fig. 7.4-11. These equations are exactly of the form required for Table 7.4-1.

c. Design

We wish to keep the state near zero to stabilize the pendulum in a vertical position with the cart at a specified position. Thus, we select the PI

$$J = \frac{1}{2} \sum_{k=0}^{\infty} (x_k^T Q x_k + \rho u_k^T u_k). \quad (12)$$

Since the reference input r_k is zero, this is a regulator problem so that the state and control deviations are equal to the actual state and control values. That is, in Table 7.4-1 we have $x_k = \tilde{x}_k$, $u_k = \tilde{u}_k$.

To find the optimal gain K in Table 7.4-1 we used the software described in Appendix A. Recall from section 7.2 that the continuous state-weighting matrices (see Example 4.2-4) should be multiplied by the sampling period T to obtain a sensible weighting matrix for discrete design. After a few design iterations using different values for Q and ρ, we found that suitable values were $Q = \text{diag}\{10, 10, 1, 1, 0.1, 0.1\}$, $\rho = 0.01$. The corresponding control gains were

$$K = -[1.294 \quad 10.02 \quad -3.648 \quad -16.94] \tag{13}$$

and the closed-loop poles were at

$$\begin{aligned} z = \;& 0.37 \;, 0.72 \\ & 0.82 \pm j0.29 \\ & 0.96 \pm j0.05 \end{aligned} \tag{14}$$

d. Controller Simulation

To simulate the digital controller one may use program DIGCTL in Appendix A. Examining Fig. 7.4-11, the difference equations describing the digital controller are

$$x_k^\theta = 0.368 x_{k-1}^\theta + \theta_{k-1} \tag{15}$$

$$x_k^p = 0.368 x_{k-1}^p + p_{k-1} \tag{16}$$

$$u_k = -(k_1 x_k^\theta + k_3 \theta_k + k_2 x_k^p + k_4 p_k). \tag{17}$$

A subroutine implementing this digital controller for use with the driver program DIGCTL appears in Fig. 7.4-12. It is also necessary to create a file for DIGCTL that contains the continuous-time dynamics of the inverted pendulum as given in Example 4.2-4.

The resulting response to an initial condition of $\theta_0 = 0.1$ rad $\approx 6°$, $p_0 = 0.1$ m is shown in Fig. 7.4-13a. The control force $u(t)$ is shown in Fig. 7.4-13b.

```
C    Digital Inverted Pendulum Controller - via direct design
C
     SUBROUTINE DIG(IT,T,X)
     REAL X(*), K(4)
     COMMON/CONTROL/u
     COMMON/OUTPUT/theta, pos, uplot
C    uplot is used for plotting purposes only
     DATA (K(I), I= 1,4)/ 1.294, 10.02, -3.648, -16.94/
C
     xt= 0.368*xt + thetam1
     xp= 0.368*xp + posm1
     u= -( K(1)*xt + K(3)*theta + K(2)*xp + K(4)*pos )
     uplot= u
C
     thetam1= theta
     posm1= pos
C
     RETURN
     END
```

Figure 7.4-12 Subroutine implementing digital controller

7.4 Discrete Design Techniques

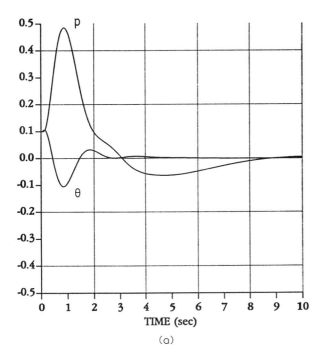

(a)

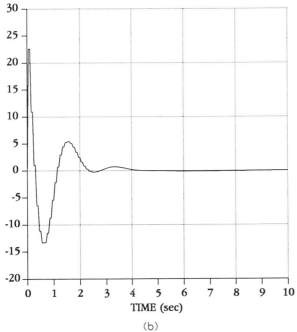

(b)

Figure 7.4-13 Response of inverted pendulum digital control system. (a) Rod angle $\theta(t)$ (rads) and cart position $p(t)$ (m). (b) Control input $u(t)$.

These results are far better than those obtained for $T = 0.1$ sec using BLT digital controls design in Example 5.2-2; indeed, they compare to the results in Example 5.3-2. However, in this example using direct discrete design we did not need to augment the plant with dynamics to approximate the ZOH, since (5) already contains the ZOH. Moreover, we did not need to perform the BLT on each component of the continuous controller to digitize it. This procedure has been replaced here by discretizing the plant using the computer program DISC to obtain (5).

It is quite interesting to note that the control magnitude required in this example is much less than that required in either Example 4.2-4, where continuous control was used, or the digital control examples in Chapter 5. This appears to be a major advantage of direct discrete-time design using large sample periods.

To verify more rigorously this digital controller, it should be simulated on the full nonlinear dynamics of the inverted pendulum (see the problems for section 2.1). This is easily accomplished using the software in Appendix A (see the problems). One would anticipate good performance on the actual nonlinear system since the excursions in the angle $\theta(t)$ are limited to ± 0.1 rad, and the nonlinearities occur in terms like $\sin\theta$ and $\cos\theta$. Note that $\sin 0.1 \approx 0.1$, so that for such small angles the linear approximation holds.

The digital controller may be implemented on an actual pendulum using the procedure given in section 6.6.

∎

Linear Quadratic Regulator with Output Feedback

In the discrete LQ regulator problem, where it is only required to drive the performance output z_k to zero, the equations in Table 7.4-1 simplify and a convergent solution algorithm exists. If the design is a sensible one and a reasonable PI is selected, the discrete LQ regulator with output feedback usually results in a stable closed-loop system with desirable poles. That is, for a multi-input/multi-output plant, the closed-loop system is stable for any reasonable choice of weighting matrices Q and R. Contrast this with classical design, where closed-loop stability in MIMO plants can be difficult and tedious to achieve.

For regulation, the reference command r_k is equal to zero and the plant becomes

$$x_{k+1} = Ax_k + Bu_k \quad (7.4.65)$$

$$y_k = Cx_k \quad (7.4.66)$$

$$z_k = Hx_k \quad (7.4.67)$$

with z_k the performance output and y_k the measured output. The control input is

$$u_k = -Ky_k \quad (7.4.68)$$

with K to be determined.

Since $r_k = 0$, according to (7.4.45), the steady-state value $\bar{x}$ is also zero and all terms involving $\bar{x}$ in Table 7.4-1 vanish. The state deviation $\tilde{x}_k$ is just the state x_k itself.

In the regulator problem, the initial state x_0 is not generally zero, for regulation means bringing nonzero initial states to zero with desirable time-response characteristics.

7.4 Discrete Design Techniques

It is reasonable to assume that the initial state is unknown, but that its initial mean-square value is known. That is

$$E\{x_0 x_0^T\} = X \qquad (7.4.69)$$

with $E\{\ \}$ the expected value and X the known initial mean-square state. The traditional assumption is that x_0 is uniformly distributed about zero so that $X = I$ [Levine and Athans 1970].

In this setting it is necessary to minimize not (7.4.51) but its expected value

$$J = \frac{1}{2} E\left[\sum_{k=0}^{\infty} (x_k^T Q x_k + u_k^T R u_k)\right] \qquad (7.4.70)$$

with $R > 0$ and $Q \geq 0$. To force z_k to zero we may select $Q = H^T H$.

Thus, the optimal value of the PI (7.4.55) may be determined using

$$J = \frac{1}{2} E\{x_0^T P x_0\} = \frac{1}{2} tr(PX) \qquad (7.4.71)$$

with P the solution to (7.4.57) and X given.

Taking all this into account, for the LQR problem the design equations of Table 7.4-1 reduce to those in Table 7.4-2.

A convergent algorithm for determining the optimal output feedback gain K is given in Halyo and Broussard [1981]. It works as long as there exists a gain K such that A_c is stable. It is similar to the continuous-time algorithm given in Table 4.1-2.

Example 7.4-4: Multivariable Digital Controller Using Discrete LQR Design

In Examples 4.1-1 and 4.2-1 we designed a continuous controller for a multivariable circuit. Let us design a digital controller for this circuit.

a. Discrete Dynamics

Using program DISC in Appendix A to discretize the continuous dynamics using step invariance with $T = 0.1$ sec yields

$$x_{k+1} = A x_k + B u_k \qquad (1)$$

$$y_k = C x_k \qquad (2)$$

with

$$A = \begin{bmatrix} 0.90031 & -0.00015 & 0.09048 & -0.00452 \\ -0.00015 & 0.90031 & 0.00452 & -0.09048 \\ -0.09048 & -0.00452 & 0.90483 & -0.09033 \\ 0.00452 & 0.09048 & -0.09033 & 0.90483 \end{bmatrix},$$

$$B = \begin{bmatrix} 0.00468 & -0.00015 \\ 0.00015 & -0.00468 \\ 0.09516 & -0.00467 \\ -0.00467 & 0.09516 \end{bmatrix} \qquad (3)$$

$$C = \begin{bmatrix} 1 & 1 & 0 & 0 \\ 0 & 1 & 0 & 0 \end{bmatrix}.$$

TABLE 7.4-2 DISCRETE LQR WITH OUTPUT FEEDBACK

System Model:

$$x_{k+1} = Ax_k + Bu_k$$

$$y_k = Cx_k$$

Control:

$$u_k = -Ky_k$$

Performance Index:

$$J = \frac{1}{2}E\left[\sum_{k=0}^{\infty}(x_k^T Q x_k + u_k^T R u_k)\right]$$

Optimal Output Feedback Gain:

$$0 = \frac{\partial H}{\partial S} = A_c^T P A_c - P + Q + C^T K^T R K C \quad (7.4.72)$$

$$0 = \frac{\partial H}{\partial P} = A_c S A_c^T - S + X \quad (7.4.73)$$

$$0 = \frac{1}{2}\frac{\partial H}{\partial K} = RKCSC^T - B^T P A_c SC^T \quad (7.4.74)$$

where:

$$A_c = A - BKC$$

$$X = E\{x_0 x_0^T\}$$

Optimal Cost:

$$J = \frac{1}{2}tr(PX)$$

The poles of this system are at

$$z = 0.9477 \pm j0.0823 \quad (4)$$
$$0.8575 \pm j0.0744.$$

It is worth verifying that these correspond to the continuous poles in Example 4.1-1 mapped according to $z = e^{sT}$.

The feedback we are interested in is of the form

$$u_k = -Ky_k. \quad (5)$$

Thus, this is a 2-input/2-output design problem with four gains to determine.

7.4 Discrete Design Techniques

b. PI and Controls Design

To speed up the response of the system and achieve good regulation, let us try the PI

$$J = \frac{1}{2} \sum_{k=0}^{\infty} (x_k^T Q x_k + u_k^T R u_k). \tag{6}$$

We used the design equations in Table 7.4-2 to determine the output feedback gain K that minimizes $E\{J\}$.

After a few design iterations we found that good responses were given using $Q = \text{diag}\{0.1, 0.2, 0, 0\}$ and $R = \text{diag}\{1.E\text{-}6, 1.E\text{-}4\}$. The corresponding control gains were

$$K = \begin{bmatrix} 1.457 & -3.953 \\ -0.009 & -8.800 \end{bmatrix} \tag{7}$$

and the closed-loop poles were

$$\begin{aligned} z &= 0.91 \pm j0.16 \\ & 0.87 \pm j0.27. \end{aligned} \tag{8}$$

c. Simulation

The subroutine in Fig. 7.4-14 contains the digital controller for use with program DIGCTL in Appendix A. The time response for nonzero initial conditions (all components of $x(0)$ were set equal to 1) is shown in Fig. 7.4-15a. The control signals are shown in Fig. 7.4-15b; note the coordinated control action that is achieved in this multivariable system using LQ design.

For comparison purposes, the open-loop response is shown in Fig. 7.4-16. It should be mentioned that this is an extremely stubborn circuit for design; it is quite difficult to improve the performance using static output feedback. However, the settling time of the system has been halved using the digital controller.

It is easy to incorporate gain element weighting in the PI exactly as in the continuous

```
C     Digital Controller for MV circuit
C
      SUBROUTINE DIG(IT,T,X)
      REAL X(*), K(2,2)
      COMMON/CONTROL/u(2)
      COMMON/OUTPUT/y(2)
      COMMON/PARAM/K
      DATA K(1,1),K(1,2),K(2,1),K(2,2)/1.457,-3.953,-.009,-8.8/
C
      u(1)= -K(1,1)*y(1) - K(1,2)*y(2)
      u(2)= -K(2,1)*y(1) - K(2,2)*y(2)
C
      RETURN
      END
```

Figure 7.4-14 Digital control subroutine for use with program DIGCTL

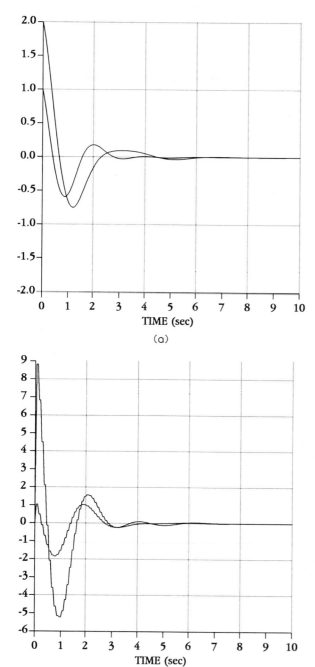

Figure 7.4-15 Response of multivariable digital control system. (a) Outputs. (b) Control inputs.

7.4 Discrete Design Techniques

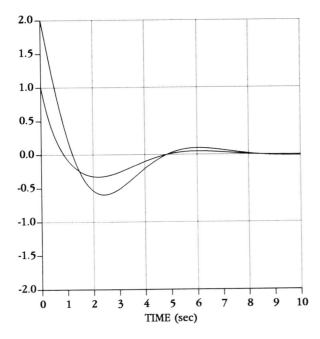

Figure 7.4-16 Open-loop response of multivariable circuit.

case (section 4.2) so that some elements of K can be set to zero, yielding more structure in the digital controller. See Table 7.4-1.

This digital controller could be implemented using the techniques in section 6.6. ∎

Linear Quadratic Regulator with State Feedback

If full state feedback

$$u_k = -Kx_k \qquad (7.4.75)$$

is allowed, then the design equations in Table 7.4-2 simplify even further and the familiar fundamental results of modern optimal control theory are recovered.

For state feedback, we may set $C = I$ in Table 7.4-2 so that equations (7.4.72) and (7.4.74) become

$$0 = A_c^T P A_c - P + Q + K^T R K \qquad (7.4.76)$$

$$0 = RKS - B^T P A_c S \qquad (7.4.77)$$

with

$$A_c = A - BK. \qquad (7.4.78)$$

Control weighting matrix R is nonsingular. If X is nonsingular and A_c is stable, then the solution S to (7.4.73) is nonsingular. Then, (7.4.77) may be solved for the feedback gain K by writing

$$0 = RK - B^TP(A - BK)$$
$$(R + B^TPB)K = B^TPA \qquad (7.4.79)$$
$$K = (R + B^TPB)^{-1}B^TPA.$$

Using (7.4.76) we may write

$$0 = (A - BK)^TP(A - BK) - P + Q + K^TRK$$
$$= A^TPA - P + Q + K^T(R + B^TPB)K - A^TPBK - K^TB^TPA,$$

whence using (7.4.79) and simplifying yields

$$0 = A^TPA - P + Q - A^TPB(R + B^TPB)^{-1}B^TPA. \qquad (7.4.80)$$

The importance of this development is that (7.4.80) does not involve the gain K. It may be solved for P, which then gives K through (7.4.79). We call (7.4.80) the *discrete algebraic Riccati equation*; it is a matrix quadratic equation.

The discrete LQR with state-variable feedback appears in Table 7.4-3.

Note that, since S is not needed to find K, it is not necessary to know X. Thus, for LQ design using full state feedback, it is not required to know any information about the initial state to find K.

Consider the recursive Riccati equation

$$P_k = A^T[P_{k+1} - P_{k+1}B(R + B^TP_{k+1}B)^{-1}B^TP_{k+1}]A + Q, \qquad (7.4.81)$$

TABLE 7.4-3 DISCRETE LQR WITH STATE FEEDBACK

System Model:
$$x_{k+1} = Ax_k + Bu_k$$

Control:
$$u_k = -Kx_k$$

Performance Index:
$$J = \frac{1}{2}\sum_{k=0}^{\infty}(x_k^TQx_k + u_k^TRu_k)$$

Optimal State Feedback Gain:
$$0 = A^TPA - P + Q - A^TPB(R + B^TPB)^{-1}B^TPA. \qquad (7.4.82)$$
$$K = (R + B^TPB)^{-1}B^TPA. \qquad (7.4.83)$$

Optimal Cost:
$$J = \frac{1}{2}x_0^TPx_0$$

Problems for Chapter 7

which develops backward in the time index k. The starting value of P_N (for some large value of $k = N$) must be selected. One of the fundamental results of modern control theory is the following, which tells when (7.4.81) can be used to determine the solution P to the algebraic Riccati equation, as well as when the closed-loop system is stable [Lewis 1986].

Theorem. Let H be any root of Q so that $Q = H^T H$. Suppose (H, A) is observable. Then (A, B) is stabilizable if and only if:
1. There exists a unique positive definite solution P to the algebraic Riccati equation.
2. There exists a unique positive definite limiting solution P_∞ to the recursive equation (7.4.81) that is independent of P_N. Moreover, P_∞ is equal to P.
3. The closed-loop plant $(A - BK)$ is asymptotically stable, where K is given by (7.4.83).

Under the hypotheses of the theorem, P may be determined by iterating the recursive Riccati equation using any starting value P_N. The identity matrix is a good choice for P_N.

The gain K is called the *Kalman gain*. The importance of this theorem is that, under reasonable assumptions, the Kalman gain stabilizes the plant. Thus, the LQR offers a convenient way to stabilize any multivariable plant. This *guaranteed stability* is an essential feature of modern control; as the design parameters Q and R are tuned in an iterative design technique to obtain suitable time responses, the closed-loop plant will always be stable.

This should be contrasted with the situation where the entries of K are treated as design parameters and tuned directly; that approach does not even guarantee stability. Indeed, for complicated multi-input/multi-output or multiloop systems, finding a stabilizing gain by varying the elements of K is a difficult task. This accounts for the limited success of root locus type techniques in multivariable design.

The strong statements of the theorem do not hold in the case of output feedback. (Some results for stochastic systems do appear in Halyo and Broussard [1981].) However, as long as $R > 0$, (H, A) is observable, $(A - BKC)$ is stabilizable by output feedback, it is found that $(A - BKC)$ is generally stable when K is determined as in Table 7.4-1.

An LQ optimal technique for computing state feedback gains that includes a computation delay of T sec is given in Mita [1985].

PROBLEMS FOR CHAPTER 7

Problems for Section 7.1

7.1-1 The state transition matrix of a time-varying state system satisfies the property $x(t) = \phi(t, \tau)x(\tau)$. Derive (7.1.16), (7.1.17).

7.1-2 Work through Example 7.1-6 to convince yourself of the claims there.

7.1-3 Discretization of Continuous-Time System. A continuous-time system is given by

$$\dot{x} = \begin{bmatrix} 0 & 1 \\ -2 & -3 \end{bmatrix} x + \begin{bmatrix} 0 \\ 1 \end{bmatrix} u, \qquad y = [1 \ 0]x.$$

   a. Discretize the system using a sample period of $T = 0.1$ sec to find A^s, B^s.
   b. Find the transfer function of the discrete system from A^s, B^s, and C. Now find the sampled transfer function from the PFE of the continuous-time transfer function.

7.1-4 Discretization of System with Delay. Repeat Problem 7.1-3 if there is a control delay of:
   a. 0.05 sec.
   b. 0.15 sec.

7.1-5 Discretization of System with Delay. A system with delay has the transfer function

$$H(s) = \frac{-6(s-1)}{(s+2)(s+3)} e^{-0.15s}.$$

Find the discretized transfer function if $T = 0.1$ sec.

7.1-6 Damped Harmonic Oscillator with Delay
   a. Evaluate the sampled plant matrices in Example 7.1-3 if $\alpha = 1$, $\omega = 2$, $T = 0.1$ sec. Verify the relation between the poles of the continuous and the discretized systems. Find the transfer functions of the continuous and discretized systems.
   b. Redo Example 7.1-3 with these parameters if there is a control delay of 0.05 sec.

Problems for Section 7.2

7.2-1 Ackermann's Formula for Discrete Design. Consider the system of Problem 7.1-3. It is desired to find a state feedback so that the closed-loop system has a percentage overshoot of 6% and a settling time of 4 sec.
   a. **Continuous State Feedback.** Find α and ζ, and hence the desired characteristic polynomial of the continuous-time closed-loop system. Given this polynomial, use Ackermann's formula to compute the state-feedback gain $u = -Kx$ required for pole placement.
   b. **Digital State Feedback.** It is now desired to design a digital state feedback $x_k = -Ku_k$ for the desired closed-loop response. Determine the required closed-loop poles of the discretized system, and hence its desired characteristic polynomial. Given this polynomial, use Ackermann's formula to compute the discrete state-feedback gain required for pole placement.

7.2-2 Digital Control of Newton's System with Delay. Repeat Example 7.2-1 if there is a computation delay of 0.05 sec.

Problems for Section 7.3

7.3-1 Use (7.3.4) to derive Table 7.3-1.

7.3-2 For the system of Problem 7.1-3, find the zero of the discretized system as a function of T. Verify that the zero approaches $z = -1$ as T becomes small.

7.3-3 For the system of Problem 7.1-3, find the observability and reachability matrices of the discretized system as a function of T. Does the system become unobservable or unreachable for any values of T?

Problems for Chapter 7

7.3-4 Newton's System. Consider the discretized Newton's system. Does it ever lose reachability for any T? Using position measurements, is observability ever lost? Using position/velocity measurements so that $y = [1 \quad 1]x$?

Problems for Section 7.4

7.4-1 Digital Controllers for Zero Steady-State Error
   a. Use the discrete final-value theorem to show that the feedforward compensator (7.4.17) results in zero steady-state error in response to a unit step reference command.
   b. Find the digital compensator required for perfect tracking of a unit ramp.

7.4-2 Discrete Classical Design. In Example 4.2-2 a continuous PI servosystem was designed for a motor. Discretize the dynamics and use root locus techniques to design a digital servo for tracking a constant commanded angular velocity. Simulate the closed-loop response.

7.4-3 Derive the design equations in Table 7.4-1.

7.4-4 Zeros of Inverted Pendulum Digital Compensator. Determine the compensator zeros in Example 7.4-3 using the control gains found there. Using these fixed zeros and fixed compensator poles of $z = 0.368$, perform a root locus design to select the control gains k_i. Are the gains found in the example using LQ design sensible?

7.4-5 Nonlinear Simulation of Inverted Pendulum Controller. The full nonlinear equations describing the inverted pendulum were given in the problems for section 2.1. Using program TRESP in Appendix A, verify the performance on this nonlinear model of the digital controller designed in Example 7.4-3.

7.4-6 Verify that the poles in Example 7.4-4 have mapped according to $z = e^{sT}$ on ZOH discretization.

7.4-7 Digital Motor Speed Controller. Redo Problem 7.4-2 using the LQ design equations in Table 7.4-1.

7.4-8 Digital Motor Speed Controller. In Example 7.4-2 regulator redesign was used to design a digital speed controller. The controller had robustness problems. Redo the design using the LQ tracker equations in Table 7.4-1. Select a PI controller for robustness.

7.4-9 Digital Controller for Ball Balancer. In the problems for section 2.1 the equations were given for a ball balancer system. In the problems for section 4.2 a continuous LQ design for this system was performed. In the problems for section 5.2 a digital controller was designed using the BLT.
   a. Using program DISC in Appendix A, discretize the ball balancer equations.
   b. Using the same control structure as in Example 7.4-3, design a digital control system using the LQ tracker equations in Table 7.4-1.

7.4-10 Digital Command Generator Tracker Design. Derive a command generator tracker (CGT) for discrete systems along the lines of the continuous CGT in section 4.4.

PART IV

FREQUENCY-DOMAIN TECHNIQUES

Up to this point in the book, we have assumed that an exact mathematical description is available of the plant to be controlled. In addition, almost all of our design techniques were given in the *time domain*, since the modern LQ performance index is a time-domain integral. We showed a variety of means for designing control systems that guarantee optimal performance of the closed-loop system.

Unfortunately, it is usual in practical design situations to have available only an approximate mathematical description of the plant, as well as to have noises and disturbances present in the system. In this part of the book we explore modern frequency-domain techniques for multi-input/multi-output systems. Using the basic notions of *multivariable transfer relations* and the *singular value*, we show how to plot the MIMO Bode magnitude plot and use it to examine closed-loop robustness to disturbances and inexact knowledge of the plant.

8

Robust Design

SUMMARY

We have shown how to design control systems using modern time-domain techniques that yield good time-response characteristics. We assumed throughout that an exact state-variable design model was available for the plant to be controlled. However, in practice there are modeling inaccuracies that can take the form of uncertain plant parameters, unmodeled high-frequency dynamics, and so on. Moreover, there may be disturbances in the system or measurement noises, all of which have an effect on the closed-loop dynamics. In this chapter we show how to take such uncertain factors into account in order to design a robust control system that gives good performance not only for the design model, but also for the actual system with these uncertainties and errors.

INTRODUCTION

All of our work so far has assumed an exact state-variable description of the plant to be controlled, whether continuous-time or discrete-time. However, such a *design model* may provide only an approximation to the actual dynamics. For instance, an inverted pendulum may have a flexible mode at high frequencies. If this mode is excited by the control signal, these *unmodeled high-frequency dynamics* can act to destabilize the closed-loop system.

On the other hand, some of the parameters in the state equations may be inexactly known. These *structured parameter variations* are a low-frequency effect that can also act to cause instability. Therefore, it is important to design controllers that have *stability robustness*, which is the ability to guarantee closed-loop stability in spite of parameter variations and high-frequency unmodeled dynamics.

It is often important to account for disturbances and sensor measurement noises. Disturbances may act to cause unsatisfactory performance in a system that has been designed without taking them into account. Therefore, it is important to design controllers that have *performance robustness*, which is the ability to guarantee acceptable performance (in terms, for instance, of percentage overshoot, settling time, and so on) even though the system may be subject to disturbances.

Robustness issues are conveniently examined in the *frequency domain*. In single-input/single-output (SISO) systems, classical control theory has been successful in robust design using tools like Bode plots, Nyquist plots, and gain and phase margins. However, it is well known that the individual gain margins, phase margins, and sensitivities of all the SISO transfer functions in a multivariable or multiloop system have little to do with its overall robustness. This is due to the coupling that generally exists between *all* inputs and *all* outputs of a MIMO system. Thus, there have been problems in extending classical robust design notions to multi-input/multi-output (MIMO) systems.

To this point, our work in modern control has been in the *time domain*, since the LQ performance index is a time-domain criterion. Modern control techniques provide a direct way to design multi-loop controllers for MIMO systems by closing all the loops simultaneously. Time-domain performance is guaranteed in terms of minimizing a quadratic performance index (PI) which, with a sensible problem formulation, generally implies closed-loop stability as well. Both continuous-time and digital controllers are easily designed using computer software like that described in Appendix A. However, all our work so far has assumed a perfectly known plant with no disturbances, and so robustness was not considered.

In this chapter we show that the classical frequency-domain robustness measures are easily extended to MIMO systems in a rigorous fashion by using the notions of *matrix transfer relations* and the *singular value* (Appendix B). In section 2.5 we introduced the multivariable loop gain and return difference, showing that they are not scalars, as in classical control, but *matrix transfer relations*. This sort of matrix transfer function approach is the basis for MIMO frequency-domain analysis.

In section 8.1 we develop a deeper understanding of the matrix transfer relations in a closed-loop system, introducing the *multivariable sensitivity and cosensitivity*. In section 8.2 we use the singular values of the loop gain to define the *multivariable Bode magnitude plot*. In terms of this plot, we present in section 8.3 bounds that *guarantee* both robust stability and robust performance for multivariable systems, deriving notions that are entirely analogous to those in classical control.

In section 8.4 we give a design technique for robust multivariable controllers using modern output-feedback theory, showing how robustness may be guaranteed. The approach is a straightforward extension of classical techniques. To yield both suitable time-domain performance and robustness, an iterative approach is described that is simple and direct using the software described in Appendix A.

This chapter applies for both continuous-time transfer functions $G(s)$ and discrete-time transfer functions $G(z)$. The latter are used for digital robust design. The corresponding frequency-domain functions will be denoted, with some abuse in notation, by $G(j\omega)$. Thus, for continuous systems $G(j\omega) = G(s)|_{s=j\omega}$, while in the discrete-time case we evaluate $G(j\omega)$ as $G(z)|_{z=e^{j\omega T}}$.

8.1 MULTIVARIABLE LOOP GAIN AND SENSITIVITY

In this section we shall examine various matrix transfer relations in a closed-loop feedback control system. These multivariable notions are important, for the individual SISO transfer relations between the individual inputs and outputs of a MIMO system mean little from the point of view of robustness.

We review the multivariable loop gain and return difference introduced in section 2.5, and then introduce the multivariable sensitivity and cosensitivity. We shall use these in the next section to introduce the multivariable Bode magnitude plot.

A Typical Feedback System

Figure 8.1-1 shows a standard feedback system of the sort that we have seen several times in our work to date. The plant is $G(s)$, and $K(s)$ is the feedback/feedforward compensator, which can be designed by any of the techniques we have covered. The plant output is $z(t) \in \mathbf{R}^q$, the plant control input is $u(t) \in \mathbf{R}^m$, and the reference input is $r(t) \in \mathbf{R}^q$.

If we are considering a digital control system, then $G(s)$ and $K(s)$ should be replaced respectively by $G(z)$ and $K(z)$, with time functions replaced by sequences. $G(z)$ is the ZOH discrete equivalent to the continuous-time plant (section 7.1), while $K(z)$ is the digital controller.

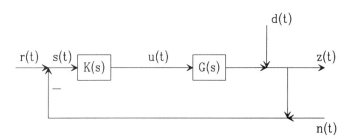

Figure 8.1-1 Standard feedback configuration

We have added a few items to the figure to characterize uncertainties. The signal $d(t)$ represents a *disturbance* acting on the system of the sort appearing in classical control. This could represent, for instance, wind gusts in an aircraft control system. The *sensor measurement noise* or errors are represented by $n(t)$. Both of these signals are generally vectors of dimension q. Typically, the disturbances occur at low frequencies, say below some ω_d, while the measurement noise $n(t)$ has its predominant effect at high frequencies, say above some value ω_n. Typical Bode plots for the magnitudes of these terms appear in Fig. 8.1-2 for the case that $d(t)$ and $n(t)$ are scalars. The reference input $r(t)$ is generally also a low-frequency signal (e.g., the unit step).

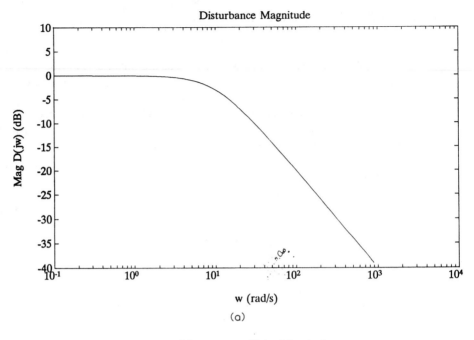

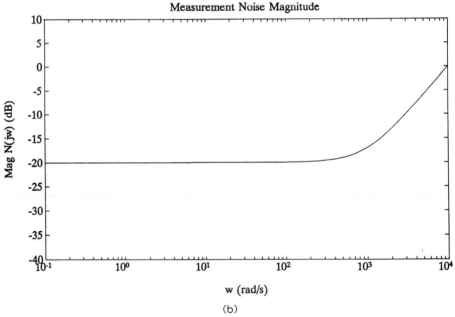

Figure 8.1-2 Typical Bode plots for the uncertain signals in the system. (a) Disturbance magnitude. (b) Measurement noise magnitude.

8.1 Multivariable Loop Gain and Sensitivity

The tracking error is

$$e(t) \equiv r(t) - z(t). \tag{8.1.1}$$

Due to the presence of $n(t)$, $e(t)$ may not be symbolized in Fig. 8.1-1. The signal $s(t)$ is in fact given by

$$s(t) = r(t) - z(t) - n(t) = e(t) - n(t). \tag{8.1.2}$$

Closed-Loop Transfer Relations

Let us perform a frequency-domain analysis on the system to see the effects of the uncertainties on system performance. This analysis holds for continuous or discrete systems, with z replacing s in the latter case. It is in the same spirit as some of the work done in section 2.5.

In terms of Laplace transforms, we may write

$$Z(s) = G(s)K(s)S(s) + D(s) \tag{8.1.3}$$

$$S(s) = R(s) - Z(s) - N(s) \tag{8.1.4}$$

$$E(s) = R(s) - Z(s). \tag{8.1.5}$$

Now we may solve for $Z(s)$ and $E(s)$, obtaining the closed-loop transfer function relations (see the problems)

$$Z(s) = (I + GK)^{-1}GK(R - N) + (I + GK)^{-1}D \tag{8.1.6}$$

$$E(s) = [I - (I + GK)^{-1}GK]R + (I + GK)^{-1}GKN - (I + GK)^{-1}D. \tag{8.1.7}$$

It is important to note that, unlike the case for SISO systems, care must be taken to perform the matrix operations in the correct order. For instance, $GK \neq KG$. The multiplications by matrix inverses must also be performed in the correct order.

We can put these equations into a more convenient form. According to the matrix inversion lemma (Appendix B), (8.1.7) may be written as

$$E(s) = (I + GK)^{-1}(R - D) + (I + GK)^{-1}GKN. \tag{8.1.8}$$

Moreover, since GK is square and invertible (even when the plant $G(s)$ is not), we can write

$$(I + GK)^{-1}GK = [(GK)^{-1}(I + GK)]^{-1} = [(GK)^{-1} + I]^{-1}$$

$$= [(I + GK)(GK)^{-1}]^{-1} = GK(I + GK)^{-1}. \tag{8.1.9}$$

Therefore, we may finally write $Z(s)$ and $E(s)$ as

$$Z(s) = GK(I + GK)^{-1}(R - N) + (I + GK)^{-1}D \tag{8.1.10}$$

$$E(s) = (I + GK)^{-1}(R - D) + GK(I + GK)^{-1}N. \tag{8.1.11}$$

Sensitivity, Cosensitivity, and Loop Gain

To simplify things a bit, define the *system sensitivity*

$$S(s) = (I + GK)^{-1} \tag{8.1.12}$$

and

$$T(s) = GK(I + GK)^{-1} = (I + GK)^{-1}GK. \tag{8.1.13}$$

Since

$$S(s) + T(s) = (I + GK)(I + GK)^{-1} = I \tag{8.1.14}$$

we call $T(s)$ the *complementary sensitivity*, or in short, the *cosensitivity*. Note that the *return difference*

$$L(s) = I + GK \tag{8.1.15}$$

is the inverse of the sensitivity. The *loop gain* is given by $G(s)K(s)$.

These expressions extend the classical notions of loop gain, return difference, and sensitivity to multivariable systems. They are generally square transfer function matrices of dimension $q \times q$.

In terms of these new quantities, we have

$$Z(s) = T(s)(R(s) - N(s)) + S(s)D(s) \tag{8.1.16}$$

$$E(s) = S(s)(R(s) - D(s)) + T(s)N(s). \tag{8.1.17}$$

Examining the latter expression, we see that to ensure small tracking errors, $S(j\omega)$ should be small at those frequencies ω where the reference input $r(t)$ and disturbance $d(t)$ are large. This will yield good *disturbance rejection*. On the other hand, for satisfactory *sensor noise rejection*, $T(j\omega)$ should be small at those frequencies ω where $n(t)$ is large.

Unfortunately, a glance at (8.1.14) reveals that $S(j\omega)$ and $T(j\omega)$ cannot simultaneously be small at any one frequency ω. According to Fig. 8.1-2, we should like to have $S(j\omega)$ small at low frequencies, where $r(t)$ and $d(t)$ dominate, and $T(j\omega)$ small at high frequencies, where $n(t)$ dominates.

These are nothing but the multivariable generalizations of the well-known SISO classical notion that a large loop gain $GK(j\omega)$ is required at low frequencies for satisfactory performance and small errors, but a small loop gain is required at high frequencies where sensor noises are present.

8.2 MULTIVARIABLE BODE PLOT

The multivariable notions just introduced are not difficult to understand on a heuristic level. Unfortunately, it is not so straightforward to determine a clear measure for the "smallness" of the sensitivity $S(j\omega)$ and cosensitivity $T(j\omega)$. These are both *square ma-*

8.2 Multivariable Bode Plot

trix transfer relations of dimension $q \times q$, with q the number of reference inputs $r(t)$. They are complex functions of the frequency ω. Clearly, the classical notion of the Bode magnitude plot, which is defined only for *scalar* complex functions of ω, must be extended to the MIMO case.

Singular Values

Some early work was done using the frequency-dependent eigenvalues of a square complex matrix as a measure of smallness [Rosenbrock 1974, MacFarlane 1970, MacFarlane and Kouvaritakis 1977]. However, note that the matrix

$$M = \begin{bmatrix} 0.1 & 100 \\ 0 & 0.1 \end{bmatrix} \quad (8.2.1)$$

has large and small components, but its eigenvalues are both at 0.1.

A better measure of the magnitude of square matrices is the *singular value* (SV) (Appendix B). Given any matrix M, we may write its *singular value decomposition* (SVD) as

$$M = U\Sigma V^*, \quad (8.2.2)$$

where U and V are square unitary matrices (i.e., $V^{-1} = V^*$, the complex conjugate transpose of V) and

$$\Sigma = \begin{bmatrix} \sigma_1 & & & & & \\ & \sigma_2 & & & & \\ & & \ddots & & & \\ & & & \sigma_r & & \\ & & & & 0 & \\ & & & & & \ddots \\ & & & & & & 0 \end{bmatrix} \quad (8.2.3)$$

with $r = \text{rank}(M)$, where σ_i are nonnegative real numbers. The singular values are the σ_i, which are ordered so that $\sigma_1 \geq \sigma_2 \geq \ldots \geq \sigma_r$. The SVD may loosely be thought of as the extension to general matrices (which may be nonsquare) of the Jordan form. If M is a function of $j\omega$, then so are U, σ_i, and V.

We note that the M given above has two singular values, namely $\sigma_1 = 100.0001$ and $\sigma_2 = 0.0001$. Thus, this measure indicates that M has a large and a small component. Indeed, note that

$$\begin{bmatrix} 0.1 & 100 \\ 0 & 0.1 \end{bmatrix} \begin{bmatrix} -1 \\ 0.001 \end{bmatrix} = \begin{bmatrix} 0 \\ 0.0001 \end{bmatrix} \quad (8.2.4)$$

while

$$\begin{bmatrix} 0.1 & 100 \\ 0 & 0.1 \end{bmatrix} \begin{bmatrix} 0.001 \\ 1 \end{bmatrix} = \begin{bmatrix} 100.0001 \\ 0.1 \end{bmatrix}. \quad (8.2.5)$$

Thus, the singular value σ_2 has the *input direction*

$$\begin{bmatrix} -1 \\ 0.001 \end{bmatrix}$$

associated with it for which the output contains the value σ_2. On the other hand, the singular value σ_1 has an associated input direction of

$$\begin{bmatrix} 0.001 \\ 1 \end{bmatrix}$$

for which the output contains the value σ_1.

There are many nice properties of the singular value which make it a suitable choice for defining the magnitude of matrix functions. Among these is the fact that the maximum singular value is an *induced matrix norm*, and norms have several useful attributes. The use of the SVs in the context of modern control was explored in Doyle and Stein [1981] and Safonov et al. [1981].

A major factor is that there are many good software packages that have good routines for computing the singular value (e.g., subroutine LSVDF in [IMSL], or MATLAB [Moler et al. 1987]). Thus, plots like those we shall present may easily be obtained by writing only a computer program to drive the available subroutines. Indeed, since the SVD uses unitary matrices, its computation is numerically stable.

Bode Magnitude Plot

The multivariable Bode magnitude plot is nothing but the plot versus frequency of the transfer function singular values. An efficient technique for obtaining the SVs of a complex matrix as a function of frequency ω is given in Laub [1981].

We note that a complete picture of the behavior of a complex matrix versus ω must take into account the SVs as well as the *multivariable phase*, which may also be obtained from the SVD [Postlethwaite et al. 1981]. Thus, complete MIMO generalizations of the Bode magnitude *and* phase plots are available. However, the theory relating to the phase portion of the plot is more difficult to use in a practical design technique, although a MIMO generalization of the Bode gain-phase relation is available [Doyle and Stein 1981]. We shall only employ plots of the SVs versus frequency, which correspond to the Bode magnitude plot for MIMO systems.

In connection with these remarks, it is worth noting that in the continuous-time case $G(j\omega) = G(s)|_{s=j\omega}$ is rational in ω, so that the Bode gain-phase relation holds. This means that, for minimum-phase plants, design may be carried out using only the Bode magnitude plot. However, for discrete-time systems $G(j\omega) = G(z)|_{z=e^{j\omega T}}$ is not rational in ω, so that this convenient feature of Bode design is lost. See section 7.4 where w-plane design is discussed. Thus, in digital design it is more important to compute the MIMO phase plot. It is found, however, that digital robust design using only the magnitude plot is often

8.2 Multivariable Bode Plot

suitable [Maciejowski 1985, Diduch and Doraiswami 1987]. The second reference shows, in fact, that digital control systems can often be *more robust* than continuous control systems.

The magnitude of a square transfer function matrix $H(j\omega)$ at any frequency ω depends on the direction of the input excitation. Inputs in a certain direction in the input space will excite only the *SVs* associated with that direction. However, for any input, the magnitude of the transfer function $H(j\omega)$ at any given frequency ω may be bounded above by its *maximum singular value*, denoted $\bar{\sigma}(H(j\omega))$, and below by its *minimum singular value*, denoted $\underline{\sigma}(H(j\omega))$. Therefore, all our results, as well as the plots we shall give, need take into account only these two constraining values of "magnitude."

Example 8.2-1: MIMO Bode Magnitude Plots

Here, we consider a simple system to make some points about the singular value plots.

a. Continuous-Time System

Consider the multivariable system

$$\dot{x} = \begin{bmatrix} -1 & -1 & 0 & 0 \\ 1 & -1 & 0 & 0 \\ 0 & 0 & -2 & 6 \\ 0 & 0 & -6 & -2 \end{bmatrix} x + \begin{bmatrix} 1 & 0 \\ 0 & 0 \\ 0 & 1 \\ 0 & 0 \end{bmatrix} u = Ax + Bu \qquad (1)$$

$$z = \begin{bmatrix} 1 & 0 & 0 & 0 \\ 0 & 0 & 1 & 0 \end{bmatrix} x = Hx \qquad (2)$$

which has a 2×2 MIMO transfer function of

$$H(s) = H(sI - A)^{-1}B = M(s)/\Delta(s) \qquad (3)$$

with

$$\Delta(s) = s^4 + 6s^3 + 50s^2 + 88s + 80$$

$$M(s) = \begin{bmatrix} 1 & 0 \\ 0 & 1 \end{bmatrix} s^3 + \begin{bmatrix} 5 & 0 \\ 0 & 4 \end{bmatrix} s^2 + \begin{bmatrix} 44 & 0 \\ 0 & 6 \end{bmatrix} s + \begin{bmatrix} 40 & 0 \\ 0 & 4 \end{bmatrix}. \qquad (4)$$

The poles are at

$$s = -1 \pm j, \qquad -2 \pm j6. \qquad (5)$$

By writing a driver program that calls standard software (e.g., subroutine LSVDF in [IMSL]) to evaluate the *SVs* of $H(j\omega)$ at closely spaced values of frequency ω, we may obtain the SV plots versus frequency shown in Fig. 8.2-1. We call this the *multivariable Bode magnitude plot* for the MIMO transfer function $H(s)$.

Since $H(s)$ is 2×2, it has two singular values. Note that although each singular value is continuous, the maximum and minimum singular values are not. This is due to the fact that the singular values can cross over each other, as the figure illustrates.

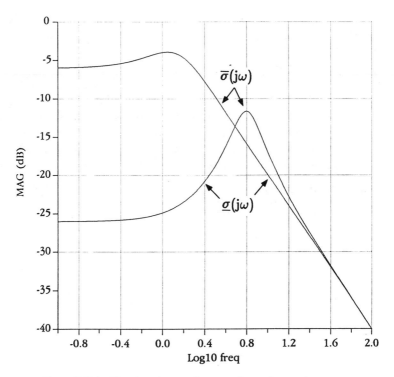

Figure 8.2-1 Singular values vs. frequency for continuous-time system

b. Discrete-Time System

Now let us sample (1) by ZOH equivalence using subroutine DISC in Appendix A. The result using a sample period of $T = 0.1$ sec is

$$x_{k+1} = \begin{bmatrix} 0.90032 & -0.09033 & 0 & 0 \\ 0.09033 & 0.90032 & 0 & 0 \\ 0 & 0 & 0.67573 & 0.46229 \\ 0 & 0 & -0.46229 & 0.67573 \end{bmatrix} x_k$$

$$+ \begin{bmatrix} 0.09501 & 0 \\ 0.00467 & 0 \\ 0 & 0.08556 \\ 0 & -0.02553 \end{bmatrix} u_k \tag{6}$$

$$z_k = \begin{bmatrix} 1 & 0 & 0 & 0 \\ 0 & 0 & 1 & 0 \end{bmatrix} x_k.$$

In the discrete-time case we determine the Bode magnitude plot by evaluating the singular values versus frequency of $H(e^{j\omega T})$, with $H(z)$ the transfer function. This is easily done using subroutines in, for instance [IMSL]. The result is shown in Fig. 8.2-2.

8.3 Frequency-Domain Performance Specifications

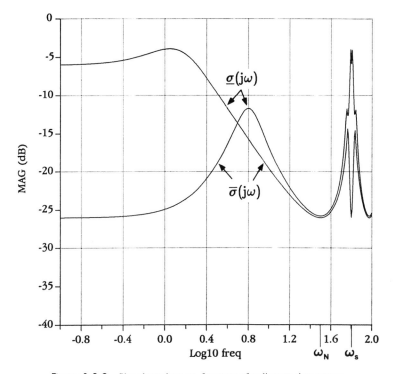

Figure 8.2-2 Singular values vs. frequency for discrete-time system

The sampling frequency is $\omega_s = 2\pi/T = 62.83 = 10^{1.8}$ rad/sec, as shown in the figure. Note that the discrete Bode plot matches the continuous Bode plot up to about $\omega_s/5 = 12.566 = 10^{1.1}$ rad/sec. Thus, for good digital designs one should select the sampling frequency at least five times greater than the maximum frequency of interest in the continuous Bode plot.

Note that the Bode plot for discrete systems is symmetric about the Nyquist frequency ω_N and periodic at the sampling frequency ω_s.

∎

8.3 FREQUENCY-DOMAIN PERFORMANCE SPECIFICATIONS

In section 8.1 we defined several square matrix transfer relations for a closed-loop multivariable servo system, including the loop gain, sensitivity, and cosensitivity. To make the error small in the closed-loop system, we saw that the sensitivity should be small at low frequencies while the cosensitivity should be small at high frequencies.

In section 8.2 we showed how to make a multivariable Bode magnitude plot of a square transfer function matrix for both continuous and discrete systems. It is now necessary to discuss *performance specifications* in the frequency domain in order to determine what a "desirable" Bode plot means in the MIMO case. The important point is that the low-frequency requirements are generally in terms of the *minimum* singular value being *large*, while the high-frequency requirements are in terms of the *maximum* singular value being *small*.

Thus, we shall discover that for robust performance the minimum singular value of the loop gain should be large at low frequencies, where disturbances are present. On the other hand, for robust stability the maximum singular value of the loop gain should be small at high frequencies, where there are significant modeling inaccuracies. We shall also see that, to guarantee stability in spite of parameter variations in the design model, the maximum singular value should be below an upper limit.

Bandwidth

First, let us point out that the classical notion of *bandwidth* holds in the MIMO case. This is the frequency ω_c for which the loop gain $GK(j\omega)$ passes through a value of 1, or 0 dB. If the bandwidth should be limited due to high-frequency noise considerations, the *largest SV* should satisfy $\bar{\sigma}(GK(j\omega_c)) = 1$ at the specified cutoff frequency ω_c.

L_2 Operator Gain

To relate frequency-domain behavior to time-domain behavior, we may take into account the following considerations [Morari and Zafiriou 1989]. Define the L_2 norm of a vector time function $s(t)$ by

$$\|s\|_2 = \left[\int_0^\infty s^T(t)s(t)\, dt \right]^{1/2}. \tag{8.3.1}$$

This is related to the total energy in $s(t)$ and should be compared to the LQ performance index.

A linear time-invariant system has input $u(t)$ and output $z(t)$ related by the convolution integral

$$z(t) = \int_{-\infty}^\infty h(t - \tau)u(\tau)\, dt, \tag{8.3.2}$$

with $h(t)$ the impulse response. The L_2 *operator gain*, denoted $\|H\|_2$, of such a system is defined as the smallest value of γ such that

$$\|z\|_2 \leq \gamma \|u\|_2. \tag{8.3.3}$$

This is just the operator norm induced by the L_2 vector norm. An important result is that the L_2 operator gain is given by

$$\|H\|_2 = \max_\omega (\bar{\sigma}(H(j\omega))), \tag{8.3.4}$$

8.3 Frequency-Domain Performance Specifications

with $H(s)$ the system transfer function. That is, $\|H\|_2$ is nothing but the *maximum value* over ω of the maximum singular value of $H(j\omega)$. Thus, $\|H\|_2$ is an *H-infinity norm* in the frequency domain.

In the discrete-time case a suitable measure of smallness is the ℓ_2 norm of a sequence given by

$$\|s_k\|_2 = \left[\sum_{k=0}^{\infty} s_k^T s_k\right]^{1/2}. \tag{8.3.5}$$

Then, the ℓ_2 operator gain of a system with transfer function $H(z)$ is again given by (8.3.4), where now, however, $H(j\omega) \equiv H(z)|_{z=e^{j\omega T}}$. This is again the maximum value over ω of the Bode magnitude plot.

These results give increased importance to $\bar{\sigma}(H(j\omega))$, for if we are interested in keeping $z(t)$ small over a range of frequencies, then we should take care that $\bar{\sigma}(H(j\omega))$ is small over that range.

It is now necessary to see how this result may be used in deriving multivariable frequency-domain performance specifications.

Some facts from Appendix B that we shall use in this discussion are

$$\underline{\sigma}(GK) - 1 \leq \underline{\sigma}(I + GK) \leq \underline{\sigma}(GK) + 1 \tag{8.3.6}$$

$$\bar{\sigma}(M) = 1/\underline{\sigma}(M^{-1}), \tag{8.3.7}$$

$$\bar{\sigma}(AB) \leq \bar{\sigma}(A)\bar{\sigma}(B) \tag{8.3.8}$$

for any matrices A, B, GK, M, with M nonsingular.

Loop gain singular values

Before we begin a discussion of performance specifications, let us note the following. If $S(j\omega)$ is small, as desired at low frequencies, then

$$\bar{\sigma}(S) = \bar{\sigma}[(I + GK)^{-1}] = 1/\underline{\sigma}(I + GK) \approx 1/\underline{\sigma}(GK). \tag{8.3.9}$$

That is, a large value of $\underline{\sigma}(GK)$ guarantees a small value of $\bar{\sigma}(S)$.

On the other hand, if $T(j\omega)$ is small, as is desired at high frequencies, then

$$\bar{\sigma}(T) = \bar{\sigma}[GK(I + GK)^{-1}] \approx \bar{\sigma}(GK). \tag{8.3.10}$$

That is, a small value of $\bar{\sigma}(GK)$ guarantees a small value of $\bar{\sigma}(T)$.

This means that specifications that $S(j\omega)$ be small at low frequencies and $T(j\omega)$ be small at high frequencies may equally well be formulated in terms of $\underline{\sigma}(GK)$ being large at low frequencies and $\bar{\sigma}(GK)$ being small at high frequencies. Thus, all of our performance specifications will be in terms of the minimum and maximum *SV*s of the loop gain $GK(j\omega)$. The practical significance of this is that we need only compute the *SV*s of

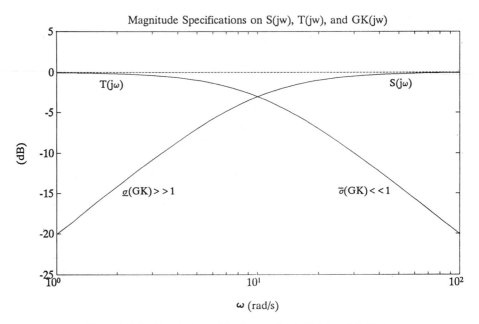

Figure 8.3-1 Magnitude specifications on $S(j\omega)$, $T(j\omega)$, and $GK(j\omega)$.

$GK(j\omega)$, and not those of $S(j\omega)$ and $T(j\omega)$. These notions are symbolized in Fig. 8.3-1, where it should be recalled that $S + T = I$.

Now, we shall first consider low-frequency specifications on the singular value plot, and then high-frequency specifications. According to our discussion relating to (8.1.17), the former will involve the reference input $r(t)$ and disturbances $d(t)$, while the latter will involve the sensor noise $n(t)$.

Low-Frequency Specifications

Several issues are important at low frequencies. Let us discuss them one at a time.

Performance robustness

For low frequencies, let us suppose that the sensor noise $n(t)$ is zero so that (8.1.17) becomes

$$E(s) = S(s)(R(s) - D(s)). \quad (8.3.11)$$

Thus, to keep $\|e(t)\|_2$ small, it is only necessary to ensure that the L_2 operator norm $\|S(j\omega)\|_2$ is small at all frequencies where $R(j\omega)$ and $D(j\omega)$ are appreciable. This may be

8.3 Frequency-Domain Performance Specifications

achieved by ensuring that, at such frequencies, $\bar{\sigma}(S(j\omega))$ is small. As we have just seen, this may be guaranteed if we select

$$\underline{\sigma}(GK(j\omega)) \gg 1, \quad \text{for } \omega \leq \omega_d, \tag{8.3.12}$$

where $D(s)$ and $R(s)$ are appreciable for $\omega \leq \omega_d$.

Thus, exactly as in the classical case [Franklin et al. 1986], we are able to specify a low-frequency performance bound that guarantees *performance robustness*; that is, good performance in the face of low-frequency disturbances. For instance, to ensure that disturbances are attenuated by a factor of 0.01, (8.3.9) shows that we should ensure $\underline{\sigma}(GK(j\omega))$ is greater than 40 dB at low frequencies $\omega \leq \omega_d$.

At this point it is worth examining Fig. 8.3-5, which illustrates the frequency-domain performance specifications we are beginning to derive. We have just derived the bound on the loop gain singular values labeled "LF conditions."

Steady-state error

Another low-frequency performance bound may be derived from steady-state error considerations. Thus, suppose that the plant is continuous with $d(t) = 0$ and reference input a step of magnitude r so that $R(s) = r/s$. Then, according to (8.3.11) and the final value theorem [Franklin et al. 1986], the steady-state error e_∞ is given by

$$e_\infty = \lim_{s \to 0} sE(s) = rS(0). \tag{8.3.13}$$

To ensure that the largest component of e_∞ is less than a prescribed small acceptable value δ_∞, we should therefore select [see (8.3.9)]

$$\underline{\sigma}(GK(0)) > r/\delta_\infty. \tag{8.3.14}$$

In the discrete case, exactly the same bound results. To show it, we may notice that the discrete step is $R(z) = r/(1 - z^{-1})$ and use the discrete final value theorem

$$\lim_{k \to \infty} e_k = \lim_{z \to 1} (1 - z^{-1})E(z). \tag{8.3.15}$$

The ultimate objective of all our concerns is to manufacture a compensator $K(s)$ in Fig. 8.1-1 that gives desirable performance. Let us now mention some low-frequency considerations that are important in the initial stages of the design of the compensator $K(s)$—they mean that certain dynamics should be built into the compensator.

In order to make the steady-state error in response to a unit step at $r(t)$ exactly equal to zero, we may ensure that there is an integrator in each path of the system $G(s)$ so that it is of Type 1 [Franklin et al. 1986]. Thus, suppose that the plant to be controlled is the continuous-time system given by

$$\begin{aligned} \dot{x} &= Ax + Bv \\ z &= Hx. \end{aligned} \tag{8.3.16}$$

To add an integrator to each control path, we may augment the dynamics so that

$$\frac{d}{dt}\begin{bmatrix} x \\ \varepsilon \end{bmatrix} = \begin{bmatrix} A & B \\ 0 & 0 \end{bmatrix}\begin{bmatrix} x \\ \varepsilon \end{bmatrix} + \begin{bmatrix} 0 \\ I \end{bmatrix} u, \quad (8.3.17)$$

$$z = [H \ 0]\begin{bmatrix} x \\ \varepsilon \end{bmatrix}$$

with ε the integrator state. See Fig. 8.3-2. The system $G(s)$ in Fig. 8.1-1 should now be taken as (8.3.17), which contains the integrators as a precompensator.

If the plant is the discrete system

$$x_{k+1} = Ax_k + Bv_k \quad (8.3.18)$$
$$z_k = Hx_k$$

then (see sections 5.2 and 7.4) we may include the modified MPZ sampled integrator by writing

$$\begin{bmatrix} x_{k+1} \\ \varepsilon_{k+1} \end{bmatrix} = \begin{bmatrix} A & B \\ 0 & I \end{bmatrix}\begin{bmatrix} x_k \\ \varepsilon_k \end{bmatrix} + \begin{bmatrix} 0 \\ IT \end{bmatrix} u_k. \quad (8.3.19)$$

with T the sample period. The BLT sampled integrator could also be used (see Example 8.3-1).

Although augmenting each control path with an integrator results in zero steady-state error, in some applications this may result in an unnecessarily complicated compensator. Note that the steady-state error may be made as small as desired without integrators by selecting $K(s)$ so that (8.3.14) holds.

Low-frequency balancing

A final concern on the low-frequency behavior of $G(s)$ needs to be addressed. It is desirable in many situations to have $\underline{\sigma}(GK)$ and $\overline{\sigma}(GK)$ close to the same value. Then, the speed of the responses will be nearly the same in all channels of the system. This is called the issue of balancing the singular values at low frequency. The SVs of G in Fig. 8.1-1 may be balanced at low frequencies as follows.

Suppose the plant has the continuous state-variable description (8.3.16), and let us add a square constant precompensator gain matrix P, so that

$$v = Pu, \quad (8.3.20)$$

is the relation between the control input $u(t)$ in Fig. 8.1-1 and the actual plant input $v(t)$.

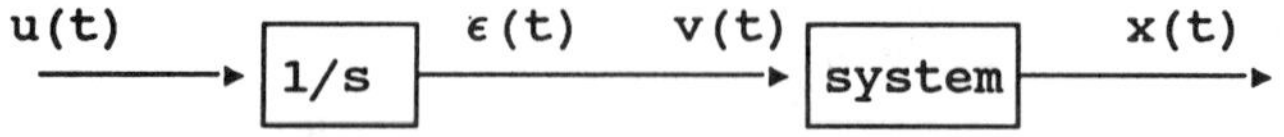

Figure 8.3-2 Plant augmented with integrators

8.3 Frequency-Domain Performance Specifications

The transfer function of the plant plus precompensator is now

$$G(s) = H(sI - A)^{-1}BP. \tag{8.3.21}$$

As s goes to zero, this approaches

$$G(0) = H(-A)^{-1}BP,$$

as long as A has no poles at the origin. Therefore, we may ensure that $G(0)$ has all SVs equal to a prescribed value of γ by selecting

$$P = \gamma[H(-A)^{-1}B]^{-1}, \tag{8.3.22}$$

for then $G(0) = \gamma I$.

The transfer function of (8.3.16) is

$$H(s) = H(sI - A)^{-1}B, \tag{8.3.23}$$

whence we see that the required value of the precompensator gain is

$$P = \gamma H^{-1}(0). \tag{8.3.24}$$

This is nothing but the (scaled) reciprocal DC gain.

In the discrete-time case (8.3.18) the precompensator for balancing is

$$P = \gamma[H(I - A)^{-1}B]^{-1} = \gamma H^{-1}(1), \tag{8.3.25}$$

with $H(z) = H(zI - A)^{-1}B$ and $H(1) = H(z)|_{z=1}$ the DC gain.

Example 8.3-1: Precompensator for Balancing and Zero Steady-State Error

Let us design a precompensator for the system in Example 8.2-1 using the notions just discussed, illustrating for both the continuous and discrete-time cases.

a. Continuous-Time Plant

Substituting the values of A, B, and H from Example 8.2-1a into (8.3.22) with $\gamma = 1$ yields

$$P = [H(-A)^{-1}B]^{-1} = \begin{bmatrix} 2 & 0 \\ 0 & 20 \end{bmatrix}. \tag{1}$$

To ensure zero-steady-state error as well as equal singular values at low frequencies, we may incorporate integrators in each input channel along with the gain matrix P by writing the augmented system

$$\frac{d}{dt}\begin{bmatrix} x \\ \varepsilon \end{bmatrix} = \begin{bmatrix} A & B \\ 0 & 0 \end{bmatrix}\begin{bmatrix} x \\ \varepsilon \end{bmatrix} + \begin{bmatrix} 0 \\ P \end{bmatrix} u \tag{2}$$

$$y = [H \quad 0]\begin{bmatrix} x \\ \varepsilon \end{bmatrix} \tag{3}$$

The singular-value plots for this plant plus precompensator appear in Fig. 8.3-3. At low frequencies there is now a slope of -20 dB/decade as well as equality of $\underline{\sigma}$ and $\bar{\sigma}$. Thus, the augmented system is both balanced and of Type 1. Compare Fig. 8.3-3 to the singular value plot of the uncompensated system in Fig. 8.2-1.

436   Chap. 8   Robust Design

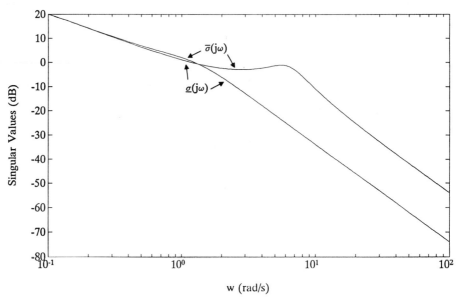

Figure 8.3-3   MIMO Bode magnitude plot for augmented continuous plant

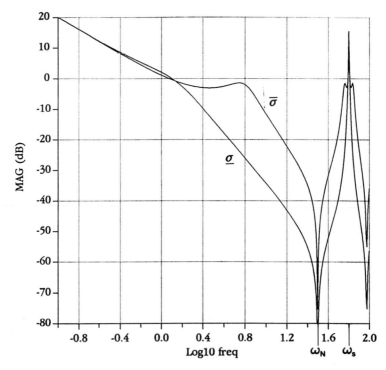

Figure 8.3-4   MIMO Bode magnitude plot for augmented discrete plant

8.3 Frequency-Domain Performance Specifications

The remaining step is the selection of the feedback gain matrix for the augmented plant (2), (3) so that the desired performance is achieved.

b. Discrete-Time Plant

Substituting the values of A, B, H from Example 8.2-1b into (8.3.25) yields exactly the same P as above (why?). To incorporate the BLT sampled integrator as well as balancing into the system, we may write the augmented system

$$\begin{bmatrix} x_{k+1} \\ \varepsilon_{k+1} \end{bmatrix} = \begin{bmatrix} A & B \\ 0 & I \end{bmatrix} \begin{bmatrix} x_k \\ \varepsilon_k \end{bmatrix} + \begin{bmatrix} BPT/2 \\ PT \end{bmatrix} u_k \qquad (4)$$

with T the sampling period. The output equation is (3). The Bode plot for this augmented system appears in Fig. 8.3-4. Compare with Fig. 8.2-2 to see the beneficial effects of balancing and integrator precompensation. ∎

High-Frequency Specifications

We now turn to a discussion of high-frequency performance specifications for robustness of the closed-loop system.

Measurement noise

The sensor noise is generally appreciable at frequencies above some known value ω_n (see Fig. 8.1-2). Thus, according to (8.1.17), to keep the tracking error norm $\|e\|_2$ small in the face of measurement noise, we should ensure that the operator norm $\|T\|_2$ is small at high frequencies above this value. By (8.3.10) this may be guaranteed if

$$\bar{\sigma}(GK(j\omega)) \ll 1, \qquad \text{for } \omega \geq \omega_n. \qquad (8.3.26)$$

See Fig. 8.3-5, where the considerations now being discussed are labelled "HF conditions." For instance, to ensure that sensor noise is attenuated by a factor of 0.1, we should guarantee that $\bar{\sigma}(GK(j\omega)) < -20$ dB for $\omega \geq \omega_n$.

Unmodeled high-frequency dynamics

One final high-frequency robustness consideration needs to be mentioned. It is unusual for the plant model to be exactly known. There are two basic sorts of modeling inaccuracies that concern us. The first is plant parameter variation in the system matrices A, B, C, H. This is a low-frequency phenomenon and will be discussed in the next subsection. The second sort of inaccuracy is due to unmodeled high-frequency dynamics; this we discuss here.

We are assuming a rigid and exact model for the purpose of controls design, and in so doing are neglecting possible flexible and vibrational modes at high frequencies. For

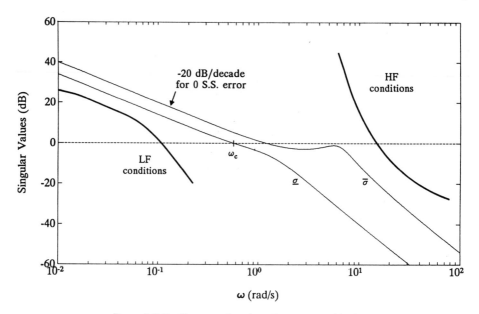

Figure 8.3-5 Frequency-domain performance specifications

instance, the inverted pendulum may be a flexible rod. Thus, although our design may guarantee closed-loop stability for the assumed mathematical model $G(s)$, stability is not assured for the actual plant $G'(s)$ with flexible modes. To guarantee *stability robustness* in the face of plant modeling uncertainty, we may proceed as follows.

The model uncertainties may be of two types. The actual plant model G' and the assumed plant model G may differ by *additive uncertainties* so that

$$G'(j\omega) = G(j\omega) + \Delta G(j\omega), \qquad (8.3.27)$$

where the unknown discrepancy satisfies a known bound

$$\bar{\sigma}(\Delta G(j\omega)) < a(\omega), \qquad (8.3.28)$$

with $a(\omega)$ known for all ω.

On the other hand, the actual plant model $G'(s)$ and the assumed plant model $G(s)$ may differ by *multiplicative uncertainties* so that

$$G'(j\omega) = [I + M(j\omega)]G(j\omega), \qquad (8.3.29)$$

where the unknown discrepancy satisfies a known bound

$$\bar{\sigma}(M(j\omega)) < m(\omega), \qquad (8.3.30)$$

with $m(\omega)$ known for all ω. We shall show a way to determine the bound $m(\omega)$ in Example

8.3 Frequency-Domain Performance Specifications

8.3-2. There, we show how to construct a *reduced-order model* for the system, which may be used for controls design. The bound $m(\omega)$ is determined from the neglected dynamics. In the next subsection, we show how to determine $m(\omega)$ in terms of plant parameter variations in the system matrices.

Since we may write (8.3.27) as

$$G'(j\omega) = [I + \Delta G(j\omega)G^{-1}(j\omega)]G(j\omega) \equiv [I + M(j\omega)]G(j\omega), \quad (8.3.31)$$

we shall confine ourselves to a discussion of multiplicative uncertainties, following Doyle and Stein [1981].

Suppose we have designed a compensator $K(s)$ or $K(z)$ so that the closed-loop system in Fig. 8.1-1 is stable. We should now like to derive a frequency-domain condition that guarantees the stability of the *actual* closed-loop system, which contains not $G(s)$, but $G'(s)$ satisfying (8.3.29), (8.3.30). For this, the multivariable Nyquist condition [Rosenbrock 1974] may be used.

Thus, it is required that the encirclement count of the map $|I + G'K|$ be equal to the negative number of unstable open-loop poles of $G'K$. By assumption, this number is the same as that of GK. Thus, the number of encirclements of $|I + G'K|$ must remain unchanged for all G' allowed by (8.3.30). This is assured if and only if $|I + G'K|$ remains nonzero as G is warped continuously toward G', or equivalently

$$0 < \underline{\sigma}[I + [I + \varepsilon M(s)]G(s)K(s)]$$

for all $0 \le \varepsilon \le 1$, all $M(s)$ satisfying (8.3.30), and all s on the standard Nyquist contour.

Since G' vanishes on the infinite radius segment of the Nyquist contour, and assuming for simplicity that no indentations are required along the $j\omega$-axis portion, this reduces to the following equivalent conditions.

$$0 < \underline{\sigma}[I + G(j\omega)K(j\omega) + \varepsilon M(j\omega)G(j\omega)K(j\omega)]$$

for all $0 \le \varepsilon \le 1, 0 \le \omega < \infty$, all M

$$iff \ 0 < \underline{\sigma}[\{I + \varepsilon MGK(I + GK)^{-1}\}(I + GK)]$$

$$iff \ 0 < \underline{\sigma}[I + MGK(I + GK)^{-1}]$$

all $0 \le \omega < \infty$, and all M

$$iff \ \bar{\sigma}[GK(I + GK)^{-1}] < 1/m(\omega) \quad (8.3.32)$$

for all $0 \le \omega < \infty$. Thus, stability robustness translates into a requirement that the cosensitivity $T(j\omega)$ be bounded above by the reciprocal of the multiplicative modeling discrepancy bound $m(\omega)$.

In the case of high-frequency unmodeled dynamics, $1/m(\omega)$ is small at high ω, so that according to (8.3.10), we may simplify (8.3.32) by writing it in terms of the loop gain as

$$\bar{\sigma}(GK(j\omega)) < 1/m(\omega), \tag{8.3.33}$$

for all ω such that $m(\omega) \gg 1$.

This bound for stability robustness is illustrated in Fig. 8.3-5.

Model Reduction and Stability Robustness

High-order systems occur in process control, aircraft and spacecraft control, control of flexible structures, and elsewhere. Many design techniques produce high-order controllers for high-order systems when a controller of smaller order would be adequate. The output-feedback design technique in Chapter 4 and section 7.4 does not necessarily have this deficiency, but the structure of the system must be understood to decide on a suitable compensator structure, and the structure of high-order systems is often difficult to analyze.

Thus, it is often important to be able to compute a *reduced-order model* of a high-order system which may then conveniently be used to design a reduced-order controller. Here we shall show a convenient technique for model reduction as well as an illustration of the stability robustness bound $m(\omega)$. The technique described here is from Athans et al. [1986].

The approach works for either continuous or discrete systems.

Model reduction by partial-fraction expansion

Suppose the actual plant is described by

$$\dot{x} = Ax + Bu \tag{8.3.34}$$

$$z = Hx. \tag{8.3.35}$$

with $x \in \mathbf{R}^n$. If A is simple with eigenvalues λ_i, right eigenvectors u_i, and left eigenvectors v_i so that

$$Au_i = \lambda_i u_i, \qquad v_i^T A = \lambda_i v_i^T, \tag{8.3.36}$$

then the transfer function

$$G'(s) = H(sI - A)^{-1}B \tag{8.3.37}$$

may be written as the partial fraction expansion (section 2.4)

$$G'(s) = \sum_{i=1}^{n} \frac{R}{s - \lambda_i}, \tag{8.3.38}$$

8.3 Frequency-Domain Performance Specifications

with residue matrices given by

$$R_i = H u_i v_i^T B. \tag{8.3.39}$$

If the value of n is large, it may be desirable to find a *reduced-order approximation* to (8.3.34), (8.3.35) for which a simplified compensator $K(s)$ in Fig. 8.1-1 may be designed. Then, if the approximation is a good one, the compensator $K(s)$ should work well when used on the actual plant $G'(s)$.

To find a reduced-order approximation $G(s)$ to the plant, we may proceed as follows.

Decide which of the eigenvalues λ_i in (8.3.38) are to be retained in $G(s)$. This may be done using engineering judgment, by omitting high-frequency modes, by omitting terms in (8.3.38) that have small residues, and so on. Let the r eigenvalues to be retained in $G(s)$ be $\lambda_1, \lambda_2, \ldots, \lambda_r$.

Define the matrix

$$Q = \text{diag}\{Q_i\},$$

where Q is an $r \times r$ matrix and the blocks Q_i are defined as

$$Q_i = 1, \quad \text{for each real eigenvalue retained}$$

$$Q_i = \begin{bmatrix} 1/2 & -j/2 \\ 1/2 & j/2 \end{bmatrix}, \quad \text{for each complex pair retained.}$$

Compute the matrices

$$V \equiv Q^{-1} \begin{bmatrix} v_1^T \\ \vdots \\ v_r^T \end{bmatrix}.$$

$$U \equiv [u_1 \ldots u_r] Q.$$

In terms of these constructions, the reduced-order system is nothing but a projection of (8.3.34) onto a space of dimension r with state defined by

$$w = Vx.$$

The system matrices in the reduced-order approximate system

$$\dot{w} = Fw + Gu \tag{8.3.40}$$

$$z = Jw + Du \tag{8.3.41}$$

are given by

$$F = VAU$$

$$G = VB \tag{8.3.42}$$

$$J = HU,$$

with the direct-feed matrix given in terms of the residues of the neglected eigenvalues as

$$D = \sum_{i=r+1}^{n} -\frac{R_i}{\lambda_i}. \tag{8.3.43}$$

The motivation for selecting such a D matrix is as follows. The transfer function

$$G(s) = J(sI - F)^{-1}G + D$$

of the reduced system (8.3.40), (8.3.41) is given as (verify!)

$$G(s) = \sum_{i=1}^{r} \frac{R_i}{s - \lambda_i} + \sum_{i=r+1}^{n} -\frac{R_i}{\lambda_i}. \tag{8.3.44}$$

Evaluating $G(j\omega)$ and $G'(j\omega)$ at $\omega = 0$, it is seen that they are equal at DC. Thus, the modeling errors induced by taking $G(s)$ instead of the actual $G'(s)$ occur at higher frequencies. Indeed, they depend on the frequencies of the neglected eigenvalues $\lambda_{r+1}, \ldots, \lambda_n$.

To determine the $M(s)$ in (8.3.29) that is induced by the order reduction, note that

$$G' = (I + M)G$$

so that

$$M = (G' - G)G^{-1}$$

or

$$M(s) = \left[\sum_{i=r+1}^{n} \frac{R_i}{\lambda_i} \frac{s}{s - \lambda_i} \right] G^{-1}(s). \tag{8.3.45}$$

Then, the high-frequency robustness bound is given in terms of

$$m(j\omega) = \bar{\sigma}(M(j\omega)). \tag{8.3.46}$$

Note that $M(j\omega)$ tends to zero as ω becomes small, reflecting our perfect certainty of the actual plant at DC.

Example 8.3-2: Model Reduction and Stability Robustness

Let us take an example to illustrate the model-reduction procedure, and show also how to compute the upper bound $m(\omega)$ in (8.3.29), (8.3.30) on the high-frequency modeling errors thereby induced. To make it easy to see what is going on, we shall take a Jordan-form system.
Let there be prescribed the MIMO system

$$\dot{x} = \begin{bmatrix} -1 & & \\ & -2 & \\ & & -10 \end{bmatrix} x + \begin{bmatrix} 1 & 0 \\ 0 & 1 \\ 2 & 0 \end{bmatrix} u = Ax + Bu \tag{1a}$$

$$z = \begin{bmatrix} 1 & 0 & 0 \\ 0 & 1 & 1 \end{bmatrix} x = Cx. \tag{1b}$$

8.3 Frequency-Domain Performance Specifications

The eigenvectors are given by $u_i = e_i$, $v_i = e_i$, $i = 1, 2, 3$, with e_i the i-th column of the 3×3 identity matrix. Thus, the transfer function is given by the partial-fraction expansion

$$G'(s) = \frac{R_1}{s+1} + \frac{R_2}{s+2} + \frac{R_3}{s+10}, \quad (2)$$

with

$$R_1 = \begin{bmatrix} 1 & 0 \\ 0 & 0 \end{bmatrix}, \quad R_2 = \begin{bmatrix} 0 & 0 \\ 0 & 1 \end{bmatrix}, \quad R_3 = \begin{bmatrix} 0 & 0 \\ 2 & 0 \end{bmatrix}. \quad (3)$$

To find the reduced-order system that retains the poles at $\lambda = -1$ and $\lambda = -2$, define

$$Q = \begin{bmatrix} 1 & 0 \\ 0 & 1 \end{bmatrix}, \quad V = \begin{bmatrix} 1 & 0 & 0 \\ 0 & 1 & 0 \end{bmatrix}, \quad U = \begin{bmatrix} 1 & 0 \\ 0 & 1 \\ 0 & 0 \end{bmatrix} \quad (4)$$

and compute the approximate system

$$\dot{w} = \begin{bmatrix} -1 & \\ & -2 \end{bmatrix} w + \begin{bmatrix} 1 & 0 \\ 0 & 1 \end{bmatrix} u = Fw + Gu \quad (5a)$$

$$z = \begin{bmatrix} 1 & 0 \\ 0 & 1 \end{bmatrix} w + \begin{bmatrix} 0 & 0 \\ 0.2 & 0 \end{bmatrix} = Jw + Du. \quad (5b)$$

This has a transfer function of

$$G(s) = \frac{R_1}{s+1} + \frac{R_2}{s+2} + D. \quad (6)$$

Singular value plots of the actual plant (1) and the reduced-order approximation (5) are shown in Fig. 8.3-6.

The multiplicative error is given by

$$M = (G' - G)G^{-1} = \begin{bmatrix} 0 & 0 \\ \frac{-0.2s(s+1)}{s+10} & 0 \end{bmatrix}, \quad (7)$$

whence

$$m(\omega) = \bar{\sigma}(M(j\omega)) = \frac{0.2\omega\sqrt{\omega^2+1}}{\sqrt{\omega^2+100}}, \quad (8)$$

and the high-frequency bound on the loop gain $GK(j\omega)$ is given by

$$\frac{1}{m(j\omega)} = \frac{5\sqrt{\omega^2+100}}{\omega\sqrt{\omega^2+1}}. \quad (9)$$

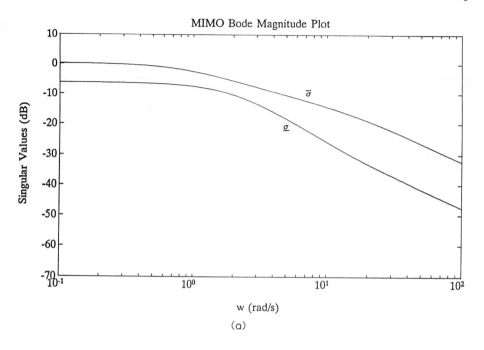

(a)

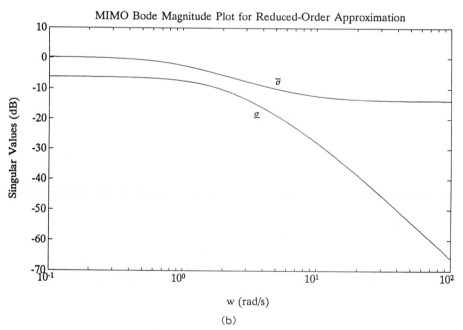

(b)

Figure 8.3-6 Singular value plots. (a) Actual plant. (b) Reduced-order approximation.

8.3 Frequency-Domain Performance Specifications

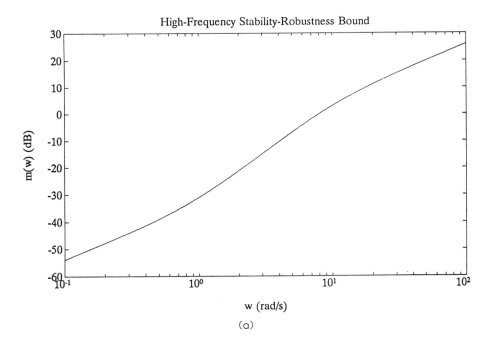

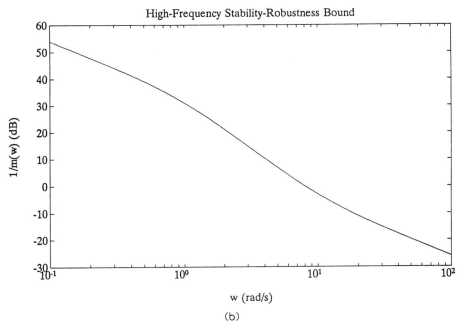

Figure 8.3-7 High-frequency stability-robustness bound. (a) $m(\omega)$. (b) $1/m(\omega)$.

This bound is plotted in Fig. 8.3-7. Note that the modeling errors become appreciable (i.e., of magnitude one) at a frequency of 8.0 rads/sec. Above this frequency, we should ensure that constraint (8.3.33) on the loop-gain magnitude holds to guarantee stability-robustness. This will be a restriction on any compensator $K(s)$ designed using the reduced-order plant (5).

■

Robustness Bounds for Plant Parameter Variations

Suppose the nominal model used for design is

$$\dot{x} = Ax + Bu$$
$$y = Cx \quad (8.3.47)$$

which has the transfer function

$$G(s) = C(sI - A)^{-1}B. \quad (8.3.48)$$

However, due to plant parameter uncertainties, the actual plant has dynamics described by

$$\dot{x} = (A + \Delta A)x + (B + \Delta B)u$$
$$y = (C + \Delta C)x, \quad (8.3.49)$$

where the plant parameter variation matrices are $\Delta A, \Delta B, \Delta C$. The variation matrices $\Delta A, \Delta B, \Delta C$ are smaller than some known bounds.

It is not difficult to show (see Stevens et al. 1987 and the problems) that the plant with variations has the transfer function

$$G'(s) = G(s) + \Delta G(s)$$

with

$$\Delta G(s) = C(sI - A)^{-1}\Delta B + \Delta C(sI - A^{-1})B$$
$$+ C(sI - A)^{-1}\Delta A(sI - A)^{-1}B, \quad (8.3.50)$$

where second-order effects have been neglected. Hence, (8.3.31) may be used to determine the multiplicative uncertainty bound $m(\omega)$ in terms of the bounds on $\Delta A, \Delta B, \Delta C$.

Any control system designed using the nominal design model must also stabilize all systems described by (8.3.49) for $\Delta A, \Delta B, \Delta C$ bounded by the given magnitudes. Then, the actual plant will be stabilized. Therefore, the loop gain should satisfy the upper bound (8.3.33) for guaranteed stability in the face of the parameter variations $\Delta A, \Delta B, \Delta C$.

Since $(sI - A)^{-1}$ has a relative degree of at least one, the high-frequency roll-off of $\Delta G(j\omega)$ is at least -20 dB/decade. Thus, plant parameter variations yield an upper bound for the cosensitivity at *low frequencies*.

8.4 ROBUST OUTPUT-FEEDBACK DESIGN

We should now like to incorporate the robustness concepts just introduced into the LQ output-feedback design procedure for continuous and digital controllers. Thus, we are interested in obtaining suitable closed-loop performance in the time domain, as well as robustness. These goals may be accomplished using the following steps.

ITERATIVE DESIGN PROCEDURE:

1. If necessary, augment the plant with compensator dynamics. These dynamics may be selected to achieve the required steady-state error behavior, to achieve balanced singular values at DC, or from other considerations.
2. Select a performance index, the PI weighting matrices Q and R, and, if applicable, the time weighting factor k in t^k.
3. Determine the optimal output feedback gain K using, for instance, Tables 4.1-1 or 4.2-1 in the continuous case or Tables 7.4-1 or 7.4-2 in the discrete case.
4. Simulate the time responses of the closed-loop system to verify that they are satisfactory. If not, select different Q, R and k and return to step 3.
5. Determine the low-frequency and high-frequency bounds required for performance robustness and stability robustness. Plot the loop gain singular values to verify that the bounds are satisfied. If they are not, select new Q, R, and k and return to step 3.

An example will illustrate the robust output-feedback design procedure. Since modern robust design is often used in aircraft control problems, we shall consider the design of an aircraft pitch rate control system.

Example 8.4-1: Aircraft Pitch Rate Control System Robust to Wind Gusts and Unmodeled Flexible Mode

This example will illustrate the design of an aircraft pitch rate control system that is robust in the presence of vertical wind gusts and the unmodeled dynamics associated with an aircraft flexible mode. We shall use the F-16 aircraft.

a. Aircraft Dynamics

An aircraft is a highly nonlinear system whose dynamics depend on a variety of environmental factors such as airspeed, altitude, air density, and so on. However, the dynamics may be linearized about an operating condition to obtain a linear time-invariant state-space model [Stevens and Lewis 1991]. For the design of a pitch rate control system it is sufficient to consider only some of the aircraft states, namely those associated with longitudinal motion in the vertical plane. Of the four longitudinal states, only two are relevant to the pitch rate control problem.

Therefore, the state of interest in this example is $x = [\alpha \quad q]^T$, with α the angle-of-attack and q the pitch rate. The angle-of-attack is the angle between the relative wind and the wing chord, while the pitch rate is the angular rate at which the aircraft nose is moving up. The dynamics of the basic F-16 aircraft in level flight, linearized about a total velocity of 502 ft/sec, 0 ft altitude, with the center of gravity at $0.35\ \bar{c}$, are given by

$$\dot{x} = \begin{bmatrix} -1.01887 & 0.90506 \\ 0.82225 & -1.07741 \end{bmatrix} x + \begin{bmatrix} -0.00215 \\ -0.17555 \end{bmatrix} \delta_e. \quad (1)$$

The control input δ_e is the elevator deflection (down is positive). This second-order dynamical model that retains only aircraft states α and q is known as the *short period approximation*.

b. Actuator and Measurements

The elevator actuator is a hydraulic motor modeled approximately as

$$H_a(s) = \frac{20.2}{s + 20.2}. \quad (2)$$

Measurements of pitch rate and angle-of-attack are taken. Unfortunately, the angle-of-attack measurements are quite noisy. Therefore, a low-pass filter of the form

$$H_F(s) = \frac{10}{s + 10} \quad (3)$$

is used to smooth out the measurements.

c. Control System Structure

The function of a pitch rate control system is to hold the aircraft pitch rate $q(t)$ at a commanded or reference value $r(t)$. A control system that accomplishes this is shown in Fig. 8.4 1, which also shows the actuator dynamics and α-smoothing filter. This is a standard classical control system with a forward-path integrator to ensure zero steady-state error. The integrator output is $\varepsilon(t)$, the actuator input is $u(t)$, and the smoothed angle-of-attack is α_F.

The tracking error is defined as

$$e = r - z \quad (4)$$

where the performance output is $z = q$, the pitch rate.

The dynamics of the aircraft, actuator, α-smoothing filter, and control system can be incorporated into one state-variable model by defining an augmented state vector as

$$x = \begin{bmatrix} \alpha \\ q \\ \delta_e \\ \alpha_F \\ \varepsilon \end{bmatrix}. \quad (5)$$

Then, the overall system dynamics in Fig. 8.4-1 are described by

$$\dot{x} = Ax + Bu + Er \quad (6)$$

$$y = Cx + Fr \quad (7)$$

$$z = Hx \quad (8)$$

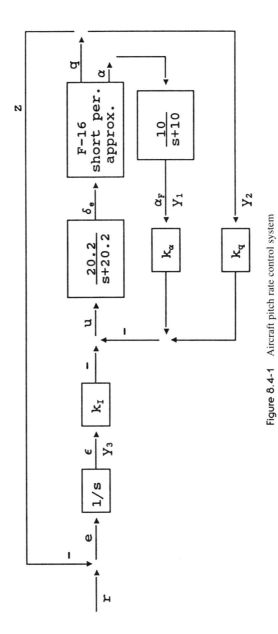

Figure 8.4-1 Aircraft pitch rate control system

with

$$A = \begin{bmatrix} -1.01887 & 0.90506 & -0.00215 & 0 & 0 \\ 0.82225 & -1.07741 & -0.17555 & 0 & 0 \\ 0 & 0 & -20.2 & 0 & 0 \\ 10.0 & 0 & 0 & -10 & 0 \\ 0 & -57.2958 & 0 & 0 & 0 \end{bmatrix}$$

$$B = \begin{bmatrix} 0 \\ 0 \\ 20.2 \\ 0 \\ 0 \end{bmatrix}, \quad E = \begin{bmatrix} 0 \\ 0 \\ 0 \\ 0 \\ 1 \end{bmatrix}$$

$$C = \begin{bmatrix} 0 & 0 & 0 & 57.2958 & 0 \\ 0 & 57.2958 & 0 & 0 & 0 \\ 0 & 0 & 0 & 0 & 1 \end{bmatrix}, \quad F = \begin{bmatrix} 0 \\ 0 \\ 0 \end{bmatrix}$$

$$H = [0 \quad 57.2958 \quad 0 \quad 0 \quad 0].$$

The factor of 57.2958 is added to convert angles from radians to degrees.

The system output is defined as $y = [\alpha_F \quad q \quad \varepsilon]^T$. Then, the control input may be expressed as

$$u = -Ky = -[k_\alpha \quad k_q \quad k_I]y = -k_\alpha \alpha_F - k_q q - k_I \varepsilon, \tag{9}$$

which is just a static output feedback of the sort discussed in Chapter 4. Note that some of the control gains are feedback gains, while some are feedforward gains. This illustrates the versatility of the output-feedback design approach, which allows the compensator to have a desired structure, and permits the closed-loop poles *and zeros* to be adjusted for optimal performance.

It is required to select the control gains to yield a good closed-loop response to a step input at r, which corresponds to a single-input/multi-output design problem. We shall assume that vertical wind gusts are present, and that the aircraft has a flexible mode which has not been accounted for in the design equations (6)–(8). Therefore, the control gains must also be selected for *robust performance*.

d. Frequency-Domain Robustness Bounds

According to [Mil. Spec. 1797] the vertical wind gust noise has a spectral density given in Dryden form as

$$\Phi_w(\omega) = 2L\sigma^2 \frac{1 + 3L^2\omega^2}{[1 + L^2\omega^2]^2}, \tag{10}$$

with ω = frequency in rad/s, σ = turbulence intensity, and L = turbulence scale length divided by true airspeed. Assuming that the vertical gust velocity is a disturbance input that changes the angle-of-attack, the software described in Stevens and Lewis [1991] can be used to find a control input matrix from gust velocity to x. Then, the magnitude of the gust disturbance versus frequency can be found. It is shown in Fig. 8.4-2. We took σ = 10 ft/sec and L = (1700 ft)/(502 ft/sec) = 3.49 sec.

8.4 Robust Output-Feedback Design

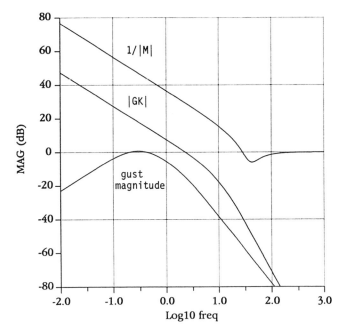

Figure 8.4-2 Frequency-domain magnitude plots and robustness bounds

Let the transfer function of the rigid dynamics (6)–(8) from $u(t)$ to $z(t)$ be denoted by $G(s)$. Then, the transfer function including the first flexible mode is given by Blakelock [1965]:

$$G'(s) = G(s)F(s) \qquad (11)$$

where

$$F(s) = \frac{\omega_n^2}{s^2 + 2\zeta\omega_n s + \omega_n^2} \qquad (12)$$

with $\omega_n = 40$ rad/sec and $\zeta = 0.3$. According to (8.3.31), therefore, the multiplicative uncertainty is given by

$$M(s) = F(s) - I = \frac{-s(s + 2\zeta\omega_n)}{s^2 + 2\zeta\omega_n s + \omega_n^2}. \qquad (13)$$

The magnitude of $1/M(j\omega)$ is shown in Fig. 8.4-2.

We should like to perform our controls design using only the rigid dynamics $G(s)$ described by (6)–(8). Then, for performance robustness in the face of the gust disturbance and stability robustness in the face of the first flexible mode, the loop gain singular values should lie within the bounds implied by the gust disturbance magnitude and $1/|M(j\omega)|$.

e. LQ Controls Design

The equations (6)–(9) are exactly of the form required for quadratic output-feedback design in Table 4.2-1. A suitable PI for selection of the control gains is

$$J = \frac{1}{2} \int_0^\infty (t^2 \bar{e}^2 + \rho \bar{u}^2) \, dt. \tag{14}$$

This is a natural PI that corresponds to the actual performance requirements of keeping the tracking error small without using too much control energy, and also has the important advantage of requiring the adjustment of only one design parameter ρ.

The software described in Appendix A was used to solve the design equations in Table 4.2-1 for several values of ρ. A good step response was found with $\rho = 1$, which yielded optimal gains of

$$K = [-0.046 \quad -1.072 \quad 3.381] \tag{15}$$

closed-loop poles of $s = -8.67 \pm j9.72, -9.85, -4.07, -1.04$, and the step response in Fig. 8.4-3.

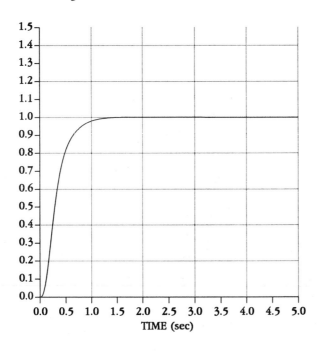

Figure 8.4-3 Optimal step response

f. Robustness Verification

To verify that the robustness bounds hold for this design, it is necessary to find the loop-gain $GK(s)$ of the closed-loop system. Thus, in Fig. 8.4-1 it is necessary to find the loop transfer function from $e(t)$ around to $e(t)$ [that is, from $e(t)$ to $-z(t)$]. With respect to this loop gain, note that some of the elements in (9) are feedforward gains while some are feedback gains.

Problems for Chapter 8

The magnitude of $GK(j\omega)$ is plotted in Fig. 8.4-2. Note that the robustness bounds are satisfied. Therefore, this design is robust in the presence of vertical turbulence velocities up to 10 ft/sec as well as the first aircraft flexible mode.

In this example the loop gain $GK(s)$ is a scalar; however, the same techniques are used when the loop gain is multivariable.

∎

PROBLEMS FOR CHAPTER 8

Problems for Section 8.1

8.1-1 Derive in detail the multivariable expressions (8.1.16) and (8.1.17) for the performance output and the tracking error.

8.1-2 Multivariable Closed-Loop Transfer Relations. In Fig. 8.1-1, the plant $G(s)$ is described by

$$\dot{x} = \begin{bmatrix} 0 & 1 & 0 \\ 0 & -3 & 0 \\ 0 & 0 & 0 \end{bmatrix} x + \begin{bmatrix} 0 & 0 \\ 1 & 0 \\ 0 & 1 \end{bmatrix} u, \quad z = \begin{bmatrix} 1 & 0 & 0 \\ 0 & 0 & 1 \end{bmatrix} x$$

and the compensator is $K(s) = 2I_2$.
a. Find the multivariable loop gain and return difference.
b. Find the sensitivity and cosensitivity.
c. Find the closed-loop transfer function from $r(t)$ to $z(t)$, and hence the closed-loop poles.

Problems for Section 8.2

8.2-1 For the continuous-time system in Example 8.2-1, plot the individual SISO Bode magnitude plots from input one to outputs one and two, and from input two to outputs one and two. Compare them to the MIMO Bode plot to see that there is no obvious relation. Thus, the robustness bounds in section 8.3 cannot be given in terms of the individual SISO Bode plots.

8.2-2 Multivariable Bode Plot. For the system in Problem 8.1-2, plot the multivariable Bode magnitude plots for:
a. the loop gain GK
b. the sensitivity S and cosensitivity T. For which frequency ranges do the plots for $GK(j\omega)$ match those for $S(j\omega)$? For $T(j\omega)$?

8.2-3 Discrete Multivariable Bode Plot. Discretize the system in Problem 8.1-2 using ZOH/step invariance. For the discrete-time system, repeat Problems 8.1-2 and 8.2-2.

8.2-4 Bode Plots for Inverted Pendulum Controller
a. Plot the multivariable Bode magnitude plot for the continuous-time inverted pendulum control system designed in Example 4.2-4.
b. Plot the multivariable Bode magnitude plot for the digital inverted pendulum control system designed in Example 5.2-2.

Problems for Section 8.3

8.3-1 Balancing and Zero Steady-State Error. Find a precompensator for balancing the SVs at low frequency and ensuring zero steady-state error for the system

$$\dot{x} = \begin{bmatrix} 0 & 1 & 0 \\ -2 & -3 & 0 \\ 0 & 0 & -3 \end{bmatrix} x + \begin{bmatrix} 0 & 0 \\ 1 & 0 \\ 0 & 1 \end{bmatrix} u, \qquad z = \begin{bmatrix} 1 & 0 & 0 \\ 0 & 0 & 1 \end{bmatrix} x.$$

Plot the SVs of the original and precompensated system.

8.3-2 Discrete Balancing and Zero Steady-State Error. Find a precompensator for balancing the SVs at low frequency and ensuring zero steady-state error for the system

$$x_{k+1} = \begin{bmatrix} 0 & 1 & 0 \\ -1/8 & -3/4 & 0 \\ 0 & 0 & -1/2 \end{bmatrix} x_k + \begin{bmatrix} 0 & 0 \\ 1 & 0 \\ 0 & 1 \end{bmatrix} x_k, \qquad z_k = \begin{bmatrix} 1 & 0 & 0 \\ 0 & 0 & 1 \end{bmatrix} x_k.$$

Plot the SVs of the original and precompensated system.

8.3-3 Prove (8.3.50). You will need to neglect any terms that contain second-order terms in the parameter variation matrices and use the fact that, for small X, $(I - X)^{-1} \approx (I + X)$.

Problems for Section 8.4

8.4-1 Model Reduction and Neglected High-Frequency Modes. An unstable system influenced by high-frequency parasitics is given by

$$\dot{x} = \begin{bmatrix} 0 & 1 & 0 \\ 1 & 0 & 1 \\ 0 & 0 & -10 \end{bmatrix} x + \begin{bmatrix} 0 & 0 \\ 0 & 1 \\ 1 & 0 \end{bmatrix} u, \qquad z = [1 \ 0 \ 0] x.$$

a. Use the technique of Example 8.3-2 to find a reduced-order model that neglects the high-frequency mode at $s = 10$ rad/sec. Find the bound $m(j\omega)$ on the magnitude of the neglected portion.

b. Using techniques like those in section 4.2, design a servo control system for the reduced-order model. In selecting your control system structure, note that the model is very similar to the angle subsystem of the inverted pendulum in Example 4.2-4. Verify the step response of the closed-loop system by performing a simulation on the reduced-order system.

c. Find the loop gain of the closed-loop system and plot its singular values. Do they fall below the bound $1/m(j\omega)$, thus guaranteeing robustness to the neglected mode? If not, return to part b and find other gains that do guarantee stability robustness.

d. Simulate your controller on the full system including the high-frequency mode. How does the step response look?

e. A better controller results if high-frequency dynamics are not neglected in the design stage. Design a servo control system for the full third-order system. It may be necessary to use a more complicated controller. Verify the step response of the closed-loop system by performing a simulation. Compare to the results of part d.

8.4-2 Ball Balancer with Flexible Mode. A ball balancer control system was designed in the problems for section 4.2. Suppose the rods are flexible so that they have a vibrational mode.
  a. Model the vibrational mode as a damped harmonic oscillator

$$\frac{\omega_n^2}{s^2 + 2\zeta\omega_n s + \omega_n^2}$$

  with $\omega_n = 20$ rad/sec, $\zeta = 0.4$, placed in parallel with the rigid system transfer function. Find and plot the magnitudes of the bounds $m(j\omega)$ and $1/m(j\omega)$.
  b. Does your control system design for the rigid system from the problems of section 4.2 satisfy the stability robustness bounds? If not, redo the design. If so, simulate your controller on the full plant including the vibrational mode to verify stability.
  c. Using the techniques in section 4.2, design a servo system for the full ball balancer including the vibrational mode. A more complicated control structure may be required. Compare to the design in part b.

8.4-3 Robust Digital Controller for Ball Balancer with Flexible Mode. In the problems for section 5.2 a digital control system was designed for the ball balancer system. Repeat Problem 8.4-2 to verify the stability robustness of this digital control system to unmodeled vibrations.

PART V

OBSERVERS, FILTERS, AND DYNAMIC REGULATORS

Up to this point in the book we have used *static* state-feedback design and output-feedback design. We say "static" because the design has amounted to finding the control gain matrix K in nondynamic expressions for the control law of the form $u = -KCx$, with $C = I$ for full state feedback. With the exception of *CGT* design (section 4.4) and model-following design (section 4.5), it was necessary to know the structure of the control system, writing the plant-plus-compensator dynamics in a form so that the only unknowns were the elements of the control gain matrix K, which were then selected using optimal LQ techniques to give the best performance possible using that specific control structure.

In many design problems, the structure of the controller is indeed known. For instance, using classical techniques it may be known that a lead compensator is needed, or that an integrator is required in each control channel to give zero steady-state error. In aircraft control, for example, experience garnered over several decades dictates the structure of the controller or autopilot.

However, for complicated modern multi-input/multi-output systems like those occurring in automotive, electronic, or aerospace design, there may be no clear guidelines for selecting the compensator structure. Fortunately, modern control theory provides many techniques for selecting compensators of suitable structure for MIMO systems. In this part of the book we deal with some of these techniques, so that one could say the prime focus here is determining the structure of the compensator.

We shall consider two basic issues. First, in Chapter 9 we show how to estimate all the states of a system given only the available measured outputs. It turns out that the state estimates are provided by a dynamical system called an observer or estimator. Moreover, the structure of the estimator is closely related to the structure of compensators that give good closed-loop performance. In Chapter 10 we examine this issue, showing how to use modern LQ techniques to design multivariable dynamic compensators for MIMO plants.

9

State Estimators

SUMMARY

In this chapter we show how to estimate or reconstruct all the internal states of a plant using measurements of only the available outputs. It turns out that the state estimates are provided by a dynamical system called an observer *or* estimator. *The estimator has two uses: first, the state estimates themselves may be required in some applications and, second, these estimates can be used for purposes of control.*

INTRODUCTION

In most practical situations it is unusual to have all the states of a dynamical system available through measurements; some states may be impossible or too expensive to measure. In this chapter we consider the problem of *estimating the state* of a plant when only partial state information is available in terms of measured outputs. The state estimates are provided by a dynamical system called an observer or estimator. The state estimation problem is often of importance in its own right since we may want to know the value of the states. For instance, in navigation, we may take noisy positional fixes using satellite or radar navigation, and the estimator can use these measurements to provide accurate estimates of current position, heading, and velocity.

The state estimates can also be used for *control* purposes. Specifically, we may design a static state-feedback control law of the form $u = -Kx$ (as in Chapter 3, for instance) assuming that all the components of the state $x(t)$ are known. Then, using the available measured outputs, we can design a *dynamic observer* to estimate the states that are not directly measured. The state estimates, denoted $\hat{x}(t)$, can then be fed back as if they were the actual states in a modified feedback law of the form $u = -K\hat{x}$. The combination of the state-feedback control law and the observer is a dynamic compensator for the plant, similar to those in classical control. The advantage to using modern techniques is that one can design dynamic compensators for complicated multi-input/multi-output plants by simply solving some matrix design equations using a digital computer. We discuss this approach to dynamic control system design in Chapter 10.

In this chapter we show several approaches for designing dynamical observers or estimators. Both discrete- and continuous-time observers are presented. In the second part of the chapter we show how to design an estimator known as the *Kalman Filter* that takes into account measurement noise and process noise. The Kalman Filter is extensively used in navigation and elsewhere.

9.1 OUTPUT-INJECTION OBSERVER DESIGN

In this section we discuss an approach to observer design that is the dual to the state-feedback design technique given for discrete systems at the end of section 7.4 and for continuous systems in section 3.3. The power of this approach is that stable observers may be designed even for multivariable systems by simply solving some standard matrix equations using commercially available software.

Digital Observer

To design a digital observer for a continuous-time plant, first discretize the plant (section 7.1) to obtain the discrete state-space model

$$x_{k+1} = Ax_k + Bu_k \tag{9.1.1}$$

$$y_k = Cx_k \tag{9.1.2}$$

with $x_k \in \mathbf{R}^n$ the state, $u_k \in \mathbf{R}^m$ the control input, and $y_k \in \mathbf{R}^p$ the available measured outputs.

Observer dynamics

Let the estimate of x_k be $\hat{x}_k$. We claim that the state observer is a dynamical system described by

$$\hat{x}_{k+1} = A\hat{x}_k + Bu_k + L(y_k - C\hat{x}_k) \tag{9.1.3}$$

or

$$\hat{x}_{k+1} = (A - LC)\hat{x}_k + Bu_k + Ly_k = A_o\hat{x}_k + Bu_k + Ly_k. \tag{9.1.4}$$

We have defined the *observer system matrix*

$$A_o = A - LC. \tag{9.1.5}$$

Thus, the observer is a system with two inputs, namely u_k and y_k, both of which are known. The output of the observer is the state estimate $\hat{x}_k$.

9.1 Output-Injection Observer Design

Since $\hat{x}_k$ is the state estimate, we could call

$$\hat{y}_k = C\hat{x}_k \qquad (9.1.6)$$

the estimated output. It is desired that $\hat{x}_k$ be close to x_k. Thus, if the observer is working properly, the quantity $y_k - \hat{y}_k$ which appears in (9.1.3) should be small. In fact

$$\tilde{y}_k = y_k - \hat{y}_k \qquad (9.1.7)$$

is the *output estimation error*.

It is worth examining Fig. 9.1-1, which depicts the state observer. Note that the observer consists of two parts: a model of the system involving (A, B, C), and an error-correcting portion that involves the output error multiplied by L. We call matrix L the *observer gain*. Since, according to the figure we are injecting the output $\hat{y}_k$ into $\hat{x}_{k+1}$, L is also called an *output injection* matrix.

To demonstrate that the proposed dynamical system is indeed an observer, it is necessary to show that it manufactures an estimate $\hat{x}_k$ that is close to the actual state x_k. For this purpose, define the *(state) estimation error* as

$$\tilde{x}_k = x_k - \hat{x}_k. \qquad (9.1.8)$$

By shifting (9.1.8) to $k + 1$ and using (9.1.1) and (9.1.4), it is seen that the estimation error has dynamics given by

$$\tilde{x}_{k+1} = (A - LC)\tilde{x}_k = A_o\tilde{x}_k. \qquad (9.1.9)$$

The initial estimation error is $\tilde{x}_0 = x_0 - \hat{x}_0$, with $\hat{x}_0$ the initial estimate, which could be taken as zero.

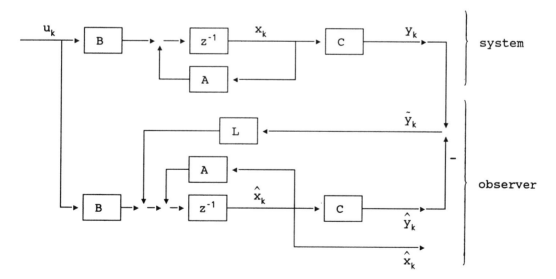

Figure 9.1-1 State observer

It is required that the estimation error vanish with time for any $\tilde{x}_0$, for then $\hat{x}_k$ will approach x_k. This will occur if $A_o = (A - LC)$ is asymptotically stable. Therefore, as long as we select the observer gain L so that $(A - LC)$ is stable, (9.1.3) is indeed an observer for the state in (9.1.1). The observer design problem is to select L so that the error vanishes suitably quickly.

Selecting the output injection matrix L

It is a well-known result of modern control theory that the poles of $(A - LC)$ may be arbitrarily assigned to desired locations by appropriate choice of L if and only if (C, A) is observable (Chapter 2). The observability of (C, A) may be checked by examining the rank of the observability matrix V. By appropriate choice of L, the observable poles of (C, A) may be assigned arbitrarily in $A_o = A - LC$. The unobservable poles of (C, A) cannot be moved and appear as poles of A_o for every L.

In fact, $(A - LC)$ can be made stable for some L as long as only the unstable poles of A are observable. This property is called detectability of (C, A) (Chapter 2) and is more difficult to test for than observability.

There are many ways to select the output injection matrix L, some of which we now give.

Selecting L by analytic manipulations. In some cases the plant A and C matrices are not too complicated. Then, we may define the *desired observer polynomial* $\Delta^D(z)$. This desired polynomial is found by selecting desired observer poles. By choosing the observer poles close to the origin in the z-plane, the estimation error $\tilde{x}_k$ can be made to vanish as quickly as desired. A general rule of thumb is that, for suitable accuracy in the state estimate $\hat{x}_k$, the slowest observer pole should have a time constant 5–10 times faster than that of the fastest plant pole.

Once $\Delta^D(z)$ has been selected, we may find the *observer characteristic polynomial*

$$\Delta_o(z) = |zI - (A - LC)|. \tag{9.1.10}$$

Then, for simple plants, we may solve $\Delta_o(z) = \Delta^D(z)$ directly for the observer gain L that results in the desired observer poles. This technique is illustrated in Example 9.1-1.

Computing L using Ackermann's formula. In the single-output case, the observability matrix V is square. Then, if the system is observable so that V is nonsingular, the observer version of *Ackermann's formula* (Chapter 2) may be used to compute L. If $\Delta^D(z)$ is the desired observer polynomial, then the observer gain required to place the observer poles as desired is given by

$$L = \Delta^D(A)V^{-1}e_n, \tag{9.1.11}$$

with $e_n = [0 \ldots 0 \ 1]^T$ the last column of the $n \times n$ identity matrix. This is the dual of the formula given in Chapter 2 for pole-placement design (see the problems).

Design using Ackermann's formula is illustrated in Example 9.1-2.

Deadbeat Observer. If the plant is observable, we can always select L so that the observer A_o has all its poles at $z = 0$. This is called a *deadbeat observer*, and guarantees

9.1 Output-Injection Observer Design

that the estimation error vanishes after n sample periods. The deadbeat observer polynomial is

$$\Delta^D(z) = z^n. \tag{9.1.12}$$

We shall illustrate deadbeat observer design in Example 9.1-1. The deadbeat observer is the dual of deadbeat control as discussed in Example 7.4-1.

Observer Design for Multi-Output Systems. If (C, A) is observable, it is not difficult even for complicated multivariable systems to select L so that the observer is stable. In fact, this is a problem we have already solved under a different guise. Let us use *duality theory* to find a suitable L.

Recall the state-feedback control law for system (9.1.1), which is

$$u_k = -Kx_k. \tag{9.1.13}$$

This results in the closed-loop system

$$x_k = (A - BK)x_k. \tag{9.1.14}$$

The state-feedback design problem is to select K for desired closed-loop properties. We have shown how this may be accomplished in Table 7.4-3. Thus, if we select the feedback gain as the Kalman gain

$$K = (R + B^TPB)^{-1}B^TPA. \tag{9.1.15}$$

with P the positive-definite solution to the discrete Algebraic Riccati equation (ARE)

$$0 = A^TPA - P + Q - A^TPB(R + B^TPB)^{-1}B^TPA, \tag{9.1.16}$$

then, if (A, B) is reachable and $(\sqrt{Q}, A)$ is observable, the closed-loop system is guaranteed to be stable. The matrices $Q \geq 0$ and $R > 0$ are design parameters that will determine the closed-loop dynamics. They are symmetric, and R can generally be taken as ρI, with I the identity matrix.

Now, compare (9.1.9) and (9.1.14). They are very similar. In fact

$$(A - LC)^T = A^T - C^TL^T, \tag{9.1.17}$$

which has the free matrix L^T to the right, exactly as in the state-feedback problem involving $(A - BK)$. This important fact is called *duality*; that is, state feedback and output injection are duals. (Note that $A - LC$ and $(A - LC)^T$ have the same poles.)

The important result of duality for us is that the same theory we have developed for selecting the state-feedback gain may be used to select the output-injection gain L. In fact, compare (9.1.17) to $(A - BK)$. Now, in the design equations (9.1.15) and (9.1.16) let us replace A, B, and K everywhere they occur by A^T, C^T, and L^T respectively. The result is

$$0 = APA^T - P + Q - APC^T(R + CPC^T)^{-1}CPA^T \tag{9.1.18}$$

$$L = APC^T(R + CPC^T)^{-1}. \tag{9.1.19}$$

These equations give a design technique for multivariable digital observers. It is only necessary to solve the first of these equations for P and then determine the observer output-injection matrix from the second equation. We call (9.1.18) the *observer algebraic Riccati equation* (ARE). Efficient algorithms for its solution are commercially available (e.g., ORACLS [Armstrong 1980] and PC-MATLAB [Moler et al. 1987]).

The next theorem shows that observer stability is guaranteed under reasonable conditions on the plant and the design parameters Q and R. It is the dual to the digital control result given for state feedback at the end of section 7.4.

Theorem. Let (C, A) be observable and $(A, \sqrt{Q})$ be reachable. Then $(A - LC)$ is stable so that the estimation error goes to zero using the gain (9.1.19) in observer (9.1.3), with P the unique positive-definite solution to the observer ARE equation (9.1.18).

Stability of the error system guarantees that the state estimate $\hat{x}_k$ will approach the actual state x_k.

The power of this theorem is that we may treat $Q \geq 0$ and $R > 0$ as design parameters that may be tuned until suitable observer behavior results using the gain computed from the observer ARE. As long as (C, A) is observable and we select $R > 0$ and $(A, \sqrt{Q})$ reachable, observer stability is assured. Thus, observer design is straightforward even for complicated MIMO systems.

An iterative design procedure for stable discrete observers would involve selecting Q and R, solving the ARE for its positive-definite solution P, and finding the observer gain L. Then, check the poles of $A_o = (A - LC)$ and/or perform some simulations. If the resulting design is not suitable, try different values for Q and R and repeat.

Observer robustness

We have assumed that the system matrices (A, B, C) are exactly known. Unfortunately, in reality this is rarely the case, since there is usually some modeling uncertainty. However, if the poles of $(A - LC)$ are selected stable enough, then the estimation error will be small in spite of uncertainties in the system matrices. That is, the observer has some built-in robustness.

Because of this robustness, the observer can often give good performance in the presence of noise. An actual continuous-time plant might be described by

$$\dot{x} = A^c x + B^c u + G^c w \qquad (9.1.20)$$

$$y = C^c x + H^c v, \qquad (9.1.21)$$

with $w(t)$ a *process noise* and $v(t)$ a *measurement noise*. Discretizing this plant and disregarding the noise terms leads to (9.1.1), (9.1.2). If an observer is designed for the discrete system using the techniques just shown, it will often perform well when used on the continuous-time plant in spite of the process and measurement noises. This is illustrated in Example 9.1-1.

In the second portion of this chapter we show how to take the process measurement noises explicitly into account during the design of the estimator so that improved performance results. The resulting estimator is called the Kalman filter.

9.1 Output-Injection Observer Design

Example 9.1-1: Digital Observer for Disk Drive Head-Positioning System

We discussed a disk drive head-positioning system in Example 4.4-1. See also Stich [1987]. The basic dynamics of a linear voice-coil actuated positioning mechanism are

$$\dot{x} = \begin{bmatrix} 0 & 1 \\ 0 & 0 \end{bmatrix} x + \begin{bmatrix} 0 \\ 1 \end{bmatrix} u + G^c w = A^c x + B^c u + G^c w, \quad (1)$$

where the state is $x = [d \; v]^T$, with $d(t)$ the position and $v(t)$ the velocity, and the control $u(t)$ is an acceleration or force-per-unit-mass input. The input is directly proportional to current, since the driver system is a current driver. This is just Newton's system since $\ddot{d} = u$ (i.e., $F = ma$).

The process noise $w(t)$ represents a random force disturbance input due, for instance, to uncertainties in the current driver or to friction effects in the voice coil.

In Example 4.4-1 we designed a control system for track following that rejects the elliptical disturbance that arises due to the fact that the disk is not quite centered. However, we used a full state-variable feedback that required knowledge of both position $d(t)$ and velocity $v(t)$. Unfortunately, only position measurements are generally feasible for the disk drive head, so that the available output is

$$y = [1 \; 0] x + v = C^c x + v. \quad (2)$$

The measurement noise $v(t)$ represents a random measurement error.

Therefore, for controls purposes let us design a digital observer that reconstructs the head velocity from the position measurements. We emphasize that pure differentiation of the position measurement will result in a very noisy signal that poorly approximates the actual velocity. On the other hand, the digital observer will be a *low-pass filter* that has noise rejection properties.

a. Deadbeat Observer for the Noise-Free Case

Assuming that the noises are absent, we may discretize the plant as in Example 7.1-2 to obtain

$$x_{k+1} = \begin{bmatrix} 1 & T \\ 0 & 1 \end{bmatrix} x_k + \begin{bmatrix} T^2/2 \\ T \end{bmatrix} u_k \equiv A x_k + B u_k \quad (3)$$

$$y_k = [1 \; 0] x_k \equiv C x_k, \quad (4)$$

with T the sample period and $x_k = [d_k \; v_k]^T$.

Using the observer gain

$$L = [L_1 \; L_2]^T, \quad (5)$$

with L_1 and L_2 scalars, and defining the output estimation error

$$\tilde{y}_k = y_k - \hat{y}_k = d_k - \hat{d}_k, \quad (6)$$

the observer is given by (9.1.3), which we may write as

$$\hat{d}_{k+1} = \hat{d}_k + T\hat{v}_k + \frac{T^2}{2} u_k + L_1 \tilde{y}_k \quad (7)$$

$$\hat{v}_{k+1} = \hat{v}_k + T u_k + L_2 \tilde{y}_k. \quad (8)$$

Let us design a deadbeat observer, which has all poles at the origin. To select L_1 and L_2, write

$$A_o = A - LC = \begin{bmatrix} 1-L_1 & T \\ -L_2 & 1 \end{bmatrix}, \qquad (9)$$

Then, the observer has characteristic polynomial given by

$$\Delta_o(z) = |zI - A_o| = z^2 - (2 - L_1)z + (L_2 T - L_1 + 1). \qquad (10)$$

For deadbeat behavior, we should like $\Delta_o(z)$ to equal the deadbeat characteristic polynomial

$$\Delta^D(z) = z^2. \qquad (11)$$

Equating $\Delta_o(z)$ and $\Delta^D(z)$ and solving for the observer gains yields

$$\begin{aligned} L_1 &= 2 \\ L_2 &= 1/T. \end{aligned} \qquad (12)$$

It is interesting to compare the deadbeat observer gains with the deadbeat control gains in Example 7.4-1. Note that, for deadbeat design there is only *one design parameter*; namely, the sample period T.

A simulation was performed using these deadbeat gains in (7) and (8). The driver program is the digital control program DIGCTLF described in Appendix A. Figure 9.1-2 shows subroutine $F(time, x, xp)$, which contains the continuous head mechanism dynamics (1), (2). The function of this subroutine is to provide a simulated measured plant output y_k for the observer. In this simulation, we set the process noise "pnoise" and the measurement noise "mnoise" to zero.

The figure also contains subroutine DIG(IT, T, x), which contains the discrete dynamics, namely, in this case the observer (7), (8). The sample period was chosen as 0.5 units. The plant initial conditions were $d(0) = -1$, $v(0) = 1$, and we set input $u(t) = 0$. The actual states as well as the estimates provided by the digital observer are shown in Fig. 9.1-3, where we selected the initial values $\hat{d}_0 = 0$, $\hat{v}_0 = 0$.

No measurement was taken at $k = 0$; the first measurement was taken at $t = T = 0.5$. Notice that, due to the deadbeat design, $A_o^2 = 0$, so that the estimates are exactly equal to the samples of $d(t)$ and $v(t)$ two sample periods after the first measurement is taken, that is, after $t = 3T = 1.5$.

b. Filtered Observer

The estimates in Fig. 9.1-3 are equal to the samples of $d(t)$ and $v(t)$ after $t = 1.5$; however, they lag behind the continuous state. They could be called (one step ahead) *predictive estimates* of the state. For controls purposes, it might be advantageous to have estimates that are more up-to-date. These are easy to obtain in this example by slightly modifying the observer.

Let us change (7), (8) to read

$$\hat{d}_{k+1} = \hat{d}_k + T\hat{v}_k + \frac{T^2}{2}u_k + L_1(y_{k+1} - \hat{d}_k) \qquad (13)$$

$$\hat{v}_{k+1} = \hat{v}_k + Tu_k + L_2(y_{k+1} - \hat{d}_k). \qquad (14)$$

Then, the current output $y_{k+1} = d_{k+1}$ is used to find x_{k+1}. Thus, the digital observer uses more up-to-date information. The up-to-date state estimates are called *filtered estimates*.

9.1 Output-Injection Observer Design

```
C    NEWTON'S SYSTEM FOR EXAMPLE 9.1-1
C    OBSERVER FOR DISK DRIVE HEAD MECH.
C    FILE EX911F.FOR

     SUBROUTINE F(time,x,xp)
     REAL x(*), xp(*), mnoise
     COMMON/CONTROL/u
     COMMON/OUTPUT/y

c    CALL RANDOM(pnoise)
c    pnoise= 10*(pnoise - 0.5)
     xp(1)= x(2)
     xp(2) = u + pnoise

c    CALL RANDOM(mnoise)
c    mnoise= 0.1*(mnoise - 0.5)
     y= x(1) + mnoise

     RETURN
     END

C    DIGITAL OUTPUT-INJECTION OBSERVER

     SUBROUTINE DIG(IT,T,x)
     REAL x(*), L1, L2
     COMMON/CONTROL/u
     COMMON/OUTPUT/y    ,d,v
c    (d,v are in OUTPUT for plotting purposes only)
     L1= 2
     L2= 1/T

     e= ym1 - d
     d= d + T*v + u*T**2/2 + L1*e
     v=      v + u*T       + L2*e
     ym1= y

     RETURN
     END
```

Figure 9.1-2 Subroutines $F(\text{time}, x, xp)$ and DIG (IT, T, x) required for driver program DIGCTLF

The simulation was repeated using this modification and the actual and estimated states are shown in Fig. 9.1-4. Note that now the estimates *lead* the value of $d(t)$.

c. Observer Robustness to Process Noise

Now, we added a significant amount of process noise, denoted "pnoise" in Fig. 9.1-2. The measurement noise "mnoise" was set to zero. The resulting simulation using the predictive observer is shown in Fig. 9.1-5. Note that, although the process noise has caused the continuous plant to drift from its unforced response, the estimates follow the actual states fairly closely. Thus, the observer is working well. This shows that the observer is robust to process noise.

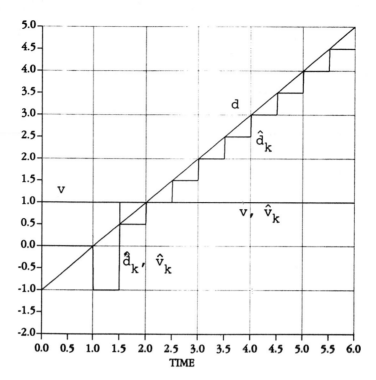

Figure 9.1-3 Actual and estimated states for head-positioning mechanism

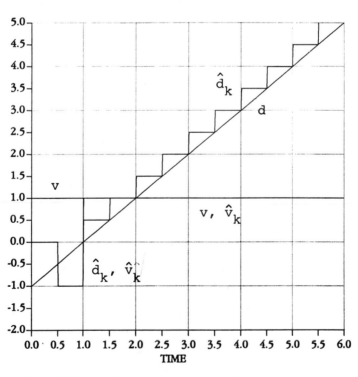

Figure 9.1-4 Actual and estimated states using filtered deadbeat observer

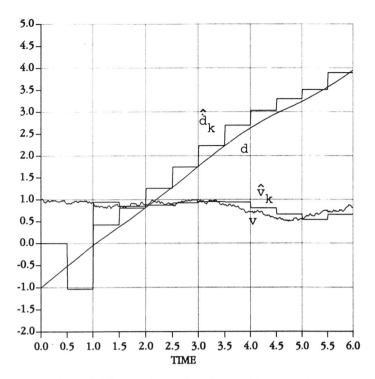

Figure 9.1-5   Actual and estimated states with process noise

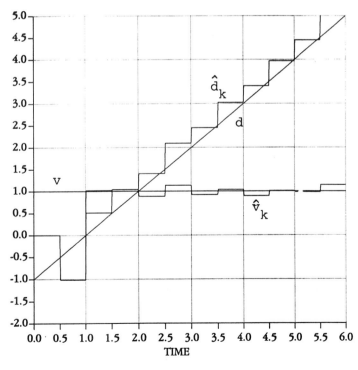

Figure 9.1-6   Actual and estimated states with measurement noise

d. Observer Robustness to Measurement Noise

Finally, we set the process noise "pnoise" to zero in Fig. 9.1-2 and injected a measurement noise "mnoise" as shown in subroutine $F(\text{time}, x, xp)$. The result using the filtered observer is shown in Fig. 9.1-6. The estimates for d_k are good, and those for v_k would be adequate for some purposes.

In the second portion of this chapter we shall discuss the Kalman filter, which has far better noise rejection properties than the deadbeat observer of this example.

∎

Continuous Observer

Let us now discuss the design of continuous-time observers. The theory is very similar to that we have just developed. The only difference is that stability is now referred to the left half of the s-plane, and not the interior of the unit circle in the z-plane. This means that we should use a different ARE for design.

A continuous-time model of the plant is

$$\dot{x} = Ax + Bu \qquad (9.1.22)$$

$$y = Cx \qquad (9.1.23)$$

with $x(t) \in \mathbf{R}^n$ the state, $u(t) \in \mathbf{R}^m$ the control input, and $y(t) \in \mathbf{R}^p$ the available measured outputs.

The state observer is a dynamical system described by

$$\dot{\hat{x}} = A\hat{x} + Bu + L(y - C\hat{x}) \qquad (9.1.24)$$

or

$$\dot{\hat{x}} = (A - LC)\hat{x} + Bu + Ly \equiv A_o\hat{x} + Bu + Ly, \qquad (9.1.25)$$

with $\hat{x}(t)$ the estimate of $x(t)$.

Using (9.1.22) and (9.1.24) we see that the estimation error $\tilde{x}(t) = x(t) - \hat{x}(t)$ has dynamics given by

$$\dot{\tilde{x}} = (A - LC)\tilde{x} = A_o\tilde{x}. \qquad (9.1.26)$$

The initial estimation error is $\tilde{x}(0) = x(0) - \hat{x}(0)$, with $\hat{x}(0)$ the initial estimate. To ensure that the estimation error vanishes with time for any $\tilde{x}(0)$, we should select the output injection L so that $A_o = (A - LC)$ is asymptotically stable. Then the estimate $\hat{x}(t)$ will closely approximate the actual state $x(t)$.

To select the observer gain, any of the techniques discussed in connection with digital observers may be used. In the single-output case, Ackermann's formula (9.1.11) may be used to choose L. It is important to note that there is no "deadbeat" observer for continuous-time systems. That is, theoretically the observer error cannot be driven exactly to zero in the continuous observer (since $z = 0$ corresponds to $s = -\infty$). Moreover, unlike the digital case, making the estimation error die out faster requires higher gains, since the poles must be farther to the left in the s-plane. Such high gains may be unreal-

9.1 Output-Injection Observer Design

izable in practice. Thus, it is clear that digital observers offer performance that cannot be matched using continuous-time observers.

For multi-output systems, where the simple design techniques are complicated to apply, to ensure that the poles of A_o are in the left-half plane, we may use the output injection matrix

$$L = PC^T R^{-1}, \qquad (9.1.27)$$

with P the positive definite solution to the continuous ARE

$$0 = AP + PA^T + Q - PC^T R^{-1} CP. \qquad (9.1.28)$$

This is the dual to the Kalman gain solution for state-feedback control given in section 3.3.

The next fundamental theorem of modern control theory shows that, under certain reasonable conditions, by using these matrix design equations it is easy to design stable observers even for complicated multivariable plants.

Theorem. Let (C, A) be observable and $(A, \sqrt{Q})$ be reachable. Then the error system (9.1.26) using the gain L given by (9.1.27), with P the unique positive definite solution to (9.1.28), is asymptotically stable.

Stability of the error system guarantees that the state estimate $\hat{x}(t)$ will approximate the actual state $x(t)$. By selecting L to place the poles of $(A - LC)$ far enough to the left in the s-plane, the estimation error $\tilde{x}(t)$ can be made to vanish as quickly as desired.

An iterative design technique for observers involves selecting the design parameters $Q \geq 0$ and $R > 0$ so that $(A, \sqrt{Q})$ is reachable and then solving the ARE for P using one of the commercially available packages (e.g., Armstrong 1980, MATLAB). Then, L is found using (9.1.27) and $A_o = A - LC$ computed. If simulation reveals the design to be unsuitable, a different Q and R may be selected and the design repeated. It is often acceptable to select Q and R as diagonal matrices.

Example 9.1-2: Continuous Observer for Disk Drive Head-Positioning System

Let us consider continuous-time observer design for the disk drive head-positioning mechanism (see Example 9.1-1). In Example 3.3-1 we discussed continuous-time state-feedback design for systems obeying Newton's laws in order to show how the design parameters Q and R influence the closed-loop behavior. This example is meant to show the influence of the design matrices Q and R on the observer poles. Therefore, we shall take an analytic approach to design here.

The plant dynamics are

$$\dot{x} = \begin{bmatrix} 0 & 1 \\ 0 & 0 \end{bmatrix} x + \begin{bmatrix} 0 \\ 1 \end{bmatrix} u = Ax + Bu, \qquad (1)$$

where the state is $x = [d \ v]^T$, with $d(t)$ the position and $v(t)$ the velocity, and the control $u(t)$ is an acceleration input. Let us take position measurements so that the measured output is

$$y = [1 \ 0] x = Cx. \qquad (2)$$

We should like to design an observer that will reconstruct the full state $x(t)$ given only position measurements. This is possible since the observability matrix is nonsingular (check it!). Let us note that simple differentiation of $y(t) = d(t)$ to obtain $v(t)$ is unsatisfactory, since differentiation increases sensor noise. In fact, the observer is a low-pass filter that provides estimates while rejecting high-frequency noise. We shall discuss two techniques for observer design.

a. Riccati Equation Design

In this example we want to analytically solve the ARE to show the relation between the design parameters Q and R and the observer poles. Such an analytic solution is not required in an actual design, since there is good software available in standard design packages for solving the observer ARE (e.g., ORACLS [Armstrong 1980] and PC-MATLAB [Moler et al. 1987]).

Selecting $R = 1$ and $Q = \text{diag}\{q_d, q_v^2\}$ with q_d, q_v nonnegative, we may assume that

$$P = \begin{bmatrix} p_1 & p_2 \\ p_2 & p_3 \end{bmatrix} \tag{3}$$

for some scalars p_1, p_2, p_3 to be determined. The observer ARE (9.1.28) becomes

$$0 = \begin{bmatrix} 0 & 1 \\ 0 & 0 \end{bmatrix}\begin{bmatrix} p_1 & p_2 \\ p_2 & p_3 \end{bmatrix} + \begin{bmatrix} p_1 & p_2 \\ p_2 & p_3 \end{bmatrix}\begin{bmatrix} 0 & 0 \\ 1 & 0 \end{bmatrix} \\ + \begin{bmatrix} q_d & 0 \\ 0 & q_v^2 \end{bmatrix} - \begin{bmatrix} p_1 & p_2 \\ p_2 & p_3 \end{bmatrix}\begin{bmatrix} 1 & 0 \\ 0 & 0 \end{bmatrix}\begin{bmatrix} p_1 & p_2 \\ p_2 & p_3 \end{bmatrix}. \tag{4}$$

which may be multiplied out to obtain the three scalar equations

$$0 = 2p_2 - p_1^2 + q_d. \tag{5a}$$

$$0 = p_3 - p_1 p_2 \tag{5b}$$

$$0 = -p_2^2 + q_v^2. \tag{5c}$$

Solving these equations gives

$$p_2 = q_v \tag{6a}$$

$$p_1 = \sqrt{2}\sqrt{q_v + q_d/2} \tag{6b}$$

$$p_3 = q_v \sqrt{2}\sqrt{q_v + q_d/2}, \tag{6c}$$

where we have selected the signs that make P positive definite.

According to (9.1.27), the observer gain is equal to

$$L = \begin{bmatrix} p_1 & p_2 \\ p_2 & p_3 \end{bmatrix}\begin{bmatrix} 1 \\ 0 \end{bmatrix} = \begin{bmatrix} p_1 \\ p_2 \end{bmatrix} \tag{7}$$

Therefore,

$$L = \begin{bmatrix} \sqrt{2}\sqrt{q_v + q_d/2} \\ q_v \end{bmatrix}. \tag{8}$$

9.1 Output-Injection Observer Design

Using (8), the error system matrix is found to be

$$A_o = (A - LC) = \begin{bmatrix} -\sqrt{2}\sqrt{q_v + q_d/2} & 1 \\ -q_v & 0 \end{bmatrix}. \tag{9}$$

Therefore, the observer characteristic polynomial is

$$\Delta_o(s) = |sI - A_o| = s^2 + 2\zeta\omega s + \omega^2, \tag{10}$$

with the observer natural frequency ω and damping ratio ζ given by

$$\omega = \sqrt{q_v}, \qquad \zeta = \frac{1}{\sqrt{2}}\sqrt{1 + q_d/2q_v}. \tag{11}$$

It is now clear how selection of Q affects the observer behavior. Note that, if $q_d = 0$ the damping ratio becomes the familiar $1/\sqrt{2}$.

The reader should verify that the system is observable, and that $(A, \sqrt{Q})$ is reachable as long as $q_v \neq 0$. Under this condition, (11) reveals that the observer is indeed stable as stated by the theorem. A comparison with Example 3.3-1, where a state feedback was designed for Newton's system, reveals some interesting aspects of duality.

b. Ackermann's Formula Design

Riccati equation observer design is useful whether the plant has only one or multiple outputs. If there is only one output, we may use Ackermann's formula (9.1.11).

Let the desired observer polynomial be

$$\Delta^D(s) = s^2 + 2\zeta\omega s + \omega^2 \tag{12}$$

for some specified damping ratio ζ and natural frequency ω. Then

$$\Delta^D(A) = A^2 + 2\zeta\omega A + \omega^2 I = \begin{bmatrix} \omega^2 & 2\zeta\omega \\ 0 & \omega^2 \end{bmatrix} \tag{13}$$

$$V = \begin{bmatrix} C \\ CA \end{bmatrix} = I \tag{14}$$

so that the observer gain is

$$L = \begin{bmatrix} 2\zeta\omega \\ \omega^2 \end{bmatrix}. \tag{15}$$

One may verify that the characteristic polynomial of $A_o = A - LC$ is indeed (12).

c. Simulation

To design an observer with a complex pole pair having damping ratio of $\zeta = 1/\sqrt{2}$ and natural frequency of $\omega = 1$ rad/sec, the observer gain was selected as

$$L = \begin{bmatrix} \sqrt{2} \\ 1 \end{bmatrix}. \tag{16}$$

The resulting time histories of the actual states and their estimates are shown in Fig. 9.1-7. The initial conditions were $d(0) = -1$, $v(0) = 1$, and the input was $u(t) = 0$. The observer

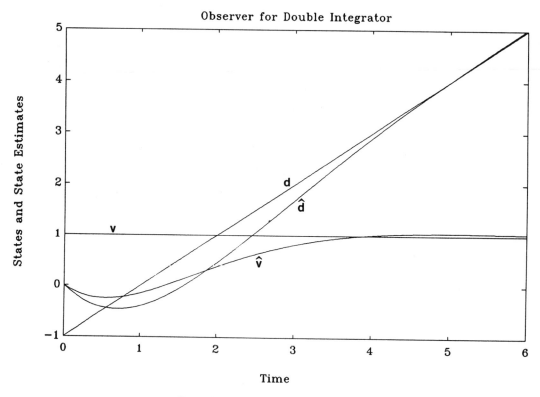

Figure 9.1-7 Actual and estimated states

was started with initial states of $\hat{d}(0) = 0$, $\hat{v}(0) = 0$. It is quite interesting to compare this simulation with that of the deadbeat digital observer in Example 9.1-1.

∎

9.2 REDUCED-ORDER OBSERVERS

In section 9.1 we showed how to design state observers using the output-injection approach. The observer had n states and so was of the same order as the plant. It is, however, possible to design *reduced-order* observers that have less than n states and so are easier to implement. In this section, we shall show how to design observers that have $n - p$ states, with p the number of measured outputs [Chen 1984].

Digital Observer

To design a digital observer for a continuous-time plant, first discretize the plant using the techniques of section 7.1 to obtain the discrete-time dynamical description

$$x_{k+1} = Ax_k + Bu_k \qquad (9.2.1)$$

9.2 Reduced-Order Observers

$$y_k = Cx_k, \qquad (9.2.2)$$

where $x_k \in \mathbf{R}^n$, $y_k \in \mathbf{R}^p$. We are interested in designing a dynamical observer to estimate the state x_k given the output y_k. Assume (C, A) is observable.

Observer dynamics

Let us propose an observer of the form

$$z_{k+1} = Fz_k + Gy_k + Hu_k \qquad (9.2.3)$$

with state $z_k \in \mathbf{R}^{n-p}$. Define an estimation error by

$$e_k = z_k - Px_k, \qquad (9.2.4)$$

where P is an $(n - p) \times n$ matrix to be determined. We should like to select the observer matrices F, G, H so that e_k vanishes with time, for then z_k will provide an estimate of Px_k.

Selecting the observer matrices F, G, and H

To examine the dynamics of e_k, shift (9.2.4) to $k + 1$ and use (9.2.1) and (9.2.3) to obtain

$$e_{k+1} = Fz_k + Gy_k + Hu_k - PAx_k - PBu_k.$$

Let us now select

$$H = PB \qquad (9.2.5)$$

and matrices F and G to satisfy the linear matrix equation

$$PA - FP = GC. \qquad (9.2.6)$$

Then, we may write

$$e_{k+1} = Fz_k - FPx_k = Fe_k, \qquad (9.2.7)$$

from which it is clear that the error will go to zero with time as long as the observer system matrix F is asymptotically stable.

Once the error e_k has vanished, we shall have approximately

$$z_k = Px_k. \qquad (9.2.8)$$

Then, we can write approximately

$$\begin{bmatrix} z_k \\ y_k \end{bmatrix} = \begin{bmatrix} P \\ C \end{bmatrix} x_k, \qquad (9.2.9)$$

so that if the coefficient matrix

$$W = \begin{bmatrix} P \\ C \end{bmatrix} \qquad (9.2.10)$$

is invertible, then

$$\hat{x}_k = \begin{bmatrix} P \\ C \end{bmatrix}^{-1} \begin{bmatrix} z_k \\ y_k \end{bmatrix} \quad (9.2.11)$$

provides an estimate for the state x_k.

The observer design hinges on solving the Lyapunov equation (9.2.6) for an F, G, and P such that F is stable and such that W is nonsingular. Given A, C, F, and G, there exists a unique P satisfying (9.2.6) if the eigenvalues of A and F are distinct. If this holds, then necessary conditions for the existence of a P such that W is nonsingular are that (A, C) be observable and (F, G) be reachable. That is, these conditions must hold, but even if they do there is no guarantee that W is nonsingular [Chen 1984].

The following design procedure for reduced-order observers is proposed:

1. Choose an $(n - p) \times (n - p)$ matrix F with desirable eigenvalues for forcing e_k to decay to zero. The eigenvalues of F should be distinct from those of the plant matrix A, with time constants about 5–10 times faster.
2. Choose G so that (F, G) is reachable.
3. Solve for the unique P in (9.2.6).
4. Check W for full rank. If W is singular, select a different G and go to step 2, or select a different F and go to step 1.

Example 9.2-1: Reduced-Order Digital Observer for Disk Drive Head-Positioning System

Let us design a reduced-order observer for the disk-drive head positioning mechanism of Example 9.1-1. The discretized system dynamics assuming position measurements are

$$x_{k+1} = \begin{bmatrix} 1 & T \\ 0 & 1 \end{bmatrix} x_k + \begin{bmatrix} T^2/2 \\ T \end{bmatrix} u_k \equiv Ax_k + Bu_k \quad (1)$$

$$y_k = \begin{bmatrix} 1 & 0 \end{bmatrix} x_k \equiv Cx_k, \quad (2)$$

with T the sample period and $x_k = [d_k \quad v_k]^T$.

a. Observer Design

The order of the observer will be $n - p = 1$. Therefore, let us select $F = a < 1$ so that the observer pole is at $z = a$ and the error e_k will decay like a^k. To make (F, G) reachable, choose $G = 1$. Then, the observer is

$$z_{k+1} = az_k + y_k + Hu_k \quad (3)$$

with $H = PB$ to be determined.
Denoting

$$P = [p_1 \quad p_2], \quad (4)$$

the Lyapunov design equation $PA - FP = GC$ becomes

$$[p_1 \quad p_2] \begin{bmatrix} 1 & T \\ 0 & 1 \end{bmatrix} - a[p_1 \quad p_2] = [1 \quad 0] \quad (5)$$

9.2 Reduced-Order Observers

or

$$[(1-a)p_1 \quad (1-a)p_2 + Tp_1] = [1 \quad 0] \tag{6}$$

whence

$$P = [p_1 \quad p_2] = \left[\dfrac{1}{1-a} \quad \dfrac{-T}{(1-a)^2}\right]. \tag{7}$$

Since

$$W = \begin{bmatrix} P \\ C \end{bmatrix} = \begin{bmatrix} 1/(1-a) & -T/(1-a)^2 \\ 1 & 0 \end{bmatrix} \tag{8}$$

is nonsingular, the choices for F and G were satisfactory. Note, however, that as T gets smaller, W becomes ill-conditioned. Thus, very small sample periods should be avoided.

For deadbeat behavior, let us select $a = 0$ so that the observer is

$$z_{k+1} = y_k - \dfrac{T^2}{2}u \tag{9}$$

and the state estimate is given by

$$\hat{x}_k = W^{-1}\begin{bmatrix} z_k \\ y_k \end{bmatrix} = \begin{bmatrix} 0 & 1 \\ -1/T & 1/T \end{bmatrix}\begin{bmatrix} z_k \\ y_k \end{bmatrix} = \begin{bmatrix} y_k \\ (y_k - z_k)/T \end{bmatrix}. \tag{10}$$

Note that

$$z_k = Px_k = d_k - Tv_k, \tag{11}$$

so that $z(t)$ provides an estimate of a linear combination of the position $d(t)$ and the velocity $v(t)$.

b. Simulation

A simulation of the actual states and their estimates may be obtained using program DIGCTLF described in Appendix A. The subroutine F(time, x, xp) to simulate the continuous dynamics is the same as in Fig. 9.1-2. Subroutine DIG(IT,T,x) for the discrete dynamics, that is, the digital observer, appears in Fig. 9.2-1. We set the process and measurement noises

```
C    DIGITAL REDUCED-ORDER OBSERVER (DEADBEAT)

     SUBROUTINE DIG(IT,T,x)
     REAL x(*)
     COMMON/CONTROL/u
     COMMON/OUTPUT/y    ,d,v
c    (d,v are in OUTPUT for plotting purposes only)

     z= yml - u*T**2/2
     yml= y
     d= y
     v= (y - z)/T

     RETURN
     END
```

Figure 9.2-1 Subroutine DIG (*IT, T, x*) required for driver program DIGCTLF

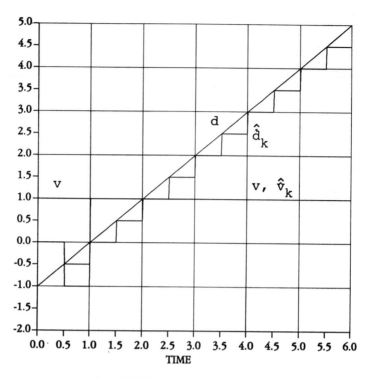

Figure 9.2-2 Actual and estimated states

to zero, the initial plant states to $d(0) = -1$, $v(0) = 1$, and the observer initial state to $z_0 = 0$. No measurement was taken at time $k = 0$. The simulation results appear in Fig. 9.2-2. Compare them to the results of Example 9.1-1a. Since the observer here has order one, the deadbeat observer results in exact estimates one sample period after the first measurement, that is, at $t = 2T = 1$ sec.

The measurement noise and process noise rejection properties of this reduced-order observer are about the same as for the observer in Example 9.1-1.

■

Continuous Observer

We are now interested in using a *continuous-time* observer to estimate $x(t)$ in

$$\dot{x} = Ax + Bu \qquad (9.2.12)$$

from measurements of the output

$$y = Cx, \qquad (9.2.13)$$

where $x \varepsilon \mathbf{R}^n$, $y \varepsilon \mathbf{R}^p$. Assume (C, A) is observable.

9.2 Reduced-Order Observers

Observer dynamics

Propose an observer of the form

$$\dot{z} = Fz + Gy + Hu, \tag{9.2.14}$$

with $z \in \mathbf{R}^{n-p}$, and define an estimation error by

$$e = z - Px, \tag{9.2.15}$$

where P is an $(n - p) \times n$ matrix to be determined. We should like to select the observer matrices F, G, H so that $e(t)$ vanishes with time, for then $z(t)$ will provide an estimate of $Px(t)$.

Selecting the observer matrices F, G, and H

To examine the dynamics of $e(t)$, differentiate (9.2.15) and use (9.2.12) and (9.2.14) to obtain

$$\dot{e} = Fz + Gy + Hu - PAx - PBu.$$

As in the discrete case, we may select H as in (9.2.5), and matrices F and G to satisfy the Lyapunov design equation (9.2.6). Then, we may write

$$\dot{e} = Fz - FPx = Fe, \tag{9.2.16}$$

from which it is clear that the error will go to zero with time as long as the observer system matrix F is asymptotically stable.

Once the error $e(t)$ has vanished, then

$$\hat{x} = \begin{bmatrix} P \\ C \end{bmatrix}^{-1} \begin{bmatrix} z \\ y \end{bmatrix} \tag{9.2.17}$$

provides an estimate for the state $x(t)$ as long as $W \equiv [P^T \quad C^T]^T$ is nonsingular.

Again, the observer design hinges on solving the Lyapunov equation (9.2.6) for an F, G, and P such that F is stable and W is nonsingular. The discussion on the existence of such solutions is identical to that in the discrete case just covered. The design technique already proposed also applies for continuous-time observers.

Example 9.2-2: Reduced-Order Continuous Observer for Disk Drive Head-Positioning System

Let us design a continuous-time reduced-order observer for the disk drive head-positioning mechanism in Example 9.1-1. The order of the observer will be $n - p = 1$. Therefore, let us select $F = -a$ so that the observer pole is at $s = -a$ and the error $e(t)$ will decay like e^{-at}. To make (F, G) reachable, choose $G = 1$. Then, the observer is

$$\dot{z} = -az + y + Hu \tag{1}$$

with $H = PB$ to be determined. Denoting

$$P = [p_1 \quad p_2], \tag{2}$$

the Lyapunov design equation $PA - FP = GC$ becomes

$$[p_1 \ p_2]\begin{bmatrix} 0 & 1 \\ 0 & 0 \end{bmatrix} + a[p_1 \ p_2] = [1 \ 0] \qquad (3)$$

or

$$[ap_1 \ ap_2 + p_1] = [1 \ 0] \qquad (4)$$

whence

$$P = [p_1 \ p_2] = [1/a \ -1/a^2]. \qquad (5)$$

Since

$$W = \begin{bmatrix} P \\ C \end{bmatrix} = \begin{bmatrix} 1/a & -1/a^2 \\ 1 & 0 \end{bmatrix} \qquad (6)$$

is nonsingular, the choices for F and G were satisfactory. Thus, the observer is

$$\dot{z} = -az + y - \frac{1}{a^2}u \qquad (7)$$

and the state estimate is given by

$$\hat{x} = W^{-1}\begin{bmatrix} z \\ y \end{bmatrix} = \begin{bmatrix} 0 & 1 \\ -a^2 & a \end{bmatrix}\begin{bmatrix} z \\ y \end{bmatrix} = \begin{bmatrix} y \\ -a^2 z + ay \end{bmatrix}. \qquad (8)$$

Note that

$$z = Px = d/a - v/a^2, \qquad (9)$$

so that $z(t)$ provides an estimate of a linear combination of the position $d(t)$ and the velocity $v(t)$. Since a has units of sec^{-1}, the units in (9) are compatible.

It is important to note that, as we increase a in an attempt to make the estimation error die out faster, the matrix W becomes ill-conditioned and problems could arise in reconstructing $\hat{x}$.

A block diagram of the reduced-order observer appears in Fig. 9.2-3. Note that no pure differentiations occur. Indeed, the observer is a low-pass filter, so that it has some noise-rejection properties.

A simulation of the actual states and their estimates is provided in Fig. 9.2-4, where

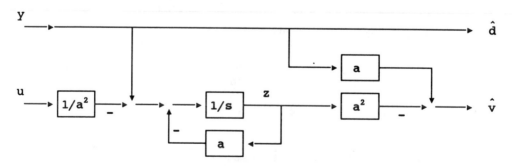

Figure 9.2-3 Continuous-time reduced-order observer for Newton's system

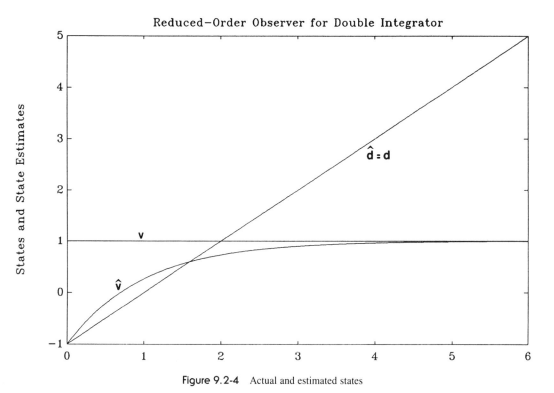

Figure 9.2-4 Actual and estimated states

we used $a = 1$. The initial conditions were $d(0) = -1$, $v(0) = 1$, and the control input was $u(t) = 0$. The observer was started with an initial condition of $z(0) = d(0) - \hat{v}(0) = 0$. Compare this to Fig. 9.1-7, where a full-order observer was used.

∎

9.3 DISCRETE KALMAN FILTER

In sections 9.1 and 9.2 we saw how to design estimators for the state assuming that an exact mathematical model was available for the system and that there is no noise present. We call these deterministic estimators or observers. Unfortunately, most practical systems have process noise, measurement noise, or both. Example 9.1-1 demonstrated that deterministic observers do have some noise rejection capability due to their inherent robustness. In the remainder of the chapter we shall show how to design state estimators that explicitly take into account the process and measurement noise to yield improved performance.

In Chapter 8 we characterized noise in terms of its spectral density, showing how to guarantee robustness using a frequency-domain approach. In this chapter we shall use a *probabilistic approach* to describe noise. Thus, it behooves the reader at this point to examine Appendix C. The estimator we shall derive is a *stochastic* estimator called the *Kalman filter*.

The Kalman filter was first introduced by R. Kalman [1960b] for discrete systems and Kalman and Bucy [1961] for continuous-time systems. These papers, along with Kalman's papers on the linear quadratic regulator [1960a], were among the key works that introduced the era of modern control theory.

In this section we shall derive the discrete-time Kalman filter given in Table 9.3-1; in section 9.4 we show how to apply it for digital filtering of continuous-time systems. In section 9.5 the continuous-time Kalman filter is derived. For more complete treatments, see Lewis [1986b], Maybeck [1979], and Gelb [1974].

Linear Discrete Stochastic Systems

A system with noise may be described by

$$x_{k+1} = Ax_k + Bu_k + Gw_k \qquad (9.3.1)$$

$$y_k = Cx_k + v_k, \qquad (9.3.2)$$

with state $x_k \in \mathbf{R}^n$, control $u_k \in \mathbf{R}^m$, measured output $y_k \in \mathbf{R}^p$. The process noise w_k represents disturbances or modeling inaccuracies. We shall assume that it is a stationary white noise process with zero mean and known covariance of Q, which we write as $w_k \sim (0, Q)$. The *measurement noise* v_k is due to sensor inaccuracy, and is assumed to be a stationary white noise process with zero mean and known covariance R, denoted $v_k \sim (0, R)$.

We assume that w_k and v_k are mutually uncorrelated so that $E\{w_j v_k^T\} = 0$ for all j and k. That is, the process and measurement noises arise in the system from independent effects, which makes sense practically. Finally, we assume that $Q \geq 0$ and $R > 0$. As covariances, Q and R are also symmetric.

We call (9.3.1), (9.3.2) a *linear discrete stochastic system*. The initial state x_0 is unknown, but we do have some knowledge of x_0 in the form of its mean value $\bar{x}_0$ and covariance P_0. Thus, $x_0 \sim (\bar{x}_0, P_0)$. It is assumed that x_0, w_j, and v_k are mutually uncorrelated for all j and k. That is, our determination of $\bar{x}_0$ and P_0 is independent of w_k and v_k, which makes physical sense.

The objective is to design an estimator that provides estimates of the state x_k taking into account the known dynamics (9.3.1) and measured data y_k (9.3.2). We shall call this estimator the Kalman filter. The difference between the Kalman filter and the deterministic observer of section 9.1 is that here we shall use our knowledge of the noise statistics to obtain improved estimates. Different formulations of the Kalman filter are given in Tables 9.3-1 through 9.3-3. The last formulation shows that it has the same structure as the deterministic observer, except that the observer gain ($L = AK_k$ in Table 9.3-3) is computed taking into account P_0, Q, and R, as well as being time-varying.

To derive the Kalman filter shown in Tables 9.3-1 through 9.3-3 we shall need to investigate two effects: the propagation of means and covariances in the stochastic system (9.3.1), and the effect of taking measurements according to (9.3.2).

Propagation of means and covariances

Let us see how the mean and covariance of the state x_k propagate under the dynamics (9.3.1). For the mean of the state, simply write

$$\bar{x}_{k+1} = \overline{Ax_k + Bu_k + Gw_k},$$

where the overbar is used to denote expected value. Now, linearity of the expectation operator implies that

$$\bar{x}_{k+1} = A\bar{x}_k + B\bar{u}_k + G\bar{w}_k,$$

however, u_k is deterministic (i.e., known) since we apply it to the system, and $\bar{w}_k = 0$, so that

$$\bar{x}_{k+1} = A\bar{x}_k + Bu_k. \tag{9.3.3}$$

To see how the state covariance $P_{x_k} \equiv E\{(x_k - \bar{x}_k)(x_k - \bar{x}_k)^T\}$ propagates, write

$$P_{x_{k+1}} = \overline{(x_{k+1} - \bar{x}_{k+1})(x_{k+1} - \bar{x}_{k+1})^T}$$

$$= \overline{[A(x_k - \bar{x}_k) + Gw_k][A(x_k - \bar{x}_k) + Gw_k]^T}$$

$$= \overline{A(x_k - \bar{x}_k)(x_k - \bar{x}_k)^T A^T} + \overline{Gw_k(x_k - \bar{x}_k)^T A^T}$$

$$+ \overline{A(x_k - \bar{x}_k)w_k^T G^T} + \overline{Gw_k w_k^T G^T}.$$

Note that x_k depends on x_0 and w_j, $j < k$. However, x_0 and w_j are mutually uncorrelated, and w_k is white so that $E\{w_j w_k^T\} = 0$, $j \neq k$. Therefore, it follows that x_k and w_k are uncorrelated for all k. Thus, the state covariance propagates according to

$$P_{x_{k+1}} = AP_{x_k}A^T + GQG^T. \tag{9.3.4}$$

This is a Lyapunov equation for P_{x_k}. The initial condition is P_0, which is assumed known.

The covariance indicates how closely distributed a random variable is about its mean, with a larger covariance indicating greater uncertainty in its value. According to (9.3.4), $P_{x_{k+1}}$ is the sum of two terms—the first term is due to the system dynamics involving the plant matrix A, and the second term is an increase in uncertainty due to the process noise w_k.

It is now desired to determine the mean and covariance of the output y_k. Using (9.3.2) and the whiteness of v_k, the output mean is given by

$$\bar{y}_k = C\bar{x}_k. \tag{9.3.5}$$

The cross-covariance between the state and output is

$$P_{x_k y_k} = \overline{(x_k - \bar{x}_k)(y_k - \bar{y}_k)^T}$$

$$= \overline{(x_k - \bar{x}_k)[C(x_k - \bar{x}_k) + v_k]^T}$$

or

$$P_{x_k y_k} = P_{x_k} C^T. \tag{9.3.6}$$

Note that

$$P_{y_k x_k} = P_{x_k y_k}^T = CP_{x_k}. \qquad (9.3.7)$$

For the covariance of the output, write

$$P_{y_k} = \overline{(y_k - \bar{y}_k)(y_k - \bar{y}_k)^T}$$

$$= \overline{[C(x_k - \bar{x}_k) + v_k][C(x_k - \bar{x}_k) + v_k]^T}$$

or

$$P_{y_k} = CP_{x_k}C^T + R, \qquad (9.3.8)$$

where we used the uncorrelatedness assumptions.

What we have done so far is determine how the state and output means and covariances at time $k+1$ may be computed from those at time k in the absence of measurements.

Kalman Filter Derivation

We want to find the best linear estimator for the state x_k that uses all the available information. The end result will be the Kalman filter in Tables 9.3-1 through 9.3-3. The best estimate $\hat{x}_k$ for the state at time k is the *conditional mean* given all previous data. The previous data consists of $\bar{x}_0$, the previous inputs u_j, $j < k$, and the measured outputs y_j, $j \leq k$, as well as the known covariances P_0, Q, and R.

To simplify the derivation of the Kalman filter, we split it into several stages.

Estimation with no measurements

Suppose that no measurements are taken. Then, the best estimate is just $\bar{x}_k$ as computed recursively by (9.3.3), which is the conditional mean given x_0 and u_j, $j < k$, P_0, and Q. That is, $\hat{x}_k = \bar{x}_k$.

Define the *estimation error* as

$$\tilde{x}_k = x_k - \hat{x}_k. \qquad (9.3.9)$$

Since the process noise has zero mean, one may take the expected value of both sides of this equation to see that $\bar{x}_k$ provides an unbiased estimate of x_k; that is, the expected value of the estimation error is equal to zero. Thus, the *error covariance* is

$$E\{\tilde{x}_k \tilde{x}_k^T\} = E\{(x_k - \hat{x}_k)(x_k - \hat{x}_k)^T\} = E\{(x_k - \bar{x}_k)(x_k - \bar{x}_k)^T\} = P_{x_k},$$

so that (9.3.4) describes the propagation of the error covariance from time k to time $k+1$ in the absence of measurements.

It is important to know the error covariance since it provides a notion of the accuracy of the estimate. A smaller covariance implies that the states x_k are more closely distributed about the mean $\hat{x}_k = \bar{x}_k$; therefore, $\bar{x}_k$ provides a better estimate of x_k. The Kalman filter computes both the state estimates and the error covariances. The error co-

9.3 Discrete Kalman Filter

variance sequence generally has P_{k+1} larger than P_k since the process noise has injected uncertainty into the estimate and degraded it [see (9.3.4)] over the time interval from k to $k + 1$.

Let us now see how the measurement at time y_k is incorporated to provide an improved estimate at time k with smaller error covariance.

Effect of measurements

It is necessary at this point to introduce some more notation. Let the estimate at time k before the measurement y_k is taken be denoted by $\hat{x}_k^-$. This is called the *a priori estimate*. The a priori estimation error at time k is

$$\tilde{x}_k^- = x_k - \hat{x}_k^- \quad (9.3.10)$$

and the associated a priori error covariance is denoted as P_k^-.

Let the estimate at time k, including the output measurement y_k, be denoted by $\hat{x}_k$. This is called the *a posteriori estimate*. The a posteriori estimation error is

$$\tilde{x}_k = x_k - \hat{x}_k \quad (9.3.11)$$

and the associated a posteriori error covariance is denoted as P_k.

The discrete Kalman filter has two steps at each time k. One is the *time update*, by which $\hat{x}_{k-1}$ is updated to $\hat{x}_k^-$, and the other is the *measurement update*, by which the measurement y_k at time k is incorporated to provide the updated estimate $\hat{x}_k$. See Fig. 9.3-1, which will help during the upcoming discussion.

The time update of the means and covariances is easy to find. According to (9.3.3) and (9.3.4)

$$\hat{x}_k^- = A\hat{x}_{k-1} + Bu_{k-1} \quad (9.3.12)$$

$$P_k^- = AP_{k-1}A^T + GQG^T, \quad (9.3.13)$$

so that the a priori estimate at time k is found from the a posteriori estimate at time $k - 1$ by propagation using the deterministic system dynamics (i.e., w_k is not included, as it is unknown). The process noise w_k degrades the estimate by injecting uncertainty, so that the a priori error covariance at time k is generally larger than the a posteriori error covariance at time $k - 1$. This is symbolized in Fig. 9.3-1.

The measurement update of the Kalman filter must now be found. At time k, we have in effect two pieces of information about the state x_k, namely, the a priori estimate $\hat{x}_k^-$, arising from the system dynamics (9.3.1) [see (9.3.12)], and the new measurement y_k in (9.3.2). We need to determine how to combine these two pieces of information in such a way as to produce an updated and improved a posteriori estimate $\hat{x}_k$. To accomplish this we proceed as follows.

Linear mean-square estimate of x_k given y_k

The optimal estimate of x_k given y_k may not be linear. However, we desire our estimator to be a linear dynamical system for ease of implementation. Moreover, if the

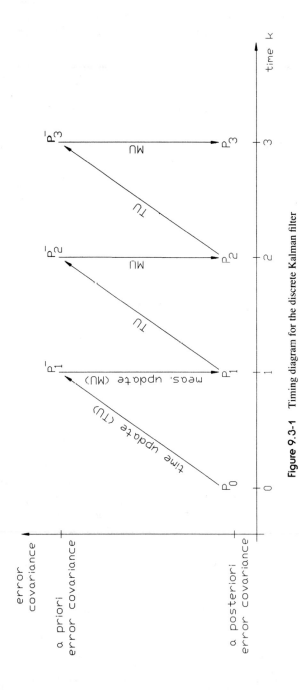

Figure 9.3-1 Timing diagram for the discrete Kalman filter

9.3 Discrete Kalman Filter

process and measurement noises are *Gaussian*, or normally distributed, the optimal estimate is linear. Therefore, let us determine the optimal *linear* estimate of x_k given y_k, assuming that

$$\hat{x}_k = F y_k + g \qquad (9.3.14)$$

for some matrix F and vector g to be determined.

To find the best choice for F and g, let us minimize the mean-square error $J = E\{\tilde{x}_k^T \tilde{x}_k\}$. Using some standard manipulations involving traces (Appendix B) and the expected value operator (Appendix C), we may write

$$J = \overline{tr(x_k - \hat{x}_k)^T(x_k - \hat{x}_k)} = \overline{tr(x_k - \hat{x}_k)(x_k - \hat{x}_k)^T}$$
$$= \overline{tr(x_k - Fy_k - g)(x_k - Fy_k - g)^T}$$
$$= \overline{tr[(x_k - \hat{x}_k^-) - (Fy_k + g - \hat{x}_k^-)][(x_k - \hat{x}_k^-) - (Fy_k + g - \hat{x}_k^-)]^T},$$

whence some straightforward work yields

$$J = tr[P_k^- + F(P_{y_k} + \bar{y}_k \bar{y}_k^T)F^T + (g - \hat{x}_k^-)(g - \hat{x}_k^-)^T$$
$$+ 2F\bar{y}_k(g - \hat{x}_k^-)^T - 2FP_{y_k x_k}].$$

(Note that we are using the definitions $\bar{y}_k = C\hat{x}_k^-$ and $P_{y_k x_k} = E\{(y_k - \bar{y}_k)(x_k - \hat{x}_k^-)\}^T$. That is, $\bar{x}_k$ is replaced everywhere by the conditional mean $\hat{x}_k^-$ given all the data through time $k - 1$.)

From Appendix B, $(d/dF)tr(FHF^T) = 2FH$ and $(d/dF)tr(DFH) = D^T H^T$ for any matrices F, D, and H. Thus, to minimize J it is required that

$$\frac{\partial J}{\partial g} = 2(g - \hat{x}_k^-) + 2F\bar{y}_k = 0$$

and

$$\frac{\partial J}{\partial F} = 2F(P_{y_k} + \bar{y}_k \bar{y}_k^T) - 2P_{x_k y_k} + 2(g - \hat{x}_k^-)\bar{y}_k^T = 0.$$

Solving the first of these equations for g yields

$$g = \hat{x}_k^- - F\bar{y}_k$$

which may be substituted into the second equation to give

$$FP_{y_k} - P_{x_k y_k} = 0$$

so that

$$F = P_{x_k y_k} P_{y_k}^{-1}.$$

Finally, using these values for g and F in (9.3.14) gives the optimal linear mean-square estimate for x_k

$$\hat{x}_k = \hat{x}_k^- + P_{x_k y_k} P_{y_k}^{-1}(y_k - \bar{y}_k). \qquad (9.3.15)$$

Note that this is a linear combination of the two pieces of information we have at time k, namely, $\hat{x}_k^-$ and the new measurement y_k.

Fortunately, we have already computed $P_{x_k y_k}$, P_{y_k}, and $\bar{y}_k$. Using (9.3.6), (9.3.8), and (9.3.5) (with the appropriate substitutions of P_k^- and $\hat{x}_k^-$) in (9.3.15) results in

$$\hat{x}_k = \hat{x}_k^- + P_k^- C^T (C P_k^- C^T + R)^{-1}(y_k - C \hat{x}_k^-). \tag{9.3.16}$$

This is the optimal linear mean-square a posteriori estimate of x_k given the a priori estimate $\hat{x}_k^-$ and the new measurement y_k. It is the estimate portion of the Kalman filter measurement update.

A posteriori error covariance

To see how accurate this a posteriori estimate is, one may compute the a posteriori error covariance P_k. Smaller error covariances imply better estimates, since they imply that the state is more closely distributed about its conditional mean. We shall now derive the covariance portion of the Kalman filter measurement update in Table 9.3-1.

Note first that the linear mean-square estimate is unbiased (i.e., $E\{\hat{x}_k\} = \bar{x}_k$, or $E\{\tilde{x}_k\} = 0$) since, using (9.3.15)

$$E\{\tilde{x}_k\} = E\{x_k - \hat{x}_k\}$$
$$= E\{(x_k - \hat{x}_k^-) - P_{x_k y_k} P_{y_k}^{-1}(y_k - \bar{y}_k)\} = E\{x_k - \hat{x}_k^-\} = E\{\tilde{x}_k^-\}.$$

Therefore, $\hat{x}_k$ provides an unbiased estimate of x_k as long as $\hat{x}_k^-$ does. However, from (9.3.12) and (9.3.1)

$$E\{\tilde{x}_k^-\} = E\{x_k - \hat{x}_k^-\} = E\{x_k - A\hat{x}_{k-1} - Bu_{k-1}\}$$
$$= AE\{x_{k-1} - \hat{x}_{k-1}\} = AE\{\tilde{x}_{k-1}\},$$

so that $\hat{x}_k^-$ provides an unbiased estimate as long as $\hat{x}_{k-1}$ does. If the initial estimate of the state is selected as

$$\hat{x}_0 = \bar{x}_0, \tag{9.3.17}$$

which is known, then the initial estimate is unbiased. Thus, all estimates of the state are unbiased.

Now, we may use (9.3.15) to write

$$P_k = E\{\tilde{x}_k \tilde{x}_k^T\} = E\{(x_k - \hat{x}_k)(x_k - \hat{x}_k)^T\}$$
$$= \overline{[(x_k - \hat{x}_k^-) - P_{x_k y_k} P_{y_k}^{-1}(y_k - \bar{y}_k)][(x_k - \hat{x}_k^-) - P_{x_k y_k} P_{y_k}^{-1}(y_k - \bar{y}_k)]^T}$$
$$= P_k^- - P_{x_k y_k} P_{y_k}^{-1} P_{y_k x_k} - P_{x_k y_k} P_{y_k}^{-1} P_{y_k x_k} + P_{x_k y_k} P_{y_k}^{-1} P_{y_k} P_{y_k}^{-1} P_{y_k x_k}$$

or

$$P_k = P_k^- - P_{x_k y_k} P_{y_k}^{-1} P_{y_k x_k}. \tag{9.3.18}$$

9.3 Discrete Kalman Filter

Note that P_k is generally less than P_k^- (i.e., $P_k^- - P_k$ is positive semidefinite). This is because the measurement y_k has been incorporated into the estimate [see (9.3.16)], improving our estimate of the state with a corresponding reduction in the error covariance.

Using (9.3.6) and (9.3.8) in (9.3.18) (with appropriate substitution by P_k^-) yields

$$P_k = P_k^- - P_k^- C^T (CP_k^- C^T + R)^{-1} CP_k^-. \qquad (9.3.19)$$

This is a matrix Riccati equation for P_k.

By employing the matrix inversion lemma (Appendix B), the Riccati equation may be written alternatively as

$$P_k = [(P_k^-)^{-1} + (C^T R^{-1} C)^{-1}]^{-1}. \qquad (9.3.20)$$

Having in mind the formula for the parallel combination of two resistors R_1 and R_2, namely $(R_1^{-1} + R_2^{-1})^{-1}$, we could interpret P_k as the parallel combination of the two matrices P_k^- and $C^T R^{-1} C$. Therefore, P_k should generally be smaller than either of these two, again corroborating the decrease in uncertainty that accompanies incorporating the measurement y_k.

At this point we have derived the measurement update portion of the Kalman filter. The estimate update is given by (9.3.16) and the covariance update is given by (9.3.19).

Kalman Filter Formulations

There are several formulations of the discrete Kalman filter. It is not difficult (see the problems) to show that

$$K_k \equiv P_k^- C^T (CP_k^- C^T + R)^{-1} = P_k C^T R^{-1}, \qquad (9.3.21)$$

so that the estimate measurement update (9.3.16) may be expressed in terms of the a posteriori error covariance as

$$\hat{x}_k = \hat{x}_k^- + P_k C^T R^{-1} (y_k - C\hat{x}_k^-). \qquad (9.3.22)$$

A filter formulation consisting of the time updates (9.3.12), (9.3.13) and the measurement updates (9.3.19), (9.3.22) appears in Table 9.3-1.

We call K_k the *Kalman gain*. Since

$$\hat{x}_k = \hat{x}_k^- + K_k (y_k - C\hat{x}_k^-), \qquad (9.3.23)$$

the Kalman gain shows the influence of the new measurement y_k in modifying the a priori estimate x_k^-. The quantity $(y_k - C\hat{x}_k^-)$ is called the *residual*. Since $C\hat{x}_k^-$ provides an estimate of the output, the residual is an (a priori) "output estimation error."

According to (9.3.19) and (9.3.21), the a posteriori error covariance may be computed in terms of the Kalman gain as

$$P_k = (I - K_k C) P_k^-. \qquad (9.3.24)$$

A Kalman filter measurement update formulation in terms of the Kalman gain appears in Table 9.3-2.

TABLE 9.3-1 DISCRETE-TIME KALMAN FILTER: TIME UPDATE/MEASUREMENT UPDATE FORMULATION

Stochastic System Model:

$$x_{k+1} = Ax_k + Bu_k + Gw_k \qquad (9.3.26)$$

$$y_k = Cx_k + v_k \qquad (9.3.27)$$

$$x_0 \sim (\bar{x}_0, P_0), \quad w_k \sim (0, Q), \quad v_k \sim (0, R)$$

Assumptions:
w_k and v_k are white noise processes mutually uncorrelated with each other and with x_0. $Q \geq 0, R > 0$.

Filter Initialization:

$$\hat{x}_0 = \bar{x}_0$$

Time Update: (Effect of system dynamics)

Error Covariance: $P_{\bar{k}+1} = AP_k A^T + GQG^T \qquad (9.3.28)$

Estimate Update: $\hat{x}_{\bar{k}+1} = A\hat{x}_k + Bu_k \qquad (9.3.29)$

Measurement Update: (Effect of measurements y_k)

Error Covariance: $P_{k+1} = P_{\bar{k}+1} - P_{\bar{k}+1} C^T (CP_{\bar{k}+1}C^T + R)^{-1} CP_{\bar{k}+1} \qquad (9.3.30)$

Estimate Update: $\hat{x}_{k+1} = \hat{x}_{\bar{k}+1} + P_{k+1}C^T R^{-1}(y_{k+1} - C\hat{x}_{\bar{k}+1}) \qquad (9.3.31)$

TABLE 9.3-2 DISCRETE-TIME KALMAN FILTER: ALTERNATIVE MEASUREMENT UPDATE EQUATIONS

$$K_{k+1} \equiv P_{\bar{k}+1}C^T(CP_{\bar{k}+1}C^T + R)^{-1} \qquad (9.3.32)$$

$$P_{k+1} = (I - K_{k+1}C)P_{\bar{k}+1} \qquad (9.3.33)$$

$$\hat{x}_{k+1} = \hat{x}_{\bar{k}+1} + K_{k+1}(y_{k+1} - C\hat{x}_{\bar{k}+1}) \qquad (9.3.34)$$

It is possible to combine the time and measurement updates to obtain the recursive formulation of the Kalman filter shown in Table 9.3-3. The estimate recursion is found by substituting (a time-shifted version of) (9.3.31) into (9.3.29) and using the definition of the Kalman gain. The error covariance recursion is found by substituting (9.3.30) into (9.3.28). It is quite interesting to notice that this formulation has exactly the same structure as the deterministic state observer in Fig. 9.1-1. The difference is that the observer gain is now $L_k = AK_k$, which is time-varying even when the plant (A, B, C) is time-invariant.

Note that in our development there is no measurement taken at time $k = 0$. That is, the first step of the Kalman filter is a time update to time $k = 1$. This may easily be modified if necessary by performing a measurement update at time $k = 0$.

9.3 Discrete Kalman Filter

TABLE 9.3-3 DISCRETE-TIME KALMAN FILTER: RECURSIVE A PRIORI FORMULATION

Kalman gain:

$$K_k \equiv P_k^- C^T (C P_k^- C^T + R)^{-1} \qquad (9.3.35)$$

Estimate Recursion:

$$\hat{x}_{k+1}^- = A\hat{x}_k^- + Bu_k + AK_k(y_k - C\hat{x}_k^-) \qquad (9.3.36)$$

Error Covariance Recursion:

$$P_{k+1}^- = A[P_k^- - P_k^- C^T (C P_k^- C^T + R)^{-1} C P_k^-]A^T + GQG^T \qquad (9.3.37)$$

The Kalman filter provides the optimal *linear* estimate of the state x_k. However, if the noises are Gaussian, then it provides the optimal estimate [Lewis 1986b].

According to Table 9.3-3, the Kalman filter is a low-pass filter with time-varying gain K_k. It therefore possesses both noise rejection and smoothing properties.

Implementation of the Kalman Filter

To implement the Kalman filter, there are two possibilities. The entire set of computations, involving both the estimate recursion and the error covariance recursion in the tables, may be performed on-line in real time. Alternatively, the Riccati equation may be solved off-line before the implementation to compute the error covariances and the Kalman gain sequence K_k, which may then be stored in computer memory. Then, in the actual implementation, it is only necessary to perform the estimate recursion using the stored values of K_k.

The Kalman filter in Table 9.3-3 provides a "predictive estimate" $\hat{x}_{k+1}^-$ of x_{k+1} in that the most current measurement y_{k+1} is not used. This automatically incorporates a delay of one sample period for computation purposes. The "filtered estimate," which includes the current measurement, is computed using

$$\hat{x}_{k+1} = \hat{x}_{k+1}^- + K_{k+1}(y_{k+1} - C\hat{x}_{k+1}^-). \qquad (9.3.25)$$

One should note that, although we assumed $R > 0$, to implement the Kalman filter only $(CP_k^- C^T + R)$, and not R itself, must be invertible.

In any actual application the matrix multiplications implied by the tables should be avoided since they can be numerically unstable in the presence of finite computer precision. There are two possibilities for achieving this. For simpler systems, it is possible to simplify the Kalman filter updates analytically to obtain equations that may be programmed. See section 9.4. For larger, more complicated systems, the *square-root algorithms* in Bierman [1977] may be used. These routines are commercially available.

In the next section we shall show an example of Kalman filter design and implementation.

Suboptimal Steady-State Kalman Filter

Even when the plant (A, B, C) is time-invariant, the optimal Kalman filter is time-varying, since the Kalman gain is a function of k. In many applications it is too much trouble to perform the calculations involved with finding the $n \times p$ matrices K_k. Even if the Kalman gain sequence is computed off-line before implementation, a great deal of memory may be required to store it. Thus, it is often satisfactory to use a simplified time-invariant filter with a constant gain K, which we shall now derive.

Examine the recursive formulation in Table 9.3-3. At *statistical steady state*, the a priori error covariance reaches a constant value which we shall call

$$P \equiv P_{k+1}^- = P_k^-. \tag{9.3.38}$$

Then, the Riccati update equation (9.3.37) in the table simplifies to the steady-state algebraic Riccati equation (ARE)

$$P = A[P - PC^T(CPC^T + R)^{-1}CP]A^T + GQG^T. \tag{9.3.39}$$

This is virtually the same ARE that was used in deterministic observer design for multivariable systems in section 9.1. The difference is that, by using a stochastic formulation, we have given meaning to the design parameters Q and R; they should simply be selected as the covariances of the measurement and process noises respectively. Then, the filter will produce optimal estimates for those noises.

The *steady-state Kalman gain* is the constant $n \times p$ matrix

$$K = PC^T(CPC^T + R)^{-1}. \tag{9.3.40}$$

The steady-state Kalman filter is the time-invariant system given by

$$\hat{x}_{k+1}^- = A\hat{x}_k^- + Bu_k + AK(y_k - C\hat{x}_k^-). \tag{9.3.41}$$

The implementation of this filter is quite simple. The constant Kalman gain matrix K may be computed off-line by solving the ARE [ORACLS, MATLAB]. Then, K is stored in memory and used to perform the estimate update (9.3.41) during the actual implementation.

To investigate the behavior of the steady-state Kalman filter, we may examine the steady-state error system. A recursion for the a priori error at statistical steady state may be found by using (9.3.26), (9.3.36), and (9.3.27) in (9.3.10) to obtain

$$\tilde{x}_{k+1}^- = A(I - KC)\tilde{x}_k^- + Gw_k - AKv_k. \tag{9.3.42}$$

It is important to realize that this steady-state error system is driven by both the process noise w_k and the measurement noise v_k. It is also clear that as long as

$$A_o = A - AKC \equiv A - LC, \tag{9.3.43}$$

is asymptotically stable, the steady-state error will be bounded as long as the noises are. The magnitude of the error depends on the degree of stability of A_o, as well as the magnitudes of w_k and v_k. A more stable A_o results in a smaller error. Note that the "observer gain" is nothing but $L = AK$.

Now recall the theorem on the stability of multivariable observers quoted in section 9.1. This theorem guarantees the stability of A_o if (C, A) is observable and $(A, G\sqrt{Q})$ is reachable. That is, the state should be observable given the measurements y_k, and the *process noise should excite all the states*. Under these circumstances, the steady-state Kalman filter provides accurate estimates of the states. In fact, all that is needed are the milder conditions of detectability of (C, A) and stabilizability of $(A, G\sqrt{Q})$; that is, the unstable states should be both observable through y_k and excited by w_k.

It is rather strange to realize that for the Kalman filter to work properly, the process noise w_k should excite all the states and the measurement noise v_k should corrupt all of the measurements (i.e., $R > 0$).

An alternative formulation of the steady-state Kalman filter (9.3.41) is

$$\hat{x}_{k+1}^- = A(I - KC)\hat{x}_k^- + Bu_k + AKy_k \tag{9.3.44}$$

which has a transfer function from the measurement y_k to the estimate $\hat{x}_k^-$ of

$$H(z) = [zI - A(I - KC)]^{-1}AK. \tag{9.3.45}$$

Since the resolvent matrix $[zI - A(I - KC)]^{-1}$ has a relative degree of one, the Kalman filter is a low-pass filter. This helps account for its noise-rejection properties. It is worth noting that (9.3.45) is nothing but the discrete-time Wiener filter.

9.4 DIGITAL FILTERING OF CONTINUOUS-TIME SYSTEMS

We have derived the Kalman filter for discrete-time systems, different formulations of which are given in Tables 9.3-1 through 9.3-3. However, most practical applications involve filtering of continuous-time systems. Due to the power and speed of digital signal processors, the fairly complex time-varying dynamics of the optimal Kalman filter are most easily implemented digitally. Therefore, let us show how to apply the discrete-time filter to continuous systems in this section.

We shall present an example showing how to design digital Kalman filters and implement them on a DSP. First, however, it is necessary to discuss two topics: discretization of stochastic systems, and modeling of nonwhite noises.

Discretization of Continuous Stochastic Systems

A continuous-time stochastic system is given by

$$\dot{x} = A^c x + B^c u + G^c w \tag{9.4.1}$$

$$y = Cx + v \tag{9.4.2}$$

with measurement noise $w(t)$ and process noise $v(t)$ both continuous-time, zero mean, white noise processes. Suppose the noise statistics are known so that $w(t) \sim (0, Q^c)$, $v(t) \sim (0, R^c)$. The statistics of the initial state are known so that $x(0) \sim (\bar{x}_0, P_0)$.

In digital filtering, the measurements are taken only at integral multiples of the sample period T. Thus, to apply the discrete Kalman filter in Tables 9.3-1–9.3-3, we must convert the continuous-time plant to the discrete-time system

$$x_{k+1} = Ax_k + Bu_k + Gw_k \qquad (9.4.3)$$

$$y_k = Cx_k + v_k \qquad (9.4.4)$$

with $w_k \sim (0, Q)$, $v_k \sim (0, R)$. The Kalman gain determined on the basis of the discretized plant and covariance matrices is the one required in the filter dynamics to provide estimates $\hat{x}_k = \hat{x}(kT)$. To perform this discretization we may proceed as follows.

The solution to the continuous dynamical equation is given by

$$x(t) = e^{A^c(t-t_0)}x(t_0) + \int_{t_0}^{t} e^{A^c(t-\tau)}B^c u(\tau)\, d\tau + \int_{t_0}^{t} e^{A^c(t-\tau)}G^c w(\tau)\, d\tau.$$

Defining $t_0 = kT$, $t = (k+1)T$, and the sampled sequence $x_k = x(kT)$, we may write

$$\begin{aligned} x_{k+1} = e^{A^c T}x_k &+ \int_{kT}^{(k+1)T} e^{A^c[(k+1)T-\tau]}B^c u(\tau)\, d\tau \\ &+ \int_{kT}^{(k+1)T} e^{A^c[(k+1)T-\tau]}G^c w(\tau)\, d\tau. \end{aligned} \qquad (9.4.5)$$

We shall include a ZOH in the digital control system by assuming that $u(t)$ is constant between the samples. With this assumption, in section 7.1 we have seen how to use a change of variables to determine A and B in (9.4.3). Indeed, it was found that

$$A = e^{A^c T}, \qquad B = \int_0^T e^{A^c \tau}B^c\, d\tau. \qquad (9.4.6)$$

Discretization of noise covariances

It remains to determine the discrete process noise covariance Q. This may be accomplished by defining the discrete stochastic sequence w_k so that (9.4.5) and (9.4.3) are equivalent. Let us define $G = I$ and

$$w_k = \int_{kT}^{(k+1)T} e^{A^c[(k+1)T-\tau]}G^c w(\tau)\, d\tau. \qquad (9.4.7)$$

This process has a mean of 0 (since $w(t)$ does) and a covariance of

$$Q = E(w_k w_k^T) = \iint_{kT}^{(k+1)T} e^{A^c[(k+1)T-\tau]}G^c E\{w(\tau)w^T(\sigma)\}(G^c)^T e^{(A^c)^T[(k+1)T-\sigma]}\, d\tau\, d\sigma.$$

However, $E[w(\tau)w^T(\sigma)] = Q^c \delta(\tau - \sigma)$, with $\delta(t)$ the Dirac delta (see Appendix C), where Q^c is a *spectral density* and $Q^c \delta(t)$ a *covariance*. Using this we may determine that

$$Q = \int_{kT}^{(k+1)T} e^{A^c[(k+1)T-\tau]}G^c Q (G^c)^T e^{(A^c)^T[(k+1)T-\tau]}\, d\tau,$$

9.4 Digital Filtering of Continuous-Time Systems

whence changing variables (twice) yields

$$Q = \int_0^T e^{A^c \tau} G^c Q (G^c)^T e^{(A^c)^T \tau} \, d\tau. \tag{9.4.8}$$

This is the equation for discretization of the process noise covariance. It is often too complicated to use, and is conveniently approximated by the series expansion

$$Q = G^c Q^c (G^c)^T T + O(2), \tag{9.4.9}$$

with O(2) denoting terms of order T^2 that may generally be neglected.

We have seen in section 7.1 that the continuous and discretized measurement matrices have the same value C. To discretize the spectral density R^c of the measurement noise $v(t)$, notice that the covariance of the discrete noise process v_k is given by

$$R_v(k) = R\delta_k, \tag{9.4.10}$$

with δ_k the Kronecker delta, which has a finite value of 1 at $k = 0$. The discrete covariance R is a finite matrix. On the other hand, the covariance of $v(t)$ is

$$R_v^c(t) = R^c \delta(t), \tag{9.4.11}$$

which is a continuous covariance matrix taking on infinite values. The matrix R^c is a spectral density matrix.

To make $R_v(k)$ approach $R_v^c(t)$ in the limit, we may use the approximation to an impulse given by

$$\delta(t) = \lim_{T \to 0} \frac{1}{T} \Pi\left(\frac{t}{T}\right) \tag{9.4.12}$$

with $\Pi(t/T)$ the unit rectangle defined as

$$\Pi(t) = \begin{cases} 1, & -\frac{1}{2} \leq t \leq \frac{1}{2} \\ 0, & \text{otherwise.} \end{cases} \tag{9.4.13}$$

Then, an appropriate definition is

$$R^c \delta(t) = \lim_{T \to 0} (RT) \frac{1}{T} \Pi\left(\frac{t}{T}\right), \tag{9.4.14}$$

so that

$$R = R^c/T. \tag{9.4.15}$$

One should clearly note the inverse effect of sampling on Q^c and R^c; for, if $T < 1$ sec, then $Q < Q^c$ while $R > R^c$. Thus, sampling *increases* the measurement noise covariance. Clearly, it is not a good idea to select T too small.

This development should be compared with the discretization of the performance index weighting matrices which was discussed in section 7.2.

Nonwhite Noise and Shaping Filters

In deriving the Kalman filter of Table 9.3-1 we assumed that the process noise w_k and measurement noise v_k were zero-mean white noise processes. Unfortunately, in practical applications it is common for this assumption to fail. In this subsection we shall demonstrate that it is straightforward to include the effects of nonwhite noises and/or random biases [Lewis 1986b].

Nonwhite process noise

If the spectral density of the process noise is not a constant with frequency, it is said to be *nonwhite* or *colored*. This situation often arises in practice where, for instance, the noise may be of low frequency or high frequency.

Consider the discrete noise process w_k. If the spectral density $\Phi_w(z)$ is rational and nonsingular for almost every z, we may factor it as

$$\Phi_w(z) = H(z)H^T(z^{-1}) \qquad (9.4.16)$$

where $H(z)$ is a square asymptotically stable rational function with zeros on or inside the unit circle. If $\Phi_w(z)$ is nonsingular for all z on the unit circle, then $H(z)$ is of minimum phase, that is, it has stable poles and zeros. The rational function $H(z)$ may be interpreted as the transfer function of a linear system which, when driven by white noise, yields the output w_k with spectral density $\Phi_w(z)$. This system is called a *shaping filter* for the noise w_k.

There are several ways to find a state-variable realization for $H(z)$ (Chapter 2), which we may symbolize as

$$\begin{aligned} x'_{k+1} &= A'x'_k + G'w'_k \\ w_k &= C'x'_k + D'w'_k. \end{aligned} \qquad (9.4.17)$$

If input noise w'_k is white with $w'_k \sim (0, I)$, then the output w_k has the required spectral density $\Phi_w(z)$. This process noise shaping filter may now be augmented to the system dynamics in Table 9.3-1 to yield

$$\begin{bmatrix} x_{k+1} \\ x'_{k+1} \end{bmatrix} = \begin{bmatrix} A & GC' \\ 0 & A' \end{bmatrix} \begin{bmatrix} x_k \\ x'_k \end{bmatrix} + \begin{bmatrix} B \\ 0 \end{bmatrix} u_k + \begin{bmatrix} GD' \\ G' \end{bmatrix} w'_k$$

$$y_k = [C \quad 0] \begin{bmatrix} x_k \\ x'_k \end{bmatrix} + v_k, \qquad (9.4.18)$$

with w'_k and v_k both white.

Now, it is only necessary to design a Kalman filter for this augmented system using Tables 9.3-1–9.3-3. The result is the optimal linear filter for the original system (9.3.26), (9.3.27) with nonwhite process noise having spectral density $\Phi_w(z)$.

9.4 Digital Filtering of Continuous-Time Systems

In practice, the spectral density of the *continuous* measurement noise process $w(t)$ would probably be determined, with the shaping filter and augmented system being determined in continuous time. In this procedure, it is necessary to factor the spectral density as

$$\Phi_w(s) = H(s)H^T(-s). \qquad (9.4.19)$$

A continuous-time realization for $H(s)$ can then be found, and appended to the system dynamics to obtain a continuous augmented system. Then, the augmented system is sampled to obtain the discrete system for digital Kalman filter design.

Some useful continuous-time spectrum shaping filters are given in Fig. 9.4-1. The random bias could be used in the filtering of ship navigation fixes, where there is a constant unknown current. Brownian motion, or the Wiener process, is useful for describing the effects of diffusion, or the effects of biases that vary with time. First-order Markov processes are valuable in modeling band-limited noises. The second-order Markov process has a periodic autocorrelation, and is useful in describing oscillatory effects such as fuel slosh or vibration.

Nonwhite measurement noise

A similar procedure may be used to deal with nonwhite measurement noise; however, the derivation yields an augmented system which has correlated process and measurement noise. The details are given in Lewis [1986b].

Design and Implementation Example

Let us now discuss an example showing how to design and simulate Kalman filters, as well as how to implement them on a digital signal processor (DSP). We shall consider the disk drive head-positioning mechanism, deriving and simulating the Kalman filter in Example 9.4-1. Then, in Example 9.4-2 we design the simpler steady-state Kalman filter, which is suitable for most applications. Finally, in Example 9.4-3 we show how to implement the Kalman filter on a DSP.

Example 9.4-1: Digital Kalman Filter for Disk-Drive Head-Positioning System

Throughout this chapter we have designed observers of various sorts for the disk drive head-positioning system. See Example 9.1-1 for a basic description of the system. A background reference is [Stich 1987]. In this example we should like to consider a more realistic model which includes an unknown bias force and process and measurement noises. The theory of the Kalman filter will allow us to incorporate the known characteristics of the noise processes into the filter design.

We shall discuss the system dynamics and noise properties. Then, the continuous system will be discretized. The digital Kalman filter will be derived, with the filter equations being simplified using scalar variables for computer programming. Finally, we will perform a computer simulation to test the filter.

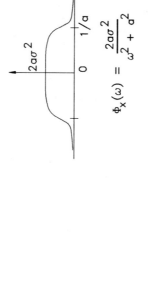

Figure 9.4-1 Some useful continuous spectrum shaping filters

9.4 Digital Filtering of Continuous-Time Systems

a. System Dynamics

A model of the disk drive head-positioning system is given by

$$\dot{x} = \begin{bmatrix} 0 & 1 & 0 \\ 0 & 0 & 1 \\ 0 & 0 & 0 \end{bmatrix} x + \begin{bmatrix} 0 \\ 1 \\ 0 \end{bmatrix} u + w = A^c x + B^c u + G^c w, \quad (1)$$

where the state is $x = [d \ v \ f]^T$, with $d(t)$ the head position and $v(t)$ the head velocity. The control input is $u(t)$, and $w(t) = [w_d \ w_v \ w_f]^T$ is process noise which disturbs the state components. We have selected the units so that the system matrices A^c, B^c, G^c have the simple elements shown (i.e., $G^c = I$).

There are several sources of constant forces on the actuator; they are modeled by the bias force f (equivalent driver input). Since the bias is unknown and must be estimated, it is necessary to include it as a state. To describe f, we have selected the Brownian motion model from Fig. 9.4-1, which is an integrator driven by white noise $w_f(t)$. This allows for a constant bias force which may vary slowly with time.

Position measurements are taken, so that the measured output is

$$y = [1 \ 0 \ 0]x + n = Cx + n \quad (2)$$

where $n(t)$ is measurement noise.

b. Noise Statistics

Using initial measurements of position we may obtain the initial position estimate $\hat{d}(0) = y(0)$. The initial velocity estimate may be computed using

$$\hat{v}(0) = (y(T) - y(0))/T, \quad (3)$$

with T the sampling period. The bias force estimate $\hat{f}(0)$ may be taken as zero. These are the three components of $\hat{x}(0) = \bar{x}(0)$ in Table 9.3-1.

The initial position covariance may be taken in the appropriate units as 1. If the initial velocity is estimated using (3), the initial velocity covariance is $1/T^2$. Since the initial force estimate is inaccurate, the initial force covariance may be taken as some large value such as 1000. Then

$$P_0 = \text{diag}\{1, 1/T^2, 1000\}. \quad (4)$$

Suppose the process noise has the spectral density

$$Q^c = \text{diag}\{0.001, 3, 0.1\}, \quad (5)$$

so that it excites primarily the second state derivative (i.e., acceleration). A reasonable value for the measurement noise spectral density is (lowercase letters denote scalars)

$$r^c = 0.1 \quad (6)$$

c. System Discretization

To design a digital filter we must first sample the continuous dynamics (1), (2). Using the techniques in section 7.1, the result is

$$x_{k+1} = \begin{bmatrix} 1 & T & T^2/2 \\ 0 & 1 & T \\ 0 & 0 & 1 \end{bmatrix} x_k + \begin{bmatrix} T^2/2 \\ T \\ 0 \end{bmatrix} u_k + w_k = Ax_k + Bu_k + w_k \quad (7)$$

$$y_k = [1 \ 0 \ 0]x_k + n_k, \quad (8)$$

where according to our work earlier in this section the discrete process and measurement noise covariances are

$$Q = Q^c T = \text{diag}\{0.001T, 3T, 0.1T\} \equiv \text{diag}\{q_d, q_v, q_t\} \quad (9)$$

$$r = r^c/T = 0.1/T. \quad (10)$$

The initial state mean and variance are given as in part b.

d. Kalman Filter Design

Let us use the recursive a priori formulation in Table 9.3-3 to design a Kalman filter for the discretized dynamics. In Stich [1987], the state estimates were used for control purposes (see Chapter 10). There, it was necessary to include a delay of one sample period T in the control input term $u(t)$, since the computation time is a significant portion of the sample period. This adds one state in the discretized system model, making it of fourth order. However, the Kalman filter in Table 9.3-3 provides a predictive estimate one sample period in the future, since the measurement y_{k+1} is not used to compute $\hat{x}_{k+1}^-$. That is, we may begin computing $\hat{x}_{k+1}^-$ at time k. Thus, the filter automatically has a computation delay of T built in, making it unnecessary to include an extra state to model a time delay in our approach.

In this part of the example we will derive and simplify the Kalman filter update at time k for programming on a digital computer. This will allow us to avoid matrix multiplies which are slow as well as inaccurate in the face of finite computer wordlength. For more complicated systems, such analytic simplification is intractable, and some square-root algorithms such as those of Bierman [1977] should be used to implement the Riccati equation in Table 9.3-3. Straightforward implementation of the Kalman filter using matrix multiplications should be avoided at all costs, since it is invariably accompanied by numerical problems.

The steps in Table 9.3-3 will now be applied to system (7), (8). The reader should fill in the missing steps of manipulation. As we proceed, we shall define intermediate scalar variables for ease of programming later. We suppress the time index k (e.g., see (11)), as all scalar variables are assumed to occur at the $(k + 1)$-th update.

Kalman gain

Denote the (symmetric) a priori error covariance at time k as

$$P_k^- = \begin{bmatrix} p_1 & p_2 & p_3 \\ p_2 & p_4 & p_5 \\ p_3 & p_5 & p_6 \end{bmatrix}, \quad (11)$$

where p_i are scalars. Referring to Table 9.3-3, the Kalman gain is computed using the following steps.

$$CP_k^- C^T + r = p_1 + r \quad (12)$$

$$K_k = \frac{1}{p_1 + r} \begin{bmatrix} p_1 \\ p_2 \\ p_3 \end{bmatrix} \equiv \begin{bmatrix} k_1 \\ k_2 \\ k_3 \end{bmatrix}. \quad (13)$$

Error covariance update

The Riccati equation error covariance update (9.3.37) may be written as

$$P_{k+1}^- = A(I - K_k C) P_k^- A^T + Q. \quad (14)$$

9.4 Digital Filtering of Continuous-Time Systems

Therefore, the error covariance update may be computed using the steps:

$$(I - K_k C)P_k^- = \begin{bmatrix} (1-k_1)p_1 & (1-k_1)p_2 & (1-k_1)p_3 \\ p_2 - k_2 p_1 & p_4 - k_2 p_2 & p_5 - k_2 p_3 \\ p_3 - k_3 p_1 & p_5 - k_3 p_2 & p_6 - k_3 p_3 \end{bmatrix} \equiv \begin{bmatrix} p_1' & p_2' & p_3' \\ p_2' & p_4' & p_5' \\ p_3' & p_5' & p_6' \end{bmatrix}. \qquad (15)$$

(Note that this is just the a posteriori error covariance P_k; as such, it is symmetric. It is a good exercise to verify this.)

$$A(I - K_k C)P_k^- = \begin{bmatrix} p_1' + p_2' T + p_3' T^2/2 & p_2' + p_4' T + p_5' T^2/2 & p_3' + p_5' T + p_6' T^2/2 \\ p_2' + p_3' T & p_4' + p_5' T & p_5' + p_6' T \\ p_3' & p_5' & p_6' \end{bmatrix}$$

$$\equiv \begin{bmatrix} q_1 & q_2 & q_3 \\ q_4 & q_5 & q_6 \\ q_7 & q_8 & q_9 \end{bmatrix} \qquad (16)$$

And, finally

$$P_{k+1}^- = A(I - K_k C)P_k^- A^T + Q$$

$$= \begin{bmatrix} q_1 + q_2 T + q_3 T^2/2 & q_2 + q_3 T & q_3 \\ q_4 + q_5 T + q_6 T^2/2 & q_5 + q_6 T & q_6 \\ q_7 + q_8 T + q_9 T^2/2 & q_8 + q_9 T & q_9 \end{bmatrix} + \begin{bmatrix} q_d & 0 & 0 \\ 0 & q_v & 0 \\ 0 & 0 & q_f \end{bmatrix}. \qquad (17)$$

Kalman filter dynamics

The Kalman filter dynamics are given by (9.3.36), which may be written in terms of scalar components as

$$\hat{d}_{k+1} = \hat{d}_k + T\hat{v}_k + \frac{T^2}{2}\hat{f}_k + \frac{T^2}{2}u_k + L_1(y_k - \hat{d}_k) \qquad (18)$$

$$\hat{v}_{k+1} = \hat{v}_k + T\hat{f}_k + Tu_k + L_2(y_k - \hat{d}_k) \qquad (19)$$

$$\hat{f}_{k+1} = \hat{f}_k + L_3(y_k - \hat{d}_k) \qquad (20)$$

where the filter gains L_i are given by

$$\begin{bmatrix} L_1 \\ L_2 \\ L_3 \end{bmatrix} \equiv AK_k = \begin{bmatrix} k_1 + k_2 T + k_3 T^2/2 \\ k_2 + k_3 T \\ k_3 \end{bmatrix}. \qquad (21)$$

Note that for notational simplicity we have suppressed the superscript "$-$" denoting that these are a priori estimates.

e. Computer Simulation

It is easy to simulate the Kalman filter on a digital computer. For this purpose, we may use the driver program DIGCTLF in Appendix A, which applies the digital Kalman filter to the *continuous system*, so that its performance may be compared at all times, even between the samples.

To use DIGCTLF for filtering of stochastic systems, three subroutines must be written. One, subroutine SYSINP(IT, Tr, x) generates the continuous-time process and measurement noise. It appears in Fig. 9.4-2. Tr is the Runge-Kutta step size and IT is an iteration counter.

```
C     SUBROUTINE TO GENERATE CONTINUOUS-TIME NOISE PROCESSES
      SUBROUTINE SYSINP(IT,Tr,x)
      REAL x(*), n
      COMMON/CONTROL/u, wd,wv,wf,n
      DATA qd,qv,qf,r/.001,3.,.1,.1/

      CALL RANDOM(anoise)
      wd= SQRT(12*qd) * (anoise-0.5)
      CALL RANDOM(anoise)
      wv= SQRT(12*qv) * (anoise-0.5)
      CALL RANDOM(anoise)
      wf= SQRT(12*qf) * (anoise-0.5)
      CALL RANDOM(anoise)
      n=  SQRT(12*r)  * (anoise-0.5)

      RETURN
      END
```

Figure 9.4-2 Subroutine to generate process and measurement noise for use with program DIGCTLF

```
C     CONTINUOUS HEAD-POSITIONING MECHANISM DYNAMICS
      SUBROUTINE F(time,x,xp)
      REAL x(*), xp(*), n
      COMMON/CONTROL/u, wd,wv,wf,n
      COMMON/OUTPUT/y

      xp(1)= x(2) + wd
      xp(2)= x(3) + wv
      xp(3) = u   + wf

      y    = x(1) + n

      RETURN
      END
```

Figure 9.4-3 Continuous system dynamics for use with program DIGCTLF

```
C     DIGITAL KALMAN FILTER

      SUBROUTINE DIG(IK,T,x)
      REAL x(*), k1,k2,k3, L1,L2,L3
      COMMON/CONTROL/u
      COMMON/OUTPUT/y    ,d,v,f
c     (d,v,f are in OUTPUT for plotting purposes only)
      DATA qd,qv,qf,r/.001,3.,.1,.1/
      DATA p1,p4,p6/1.,1.,1000./

C     INITIALIZE COVARIANCES AND ESTIMATE
```

Figure 9.4-4 Digital Kalman filter dynamics for use with program DIGCTLF

```
            IF(IK.EQ.0) THEN
               p4= p4/T**2
               qd= qd*T
               qv= qv*T
               qf= qf*T
               r = r/T
               d= y
            END IF

C     KALMAN FILTER GAIN AND ERROR COVARIANCE UPDATE

C        Kalman gains:
         k1= p1/(p1+r)
         k2= p2/(p1+r)
         k3= p3/(p1+r)

C        Error covariance update:
         p1p= (1-k1)*p1
         p2p= (1-k1)*p2
         p3p= (1-k1)*p3
         p4p= p4 - k2*p2
         p5p= p5 - k2*p3
         p6p= p6 - k3*p3

         q1=  p1p + p2p*T + p3p*T**2/2
         q2=  p2p + p4p*T + p5p*T**2/2
         q3=  p3p + p5p*T + p6p*T**2/2
         q4=  p2p + p3p*T
         q5=  p4p + p5p*T
         q6=  p5p + p6p*T
         q7=  p3p
         q8=  p5p
         q9=  p6p

         p1=  q1 + q2*T + q3*T**2/2 + qd
         p2=  q2 + q3*T
         p3=  q3
         p4=  q5 + q6*T              + qv
         p5=  q6
         p6=  q9                     + qf

C        Filter gains:

         L1= k1 + k2*T + k3*T**2/2
         L2=      k2   + k3*T
         L3=             k3

C     KALMAN FILTER DYNAMICS
         e= ym1 - d
         d= d + T*v + f*T**2/2 + u*T**2/2 + L1*e
         v=     v   + f*T      + u*T      + L2*e
         f=           f                   + L3*e
         ym1= y

         RETURN
         END
```

Figure 9.4-4 (*continued*)

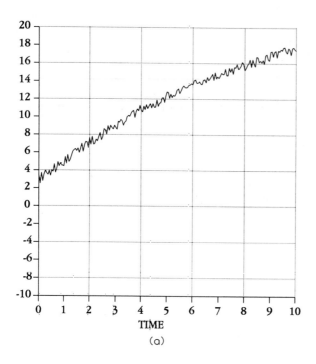

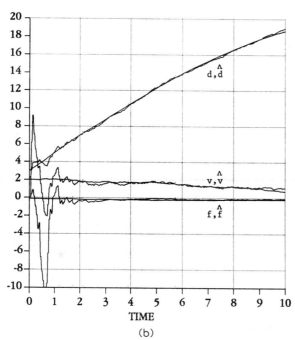

Figure 9.4-5 Kalman filter simulation results. (a) Measured output $y(t)$. (b) Actual and estimated states.

The FORTRAN V random number generator RANDOM( ) produces white noise uniformly distributed between 0 and 1. The subroutine uses RANDOM( ) to manufacture white noise of zero mean with the correct covariance. First, it subtracts 0.5 to yield a zero-mean process uniformly distributed between -0.5 and 0.5, which has a variance of $1/12$. (Recall that a uniform process distributed between $\pm a$ has a variance of $a^2/3$.) To produce a process with a variance of q_d, for instance, it is thus necessary to multiply the noise values by the standard deviation factor $(12q_d)^{1/2}$.

Note that SYSINP produces uniformly distributed white noise. Thus, the Kalman filter is the best *linear* estimator. If the noise were Gaussian, the Kalman filter would be *the* optimal estimator.

Subroutine F(time, x, xprime), provides the continuous-time dynamics (1), (2) and injects the noises. These dynamics are integrated using a Runge-Kutta integrator. The result is a simulated system output $y(t)$. Subroutine F appears in Fig. 9.4-3.

The third subroutine, DIG(IK, T, x), provides the discrete Kalman filter dynamics. It is shown in Fig. 9.4-4. T is the sample period and IK is an iteration counter. DIG is called only at the sample times, so that it uses the sampled output y_k. This setup should be compared to the digital control simulation procedure in section 5.1.

Note that subroutine DIG(IK, T, x) is structured exactly as is part d of this example. All of the intermediate scalar variables we used appear in the program. No matrix multiplies are used since the program is constructed around our analytical simplifications.

The simulation results, using $T = 0.05$ units, are shown in Fig. 9.4-5. Fig. 9.4-5a shows the measured output and Fig. 9.4-5b shows the actual and estimated states. Note that the estimated value of position $\hat{d}(t)$ is considerably better than the measured output $y(t) = d(t) + n(t)$. This is due to the noise rejection property of the Kalman filter. The Kalman filter is a low-pass filter that rejects noise.

∎

Example 9.4-2: Steady-State Kalman Filter

Although the Kalman filter designed in Example 9.4-1 performs very well, it is complicated to implement due to the time-varying nature of the error covariance and the Kalman gain. A much simpler filter can be designed using the steady-state Kalman gain.

a. Steady-State Kalman Filter Design

The steady-state Kalman filter is given by

$$\hat{x}_{k+1}^- = A\hat{x}_k^- + Bu_k + L(y_k - C\hat{x}_k^-) \qquad (1)$$

with constant filter gain $L = AK$, and the steady-state Kalman gain K given by (9.3.40) with P the positive definite solution to the ARE (9.3.39). There are several ways to determine the positive definite solution to the ARE, including routines in various commercially available software packages such as ORACLS [Armstrong 1980] and PC-Matlab [Moler et al. 1987]. One way is to simply iterate the time-varying Riccati equation in Table 9.3-3 until the solution converges. Let us take that approach here, since in Example 9.4-1 we have already done all of the work required.

For the simulation performed in Example 9.4-1, the error covariance diagonal elements p_1, p_4, and p_6 are shown in Fig. 9.4-6. Note that they reach a steady-state value in

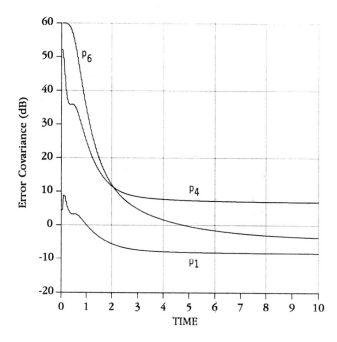

Figure 9.4-6 Diagonal elements of the error covariance

about 10 time units. The time-varying optimal Kalman gain elements k_1, k_2, k_3 are shown in Fig. 9.4-7. They reach steady-state values of

$$K = \begin{bmatrix} k_1 \\ k_2 \\ k_3 \end{bmatrix} = \begin{bmatrix} 0.161 \\ 0.281 \\ 0.050 \end{bmatrix} \qquad (2)$$

after about 6 time units. The time-varying filter gains $L_k = AK_k$ are virtually identical to Fig. 9.4-7; they reach steady-state values of

$$L = \begin{bmatrix} L_1 \\ L_2 \\ L_3 \end{bmatrix} = \begin{bmatrix} 0.175 \\ 0.283 \\ 0.050 \end{bmatrix}. \qquad (3)$$

The steady-state Kalman filter is just (1) with these filter gains. It may be written in scalar form as equations (18)–(20) in Example 9.4-1.

b. Simulation

A simulation of the steady-state Kalman filter may be performed using program DIGCTLF in Appendix A exactly as in Example 9.4-1. Subroutines SYSINP and F are the same as in that example. However, the Kalman filter subroutine DIG is far simpler, as the recursive updates of the error covariance and the Kalman gain are not needed.

The steady-state Kalman filter subroutine is shown in Fig. 9.4-8; it only contains the filter dynamics (18)–(20) with the time-invariant gains L_i in (3), and should be compared to the complicated optimal time-varying filter in Fig. 9.4-4. The simulation results using the

9.4 Digital Filtering of Continuous-Time Systems

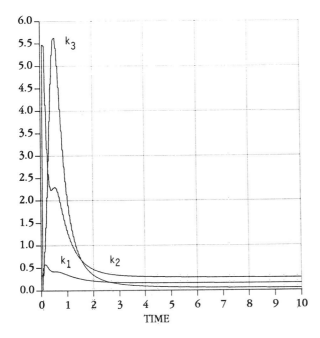

Figure 9.4-7 Time-varying Kalman gains

time-invariant steady-state Kalman filter with $T = 0.05$ appear in Fig. 9.4-9. The estimates shown there are *suboptimal estimates* since the optimal time-varying Kalman gain is not used in the filter. However, they are quite good when compared with the optimal estimates in Fig. 9.4-5b. The major deterioration resulting from use of the constant steady-state gains is the long time of convergence of the bias estimate $\hat{f}$. Indeed, convergence has not yet occurred

```
C     DIGITAL STEADY-STATE KALMAN FILTER

      SUBROUTINE DIG(IK,T,x)
      REAL x(*), L1,L2,L3
      COMMON/CONTROL/u
      COMMON/OUTPUT/y    ,d,v,f
c   (d,v,f are in OUTPUT for plotting purposes only)
      DATA L1,L2,L3/.175,.283,.05/

      e= ym1 - d
      d= d + T*v + f*T**2/2 + u*T**2/2 + L1*e
      v=      v + f*T       + u*T      + L2*e
      f=          f                    + L3*e
      ym1= y

      RETURN
      END
```

Figure 9.4-8 Subroutine implementing steady-state Kalman filter

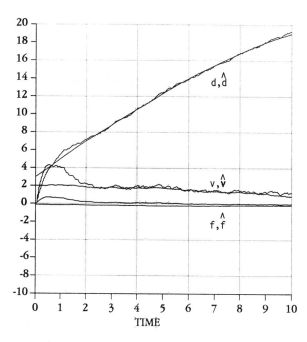

Figure 9.4-9 Simulated actual and estimated states using the steady-state Kalman filter

in Fig. 9.4-9 even within 10 time units. However, for most applications these estimates would be quite adequate.

∎

Example 9.4-3: DSP Implementation of the Kalman Filter

Let us now demonstrate the implementation of the Kalman filter we have just designed on the TI TMS320 C25 fixed-point digital signal processor (DSP). The hardware used was the Chimera board made by Atlanta Signal Processors, Inc., which is especially built for processing time-dependent signals (see Appendix D). The optimal time-varying filter of Example 9.4-1 and the simplified steady-state filter of Example 9.4-2 are both implemented in exactly the same fashion, except that the code required for the time-varying filter is much longer due to the error covariance and filter gain update recursion. To demonstrate the technique without obfuscation, let us therefore implement here the steady-state filter.

To implement the steady-state Kalman filter, it is necessary to translate the filter code in Fig. 9.4-8 into the DSP machine language. This DSP code is shown in Fig. 9.4-10.

The C25 implementation of the Kalman filter consists of three subroutines. One subroutine, FILTER__INIT, places the filter coefficients in data memory. This subroutine also clears a flag which will be used to indicate the first and second iterations of the filter so that the initial position $d(0)$ and initial velocity $v(0)$ can be estimated as in Example 9.4-1b.

FILTER__INPUT obtains the position measurements y_k and control inputs u_k and stores them in data memory. This subroutine is not included in Fig. 9.4-10 because it is hardware dependent. In this implementation, we used the data samples y_k that were generated by the simulation of Example 9.4-2 using subroutine $F(\text{time}, x, \dot{x})$ with program DIGCTLF. In a real-time control application, y_k might be obtained from an A/D converter.

```
;       ****************************************************************
;                               KALMAN.ASM
;               Kalman filter for a disk drive head positioning system.
;       ****************************************************************
;
;       ****************************************************************
;       Kalman filter initialization subroutine
;       The FILTER_INIT subroutine stores the filter coefficients in
;       data memory and clears the flags used by the FILTER subroutine.
;       FILTER_INIT should be called only once, prior to the first
;       execution of the FILTER subroutine.
;       ****************************************************************

                ;allocate data memory for filter coefficients
                ;all filter coefficients are stored in Q15 format
T               .set    300h            ;sample period
T2              .set    301h            ;T2 := T^2/2
L1              .set    302h            ;gain L1
L2              .set    303h            ;gain L2
L3              .set    304h            ;gain L3
                ;allocate data memory for the sampling frequency 1/T, which is used
                ; in the calculation of the initial velocity estimate
Tinv            .set    305h            ;Tinv := 1/T  (16-bit signed integer)
                                        ;note that 1/T>1 (assuming T<1), hence
                                        ; it cannot be stored as a fraction
                ;allocate data memory for inputs to the Kalman filter
                ;all filter inputs are stored in Q15 format
y               .set    306h            ;position measurement y[k]
u               .set    307h            ;control input u[k]
                ;allocate data memory for the state variables of the Kalman filter
                ;all state variables are stored in Q15 format
d               .set    308h            ;position estimate d[k] on entry,
                                        ; d[k+1] on exit
v               .set    309h            ;velocity estimate v[k] on entry,
                                        ; v[k+1] on exit
f               .set    30ah            ;bias force estimate f[k] on entry,
                                        ; f[k+1] on exit
;               allocate additional data memory required to implement filter
e               .set    30bh            ;position estimate error (Q15),
                                        ; e[k] = y[k] - d[k]
y0              .set    30ch            ;initial position measurement y[0],
                                        ; which must be saved in order to
                                        ; estimate v[1]
flag            .set    30dh            ;flag:  bit 0 is set the first time
                                        ;           the filter subroutine is
                                        ;           executed
                                        ;       bit 1 is set the second time
                                        ;           the filter subroutine is
                                        ;           executed

                .text
filter_init
                ldpk            300h>>7  ;set data page pointer to block B1
                                         ; (dp=300h shifted right 7 times)
                ;initialize coefficients (all coefficients are stored in Q15 format)
                lalk    1638
                sacl    T                ;sample period T=0.05 s
                lack    41
                sacl    T2               ;T^2/2=0.00125 s^2
                lalk    5734
```

Figure 9.4-10 TI TMS320C25 DSP code for implementing the steady-state Kalman filter on a disk drive head-positioning mechanism

```
                sacl    L1                      ;gain L1=0.175
                lalk    9273
                sacl    L2                      ;gain L2=0.283
                lalk    1638
                sacl    L3                      ;gain L3=0.05
;initialize the sampling frequency 1/T (16-bit signed integer)
                lack    20
                sacl    Tinv                    ;1/T = 20 Hz
;clear the flag bits
                zac                             ;zero acc
                sacl    flag                    ;zero the flag bits
                ret
;
;       ***************************************************************
;       Kalman filter subroutine
;       The FILTER subroutine implements the Kalman filter equations:
;       d[k+1] = d[k] + T*v[k] + T^2/2*f[k] + T^2/2*u[k] + L1*(y[k]-d[k]),
;       v[k+1] =        v[k] +   T*f[k] +     T*u[k] + L2*(y[k]-d[k]),
;       f[k+1] =                 f[k]                + L3*(y[k]-d[k]).
;       The inputs u[k] and y[k] to the filter are provided by
;       subroutine FILTER_INPUT. The state variables d[k+1], v[k+1],
;       and f[k+1] are stored in data memory locations 308h-30ah upon
;       exit and must be unchanged upon the next entry. Subroutine
;       FILTER_INIT must be called prior to the first execution of
;       FILTER.
;       ***************************************************************

                .text
filter
                call    filter_input            ;get the inputs to the Kalman filter
;The FILTER_INPUT subroutine provides the inputs y[k] and u[k]
;to the Kalman filter. In practice, the position measurement
;y[k] might come from an A/D converter, and the control input
;u[k] would be the last control input calculated by the control
;algorithm. FILTER_INPUT must place y[k] and u[k] (in Q15
;format) in data memory locations 306h and 307h.
                ssxm                            ;set sign extension mode
                sovm                            ;enable overflow mode
                ldpk    300h>>7                 ;data page is first half of block B1
                bit     flag,15                 ;test the LSB of the flag to see if
                                                ; this is the first sample
                bbnz    filter1                 ;branch if not the first sample

                lac     y                       ;get the initial position measurement
                                                ; y[0] in the acc
                sacl    y0                      ;save y[0] so it can be used to
                                                ; calculate v[0] after the next
                                                ; sample
                sacl    d                       ;the initial position estimate d[0]
                                                ; is y[0]
                zac                             ;zero acc
                sacl    f                       ;the initial bias force estimate f[0]
                                                ; is zero
                sacl    v                       ;the velocity cannot be estimated
                                                ; until a second position measurement
                                                ; is made, so we will return a zero
                                                ; for the velocity estimate v[0]
                lack    1                       ;acc=1
                sacl    flag                    ;set the first sample flag
                b       filter2                 ;branch to the filter equations
```

Figure 9.4-10 (*continued*)

```
filter1          bit      flag,14          ;test bit 1 of the flag to see if
                                           ; this is the second sample
                 bbnz     filter2          ;branch to filter equations if not the
                                           ; second sample

        ;We can now calculate a velocity estimate v[1] using
        ;v[1] = 1/T * (y[1] - y[0]). We will replace the estimate v[1]
        ;calculated by the Kalman filter after the first sample with this
        ;new estimate.
                 spm      0                ;set product mode for no shifting
                                           ; between product register and acc
                 lt       Tinv             ;load the T register with the sampling
                                           ; frequency (16-bit signed integer)
                 mpy      y                ;P = 1/T*y[1]   (Q15)
                 pac                       ;acc=P
                 mpy      y0               ;P = 1/T*y[0]   (Q15)
                 spac                      ;acc = 1/T*(y[1] - y[0])  (Q15)
                 sacl     v                ;replace previous v[1] with new v[1]
                 lack     3                ;acc=3
                 sacl     flag             ;set the first and second sample flags
        ;Kalman filter equations
filter2          spm      1                ;set product mode for 1 left shift
                                           ; between product register and acc
                 lac      y                ;acc = y[k]   (Q15)
                 sub      d                ;acc = y[k] - d[k]   (Q15)
                 sacl     e                ;e[k] = y[k] -d[k]   (Q15)
                 zalh     d                ;acc = d[k]   (Q31)
                 lt       T                ;T register = sample period   (Q15)
                 mpy      v                ;P = T*v[k]   (Q30)
                 lta      T2               ;acc = acc + P<<1   (Q31)
                                           ;T register = T^2/2   (Q15)
                 mpy      f                ;P = T^2/2*f[k]   (Q30)
                 mpya     u                ;acc = acc + P<<1   (Q31)
                                           ;P = T^2/2*u[k]   (Q30)
                 lta      L1               ;acc = acc + P<<1   (Q31)
                                           ;T register = L1   (Q15)
                 mpy      e                ;P = L1*e[k]   (Q30)
                 apac                      ;acc = acc + P<<1   (Q31)
                 sach     d                ;store d[k+1]   (Q15)
                 zalh     v                ;acc = v[k]   (Q31)
                 lt       T                ;T register = sample period   (Q15)
                 mpy      f                ;P = T*f[k]   (Q30)
                 mpya     u                ;acc = acc + P<<1   (Q31)
                                           ;P = T*u[k]   (Q30)
                 lta      L2               ;acc = acc + P<<1   (Q31)
                                           ;T register = L2   (Q15)
                 mpy      e                ;P = L2*e[k]   (Q30)
                 apac                      ;acc = acc + P<<1   (Q31)
                 sach     v                ;store v[k+1]   (Q15)
                 zalh     f                ;acc = f[k]   (Q31)
                 lt       L3               ;T register = L3
                 mpy      e                ;P = L3*e[k]   (Q30)
                 apac                      ;acc = acc + P<<1   (Q31)
                 sach     f                ;store f[k+1]   (Q15)
                 ret

                 .end
```

Figure 9.4-10 (*continued*)

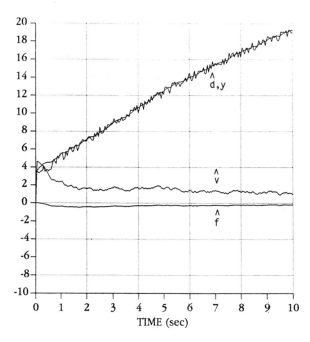

Figure 9.4-11 Output measurements and estimated states for the head-positioning mechanism using the TI TMS320C25 DSP

Subroutine FILTER implements the Kalman filter equations. The initial estimates are obtained as $\hat{d}(0) = y_0$, $\hat{v}(0) = (y_1 - y_0)/T$, so that the initial velocity estimate cannot be computed until the second iteration. Note that since the sampling frequency $1/T$ is stored as an integer, the product of the sampling frequency and the difference of the first two position measurements is obtained from the low accumulator rather than the high accumulator.

The results of the steady-state Kalman filter implementation on the C25 DSP are shown in Fig. 9.4-11. The position measurements y_k used in the simulation of Example 9.4-2 were multiplied by a factor of 1/20 to bring them into the range [−1, 1) required for Q15 representation. The output of the Kalman filter was multiplied by 20 to restore the original scale.

The results of the C25 implementation in Fig. 9.4-11 may be compared to the simulation results in Fig. 9.4-9. The C25 implementation converges faster, probably due to the fact that in the simulation of Example 9.4-2 we took $\hat{v}(0) = 0$, while in this DSP implementation we used the initialization procedure of Example 9.4-1b. ∎

9.5 CONTINUOUS KALMAN FILTER

The discrete-time Kalman filter given in Tables 9.3-1 through 9.3-3 is useful for digital signal processor (DSP) implementations of a state estimator, as we saw in section 9.4.

9.5 Continuous Kalman Filter

However, the continuous-time Kalman filter is important since it gives some additional insight. Therefore, we discuss it briefly here.

Let there be prescribed the linear time-invariant continuous-time stochastic system

$$\dot{x} = Ax + Bu + Gw \qquad (9.5.1)$$

$x \in \mathbf{R}^n$, with measurements

$$y = Cx + v, \qquad (9.5.2)$$

$y \in \mathbf{R}^p$. Assume the process noise $w(t)$ and measurement noise $v(t)$ are white noise processes, mutually uncorrelated with each other and with the initial state $x(0)$. Let $w(t) \sim (0, Q)$, $v(t) \sim (0, R)$, $x(0) \sim (\bar{x}_0, P_0)$, with known spectral densities $Q \geq 0$, $R > 0$, and covariance $P_0 \geq 0$.

If $w(t)$ and/or $v(t)$ is not white, then a noise shaping filter should be added to the system model as described in section 9.4.

It is desired to find a linear dynamical filter that provides optimal state estimates $\hat{x}(t)$ at all times given the data $y(t)$ and the noise statistics. This is known as the Kalman filter, and it is given in Table 9.5-1. Notice that it is of exactly the same form as the continuous observer discussed in section 9.1, except that the observer gain K is computed using the noise statistics. Let us now derive the filter in the table.

Kalman Filter Derivation

We shall derive the continuous-time Kalman filter by "unsampling" the discrete-time Kalman filter in Table 9.3-1 [Lewis 1986b].

The Euler's approximation (section 7.1) to the continuous-time dynamics, assuming sampling period T, is given by

$$x_{k+1} = (I + AT)x_k + BTu_k + Gw_k \qquad (9.5.3)$$

$$y_k = Cx_k + v_k \qquad (9.5.4)$$

with (section 9.4) $w_k \sim (0, QT)$, $v_k \sim (0, R/T)$, $x_0 (\bar{x}_0, P_0)$.

The discrete Kalman filter for this system is given by Tables 9.3-1, 9.3-2. Thus, the error covariance update equations are

$$P^-_{k+1} = (I + AT)P_k(I + AT)^T + GQG^T T \qquad (9.5.5)$$

$$K_{k+1} = P^-_{k+1}C^T(CP^-_{k+1}C^T + R/T)^{-1} \qquad (9.5.6)$$

$$P_{k+1} = (I - K_{k+1}C)P^-_{k+1}. \qquad (9.5.7)$$

By letting T go to zero, we shall discover the continuous-time error covariance update dynamics.

Let us examine the behavior of the Kalman gain K_k as T goes to zero. According to (9.5.6)

$$\frac{1}{T}K_k = P^-_k C^T(CP^-_k C^T T + R)^{-1}, \qquad (9.5.8)$$

and taking the limit as $T \to 0$ yields

$$\lim_{T \to 0} \frac{1}{T} K_k = P_k^- C^T R^{-1}. \qquad (9.5.9)$$

It is important to realize this means that

$$\lim_{T \to 0} K_k = 0, \qquad (9.5.10)$$

which should be remembered when selecting the sample period in digital filtering.

Now examine (9.5.5), which says

$$P_{k+1}^- = P_k^- + (AP_k^- + P_k^- A^T + GQG^T)T + o(T^2),$$

where $o(T^2)$ denotes terms of order T^2. Substituting (9.5.7) into this equation yields

$$P_{k+1}^- = (I - K_k C)P_k^- + [A(I - K_k C)P_k^- + (I - K_k C)P_k^- A^T + GQG^T]T + o(T^2),$$

whence, dividing by T

$$\frac{1}{T}(P_{k+1}^- - P_k^-) = (AP_k^- + P_k^- A^T + GQG^T - AK_k C P_k^-$$

$$- K_k C P_k^- A^T) - \frac{1}{T} K_k C P_k^- + o(T).$$

Letting $T \to 0$ and defining the continuous-time error covariance as

$$P(kT) = P_k^- \qquad (9.5.11)$$

results in

$$\dot{P}(t) = AP + PA^T + GQG^T - PC^T R^{-1} CP, \qquad (9.5.12)$$

where we used (9.5.9), (9.5.10). This is the continuous-time error covariance dynamics. It is a *continuous-time Riccati equation* whose discrete counterpart appears in Table 9.3-3. The Riccati equation should be initialized using $P(0) = P_0$, the initial state covariance.

It is now necessary to find the continuous-time estimate dynamics. Using (9.3.29) in (9.3.31), we may write the discrete Kalman filter for the Euler's system in terms of the a posteriori estimate as

$$\hat{x}_{k+1} = (I + AT)\hat{x}_k + BTu_k + K_{k+1}[y_{k+1} - C(I + AT)\hat{x}_k - CBTu_k],$$

which may be divided by T to yield

$$\frac{\hat{x}_{k+1} - \hat{x}_k}{T} = A\hat{x}_k + Bu_k + \frac{K_{k+1}}{T}[y_{k+1} - C\hat{x}_k - C(A\hat{x}_k + Bu_k)T].$$

Since $\hat{x}(kT) = \hat{x}_k$, as $T \to 0$ this results in

$$\dot{\hat{x}}(t) = A\hat{x}(t) + Bu(t) + P(t)C^T R^{-1}[y(t) - C\hat{x}(t)]. \qquad (9.5.13)$$

This is the continuous-time Kalman filter dynamics. It should be initialized using $\hat{x}(0) = \bar{x}_0$.

9.5 Continuous Kalman Filter

Defining the continuous-time Kalman gain by

$$K(kT) = \frac{1}{T}K_k, \qquad (9.5.14)$$

we see that, as $T \to 0$

$$K(t) = P(t)C^T R^{-1} \qquad (9.5.15)$$

and

$$\dot{\hat{x}}(t) = A\hat{x}(t) + Bu(t) + K(t)[y(t) - C\hat{x}(t)]. \qquad (9.5.16)$$

The continuous-time Kalman filter equations are collected in Table 9.5-1 for easy reference.

TABLE 9.5-1 CONTINUOUS-TIME KALMAN FILTER

Stochastic System and Measurement Model:

$$\dot{x} = Ax + Bu + Bw \qquad (9.5.17)$$

$$y = Cx + v \qquad (9.5.18)$$

$$x(0) \sim (\bar{x}_0, P_0), \; w(t) \sim (0, Q), \; v(t) \sim (0, R)$$

Assumptions:

$w(t)$ and $v(t)$ are white noise processes mutually uncorrelated with each other and with the initial state. $Q \geq 0, R > 0$.

Error Covariance Propagation:

$$\dot{P} = AP + PA^T + GQG^T - PC^T R^{-1} CP, \; P(0) = P_0 \qquad (9.5.19)$$

Kalman Gain:

$$K(t) = PC^T R^{-1} \qquad (9.5.20)$$

Kalman Filter Dynamics:

$$\dot{\hat{x}} = A\hat{x} + Bu + K(y - C\hat{x}), \; \hat{x}(0) = \bar{x}_0 \qquad (9.5.21)$$

Filter Properties and Implementation

It is important to note that, while, in the limit $T \to 0$, the discrete error covariance sequence P_k^- is a sampled version of the continuous error covariance $P(t)$, the same is not true for the Kalman gains. In fact, K_k is a sampled version of $TK(t)$.

While in the discrete-time case only the invertibility of $(CP_k^- C^T + R)$ is required to implement the Kalman filter, the continuous filter requires R to be invertible. If R is singular, a filter known as the *Deyst filter* must be used [Lewis 1986b].

The Kalman filter provides the optimal *linear* estimate $\hat{x}(t)$, but if all noises are Gaussian, it provides *the* optimal estimator. Note that the Kalman filter has the same structure as the continuous-time observer in section 9.1. It has two parts—a model of the system dynamics (A, B, C) and an error-correcting part $K(y - C\hat{x})$. Unlike the discrete Kalman filter, the continuous filter cannot be split up into a "time update" and a "measurement update."

For implementation purposes, there are two possibilities. The error covariance $P(t)$ and Kalman gain $K(t)$ may be computed off-line and stored, or they may be computed on-line in real time as the measurements are taken. Computer simulations may be performed using a program like TRESP in Appendix A which is based on a Runge-Kutta integrator. See Example 9.5-1.

Steady-State Kalman Filter

The Kalman filter is *time-varying* even if the system is time-invariant. A suboptimal *steady-state* filter with a constant Kalman gain version is easier to implement, and is usually satisfactory for most applications.

To derive the continuous-time steady-state Kalman filter, note that at statistical steady-state, $dP/dt = 0$, that is, the error covariance is constant. Therefore, the Riccati equation becomes the *algebraic Riccati equation* (ARE)

$$0 = AP + PA^T + GQG^T - PC^TR^{-1}CP. \tag{9.5.22}$$

The steady-state Kalman gain is the *constant $n \times p$* matrix

$$K = PC^TR^{-1}. \tag{9.5.23}$$

Implementation of the steady-state filter is much easier than the optimal time-varying filter. The constant Kalman gain may be computed once off-line and stored, and then used in the filter on-line when the measurements are taken. It is still, however, necessary to implement the continuous-time dynamics of the filter. Therefore, for DSP applications digital filtering using the discrete Kalman filter is preferred.

Note that the ARE is the same equation that was used for design of multivariable continuous-time observers in section 9.1. The difference is that here we use the noise statistics Q and R to compute P and hence the Kalman gain K in an optimal fashion for the given noise.

The Kalman filter may be written as

$$\dot{\hat{x}} = (A - KC)\hat{x} + Bu + Ky, \tag{9.5.24}$$

where in the steady-state filter K is constant. In this case, the transfer function from the measurements $y(t)$ to the state estimate $\hat{x}(t)$ is given by

$$H(s) = [sI - (A - KC)]^{-1}K, \tag{9.5.25}$$

so that the Kalman filter is a low-pass filter. This accounts for its noise-rejection properties. It is interesting to realize that (9.5.25) is the Wiener filter for the system (9.5.17).

The theorem we presented in section 9.1 is relevant here. It states that, if (C, A) is observable and $(A, \sqrt{Q})$ is reachable, then there is a unique positive definite solution P

9.5 Continuous Kalman Filter

to the ARE, and the steady-state Kalman filter is guaranteed to be asymptotically stable. Thus, it produces asymptotically accurate estimates.

Again, we have reached the rather strange conclusion that, for the Kalman filter to work properly, the process noise must excite all the states and the measurement noise must corrupt all the measurements (i.e., $R > 0$).

Example 9.5-1: Continuous Filtering for Damped Harmonic Oscillator

Suppose the plant is the damped harmonic oscillator with velocity measurements described by

$$\dot{x} = \begin{bmatrix} 0 & 1 \\ -\omega_n^2 & -2\alpha \end{bmatrix} x + \begin{bmatrix} 0 \\ 1 \end{bmatrix} w = Ax + Gw \quad (1)$$

$$y = [0 \quad 1]x + n = Cx + n, \quad (2)$$

where the state is $x = [d \quad v]^T$, with $d(t)$ the position and $v(t)$ the velocity. See Example 2.1-5. The natural frequency is ω_n and the real part of the poles is $-\alpha$. Let the process noise be white with $w(t) \sim (0, q)$ and the measurement noise be white with $n(t) \sim (0, r)$, $r \neq 0$. The initial state is random with $x(0) \sim (\bar{x}_0, P_0)$.

We shall show here how to design and simulate the continuous Kalman filter for this system.

a. Riccati Equation Simplification

To compute the error covariance, let us simplify the Riccati equation in Table 9.5-1. Since $P(t)$ is symmetric, define

$$P = \begin{bmatrix} p_1 & p_2 \\ p_2 & p_4 \end{bmatrix}, \quad (3)$$

with $p_i(t)$ time-varying scalars. Then, simplification of the Riccati equation yields the three coupled scalar differential equations (verify this)

$$\dot{p}_1 = 2p_2 - p_2^2/r \quad (4)$$

$$\dot{p}_2 = p_4 - \omega_n^2 p_1 - 2\alpha p_2 - p_2 p_4/r \quad (5)$$

$$\dot{p}_4 = -2\omega_n^2 p_2 - 4\alpha p_4 + q - p_4^2/r. \quad (6)$$

These equations describe the dynamics of the error covariance.

b. Kalman Filter

The Kalman filter dynamics may be computed using Table 9.5-1. Thus, the Kalman gain is found to be

$$K = PC^T/r = \frac{1}{r}\begin{bmatrix} p_2(t) \\ p_4(t) \end{bmatrix} \equiv \begin{bmatrix} k_1 \\ k_2 \end{bmatrix}. \quad (7)$$

The filter dynamics in Table 9.5-1 may be simplified to yield the scalar equations

$$\dot{\hat{d}} = \hat{v} + k_1(y - \hat{v}) \quad (8)$$

$$\dot{\hat{v}} = -\omega_n^2 \hat{d} - 2\alpha \hat{v} + k_2(y - \hat{v}). \quad (9)$$

c. Computer Simulation of the Continuous Kalman Filter

A computer simulation of the Kalman filter may be carried out using program DIGCTLF in Appendix A. Three subroutines are needed; they are shown in Fig. 9.5-1.

Subroutine SYSINP(IT, Tr, x) is called once each Runge-Kutta time period Tr. It computes the continuous-time noise processes $w(t)$ and $n(t)$. See the discussion in Example 9.4-1.

DIGCTLF needs a subroutine F(time, X, $\dot{X}$) to provide the continuous dynamics required by the Runge-Kutta integrator. In this example, the continuous dynamics have two parts—the plant and the Kalman filter. Thus, subroutine F(time, X, $\dot{X}$) in Fig. 9.5-1 calls two other subroutines. The Runge-Kutta state vector $X(t)$ contains all the variables which are described using dynamical equations, including the plant state $d(t)$, $v(t)$, the error covariance elements $p_1(t)$, $p_2(t)$, $p_4(t)$, and the estimates $\hat{d}(t)$, $\hat{v}(t)$.

Subroutine SYSTEM(time, X, $\dot{X}$) provides the harmonic oscillator dynamics (1), (2). It uses $X(1)$ and $X(2)$ to represent the plant states d and v. The function of this subroutine is to provide simulated plant measured outputs.

Subroutine KALMAN(time, X, $\dot{X}$) contains the error covariance and Kalman filter dynamics (4)–(9). It assigns p_1, p_2, and p_4 respectively to components $X(3)$, $X(4)$, and $X(5)$ of the dynamical vector. The state estimates $\hat{d}$ and $\hat{v}$ are assigned to $X(6)$, $X(7)$.

Note how the structure of the subroutines exactly duplicates the equations we have derived from Table 9.5-1.

The simulation results are shown in Fig. 9.5-2. Figure 9.5-2a shows the states $d(t)$, $v(t)$, as well as the measured output $y(t)$. Fig. 9.5-2b shows the states and their estimates $\hat{d}(t)$, $\hat{v}(t)$. Note that the estimates are quite good even though the measured output is severely corrupted by noise.

We used $d(0) = 1$, $v(0) = 0$. The initial estimates were taken equal to zero. The error covariance was initialized with $P(0) = 10I$, with I the identity. The motivation for selecting a large initial error covariance is that this makes the estimates converge more quickly to the states. We used a Runge-Kutta step size of 10 msec.

Plots of the error covariance elements (in dB) are shown in Fig. 9.5-3. The position variance $p_1(t)$ and velocity variance $p_4(t)$ reach steady-state values in about 12 sec, and the position-velocity cross-covariance term $p_2(t)$ goes to zero with time.

d. Steady-State Kalman Filter

The optimal filter is time-varying, which means that the error covariance dynamics (4)–(6) must be used to compute the optimal time-varying Kalman gain $K(t)$. See Fig. 9.5-1. In some applications, this could be too much trouble, as well as unneccessary.

To find the suboptimal time-invariant filter with constant Kalman gain, we may find the steady-state solution to the ARE. Setting $\dot{p}_i = 0$ in (4)–(6) yields three coupled nonlinear *algebraic* equations. These equations are easily solved (as the reader should verify) to obtain the steady-state error covariance elements

$$p_1 = \frac{2\alpha r}{\omega_n^2}[(1 + q/4\alpha^2 r)^{1/2} - 1] \tag{10}$$

$$p_2 = 0 \tag{11}$$

$$p_4 = 2\alpha r[(1 + q/4\alpha^2 r)^{1/2} - 1]. \tag{12}$$

We have selected positive roots so that P is positive definite. The quantity $\lambda = q/\alpha r$ may be interpreted as a signal-to-noise ratio.

```
C     SUBROUTINE TO GENERATE CONTINUOUS-TIME NOISE PROCESSES
      SUBROUTINE SYSINP(IT,Tr,x)
      REAL x(*), n
      COMMON/CONTROL/u, w,n
      DATA q,r/1.,.05/

      CALL RANDOM(anoise)
      w= SQRT(12*q) * (anoise-0.5)
      CALL RANDOM(anoise)
      n= SQRT(12*r) * (anoise-0.5)

      RETURN
      END

C     CONTINUOUS-TIME DYNAMICS
      SUBROUTINE F(time,x,xp)
      REAL x(*), xp(*)
      CALL SYSTEM(time,x,xp)
      CALL KALMAN(time,x,xp)
      RETURN
      END

C     CONTINUOUS HARMONIC OSCILLATOR DYNAMICS
      SUBROUTINE SYSTEM(time,x,xp)
      REAL x(*), xp(*), n
      COMMON/CONTROL/u, w,n
      COMMON/OUTPUT/y
      DATA w2,a/.64,.16/

      xp(1)=               x(2)
      xp(2)= -w2*x(1) - 2*a*x(2) + w

      y    =   x(2)  + n

      RETURN
      END

C     CONTINUOUS KALMAN FILTER
      SUBROUTINE KALMAN(time,x,xp)
      REAL x(*), xp(*), k1,k2
      COMMON/CONTROL/u
      COMMON/OUTPUT/y
      DATA w2,a,q,r/.64,.16,1.,.05/

C     Error Covariance Dynamics [p1= x(3), p2= x(4), p4= x(5)]
      xp(3)= 2*x(4)                          - x(4)**2/r
      xp(4)= x(5) - w2*x(3) - 2*a*x(4) - x(4)*x(5)/r
      xp(5)= -2*w2*x(4) - 4*a*x(5) + q - x(5)**2/r

C     Kalman Gains
      k1= x(4)/r
      k2= x(5)/r

C     Kalman Filter Dynamics [dhat= x(6), vhat= x(7)]
      e   = y - x(7)
      xp(6)=            x(7)   + k1*e
      xp(7)= -w2*x(6) -2*a*x(7) + k2*e

      RETURN
      END
```

Figure 9.5-1 Subroutines used with DIGCTLF for continuous Kalman filter simulation

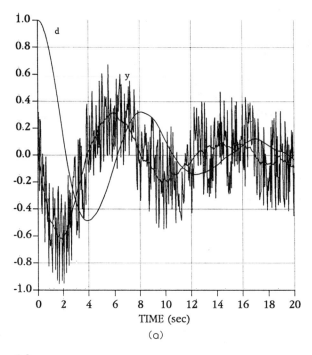

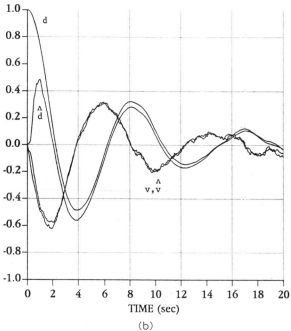

Figure 9.5-2 Kalman filter simulation results. (a) Plant states and measured output ($d(0)=1$, $v(0)=0$). (b) States and estimates.

9.5 Continuous Kalman Filter

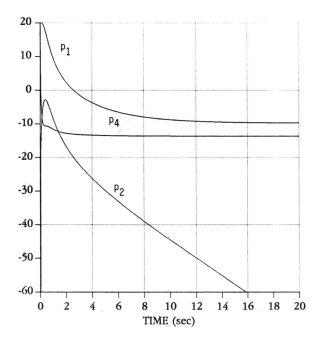

Figure 9.5-3 Plots of error covariance elements vs. time (in dB)

The steady-state Kalman gain has the elements

$$k_1 = p_2/r = 0 \tag{13}$$

$$k_2 = p_4/r = 2\alpha[(1 + q/4\alpha^2 r)^{1/2} - 1]. \tag{14}$$

Note that (C, A) is observable and $(A, G\sqrt{q})$ is reachable, so asymptotic accuracy of the estimates is assured.

Using the values for α, ω_n^2, q, and r given in Fig. 9.5-1, the steady-state error covariance is computed to be

$$P = \begin{bmatrix} 0.3273 & 0 \\ 0 & 0.2082 \end{bmatrix}. \tag{15}$$

This corresponds to the steady-state values in Fig. 9.5-3 (note the scale in the figure is in dB). In fact, an alternative to computing the steady-state error covariance analytically as we have done is simply to integrate the Riccati equation numerically until the solution reaches a constant value. This approach may be used for complicated systems.

The steady-state Kalman gains are

$$k_1 = 0, \qquad k_2 = 4.164. \tag{16}$$

The software for simulation of the steady-state filter is shown in Fig. 9.5-4, where subroutine KALMAN(time, X, $\dot{X}$) appears. The other subroutines needed by the driver program DIGCTLF in Appendix A are the same as in Fig. 9.5-1. Note that, now, the error covariance updates are not needed so the filter is simpler to simulate.

C   STEADY-STATE CONTINUOUS KALMAN FILTER

```
      SUBROUTINE KALMAN(time,x,xp)
      REAL x(*), xp(*), k1,k2
      COMMON/CONTROL/u
      COMMON/OUTPUT/y
      DATA w2,a/.64,.16/

      k1= 0
      k2= 4.164
```
C   Kalman Filter Dynamics [dhat= x(3), vhat= x(4)]
```
      e    = y - x(4)
      xp(3)=              x(4)    + k1*e
      xp(4)= -w2*x(3) - 2*a*x(4) + k2*e

      RETURN
      END

      SUBROUTINE DIG(IK,T,X)
      RETURN
      END
```

Figure 9.5-4 Subroutine for use with DIGCLTF for steady-state continuous Kalman filter

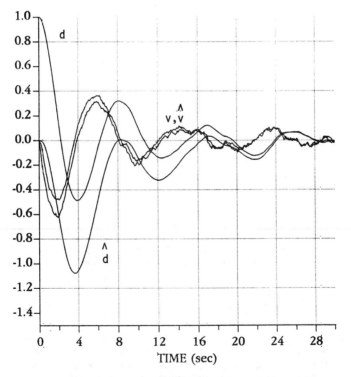

Figure 9.5-5 States and estimates using the steady-state Kalman filter

The simulation results using the steady-state Kalman filter are provided in Fig. 9.5-5. Note that the performance is not as good as that of the optimal time-varying filter in this case. In fact, it takes 30 sec for the position estimate $\hat{d}$ to converge; this is not surprising since the steady-state Kalman gain element k_1 is equal to zero.

∎

PROBLEMS FOR CHAPTER 9

Problems for Section 9.1

9.1-1 Duality and Ackermann's Formula. Substitute (A^T, B^T) for (A, C) in the observer Ackermann's formula (9.1.11). Verify that the result is the controller Ackermann's formula in Chapter 2. Thus, these two versions of Ackermann's formula are *duals*.

9.1-2 Duality and the Continuous-Time ARE. Replace (A, B, C) in (9.1.27), (9.1.28) by (A^T, C^T, B^T) and verify that the result is the ARE used for LQR design in section 3.3.

9.1-3 Continuous Observer with Noise. In Example 9.1-1 a digital observer was simulated with process and measurement noises to demonstrate its robustness. Perform similar noisy simulations for the continuous-time observer of Example 9.1-2.

9.1-4 Continuous Observer for Armature-Controlled DC Motor. In Example 2.1-3 a state-space description was presented for an armature-controlled DC motor. It is common to measure the angular velocity of the motor, but feedback of the current as well is sometimes needed for control purposes. Therefore, let us take the output as $\omega(t)$ and design an observer to reconstruct $i(t)$. Use a control input of 100 V.
  a. Design an observer for the motor that has $\zeta = 1/2$ and $\omega = 2$ rad/sec. Simulate the observer with and without noise.
  b. Design an observer using the ARE. Select various values for Q and R. You may need to use ORACLS [Armstrong 1980], PC-MATLAB [Moler et al. 1987], or MATRIX$_x$ [1989] to solve the ARE and find the observer gain. Simulate the observer with and without noise.

9.1-5 Digital Observer for Armature-Controlled DC Motor. Repeat Problem 9.1-4 with a digital observer. Use various sample periods between about 0.01 sec–0.1 sec. Use program DISC in Appendix A to discretize the system. Try deadbeat design as well as some other choices.

9.1-6 Multivariable Continuous Observer. Design an observer for the MIMO circuit in Examples 2.5-2 and 4.1-1. A design approach using the ARE easily handles this two-output system. Select various values for the design parameters Q and R, in each case finding the observer poles, and performing simulations with and without noise.

9.1-7 Multivariable Digital Observer. Repeat Problem 9.1-6 using a digital observer.

9.1-8 Continuous Observer for Inverted Pendulum. Design an observer for the inverted pendulum in Examples 4.2-4, 5.2-2, 5.3-1, 7.4-3. Use process and measurement noises. The multi-output nature of this system means that the ARE provides a convenient design technique.

9.1-9 Digital Observer for Inverted Pendulum. Repeat Problem 9.1-8 using a digital observer. Try various sample periods.

Problems for Section 9.2

9.2-1 Robustness of Deadbeat Observer to Noise. In Example 9.2-1 a reduced-order digital observer was designed for Newton's system.
   a. Perform simulations in the presence of process and measurement noise as in Example 9.1-1.
   b. Increasing the noise too much results in bad estimates. Select a small nonzero value of the design parameter a and compute $\hat{x}_k$ in terms of y_k and z_k. Perform noisy simulations and compare the performance to the deadbeat case.

9.2-2 Reduced-Order Observer for Armature-Controlled DC Motor. Design a reduced-order observer for the motor of Problem 9.1-4. Select the observer pole at $s = -5$. Note that a problem with the design procedure is finding a suitable solution to the Lyapunov design equation.

9.2-3 Digital Reduced-Order Observer for Armature-Controlled DC Motor. Repeat Problem 9.2-2 using a digital observer. Try various sample periods.

Problems for Section 9.3

9.3-1 Demonstrate the equality (9.3.21).

9.3-2 Derive the Kalman filter formulation in Table 9.3-3.

9.3-3 Software for Digital Kalman Filter. Use the UD factorization subroutine given in [Bierman 1977] to write a computer program that implements the time-varying Kalman filter of Table 9.3-1. Your program should read in the plant matrices (A, B, C, G) and the noise covariance matrices Q, R, P_0. It should request the initial estimate $\bar{x}_0$. Its input should be the data y_k from the plant and it should output the state estimates and selected elements of the error covariance for plotting.

To obtain the data y_k for a filter simulation run, one can simulate the plant with noise inputs using a digital computer exactly as was done in the examples of section 9.1. One way to achieve all of this is to use the Bierman UD factorization in subroutine DIG of program DIGCTLF in Appendix A.

Problems for Section 9.4

9.4-1 Digital Kalman Filter for Armature-Controlled DC Motor. In Example 2.1-3 the equations were given for an armature-controlled DC motor.
   a. Using the software written in Problem 9.3-3 simulate the optimal Kalman filter. Assume that $G = I$, $Q = I$, $R = 0.1I$, $P_0 = 100I$, $\bar{x}_0 = 0$. Use zero initial conditions on the motor and a motor control voltage of 100 V.
   b. Simulate the steady-state Kalman filter for the motor.

9.4-2 Digital Kalman Filter for Damped Harmonic Oscillator. It is desired to estimate $z(t)$ and $z'(t)$ for the harmonic oscillator

$$z'' + 2\zeta\omega z' + \omega^2 z = \omega^2 u + w,$$

where $w(t) \sim (0, 0.1)$ is a process noise. Take $\zeta = 1/2$, $\omega = 2$. The measurements are the corrupted position measurements

$$y = z + v,$$

with measurement noise $v(t) \sim (0, 0.1)$.

a. Write a state-space realization that has z, z' as the state components. Discretize using $T = 0.05$ sec.

b. Simulate the optimal Kalman filter on the plant using $z(0) = z'(0) = 0$, with $u(t)$ the unit step. For filter initialization use zero initial estimates and a large initial error covariance. Examine the effect of different P_0 on the filter performance.

c. Simulate the steady-state Kalman filter.

9.4-3 Multivariable Kalman Filter. A 2-input/2-output circuit was given in Examples 2.5-2 and 4.1-1. Take a noisy version of that circuit with $G = B$, $Q = 0.1I$, $R = 0.1I$.

a. Discretize the plant and noise covariance matrices.

b. Simulate the digital Kalman filter. Select zero plant initial conditions and set both control inputs equal to one. Use zero initial estimates and a large initial error covariance to achieve rapid convergence.

c. Simulate the steady-state Kalman Filter.

Problems for Section 9.5

9.5-1 Kalman Filter for Newton's System. Repeat Example 9.5-1 for Newton's system with position measurements

$$\dot{d} = v, \qquad \dot{v} = u + w, \qquad y = d + n,$$

with process noise $w \sim (0, 0.1)$ and measurement noise $v \sim (0, 0.1)$. Derive the optimal and the steady-state Kalman filters from scratch. (Your results can be checked by setting $\alpha = 0$, $\omega_n = 0$ in Example 9.5.1.) For the simulation, use initial conditions of $d(0) = -1$, $v(0) = 1$ in the plant and take $u(t)$ as the unit step. Take the initial estimates as zero and a large initial error covariance.

9.5-2 Kalman Filter for Armature-Controlled DC Motor. Simulate the continuous-time Kalman filter for the system described in Problem 9.4-1. Derive analytic scalar expressions for the covariance and estimate dynamics. Compare to the results using the digital Kalman filter.

10
Multivariable Dynamic Compensator Design

SUMMARY

Our focus in this book has thus far been on controls design when the dynamical structure of the compensator is known using engineering judgement. However, modern multivariable systems, such as jet engines, may be so complicated that it is not obvious which compensator structures are suitable. In this chapter we deal with design techniques that automatically give the dynamic compensator structure required to achieve performance specifications for multi-input/multi-output systems.

INTRODUCTION

The central theme in this book so far has been controls design using partial state, or *output* feedback. A major motivation for output-feedback design is that the compensator structures obtained are simple and intuitive. We saw in section 4.2 that by using output feedback a compensator of any desired structure may be used, with the feedback gains being selected by modern LQ techniques. In many practical situations, the dynamical structure of the compensator is indeed known using engineering judgment. For instance, in aircraft controls, decades of experience dictate when washout filters are needed, when integral action is required, and so on. In such situations, output-feedback controls design is very suitable.

On the other hand, many modern MIMO systems (e.g., jet engines) are so complicated that it is impossible to specify a compensator structure suitable for attaining the performance objectives. Indeed, the performance objectives themselves may be conflicting, involving requirements on the time response as well as on robustness.

We have seen some controls design techniques that automatically yield the dynamical structure of the compensator. In CGT design (section 4.4) the compensator structure depends on the dynamics of the reference command or the disturbance. In explicit model-following design (section 4.5) the compensator structure depends on the dynamics of the model with desired performance characteristics.

In this chapter we give some additional modern controls design techniques that automatically give the dynamical structure of a compensator which results in the fulfillment of the design objectives. A major advantage of these techniques is that the compensator structure and gains are obtained, even for complicated multivariable systems, by simply solving some matrix design equations, for which good software is available commercially.

We shall cover compensator design using state-space techniques. We begin with linear-quadratic-Gaussian (LQG) design, which uses full state-feedback design followed by a Kalman filter. The powerful LQG/loop-transfer recovery method is next presented.

The techniques of this chapter apply for both continuous-time and digital controls design.

10.1 LINEAR-QUADRATIC-GAUSSIAN DESIGN

The basic approach in modern control theory for the design of multivariable dynamic compensators is the *linear-quadratic-Gaussian* (LQG) approach. In LQG design, all the theory of full state-variable feedback (sections 3.3 and 7.4) and the Kalman filter (sections 9.3–9.5) is relevant. This is important from several points of view, as we now discuss.

Full state-feedback design has some important advantages over output-feedback design. In sections 3.3 and 7.4 we saw that the steady-state design equations for full state-variable feedback are simpler than those for output feedback. In fact, in state-variable design it is only necessary to solve the matrix Riccati equation, for which there are good routines in several commercially available software packages (e.g., ORACLS [Armstrong 1980], PC-MATLAB [Moler et al. 1987], and MATRIX$_x$ [1989]). By contrast, in output-feedback design it is necessary to solve coupled nonlinear equations (see Tables 4.1-1, 4.2-1, 7.4-1), which must generally be done using iterative algorithms [Moerder and Calise 1985, Press et al. 1986].

Moreover, in the case of full state feedback, if the system (A, B) is reachable and $(\sqrt{\bar{Q}}, A)$ is observable (with Q the state weighting in the PI), then the steady-state Kalman control gain is guaranteed to stabilize the plant and yield a global minimum value for the PI. This is a fundamental result of modern control theory, and no such result yet exists for output feedback.

Another issue is that the LQ regulator with full state feedback enjoys some important *robustness properties* that are not guaranteed using output feedback. Specifically, as we shall see in section 10.2, it has an infinite gain margin and 60° of phase margin.

Similar statements hold for Kalman filter design. The filter Riccati equation is straightforward to solve, and we saw in sections 9.3 and 9.5 that under some observability and reachability conditions, the Kalman filter is guaranteed to provide stable estimates of the states.

In LQG design, many of these separate advantages of state feedback and the Kalman filter are retained.

LQG Formulation

LQG design relies on the fact that the linear-quadratic regulator (LQR) with full state-variable feedback (sections 3.3 and 7.4) and the Kalman filter (sections 9.3–9.5) can be used together to design a dynamic regulator. An important advantage of LQG design is that the compensator structure is automatically given by the procedure, so that it need not be known beforehand. In fact, it is closely related to the structure of the observers and filters in Chapter 9. This makes LQG design useful in the control of complicated modern-day systems such as large-scale space structures and aircraft engines, where an appropriate compensator structure may be difficult to decide on using basic engineering intuition.

LQG design may be used for either continuous-time or digital controls design. We shall discuss the procedure using the continuous-time framework to avoid a proliferation of subscripts in the time index. A digital design example will then be given. The LQG approach also applies to the design of time-varying feedbacks and filters. However, we shall discuss the time-invariant or steady-state LQG dynamic regulator.

Suppose the plant and measured output are given by

$$\dot{x} = Ax + Bu + Gw \qquad (10.1.1)$$

$$y = Cx + v \qquad (10.1.2)$$

with $x(t) \in \mathbf{R}^n$, $u(t)$ the control input, $w(t)$ the process noise, and $v(t)$ the measurement noise. Suppose that the full state-feedback control

$$u = -Kx + r \qquad (10.1.3)$$

has been designed, with $r(t)$ the command or reference input. That is, the state feedback gain K has been selected by some technique, such as the LQR technique in section 3.3. LQR design for the steady-state control gain K involves solving the control algebraic Riccati equation (ARE), which is easily accomplished, even for complicated MIMO systems, using available software packages such as those previously mentioned.

If the control (10.1.3) is substituted into (10.1.1), the closed-loop system is found to be

$$\dot{x} = (A - BK)x + Br + Gw. \qquad (10.1.4)$$

Full-state feedback design is attractive because if the reachability and observability conditions in section 3.3 hold, the closed-loop system is guaranteed stable. Such a strong result has not yet been shown for output feedback. Moreover, using full state feedback all the poles of $(A - BK)$ may be placed arbitrarily as desired. However, the control law (10.1.3) cannot be implemented in most practical situations since all the states are usually not available as measurements.

Now, suppose that an observer or Kalman filter

$$\dot{\hat{x}} = (A - LC)\hat{x} + Bu + Ly \qquad (10.1.5)$$

10.1 Linear-Quadratic-Gaussian Design

has been designed. That is, the filter gain L has been selected by any of the techniques discussed in Chapter 9 to provide stable state estimates $\hat{x}(t)$. If Kalman filter design is selected, the filter gain is easily determined by solving the filter ARE using software packages such as those mentioned above.

Since all the states are not measurable and the control (10.1.3) cannot be implemented in practice, we propose to *feed back the estimate* $\hat{x}(t)$ instead of the actual state $x(t)$. That is, let us examine the feedback law

$$u = -K\hat{x} + r. \qquad (10.1.6)$$

The closed-loop structure using this controller is shown in Fig. 10.1-1. Due to the fact that the observer is a dynamical system, the proposed controller is nothing but a *dynamical regulator or compensator* of the sort seen in classical control theory. However, in contrast to classical design the LQG theory makes it easy to design multivariable regulators with guaranteed stability even for complicated MIMO systems.

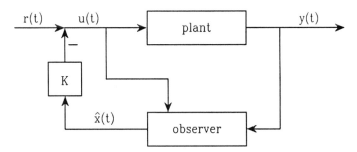

Figure 10.1-1 Dynamic regulator using observer and full state feedback

The feedback gain K and observer gain L can be selected using any techniques that guarantee suitable properties of $(A - BK)$ and $(A - LC)$. In LQG design, K is selected using the LQR ARE in section 3.3 and L is selected using the Kalman filter ARE in section 9.5. The design equations for the continuous-time LQG regulator appear in Table 10.1-1.

We propose to show that using the dynamic LQG controller in Table 10.1-1:

1. The closed-loop poles are the same as if the full state feedback (10.1.3) had been used.
2. The transfer function from $r(t)$ to $y(t)$ is the same as if (10.1.3) had been used.

A few seconds of thought should make one realize that these results are quite remarkable.

The importance of these results is that the state feedback K and the observer gain L may be designed *separately* to yield desired closed-loop plant behavior and observer behavior. This is the *Separation Principle* which is at the heart of modern controls design.

TABLE 10.1-1 CONTINUOUS-TIME LQG DYNAMIC REGULATOR

LQR State Feedback Design

$$0 = A^T S + SA + Q_c - SBR_c^{-1}B^T S \quad (10.1.15)$$

$$K = R_c^{-1}B^T S \quad (10.1.16)$$

where Q_c and R_c are PI weighting matrices

Kalman Filter Gain Design

$$0 = AP + PA^T + GQ_f G^T - PC^T R_f^{-1} CP \quad (10.1.17)$$

$$L = PC^T R_f^{-1} \quad (10.1.18)$$

where Q_f and R_f are noise spectral density matrices

LQG Dynamic Regulator

$$\dot{\hat{x}} = A\hat{x} + Bu + L(y - C\hat{x}) \quad (10.1.19)$$

$$u = -K\hat{x} + r \quad (10.1.20)$$

Two important ramifications of the separation principle are that closed-loop stability is guaranteed under the controllability and observability assumptions, and good software is available to solve the matrix design equations that yield K and L.

The Separation Principle

To show the two important results just mentioned, define the estimation error

$$\tilde{x}(t) = x(t) - \hat{x}(t). \quad (10.1.7)$$

Using (10.1.1) and (10.1.5) we may derive the error dynamics to be

$$\dot{\tilde{x}} = (A - LC)\tilde{x} + Gw - Lv \equiv A_o \tilde{x} + Gw - Lv. \quad (10.1.8)$$

In terms of $\tilde{x}(t)$, we may write (10.1.6) as

$$u = -Kx + K\tilde{x} + r \quad (10.1.9)$$

which, when used in (10.1.1) yields

$$\dot{x} = (A - BK)x + BK\tilde{x} + Br + Gw. \quad (10.1.10)$$

This equation demonstrates that the closed-loop system is driven by the process noise $w(t)$, the reference command $r(t)$, and the estimation error $\tilde{x}(t)$.

Now, write (10.1.10) and (10.1.8) as the augmented system

$$\frac{d}{dt}\begin{bmatrix} x \\ \tilde{x} \end{bmatrix} = \begin{bmatrix} A - BK & BK \\ 0 & A - LC \end{bmatrix}\begin{bmatrix} x \\ \tilde{x} \end{bmatrix} + \begin{bmatrix} B \\ 0 \end{bmatrix} r + \begin{bmatrix} G \\ G \end{bmatrix} w - \begin{bmatrix} 0 \\ L \end{bmatrix} v \quad (10.1.11)$$

10.1 Linear-Quadratic-Gaussian Design

$$y = [C \quad 0] \begin{bmatrix} x \\ \tilde{x} \end{bmatrix} + v. \qquad (10.1.12)$$

This equation describes the complete closed-loop dynamics; namely, those of the plant and those of the error. Since the augmented system is block triangular, the closed-loop characteristic equation is

$$\Delta(s) = |sI - (A - BK)| \cdot |sI - (A - LC)| = 0. \qquad (10.1.13)$$

That is, the closed-loop poles are nothing but the plant poles that result by choosing K, plus the desired observer poles that result by choosing L. Thus, the state-feedback gain K and observer gain L may be selected separately for desirable closed-loop behavior.

The closed-loop transfer function from $r(t)$ to $y(t)$ is given by

$$H_c(s) = [C \quad 0] \begin{bmatrix} A - BK & BK \\ 0 & A - LC \end{bmatrix}^{-1} \begin{bmatrix} B \\ 0 \end{bmatrix},$$

and the triangular form of the system matrix makes it easy to see that

$$H_c(s) = C[I - (A - BK)]^{-1}B. \qquad (10.1.14)$$

This, however, is exactly what results if the full state feedback (10.1.3) is used.

These remarkable results show that, using feedback of the state estimate, the performance is in many respects the same as if full state feedback is used.

Of course, the initial conditions also affect the output $y(t)$. However, since the observer is stable, the effects of the initial error $\tilde{x}(0)$ will vanish with time. The observer poles [i.e., those of $(A - LC)$] should be chosen faster than the desired closed-loop plant poles [i.e., those of $(A - BK)$] for good closed-loop behavior. A rule of thumb is that the observer time constants should be about 5–10 times faster than the closed-loop plant time constants.

Transfer Function Description of the LQG Regulator

From our point of view, when possible it is usually better to design compensators using output feedback as we have demonstrated in previous chapters than to use separation-principle design. To see why, let us examine the structure of the dynamic compensator in Fig. 10.1-1 and Table 10.1-1 in more detail.

The control input $u(t)$ may be expressed as

$$U(s) = H_y(s)Y(s) + H_u(s)U(s) + R(s) \qquad (10.1.21)$$

where, according to Table 10.1-1, the transfer function from $y(t)$ to $u(t)$ is

$$H_y(s) = -K[sI - (A - LC)]^{-1}L \qquad (10.1.22)$$

and the transfer function from $u(t)$ to $u(t)$ is

$$H_u(s) = -K[sI - (A - LC)]^{-1}B. \qquad (10.1.23)$$

Now, note that the dynamic compensator designed by this technique has order equal to the order n of the plant. This means that it generally has an order that is higher than necessary to obtain suitable closed-loop performance. Moreover, it has no special structure and so is difficult to understand on an intuitive level.

It is possible to design *reduced-order* compensators using the separation principle. Three possible approaches are:

1. First find a reduced-order model of the plant, then design a compensator for this reduced-order model.
2. First design a compensator for the full plant, then reduce the order of the compensator.
3. Design the reduced-order compensator directly from the full-order plant.

One technique for order-reduction is the partial-fraction-expansion technique in Example 8.3-2. Other techniques include principal component analysis [Moore 1982] and the frequency-weighted technique in Anderson and Liu [1989]. A very convenient approach is given in Ly, Bryson, and Cannon [1985].

It is important to realize that, even if the plant is minimal (i.e., reachable and observable), the LQ regulator may not be. That is, it may have unreachable or unobservable states. A technique for reducing the regulator to minimal form is given in Yousuff and Skelton [1984].

Digital LQG Regulator

In LQG design for discrete-time systems like

$$x_{k+1} = Ax_k + Bu_k + GW_k \tag{10.1.24}$$

$$y_k = Cx_k + v_k, \tag{10.1.25}$$

there are two possibilities. According to Tables 9.3-1 through 9.3-3, the steady-state Kalman filter may be written as the *a priori recursion*

$$\hat{x}_{k+1}^- = A\hat{x}_k^- + Bu_k + AK_f(y_k - C\hat{x}_k^-) \tag{10.1.26}$$

or the *a posteriori recursion*

$$\hat{x}_{k+1} = A\hat{x}_k + Bu_k + K_f[y_{k+1} - C(A\hat{x}_k + Bu_k)]. \tag{10.1.27}$$

The a priori state estimate $\hat{x}_k^-$ does not include the most recent measurement y_k, while the a posteriori estimate $\hat{x}_k$ does include y_k. The steady-state Kalman gain is K_f.

The feedback control may be taken either as

$$u_k = -K\hat{x}_k^- + r_k \tag{10.1.28}$$

10.1 Linear-Quadratic-Gaussian Design

or as

$$u_k = -K\hat{x}_k + r_k. \tag{10.1.29}$$

An advantage to using the a priori feedback (10.1.28) is that the filter dynamics (10.1.26) have the same form as (10.1.5). Note that the discrete filter gain is

$$L = AK_f. \tag{10.1.30}$$

Thus, the results just derived for continuous-time systems hold here also. On the other hand, selecting the a posteriori feedback (10.1.29) yields different results (see the problems). Another advantage is that u_k can be computed at time $k - 1$. This allows some time for computation delays.

The feedback gain K and observer gain can be selected by any technique that yields suitable behavior of $(A - BK)$ and $(A - LC)$. In digital LQG design, the feedback gain is selected using the discrete LQR ARE in Table 7.4-3, and the observer gain is selected using the discrete Kalman filter ARE at the end of section 9.3. The digital LQG regulator design equations appear in Table 10.1-2.

Example 10.1-1: Digital LQG Regulator for Inverted Pendulum

We have designed controllers for the inverted pendulum in Examples 4.2-4, 5.2-2, 5.3-1, 5.3-2, 6.6-1, 7.4-3. In every case, we selected a lead compensator in each channel. In this example, we design a LQG regulator, so that the compensator dynamics will be those of the a priori Kalman filter in Table 10.1-2.

TABLE 10.1-2 DIGITAL LQG DYNAMIC REGULATOR

LQR State Feedback Design

$$0 = A^TPA - P + Q_c - A^TPB(R_c + B^TPB)^{-1}B^TPA. \tag{10.1.31}$$

$$K = (R_c + B^TPB)^{-1}B^TPA, \tag{10.1.32}$$

where Q_c and R_c are PI weighting matrices

Kalman Filter Gain Design

$$0 = APA^T - P + GQ_fG^T - APC^T(CPC^T + R_f)^{-1}CPA^T. \tag{10.1.33}$$

$$L = APC^T(CPC^T + R_f)^{-1}. \tag{10.1.34}$$

where Q_f and R_f are noise covariance matrices

LQG Dynamic Regulator

$$\hat{x}_{k+1}^- = A\hat{x}_k^- + Bu_k + L(y_k - C\hat{x}_k^-) \tag{10.1.35}$$

$$u_k = -K\hat{x}_k^- + r_k \tag{10.1.36}$$

a. Digital State-Variable Feedback

The dynamics of the pendulum used in this example are slightly different than those in the other examples due to a different cast mass. Using a sample period of $T = 0.001$ sec the discretized inverted pendulum dynamics are

$$x_{k+1} = \begin{bmatrix} 1.00002 & 0.001 & 0 & 0 \\ 0.047 & 1.00002 & 0 & 0 \\ -1.e-6 & 0 & 1 & 0.001 \\ -0.0025 & -1.e-6 & 0 & 1 \end{bmatrix} x_k + \begin{bmatrix} -0.0015 \\ -3.02 \\ 0.0003 \\ 0.5951 \end{bmatrix} u_k$$

$$\equiv Ax_k + Bu_k \tag{1}$$

$$y_k = \begin{bmatrix} 0.0222222 & 0 & 0 & 0 \\ 0 & 0 & 0.16 & 0 \end{bmatrix} x_k \equiv Cx_k. \tag{2}$$

The state is $x_k = [\theta_k \ \dot{\theta}_k \ p_k \ \dot{p}_k]^T$, with units of deg, deg/sec, in., and in./sec, respectively. The numbers in (1) and (2) are given to the accuracy with which they can be represented in the fixed-point Texas Instruments C25 DSP, which will be used for implementation purposes in Example 10.1-2.

The measured output is $y_k = [\bar{\theta}_k \ \bar{p}_k]^T$, where $\bar{\theta}_k$ and $\bar{p}_k$ are both normalized to the range $[-1, 1]$. This is for ease of implementation on a fixed-point DSP in Example 10.1-2. The control input u_k is also normalized to the range $[-1, 1]$, where a value of 1 corresponds to an acceleration of 595 in./sec^2.

The continuous system has poles at $s = 0, 0, 6.85, -6.85$. The discrete system has poles at

$$z = 1, 1, 1.00687, 0.99317. \tag{3}$$

which corresponds to the transformation $z = e^{sT}$.

Using MATLAB [Moler et al. 1987], the discrete LQR ARE in Table 10.1-2 was solved using $Q_c = \text{diag}\{1, 0.0002, 52, 3\}$, $R_c = 10{,}000$ to obtain the Kalman control gain

$$K = [-0.11081 \ -0.018145 \ -0.07117 \ -0.04801]. \tag{4}$$

Using this gain, the state feedback

$$u_k = -Kx_k \tag{5}$$

was applied to the continuous-time dynamics using program DIGCTL in Appendix A. The simulation results for this digital state-feedback controller are shown in Fig. 10.1-2. The poles of the closed-loop system $(A - BK)$ are at

$$z = 0.99605 \pm j0.00132, 0.98985, 0.99177. \tag{6}$$

The values for Q_c and R_c were obtained after several iterations, examining in each case the simulation results. The normalization scaling is also a factor in the selection of these design weighting matrices.

b. Digital LQG Regulator

All the states are not available as measurements, so that the state-feedback control (5) cannot be implemented in practice. In fact, the outputs are given by (2). Therefore, it is necessary to design an observer to estimate the states.

10.1 Linear-Quadratic-Gaussian Design

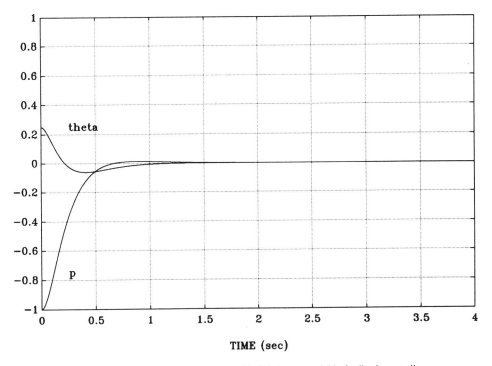

Figure 10.1-2 Inverted pendulum response with digital state-variable feedback controller

Using MATLAB, the discrete filter ARE in Table 10.1-2 was solved to obtain the steady-state Kalman filter gain

$$K_f = \begin{bmatrix} 5.048 & -0.003 \\ 300.0 & 0. \\ -0.0005 & 0.5982 \\ -0.110 & 30.071 \end{bmatrix} \quad (7)$$

The process noise covariance was selected as $Q_f = \text{diag}\{0, 100, 0, 1\}$ and the measurement noise covariance was $R_f = \text{diag}\{0.001, 0.001\}$. Notice that the process noise enters the system as an acceleration disturbance, influencing $\ddot{\theta}$ and $\ddot{p}$.

The filter gain was computed as

$$L = AK_f = \begin{bmatrix} 5.348 & -0.003 \\ 300.2 & 0. \\ -0.0006 & 0.6283 \\ -0.123 & 30.071 \end{bmatrix}. \quad (8)$$

C DIGITAL LQG REGULATOR FOR INVERTED PENDULUM

```
      SUBROUTINE DIG(IK,T,x)
      REAL x(*), k(4), L(4,2)
      COMMON/CONTROL/u
      COMMON/OUTPUT/y(2) , th, thd, p, pd
      DATA (k(I), I= 1,4) / -.11081,  -.018145, -.07117, -.04801/
      DATA ((L(I,J), J= 1,2), I= 1,4) / 5.348,  -.003, 300.2, 0.,
     &  -.0006, .6283, -.123, 30.071/

C  Store Previous Values

      thm=   th
      thdm=  thd
      pm=    p
      pdm=   pd

C  Steady-State Kalman Filter

      eth= y(1) - thm/45
      ep = y(2) - pm/6.25

      th=  1.00002*thm +  .001*thdm - .0015*u  + L(1,1)*eth + L(1,2)*ep
      thd=  .047*thm + 1.00002*thdm - 3.02*u   + L(2,1)*eth + L(2,2)*ep
      p= -1.e-6*thm + pm + .001*pdm + 3.e-3*u  + L(3,1)*eth + L(3,2)*ep
      pd= -.0025*thm - 1.e-6*thdm + pdm + .5951*u
     &                                         + L(4,1)*eth + L(4,2)*ep

C  Estimate Feedback Control

      u= - ( k(1)*th + k(2)*thd + k(3)*p + k(4)*pd )

      RETURN
      END
```

Figure 10.1-3 LQG digital control subroutine for use with driver program DIGCTL

The poles of the filter system matrix $(A - LC)$ are given by

$$z = 0.94060 \pm j0.05562, \; 0.94973 \pm j0.04780, \tag{9}$$

which are considerably faster than the poles of the closed-loop plant $(A - BK)$.

The digital LQG regulator

$$\hat{x}_{k+1}^- = A\hat{x}_k^- + Bu_k + L(y_k - C\hat{x}_k^-) \tag{10}$$

$$u = -K\hat{x}_k^- \tag{11}$$

was applied to the continuous-time plant using program DIGCTL. The required controller subroutine appears in Fig. 10.1-3. The simulation results are identical to those using full state feedback in Fig. 10.1-2. This is because the filter poles are so much faster than the closed-loop plant dynamics.

■

10.1 Linear-Quadratic-Gaussian Design

Example 10.1-2: DSP Implementation of Digital LQG Regulator

We now show how to implement the LQG regulator just designed for the inverted pendulum on a Texas Instruments TMS320 C25 digital signal processor (DSP). The hardware used was the Atlanta Signal Processors, Inc. Chimera board. See Appendix D for a description of the C25 chip.

a. DSP Implementation

The inverted pendulum hardware was described in Example 6.6-1, where a second-order digital controller was implemented. It is also worth glancing at Example 9.4-3, where a Kalman filter was implemented. To implement the LQG regulator, it is necessary to translate the code in Fig. 10.1-3 to the C25 assembly language. This C25 code is shown in Fig. 10.1-4. The program is very long, since all the details of memory allocation and coefficient initialization are shown. The actual regulator occupies somewhat more that half the total length.

The pendulum angle must remain in the range $[-45°, 45°]$ and the cart position in the range $[-6.25$ in., 6.25 in.] to avoid overflow (or saturation) of the measurement variables.

The DSP implementation of the LQG regulator requires 55 words of data memory and 284 words of program memory. The worst-case execution time is 39 microseconds for a single iteration at a C25 clock frequency of 40 MHz. Therefore, the processor is free to perform other tasks for 961 out of the 1000 microseconds in a sample period.

b. Friction Compensation

Through experimentation with the state-feedback gains and the filter gains, it was discovered that the unmodeled friction inherent in the drive mechanism of the pendulum cart prevents LQG regulators with feedback gains of small magnitude from successfully balancing the pendulum. The magnitude of the feedback gains must be sufficiently large to overcome the model mismatch. Large feedback gains result in fast dynamics for the closed-loop plant, which in turn require that the filter dynamics be correspondingly faster in order for the filter states to accurately track the rapidly changing plant states. Rapid filter dynamics, in turn, result in poor noise rejection. Because the angle and position measurements for our pendulum are extremely noisy, the pendulum is repeatedly forced far away from equilibrium when an LQG regulator with feedback gains large enough to overcome the friction is employed.

To correct this, we can use *feedforward compensation* in order to allow the magnitude of the feedback gains to be reduced. That is, we can feed forward a compensation term for the friction in the drive mechanism so that the plant effectively has no friction, and is close to the linear model used in Example 10.1-1 for the regulator design.

A model for the inverted pendulum with friction in the drive mechanism is

$$x_{k+1} = Ax_k + B(u_k + f_k) \equiv Ax_k + Bu_k^c, \quad (1)$$

where the A and B matrices are given in Equation (1) of Example 10.1-1 and u_k is the control signal designed in that example. The nonlinear friction term f_k can be modeled by

$$f_k = k_v \dot{p}_k + \text{sgn}(\dot{p}_k)(f_s + k_s p_k), \quad (2)$$

with k_v the coefficient of viscous friction and the second term modeling dynamic friction. Identification methods were used to fit step-response data from the inverted pendulum to this model, and the friction parameters were determined to be $k_v = 0.005081$, $f_s = 0.3487$, and $k_s = 0.02627$.

```
;       ***************************************************************
;                               PENDULUM.ASM
;       LQG regulator for inverted pendulum.
;       ***************************************************************

        ;define subroutines
                .def    con_init        ;controller initialization
                .def    cntrl           ;control algorithm

        ;define constants
DB0             .set    200h            ;address of internal data block B0

        ;notation: Filter coefficients and variables have units [units] and
        ;          are in Q15 format with a scale factor of (2^n).
        ;          [ntu]  normalized theta (angle) units
        ;          [npu]  normalized position units
        ;          [nau]  normalized acceleration units
        ;          [in]   inches
        ;          [deg]  degrees
        ;          [s]    seconds
        ; e.g. If variable x has units [in] and a scale factor of (2^3), then
        ;      the range of x is 2^3=8 inches and the resolution is
        ;      2^-12=0.0002 inches.

        ;allocate data memory for observer matrices
a11             .usect  "B0",1  ;(2^0)
a12             .usect  "B0",1  ;(2^-5) [s]
a21             .usect  "B0",1  ;(2^5) [1/s]          [ 2*a11   a12    0    0  ]
a22             .usect  "B0",1  ;(2^0)                [  a21  2*a22    0    0  ]
a31             .usect  "B0",1  ;(2^-3) [in/deg]   A=[  a31    a32    1   a34 ]
a32             .usect  "B0",1  ;(2^-8) [in-s/deg]    [  a41    a42    0    1  ]
a34             .usect  "B0",1  ;(2^-4) [s]
a41             .usect  "B0",1  ;(2^1) [in/deg-s]
a42             .usect  "B0",1  ;(2^-4) [in/deg]

b1              .usect  "B0",1  ;(2^3) [deg/nau]      [ b1 ]
b2              .usect  "B0",1  ;(2^8) [deg/nau-s]    [ b2 ]
b3              .usect  "B0",1  ;(2^0) [in/nau]    B=[ b3 ]
b4              .usect  "B0",1  ;(2^4) [in/nau-s]     [ b4 ]

c11             .usect  "B0",1  ;(2^-6) [ntu/deg]     [ 2*c11   0      0    0 ]
c23             .usect  "B0",1  ;(2^-3) [npu/in]   C=[   0     0    2*c23   0 ]

        ;allocate data memory for coefficients needed to calculate initial
        ; values for the state estimates
Tinv            .usect  "B0",1  ;(2^10) [1/s]         1/T (sampling frequency)
c11inv          .usect  "B0",1  ;(2^6) [deg/ntu]      1/(2*c11)
c23inv          .usect  "B0",1  ;(2^3) [in/npu]       1/(2*c23)

        ;allocate data memory for observer gains
l11             .usect  "B0",1  ;(2^6) [deg/ntu]
l12             .usect  "B0",1  ;(2^6) [deg/npu]
l21             .usect  "B0",1  ;(2^11) [deg/ntu-s]   [ l11   l12 ]
l22             .usect  "B0",1  ;(2^11) [deg/npu-s]   [ l21   l22 ]
```

Figure 10.1-4 C25 code for implementation of digital LQG regulator

```
l31             .usect  "B0",1  ;(2^3) [in/ntu]        L=[ l31   l32 ]
l32             .usect  "B0",1  ;(2^3) [in/npu]          [ l41   l42 ]
l41             .usect  "B0",1  ;(2^7) [in/ntu-s]
l42             .usect  "B0",1  ;(2^7) [in/npu-s]

                ;allocate data memory for state feedback gains
k1              .usect  "B0",1  ;(2^0) [nau/deg]
k2              .usect  "B0",1  ;(2^-3) [nau-s/deg]
k3              .usect  "B0",1  ;(2^0) [nau/in]        K=[ k1   k2   k3   k4]
k4              .usect  "B0",1  ;(2^-1) [nau-s/in]

                ;allocate data memory for friction compensation terms
fsp             .usect  "B0",1  ;(2^3) [nau]
fsm             .usect  "B0",1  ;(2^3) [nau]
ks              .usect  "B0",1  ;(2^0) [nau/in]
kv              .usect  "B0",1  ;(2^-4) [nau-s/in]

                ;allocate data memory for controller state variables
th              .usect  "B0",1  ;(2^6) [deg]     angle k on entry, k+1 on exit
th1             .usect  "B0",1  ;(2^6) [deg]     angle k-1 on entry, k on exit
thd             .usect  "B0",1  ;(2^11) [deg/s]  ang. vel. k on entry,k+1 on exit
thd1            .usect  "B0",1  ;(2^11) [deg/s]  ang. vel. k-1 on entry,k on exit
p               .usect  "B0",1  ;(2^3) [in]      position k on entry, k+1 on exit
p1              .usect  "B0",1  ;(2^3) [in]      position k-1 on entry, k on exit
pd              .usect  "B0",1  ;(2^7) [in/s]    velocity k on entry, k+1 on exit
pd1             .usect  "B0",1  ;(2^7) [in/s]    velocity k-1 on entry, k on exit

                ;allocate data memory for controller inputs and outputs
ang             .usect  "B0",1  ;(2^0) [ntu]     angle measurement k
pos             .usect  "B0",1  ;(2^0) [npu]     position measurement k
u               .usect  "B0",1  ;(2^3) [nau]     control signal k on entry,
                ;                                k+1 on exit
uc              .usect  "B0",1  ;(2^0) [nau]     control signal adjusted to
                ;                                offset the effects of friction

                ;allocate additional data memory needed to implement controller
eth             .usect  "B0",1  ;(2^6) [deg]     angle error k
ep              .usect  "B0",1  ;(2^3) [in]      position error k
flag            .usect  "B0",1  ;bit 0 is set the first time the control
                ;                 algorithm is executed
                ;              ;bit 1 is set the second time the control
                ;                 algorithm is executed
round           .usect  "B0",1  ;used to round Q31 numbers to Q15 numbers
acch            .usect  "B0",1  ;temporary storage for high accumulator
accl            .usect  "B0",1  ;temporary storage for low accumulator
scnt            .usect  "B0",1  ;shift counter for shift subroutine

;       ****************************************************************
;       Controller initialization
;       ****************************************************************

                .data
;control algorithm coefficients
```

Figure 10.1-4 *(continued)*

```
coeff
            .word   4000h       ;a11 = 1.00002/2
            .word   0419h       ;a12 = 0.001000 s
            .word   0030h       ;a21 = 0.047 /s
            .word   4000h       ;a22 = 1.00002/2
            .word   00000h      ;a31 = -0.000001 in/deg
            .word   00000h      ;a32 = -0.0000000 in-s/deg
            .word   020Ch       ;a34 = 0.001000 s
            .word   0FFD7h      ;a41 = -0.0025 in/deg-s
            .word   0FFFFh      ;a42 = -0.000001 in/deg
            .word   0FFFAh      ;b1  = -0.0015 deg/nau
            .word   0FE7Eh      ;b2  = -3.02 deg/nau-s
            .word   000Ah       ;b3  = 0.00030 in/nau
            .word   04C3h       ;b4  = 0.5951 in/nau-s
            .word   5B06h       ;c11 = 0.0222222/2 ntu/deg
            .word   51ECh       ;c23 = 0.160000/2 npu/in
            .word   7D00h       ;Tinv= 1000.00 /s
            .word   5A00h       ;c11inv= 45.000 deg/ntu
            .word   6400h       ;c23inv= 6.2500 in/npu
            .word   0AB2h       ;l11 = 5.348 deg/ntu
            .word   0FFFEh      ;l12 = -0.003 deg/npu
            .word   12C3h       ;l21 = 300.2 deg/ntu-s
            .word   0001h       ;l22 = 0.0 deg/npu-s
            .word   0FFFEh      ;l31 = -0.0006 in/ntu
            .word   0A0Eh       ;l32 = 0.6283 in/npu
            .word   0FFE1h      ;l41 = -0.123 in/ntu-s
            .word   1E12h       ;l42 = 30.071 in/npu-s
            .word   0F1D1h      ;k1  = -0.11081 nau/deg
            .word   0ED6Bh      ;k2  = -0.018145 nau-s/deg
            .word   0F6E4h      ;k3  = -0.07117 nau/in
            .word   0F3B6h      ;k4  = -0.04801 nau-s/in
            .word   0594h       ;fsp = 0.3487 nau
            .word   04D4h       ;fsm = 0.3017 nau
            .word   035Dh       ;ks  = 0.02627 nau/in
            .word   0A68h       ;kv  = 0.005081 nau-s/in
last_coeff  .set    $-1

            .text
con_init
        ;zero data memory block B0
            lrlk    ar1,DB0         ;ar1 points to B0
            larp    ar1
            zac
            rptk    255             ;256 words to zero
            sacl    *+              ;zero block B0
        ;initialize coefficients in data memory
            ldpk    DB0>>7          ;controller data page
            lrlk    ar1,a11         ;ar1 points to coefficient table in
                                    ; data memory
            lalk    coeff           ;low accumulator points to coefficient
                                    ; table in program memory
            rptk    last_coeff-coeff    ;(last_coeff-coeff+1) words
                                        ; to move
            tblr    *+              ;move the table
```

Figure 10.1-4 (*continued*)

```
                lalk    8000h               ;acc = 1000 0000 0000 0000b
                sacl    round
                ret

;       ****************************************************************
;       Control algorithm
;       The sampling period of 1ms must be enforced by a mechanism
;       external to this subroutine.  Prior to calling this subroutine,
;       the angle[k] and position[k] measurements must be placed in
;       data memory locations ang and pos, respectively.  On exit, the
;       control signal uc[k+1] is in data memory location uc, ready to
;       be sent to the output device simultaneously with the next
;       sampling of angle and position.
;       ****************************************************************

                .text
cntrl
                ssxm                        ;set sign extension mode
                sovm                        ;enable overflow mode
                spm     1                   ;set product mode for one left shift
                                            ; between product register and acc
                ldpk    DB0>>7              ;data page is first half of block B0
                larp    ar6                 ;ar6 is used by saturating shift sub.
                bit     flag,15             ;test the LSB of the flag to see if
                                            ; this is the first sample
                bbnz    cntrl1              ;branch if not the first sample

        ;calculate initial values for the angle and position estimates
        ;th[0] = c11inv*ang[0]
                lt      c11inv              ;T = c11inv
                mpy     ang                 ;P = c11inv*ang (Q30*2^6)
                ltp     c23inv              ;T = c23inv, acc = P<<1 (Q31*2^6)
                adds    round               ;round high acc
                sach    th                  ;store th[0]
        ;p[0] = c23inv*pos[0]
                mpy     pos                 ;P = c23inv*pos (Q30*2^3)
                pac                         ;acc = P<<1 (Q31*2^3)
                adds    round               ;round high acc
                sach    p                   ;store p[0]
                zac                         ;zero acc
        ;thd[0] = 0
                sacl    thd                 ;store thd[0]
        ;pd[0] = 0
                sacl    pd                  ;store pd[0]
                lack    1                   ;acc=1
                sacl    flag                ;set the first sample flag
                b       cntrl2

cntrl1
                bit     flag,14             ;test bit 1 of the flag to see if
                                            ; this is the second sample
                bbnz    cntrl2              ;branch if not the second sample
```

Figure 10.1-4 (*continued*)

```
        ;calculate initial values for all of the state estimates
        ;th[1] = c11inv*ang[1]
                lt      c11inv          ;T = c11inv
                mpy     ang             ;P = c11inv*ang (Q30*2^6)
                ltp     c23inv          ;T = c23inv, acc = P<<1 (Q31*2^6)
                adds    round           ;round high acc
                sach    th              ;store th[1]
        ;p[1] = c23inv*pos[1]
                mpy     pos             ;P = c23inv*pos (Q30*2^3)
                ltp     Tinv            ;T = Tinv, acc = P<<1 (Q31*2^6)
                adds    round           ;round high acc
                sach    p               ;store p[1]
        ;thd[1] = Tinv*th[1] - Tinv*th[0]
                mpy     th              ;P = Tinv*th (Q30*2^16)
                pac                     ;acc = P<<1 (Q31*2^16)
                mpy     th1             ;P = Tinv*th1 (Q30*2^16)
                spac                    ;acc = acc - P<<1 (Q31*2^16)
                lark    ar6,5
                call    shift           ;acc = acc<<2 (Q31*2^11)
                adds    round           ;round high acc
                sach    thd             ;store thd[1]
        ;pd[1] = Tinv*p[1] - Tinv*p[0]
                mpy     p               ;P = Tinv*p (Q30*2^13)
                pac                     ;acc = P<<1 (Q31*2^13)
                mpy     p1              ;P = Tinv*p1 (Q30*2^13)
                spac                    ;acc = acc -P<<1 (Q31*2^13)
                lark    ar6,6
                call    shift           ;acc = acc<<3 (Q31*2^7)
                adds    round           ;round high acc
                sach    pd              ;store pd[1]
                lack    3               ;acc = 3
                sacl    flag            ;set the first and second sample flags
cntrl2
        ;age the states
                dmov    th              ;th1 = th[k]
                dmov    thd             ;thd1 = thd[k]
                dmov    p               ;p1 = p[k]
                dmov    pd              ;pd1 = pd[k]
        ;calculate eth and ep
        ;eth[k] = ang[k] - c11*th[k] - c11*th[k]
                zalh    ang             ;acc = ang (Q31*2^0)
                lt      c11             ;T = c11
                mpy     th1             ;P = c11*th1 (Q30*2^0)
                spac                    ;acc = acc - P<<1 (Q31*2^0)
                lts     c23             ;T = c23, acc = acc - P<<1 (Q31*2^0)
                adds    round           ;round high acc
                sach    eth             ;store angle error eth[k]
        ;ep[k] = pos[k] - c23*p[k] - c23*p[k]
                zalh    pos             ;acc = pos (Q31*2^0)
                mpy     p1              ;P = c23*p1 (Q30*2^0)
                spac                    ;acc = acc - P<<1 (Q31*2^0)
                lts     a11             ;T = a11, acc = acc -P<<1 (Q31*2^0)
                adds    round           ;round high acc
                sach    ep              ;store position error ep[k]
```

Figure 10.1-4 (*continued*)

```
;calculate th[k+1]
;th[k+1] = a11*th[k] + a11*th[k] + a12*thd[k] + b1*u[k] +
;          + l11*eth[k] + l12*ep[k]
        mpy     th1             ;P = a11*th1 (Q30*2^6)
        pac                     ;acc = P<<1 (Q31*2^6)
        lta     a12             ;T = a12, acc = acc + P<<1 (Q31*2^6)
        mpy     thd1            ;P = a12*thd1 (Q30*2^6)
        lta     b1              ;T = b1, acc = acc + P<<1 (Q31*2^6)
        mpy     u               ;P = b1*u (Q30*2^6)
        lta     l11             ;T = l11, acc = acc + P<<1 (Q31*2^6)
        mpy     eth             ;P = l11*eth (Q30*2^6)
        lta     l12             ;T = l12, acc = acc + P<<1 (Q31*2^6)
        mpy     ep              ;P = l12*ep (Q30*2^6)
        lta     a21             ;T = a21, acc = acc + P<<1 (Q31*2^6)
        adds    round           ;round high acc
        sach    th              ;store angle estimate th[k+1]
;calculate thd[k+1]
;thd[k+1] = a21*th[k] + a22*thd[k] + a22*thd[k] + b2*u[k] +
;          + l21*eth[k] + l22*ep[k]
        mpy     th1             ;P = a21*th1 (Q30*2^11)
        ltp     a22             ;T = a22, acc = P<<1 (Q31*2^11)
        mpy     thd1            ;P = a22*thd1 (Q30*2^11)
        apac                    ;acc = acc + P<<1 (Q31*2^11)
        lta     b2              ;T = b2, acc = acc + P<<1 (Q31*2^11)
        mpy     u               ;P = b2*u (Q30*2^11)
        lta     l21             ;T = l21, acc = acc + P<<1 (Q31*2^11)
        mpy     eth             ;P = l21*eth (Q30*2^11)
        lta     l22             ;T = l22, acc = acc + P<<1 (Q31*2^11)
        mpy     ep              ;P = l22*ep (Q30*2^11)
        lta     a31             ;T = a31, acc = acc + P<<1 (Q31*2^11)
        adds    round           ;round high acc
        sach    thd             ;store ang. vel. estimate thd[k+1]
;calculate p[k+1]
;p[k+1] = a31*th[k] + a32*thd[k] + p[k] + a34*pd[k] + b3*u[k] +
;          + l31*eth[k] + l32*ep[k]
        mpy     th1             ;P = a31*th1 (Q30*2^3)
        ltp     a32             ;T = a32, acc = P<<1 (Q31*2^3)
        mpy     thd1            ;P = a32*thd1 (Q30*2^3)
        lta     a34             ;T = a34, acc = acc + P<<1 (Q31*2^3)
        addh    p1              ;acc = acc + p1 (Q31*2^3)
        mpy     pd1             ;P = a34*pd1 (Q30*2^3)
        lta     b3              ;T = b3, acc = acc + P<<1 (Q31*2^3)
        mpy     u               ;P = b3*u (Q30*2^3)
        lta     l31             ;T = l31, acc = acc + P<<1 (Q31*2^3)
        mpy     eth             ;P = l31*eth (Q30*2^3)
        lta     l32             ;T = l32, acc = acc + P<<1 (Q31*2^3)
        mpy     ep              ;P = l32*ep (Q30*2^3)
        lta     a41             ;T = a41, acc = acc + P<<1 (Q31*2^3)
        adds    round           ;round high acc
        sach    p               ;store position estimate p[k+1]
;calculate pd[k+1]
;pd[k+1] = a41*th[k] + a42*thd[k] + pd[k] + b4*u[k] +
;          + l41*eth[k] + l42*ep[k]
        mpy     th1             ;P = a41*th1 (Q30*2^7)
```

Figure 10.1-4 *(continued)*

```
        ltp     a42             ;T = a42, acc = acc + P<<1 (Q31*2^7)
        mpy     thd1            ;P = a42*thd1 (Q30*2^7)
        lta     b4              ;T = b4, acc = acc + P<<1 (Q31*2^7)
        addh    pd1             ;acc = acc + pd1 (Q31*2^7)
        mpy     u               ;P = b4*u (Q30*2^7)
        lta     l41             ;T = l41, acc = acc + P<<1 (Q31*2^7)
        mpy     eth             ;P = l41*eth (Q30*2^7)
        lta     l42             ;T = l42, acc = acc + P<<1 (Q31*2^7)
        mpy     ep              ;P = l42*ep (Q30*2^7)
        lta     k1              ;T = k1, acc = acc + P<<1 (Q31*2^7)
        adds    round           ;round high acc
        sach    pd              ;store velocity estimate pd[k+1]
;calculate control signal u[k+1]
;u[k+1] = -(k1*th[k+1] + k2*thd[k+1] + k3*p[k+1] + k4*pd[k+1])
        zac                     ;acc = 0
        mpy     th              ;P = k1*th (Q30*2^6)
        lts     k2              ;T = k2, acc = acc - P<<1 (Q31*2^6)
        lark    ar6,3
        call    shift           ;acc = acc<<1 (Q31*2^3)

        sach    acch            ;save high acc
        sacl    accl            ;save low acc
        zac                     ;acc = 0
        mpy     thd             ;P = k2*thd (Q30*2^8)
        lts     k3              ;T = k3, acc = acc - P<<1 (Q31*2^8)
        lark    ar6,5
        call    shift           ;acc = acc<<4 (Q31*2^3)

        adds    accl            ;restore low partial sum
        addh    acch            ;restore high partial sum
        mpy     p               ;P = k3*p (Q30*2^3)
        lts     k4              ;T = k4, acc = acc - P<<1 (Q31*2^3)
        sach    acch            ;save high acc
        sacl    accl            ;save low acc
        zac                     ;acc = 0
        mpy     pd              ;P = k4*pd (Q30*2^6)
        spac                    ;acc = P<<1 (Q31*2^6)
        lark    ar6,3
        call    shift           ;acc = acc<<1 (Q31*2^3)

        adds    accl            ;restore low partial sum
        addh    acch            ;restore high partial sum, acc = u[k+1]
;add terms to compensate for friction
        bit     pd,0            ;test the sign bit of pd
        bbnz    cntrl3          ;branch if pd is negative

;pd[k+1]>=0: uc[k+1] = u[k+1] + fsp + ks*p[k+1] + kv*pd[k+1]
        addh    fsp             ;add fsp
        lt      ks              ;T = ks
        mpy     p               ;P = ks*p (Q30*2^3)
        lta     kv              ;T = kv, acc = acc + P<<1 (Q31*2^3)
        mpy     pd              ;P = kv*pd (Q30*2^3)
        apac                    ;acc = acc + P<<1 (Q31*2^3)
        b       cntrl4
```

Figure 10.1-4 (*continued*)

```
cntrl3
        ;pd[k+1]<0: uc[k+1] = u[k+1] - fsm - ks*p[k+1] + kv*pd[k+1]
                subh    fsm             ;subtract fsm
                lt      ks              ;T = ks
                mpy     p               ;P = ks*p (Q30*2^3)
                lts     kv              ;T = kv, acc = acc - P<<1 (Q31*2^3)
                mpy     pd              ;P = kv*pd (Q30*2^3)
                apac                    ;acc = acc + P<<1 (Q31*2^3)
cntrl4
                lark    ar6,3
                call    shift           ;acc = acc<<3 (Q31*2^0)

                adds    round           ;round high acc
                sach    uc              ;store uc[k+1]
                zalh    uc              ;acc = uc (Q31*2^0)
                rptk    2
                sfr                     ;acc = acc>>3 (Q31*2^3)
        ;remove friction compensation terms from u
                bit     pd,0            ;test the sign bit of pd
                bbnz    cntrl5          ;branch if pd is negative

        ;pd[k+1]>=0: u[k+1] = uc[k+1] - fsp - ks*p[k+1] - kv*pd[k+1]
                subh    fsp             ;subtract fsp
                lt      ks              ;T = ks
                mpy     p               ;P = ks*p (Q30*2^3)
                lts     kv              ;T = kv, acc = acc - P<<1 (Q31*2^3)
                mpy     pd              ;P = kv*pd (Q30*2^3)
                spac                    ;acc = acc - P<<1
                b       cntrl6
cntrl5
        ;pd[k+1]<0: u[k+1] = uc[k+1] + fsm + ks*p[k+1] - kv*pd[k+1]
                addh    fsm             ;add fsm
                lt      ks              ;T = ks
                mpy     p               ;P = ks*p (Q30*2^3)
                lta     kv              ;T = kv, acc = acc + P<<1
                mpy     pd              ;P = kv*pd (Q30*2^3)
                spac                    ;acc = acc - P<<1
cntrl6
                adds    round           ;round high acc
                sach    u               ;store u[k+1]
                ret

;       ***************************************************************
;       Saturating shifter
;       Entry:  auxiliary register AR6 contains number of times to
;               shift accumulator left (0-255)
;       Exit :  auxiliary register AR6 is unchanged
;               AR6 is selected auxiliary register
;       ***************************************************************
shift
                larp    AR6             ;make AR6 current
```

Figure 10.1-4 (*continued*)

```
                bnz     shift1          ;branch if acc<>0

                ret                     ;acc=0; no need to shift
shift1
                banz    shift2,*        ;branch if a shift > 0  bits is
                                        ;requested

                ret                     ;finished
shift2
                mar     *-              ;decrement auxiliary register
                sar     ar6,scnt        ;save shift count
                mar     *+              ;restore auxiliary register
                rpt     scnt            ;repeat norm instruction scnt+1 times
                norm    *-              ;shift acc left and decrement aux reg
                                        ; unless overflow occurs

                banz    shift3          ;branch if overflow occurred

                b       shift6          ;branch to exit sequence
shift3
                rsxm
                bgz     shift4          ;branch if positive overflow

                lalk    8000h,15        ;negative overflow
                sfl                     ;acc = 8000 0000h
                b       shift5          ;branch to exit sequence
shift4
                lalk    7FFFh,15        ;positive overflow
                sfl
                adlk    0FFFFh          ;acc = 7FFF FFFFh
shift5
                ssxm
shift6
                lar     ar6,scnt
                mar     *+              ;restore auxiliary register
                ret
```

Figure 10.1-4 (*continued*)

Experimental results show that adding the friction compensation term f_k to the control u_k from Example 10.1-1 does indeed allow feedback gains of smaller magnitude to balance the pendulum.

c. Discussion

The C25 assembly language program in Fig. 10.1-4 implements the LQG controller of Example 10.1-1 with the addition of the friction compensation as just described. The response of the pendulum with this LQG regulator is plotted in Fig. 10.1-5. These results may be compared to the simulation results for similar initial conditions shown in Fig. 10.1-2. The

10.1 Linear-Quadratic-Gaussian Design

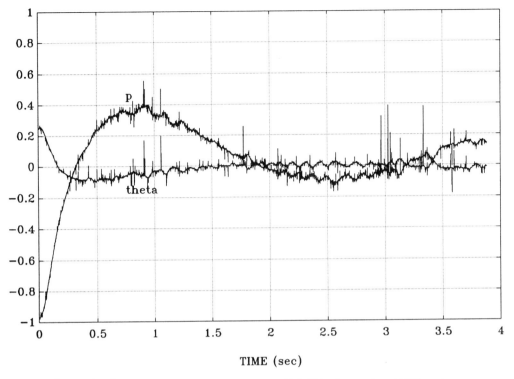

Figure 10.1-5 Actual inverted pendulum response with DSP implementation of LQG regulator

angle response is very similar to the simulation, but the position response has more overshoot and a longer settling time.

The DSP code for the LQG regulator includes several refinements over the previous implementation examples in this book. All 16-bit results have been rounded rather than truncated. The saturating shift subroutine now makes use of the NORM instruction to detect overflow. The coefficients have been scaled so that shifting of intermediate products to align the binary points for addition is only necessary when summing the products of the feedback gains with the filter states. Shifting with saturation is unavoidable at this stage, since the magnitude of the control signal can be larger than the maximum allowable value of 1 when the pendulum is far from equilibrium. (A control signal of magnitude 1 corresponds to the largest force that the actuator is capable of producing.)

Notice that the range of the control signal internal to the algorithm is $[-2^3, 2^3]$. The control signal is scaled down to the range $[-1, 1]$ just before it is stored. The purpose of the three extra bits of "headroom" is to prevent a single large product in the expression

$$u_k = -k_1 \hat{\theta}_k^- - k_2 \hat{\dot{\theta}}_k^- - k_3 \hat{p}_k^- - k_4 \hat{\dot{p}}_k^- \tag{3}$$

from saturating the control signal before all four products have been summed.

∎

10.2 LQG/LOOP-TRANSFER RECOVERY ROBUST DESIGN

In section 10.1 we saw how to design a multivariable dynamic compensator for MIMO plants using the separation principle. In this LQG approach, the feedback gain K is selected by solving the LQR algebraic Riccati equation, and the compensator dynamics are given in terms of the observer gain designed using the Kalman filter ARE. The design equations for the continuous-time LQG dynamic regulator appear in Table 10.1-1, and those for the digital LQG regulator appear in Table 10.1-2.

It is now necessary to consider *robust design* of compensators using the LQG approach. The issue is that in any actual situation the plant dynamics may not be exactly known, and there may exist disturbances or measurement noises in the system. We would like the compensator to provide not only good performance, but also *performance robustness* in the face of disturbances and *stability robustness* in the presence of unmodeled plant dynamics. In general, the LQG dynamic regulator has no guaranteed robustness.

In Chapter 8 we showed how to extend classical notions of robust design to multivariable systems using singular value plots versus frequency. In section 8.4 we demonstrated how a given compensator can be checked for robustness. In that section we specifically covered the case of compensators designed using output-feedback techniques; however, all those results apply also for LQG design.

In Tables 10.1-1 and 10.1-2, the matrices Q_c, R_c, Q_f, R_f are *design parameters*. In this section we shall show how they may be selected to guarantee robustness in the closed-loop system. The approach presented for selecting these design matrices is the *linear-quadratic Gaussian/loop-transfer recovery* (LQG/LTR) design technique for robust controllers. This approach is quite popular in the current literature and has been used extensively by Honeywell and others to design multivariable aircraft flight control systems [Doyle and Stein 1981, Athans 1986]. It is based on the fact that the static linear-quadratic regulator using state-variable feedback has certain *guaranteed robustness properties*.

We shall work with continuous-time systems, though similar results hold for digital design. Thus, suppose a state-feedback gain K has been computed using the LQR ARE as in Table 10.1-1. This state feedback cannot be implemented since all of the states are not available as measurements; however, it can be used as the basis for the design of the dynamic LQG regulator in Table 10.1-1 by using a Kalman filter to provide state estimates for feedback purposes. We would like to discuss two issues. First, we shall show that state feedback, in contrast to output feedback, has certain guaranteed robustness properties in terms of gain and phase margins. Then, we shall see that the Kalman filter design matrices Q_f and R_f may be selected so that the dynamic regulator *recovers the desirable robustness properties of full state feedback*.

The final result of our work will be the two LQG/LTR design procedures in Table 10.2-1. Let us now derive these algorithms.

Guaranteed Robustness of the Linear-Quadratic Regulator

In Chapter 8 we discussed conditions for performance robustness and stability robustness for a general feedback configuration of the form shown in Fig. 10.2-1, where $G(s)$ is the

10.2 LQG/Loop-Transfer Recovery Robust Design

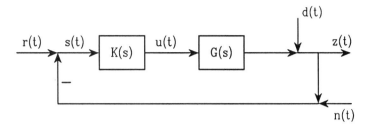

Figure 10.2-1 Standard feedback configuration

plant and $K(s)$ is the compensator. The static linear-quadratic regulator using *full state feedback* has many important properties, as we have seen in section 3.3. In this subsection we should like to return to the state-feedback LQR to show that it has certain *guaranteed robustness properties* that make it even more useful [Safonov and Athans 1977]. Specifically, the LQR with state feedback has an infinite gain margin and 60° of phase margin.

LQR return difference relation

Suppose that in Fig. 10.2-1 $K(s) = K$, the constant optimal LQ state-feedback gain determined using the LQR ARE as in Table 10.1-1. Suppose moreover that

$$G(s) = (sI - A)^{-1}B \tag{10.2.1}$$

is a plant in state-variable formulation.

For this subsection, it will be necessary to consider the loop gain referred to the control input $u(t)$ in Fig. 10.2-1. This is in contrast to the work in section 8.1, where we referred the loop gain to the output $z(t)$, or equivalently to the signal $s(t)$ in the figure. Breaking the loop at $u(t)$ yields the loop gain

$$KG(s) = K(sI - A)^{-1}B. \tag{10.2.2}$$

Our discussion will be based on the *optimal return difference relation* that holds for the static LQR with state feedback [Lewis 1986, Grimble and Johnson 1988, Kwakernaak and Sivan 1972], namely

$$[I + K(-sI - A)^{-1}B]^T[I + K(sI - A)^{-1}B]$$
$$= I + \frac{1}{\rho}B^T(-sI - A)^{-T}Q_c(sI - A)^{-1}B. \tag{10.2.3}$$

We have selected $R_c = I$.

Denoting the i-th singular value (section 8.2) of a matrix M as $\sigma_i(M)$, we note that by definition

$$\sigma_i(M) = \sqrt{\lambda_i(M^*M)}, \tag{10.2.4}$$

with $\lambda_i(M^*M)$ the i-th eigenvalue of matrix M^*M and M^* the complex conjugate transpose of M. Therefore, according to (10.2.3) there results

$$\sigma_i[I + KG(j\omega)] = \left[\lambda_i\left[I + \frac{1}{\rho}B^T(-j\omega I - A)^{-T}Q_c(j\omega I - A)^{-1}B\right]\right]^{1/2}$$

$$= \left[1 + \frac{1}{\rho}\lambda_i[B^T(-j\omega I - A)^{-T}Q_c(j\omega I - A)^{-1}B]\right]^{1/2}$$

or

$$\sigma_i[I + KG(j\omega)] = \left[1 + \frac{1}{\rho}\sigma_i^2[H(j\omega)]\right]^{1/2} \quad (10.2.5)$$

with

$$H(s) = H(sI - A)^{-1}B \quad (10.2.6)$$

and matrix H defined by $Q_c = H^T H$.

We could call (10.2.5) the *optimal singular value relation* of the LQR. It is important due to the fact that the right-hand side is known in terms of *open-loop quantities* before the optimal feedback gain is found by solution of the ARE, while the left-hand side is the closed-loop return difference. Thus, exactly as in classical control, we are able to derive properties of the closed-loop system in terms of properties of the open-loop system.

According to this relation, for all ω the minimum singular value satisfies the LQ optimal singular value constraint

$$\underline{\sigma}[I + KG(j\omega)] \geq 1. \quad (10.2.7)$$

Thus, the static LQ regulator always results in a *decreased sensitivity*.

Gain margin

Some important conclusions on the guaranteed robustness of the LQR may now be discovered using the *multivariable Nyquist criterion* [Postlethwaite et al. 1981], which we shall refer to the polar plot of the return difference $I + KG(s)$, where the origin is the critical point [Grimble and Johnson 1988]. (Usual usage is to refer the criterion to the polar plot of the loop gain $KG(s)$, where -1 is the critical point.)

A typical polar plot of $\underline{\sigma}[I + KG(j\omega)]$ is shown in Fig. 10.2-2, where the optimal singular value constraint appears as the condition that all the singular values remain outside the unit disc. To see how the end points of the plots were discovered, note that, since $K(sI - A)^{-1}B$ has relative degree of at least one, its limiting value for $s = j\omega$ as $\omega \to \infty$ is zero. Thus, in this limit $I + KG(j\omega)$ tends to I. On the other hand, as $\omega \to 0$, the limiting value of $I + KG(j\omega)$ is determined by the DC loop gain, which should be large.

The multivariable Nyquist criterion says that the closed-loop system is stable if none of the singular value plots of $I + KG(j\omega)$ encircle the origin in the figure. Clearly, due

10.2 LQG/Loop-Transfer Recovery Robust Design

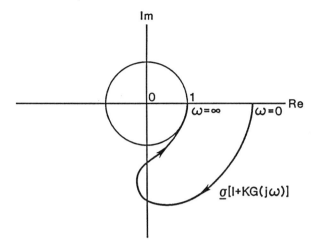

Figure 10.2-2 Typical polar plot for optimal LQ return difference (referred to the plant input)

to the optimal singular value constraint, no encirclements are possible. This constitutes a proof of the *guaranteed stability* of the LQR.

Multiplying the optimal feedback K by any positive scalar gain k results in a loop gain of $kKG(s)$, which has a minimum singular value plot identical to the one in Fig. 10.2-2 except that it is scaled outward. That is, the $\omega \to 0$ limit (i.e., the DC gain) will be larger, but the $\omega \to \infty$ limit will still be 1. Thus, the closed-loop system will still be stable. In classical terms, the LQ regulator with full state feedback has *an infinite gain margin*.

Phase margin

The *phase margin* may be defined for multivariable systems as the angle marked "PM" in Fig. 10.2-3. As in the classical case, it is the angle through which the polar plot

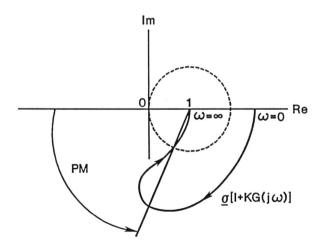

Figure 10.2-3 Definition of multivariable phase margin

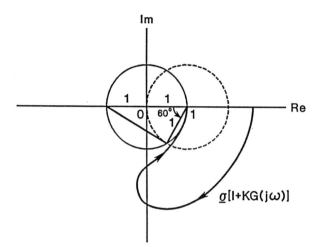

Figure 10.2-4 Guaranteed phase margin of the LQR

of $\underline{\sigma}[I + KG(j\omega)]$ must be rotated (about the point 1) clockwise to make the plot go through the critical point.

Figure 10.2-4 combines Fig. 10.2-2 and Fig. 10.2-3. By using some simple geometry, we may find the value of the angle indicated as 60°. Therefore, due to the LQ singular value constraint, the plot of $\underline{\sigma}[I + KG(j\omega)]$ must be rotated through at least 60° to make it pass through the origin. The LQR with full state feedback thus has a *guaranteed phase margin of at least 60°*.

This means that a phase shift of up to 60° may be introduced in any of the m paths in Fig. 10.2-1, or in all paths simultaneously as long as the paths are not coupled to each other in the process.

This phase margin is excessive; it is higher than that normally required in classical control system design. This overdesign means that, in other performance aspects, the LQ regulator may have some deficiencies. One of these turns out to be that, at the crossover frequency (loop gain = 1), the slope of the multivariable Bode plot is -20 dB/decade, which is a relatively slow attenuation rate [Doyle and Stein 1981]. By allowing a Q weighting matrix in the PI that is not positive semidefinite, it is possible to obtain better LQ designs that have higher rolloff rates at high frequencies [Shin and Chen 1974; Ohta, Nakinuma, and Nikiforuk 1990]. (See also AL-Sunni, Stevens, and Lewis 1991.)

Stability with multiplicative uncertainty

A stability robustness bound like (8.3.32) may be obtained for the loop gain referred to the input $u(t)$. It is

$$\bar{\sigma}[KG(I + KG)^{-1}] < 1/m(\omega) \qquad (10.2.8)$$

10.2 LQG/Loop-Transfer Recovery Robust Design

The inverse of this is

$$m(\omega) < \frac{1}{\overline{\sigma}[KG(I + KG)^{-1}]} = \underline{\sigma}[I + (KG)^{-1}]. \tag{10.2.9}$$

It can be shown (see the problems) that (10.2.7) implies that

$$\underline{\sigma}[I + (KG(j\omega))^{-1}] \geq \frac{1}{2}. \tag{10.2.10}$$

Therefore, the LQR with state feedback remains stable for all multiplicative uncertainties in the plant transfer function which satisfy $m(\omega) < \frac{1}{2}$.

Recovery of Robust Loop Gain at the Input

In most real design situations state feedback cannot be used, since it is rare to have all the states available as measurements. If the LQG regulator in Table 10.1-1 is used to control the plant, we saw in section 10.1 that, due to the separation principle, the closed-loop system has the same transfer function as the state-feedback controller. However, it is an unfortunate occurrence that none of the robustness properties just derived for the static LQR with full state feedback is guaranteed to hold for the static LQG regulator [Doyle 1978]. Therefore, we next address the issue of robust LQG compensator design. The results will be the first LQG/LTR design procedure summarized in Table 10.2-1.

The design equations for the LQG regulator in Tables 10.1-1 and 10.1-2 involve several design parameters—namely, the PI weighting matrices Q_c, R_c and the noise covariances Q_f, R_f. We shall show here that the key to attaining a robust LQG compensator lies in the correct selection of these design matrices.

Let us study the plant

$$\dot{x} = Ax + Bu + Gw \tag{10.2.11}$$

$$y = Cx + v, \tag{10.2.12}$$

with process noise $w(t) \sim (0, Q_f)$ and measurement $n(t) \sim (0, \nu^2 R_f)$ both white, $Q_f > 0$, $R_f > 0$, and ν a scalar parameter.

State-feedback loop gain

Consider first the full state feedback

$$u = -Kx. \tag{10.2.13}$$

According to Fig. 10.2-5a, where the plant transfer function is

$$\phi(s)B = (sI - A)^{-1}B, \tag{10.2.14}$$

the loop gain for static full state feedback, breaking the loop at the input $u(t)$, is

$$L_s(s) = K\phi B. \tag{10.2.15}$$

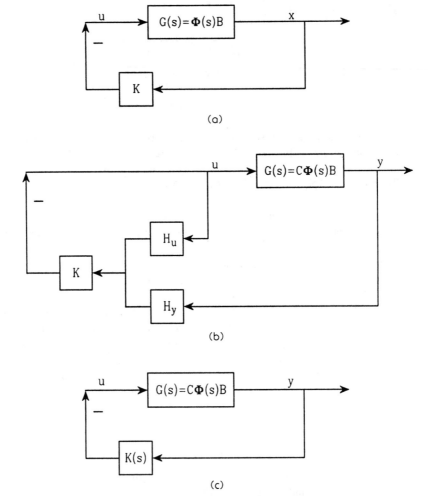

Figure 10.2-5 (a) Loop gain with full state feedback. (b) Regulator using observer and estimate feedback. (c) Regulator loop gain.

LQG regulator transfer function

Using the dynamic LQG Regulator in Table 10.1-1, the closed-loop system appears in Fig. 10.2-5b, where the regulator is given by (section 10.1)

$$U(s) = -K[sI - (A - LC)]^{-1}BU(s) - K[sI - (A - LC)]^{-1}LY(s)$$
$$\equiv -H_u(s)U(s) - H_y(s)Y(s) \qquad (10.2.16)$$

and L is the observer gain. Denoting the observer resolvent matrix as

$$\phi_o(s) = [sI - (A - LC)]^{-1} \qquad (10.2.17)$$

10.2 LQG/Loop-Transfer Recovery Robust Design

we write

$$H_u = K\phi_o B, \qquad H_y = K\phi_o L. \qquad (10.2.18)$$

To find an expression for $K(s)$ in Fig. 10.2-5c using the regulator, note that $(I + H_u)U = -H_y Y$, so that

$$U = -(I + H_u)^{-1} H_y Y = -K(s)Y. \qquad (10.2.19)$$

However

$$\begin{aligned}(I + H_u)^{-1} K &= [I + K(sI - (A - LC))^{-1} B] K \\ &= [I - K(sI - (A - BK - LC))^{-1} B] K \\ &= K(sI - (A - BK - LC))^{-1} [(sI - (A - BK - LC)) - BK] \\ &= K(sI - (A - BK - LC))^{-1} \phi_o^{-1},\end{aligned}$$

where the matrix inversion lemma was used in the second step. Therefore

$$K(s) = (I + H_u)^{-1} H_y = K[sI - (A - BK - LC)]^{-1} \phi_o^{-1} \phi_o L$$

or

$$K(s) = K[sI - (A - BK - LC)]^{-1} L \equiv K\phi_r L, \qquad (10.2.20)$$

with $\phi_r(s)$ the regulator resolvent matrix.

Recovery of state-feedback loop gain

It is important to note that either the control gain K or the observer gain L may be determined first in the LQG regulator. In this subsection we shall assume that a state-feedback gain K has already been determined using the LQR ARE design equations in Table 10.1-1. As we have just seen, this K yields suitable robustness properties of $K\phi B$. We call $K\phi B$ the *target feedback loop* for loop-gain recovery at the input. The gain K should also be selected for suitable time-response characteristics.

We should now like to present a technique for designing a Kalman filter that results in a regulator that *recovers* the guaranteed robustness properties of the static full-state feedback control law as the design parameter ν goes to zero. The technique is called LQG/loop-transfer recovery (LQG/LTR), since the loop gain (i.e., loop transfer function) $K\phi B$ of static full state feedback is recovered in the regulator as $\nu \to 0$.

As we shall see, the key to robustness using a dynamic LQG regulator is in the selection of the noise spectral densities Q_f and R_f.

We shall now show how to make the loop gain (at the input) using the regulator

$$L_r(s) = K(s)G(s) = K\phi_r LC\phi B \qquad (10.2.21)$$

approach the loop gain $L_s(s) = K\phi B$ using full state feedback, which is guaranteed to be robust. The references for this subsection are [Doyle and Stein 1979, 1981; Athans 1986; Stein and Athans 1987; Birdwell 1989].

To design the Kalman filter so that the regulator loop gain at the input $L_r(s)$ is the same as the state feedback loop gain $L_s(s)$, we shall need to assume that the plant $C\phi B$ is *minimum-phase* (i.e., with stable zeros), with B and C of full rank and $\dim(u) = \dim(y)$.

Let us propose $G = I$ and the process noise spectral density matrix

$$Q_f = \nu^2 Q_0 + BB^T, \tag{10.2.22}$$

with $Q_0 > 0$. Then, according to Table 10.1-1

$$L = PC^T(\nu^2 R_f)^{-1} \tag{10.2.23}$$

and the Kalman filter ARE becomes

$$0 = AP + PA^T + (\nu^2 Q_0 + BB^T) - PC^T(\nu^2 R_f)^{-1}CP. \tag{10.2.24}$$

According to Kwakernaak and Sivan [1972], if the aforementioned assumptions hold, then $P \to 0$ as $\nu \to 0$, so that the ARE shows

$$L(\nu^2 R_f)L^T = PC^T(\nu^2 R_f)^{-1}CP \to BB^T.$$

The general solution of this equation is

$$L \to \frac{1}{\nu} BUR_f^{-1/2}, \tag{10.2.25}$$

with U any unitary matrix.

We claim that in this situation $L_r(s) \to L_s(s)$ as $\nu \to 0$. Indeed, defining the full-state-feedback closed-loop resolvent as

$$\phi_c(s) = (sI - (A - BK))^{-1} \tag{10.2.26}$$

we may write

$$L_r(s) = K(s)G(s) = K[sI - (A - BK - LC)]^{-1}LC\phi B$$

$$= K[\phi_c^{-1} + LC]^{-1}LC\phi B$$

$$= K[\phi_c - \phi_c L(I + C\phi_c L)^{-1}C\phi_c]LC\phi B$$

$$= K\phi_c L[I - (I + C\phi_c L)^{-1}C\phi_c L]C\phi B$$

$$= K\phi_c L[(I + C\phi_c L) - C\phi_c L](I + C\phi_c L)^{-1}C\phi B \tag{10.2.27}$$

$$= K\phi_c L(I + C\phi_c L)^{-1}C\phi B$$

$$\to K\phi_c B(C\phi_c B)^{-1}C\phi B$$

$$= K\phi B(I + K\phi B)^{-1}[C\phi B(I + K\phi B)^{-1}]^{-1}C\phi B$$

$$= [K\phi B(C\phi B)^{-1}] C\phi B = K\phi B.$$

The matrix inversion lemma was used in going from line 2 to line 3, and from line 7 to 8. The limiting value (10.2.25) for L was used at the arrow.

What we have shown is that, using $G = I$ and the process noise given by (10.2.22),

10.2 LQG/Loop-Transfer Recovery Robust Design

as $\nu \to 0$ the dynamic regulator loop gain using a Kalman filter approaches the target feedback loop gain $K\phi B$ using full state feedback. This means that, as $\nu \to 0$, all the robustness properties of the full state feedback control law are recovered in the dynamic regulator. It also means that the time response of the LQG regulator should approach the time response of the full state feedback law as $\nu \to 0$.

The *LQG/LTR design procedure* is thus as follows. The procedure is summarized in the first design algorithm in Table 10.2-1.

TARGET FEEDBACK LOOP DESIGN

1. Use the control ARE in Table 10.2-1 to design a state feedback gain K giving desirable robustness and time-response properties. This may involve iterative design varying the PI weighting matrices Q_c and R_c.

TABLE 10.2-1 CONTINUOUS-TIME LQG/LTR DESIGN ALGORITHMS

LQG Dynamic Regulator

$$\dot{\hat{x}} = (A - LC)\hat{x} + Bu + Ly$$

$$y = Cx$$

$$u = -K\hat{x}$$

where K and L are found by one of the following algorithms:

Algorithm 1: LQG/LTR at the Input

LQR State-Feedback Design for Target Feedback Loop $K\phi B$

$$0 = A^T S + SA + Q_c - SBR_c^{-1}B^T S$$

$$K = R_c^{-1}B^T S$$

Kalman Filter Design for Loop-Transfer Recovery

$$0 = AP + PA^T + Q_f - PC^T(\nu^2 R_f)^{-1} CP$$

$$L = PC^T(\nu^2 R_f)^{-1}$$

where $Q_f = \nu^2 Q_0 + BB^T$, $Q_0 \geq 0$

Algorithm 2: LQG/LTR at the Output

Kalman Filter Design for Target Feedback Loop $C\phi L$

$$0 = AP + PA^T + GQ_f G^T - PC^T R_f^{-1} CP$$

$$L = PC^T R_f^{-1}$$

LQR State-Feedback Design for Loop-Transfer Recovery

$$0 = A^T S + SA + Q_c - SB(\rho^2 R_c)^{-1} B^T S$$

$$K = (\rho^2 R_c)^{-1} B^T S$$

where $Q_c = \rho^2 Q_0 + C^T C$, $Q_0 \geq 0$

LOOP-GAIN RECOVERY DESIGN

2. Select $G = I$, process noise spectral density $Q_f = \nu^2 Q_0 + BB^T$ and noise spectral density $\nu^2 R_f$ for some $Q_0 > 0$ and $R_f > 0$. Fix the design parameter ν and use the Kalman filter ARE to solve for the Kalman gain L.

3. Plot the maximum and minimum singular values of the regulator loop gain $L_r(s)$ and verify that the robustness bounds are satisfied. Verify that the time response using the LQG regulator is suitable. If the results are not acceptable, decrease ν and return to 2.

A *reduced-order* regulator with suitable robustness properties may be designed by the LQG/LTR approach using the notions in section 10.1. That is, either a regulator may be designed for a reduced-order model of the plant, or the regulator designed for the full-order plant may then have its order reduced. In using the first approach, a high-frequency bound characterizing the unmodeled dynamics should be used to guarantee stability robustness.

An interesting aspect of the LQR/LTR approach is that the recovery process may be viewed as a *frequency-domain linear-quadratic* technique that trades off the smallness of the sensitivity $S(j\omega)$ and the cosensitivity $T(j\omega)$ at various frequencies. These notions are explored in Stein and Athans [1987] and Safonov et al. [1981].

Nonminimum-Phase Plants and Parameter Variations

The limiting value of $K(s)$ is given by the bracketed term in (10.2.27). Clearly, as $\nu \to 0$, the regulator inverts the plant transfer function $C\phi B$. If the plant is of minimum-phase, with very stable zeros, the LQG/LTR approach generally gives good results. On the other hand, if the plant is nonminimum-phase or has stable zeros with large time constants, the approach can be unsuitable.

In some applications, however, even if the plant is nonminimum-phase the LQG/LTR technique can produce satisfactory results [Athans 1986]. In this situation, better performance may result if the design parameter ν is not nearly zero. If the right-half plane zeros occur at high frequencies where the loop gain is small, the LQG/LTR approach works quite well.

An additional defect of the LQG/LTR approach appears when there are plant parameter variations. Stability in the presence of parameter variations requires that the loop gain singular values be below some upper bound at low frequencies. However, this bound is not taken into account in the LQG/LTR derivation. Thus, LQG/LTR can yield problems for aircraft controls design, where gain scheduling is required. The H-infinity design approach [Francis et al. 1984, Doyle et al. 1989] has been used with success to overcome this problem.

Recovery of Robust Loop Gain at the Output

We have shown that, by designing the state feedback K first and then computing the Kalman filter gain L using the specific choice of noise spectral densities given in the first

10.2 LQG/Loop-Transfer Recovery Robust Design

LQG/LTR design algorithm given in Table 10.2-1, the LQG regulator loop gain $K(s)G(s)$ in Fig. 10.2-5, which is *referred to the input*, recovers the robustness of the target feedback loop $K\phi B$ using full state feedback.

However, in section 8.3 we saw that for a small tracking error the robustness should be studied in terms of the loop gain $G(s)K(s)$ referred to the error, or equivalently to the system *output*. Therefore, we should now like to show how to design a stochastic regulator that recovers a robust loop gain $G(s)K(s)$. The result will be the second LQG/LTR design procedure shown in Table 10.2-1.

Thus, suppose we first design a Kalman filter with gain L using Table 10.1-1. By duality theory, one may see that the Kalman filter loop gain

$$L_k(s) = C\phi L \tag{10.2.28}$$

enjoys exactly the same guaranteed robustness properties as the state-feedback loop gain $K\phi B$ that were described earlier in this section. The Kalman filter should be designed so that the robustness and time response of this target feedback loop $C\phi L$ are suitable.

The regulator loop gain referred to the output is

$$L_r^o(s) = G(s)K(s) = C\phi BK\phi_r L. \tag{10.2.29}$$

Thus, we should like to determine how to design a state-feedback gain K so that $L_r^o(s)$ approaches the target feedback loop $C\phi L$ for loop-gain recovery at the output. As we shall see, the key to this is in the selection of the PI weighting matrices Q_c and R_c in Table 10.1-1.

To determine K, let us propose the PI

$$J = \frac{1}{2} \int_0^\infty (x^T Q_c x + \rho^2 u^T R_c u)\, dt \tag{10.2.30}$$

with

$$Q_c = \rho^2 Q_0 + C^T C, \tag{10.2.31}$$

with $Q_0 \geq 0$. By using techniques dual to those above, we may demonstrate that as $\rho \to 0$, the state feedback gain determined using Table 10.1-1 approaches

$$K \to \frac{1}{\rho} R_c^{-1/2} WC, \tag{10.2.32}$$

with W a unitary matrix. Using this fact, it may be shown that

$$L_r^o(s) = G(s)K(s) \to C\phi L. \tag{10.2.33}$$

The LQG/LTR design technique for loop gain recovery at the output is therefore exactly dual to that for recovery at the input. Specifically, the filter gain L is first determined using Table 10.1-1 for desired robustness and time response properties. Then, Q_c and R_c are selected, with Q_c of the special form (10.2.31). For a small value of ρ, the state-feedback gain K is determined using Table 10.1-1. If the singular value Bode plots of $L_r^o(s)$ do not show acceptable robustness, then ρ is decreased and a new K is determined.

The LQG/LTR algorithm for loop-gain recovery at the output is summarized in Table 10.2-1.

If the plant $C\phi B$ is minimum-phase, all is well as ρ is decreased. However, if there are zeros in the right-half plane there could be problems as ρ becomes too small, although with care the LQG/LTR technique often still produces good results for suitable (i.e. larger) ρ.

Example 10.2-1: LQG/LTR Design of Aircraft Lateral Control System

Since LQG/LTR has been quite popular for aircraft controls design, we shall illustrate the technique using an F-16 aircraft example. All computations, including solving for the state feedback gains and Kalman filter gains, were carried out very easily using MATLAB [Moler et al. 1987].

a. Control Objective

The tracking control system shown in Fig. 10.2-6 is meant to provide coordinated turns by causing the bank angle $\phi(t)$ to follow a desired command while maintaining the sideslip angle $\beta(t)$ at zero. It is a two-channel system with control input $u = [u_\phi \ u_\beta]^T$.

The reference command is $r = [r_\phi \ r_\beta]^T$. The control system should hold ϕ at the commanded value of r_ϕ and $\beta(t)$ at the commanded value of r_β, which is equal to zero. The tracking error is $e = [e_\phi \ e_\beta]^T$ with

$$e_\phi = r_\phi - \phi$$
$$e_\beta = r_\beta - \beta. \tag{1}$$

The negatives of the errors appear in the figure since a minus sign appears in $u = -K\hat{x}$ as is standard for LQG design.

b. State Equations of Aircraft and Basic Compensator Dynamics

To obtain the basic aircraft dynamics, a nonlinear F-16 model was linearized at a nominal flight condition having total velocity of 502 ft/sec, 0 ft altitude, 300 psf dynamic pressure, and center of gravity at $0.35\bar{c}$ [Stevens and Lewis 1991]. Since only the lateral dynamics are of interest here, the states are sideslip β, bank angle ϕ, roll rate p, and yaw rate r. Additional

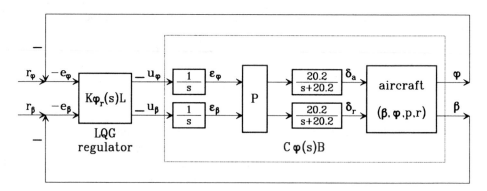

Figure 10.2-6 Aircraft turn coordinator control system

10.2 LQG/Loop-Transfer Recovery Robust Design

states δ_a and δ_r are introduced by the aileron and rudder actuators, both of which are modeled as having approximate transfer functions of $20.2/(s + 20.2)$. The aileron deflection is δ_a and the rudder deflection is δ_r.

The singular values versus frequency of the basic aircraft with actuators are shown in Fig. 10.2-7. Clearly, the steady-state error will be large in closed-loop since the loop gain has neither integrator behavior nor large singular values at DC. Moreover, the singular values are widely separated at DC, so that they are not balanced.

To correct these deficiencies we may use the techniques of Example 8.3-1. The DC gain of the system is given by

$$H(0) = \begin{bmatrix} -727.37 & -76.94 \\ -2.36 & 0.14 \end{bmatrix}. \quad (2)$$

First, the dynamics are augmented by integrators in each control channel. We denote the integrator outputs by ε_ϕ, ε_β. The singular value plots including the integrators are shown in Fig. 10.2-8. The DC slope is now -20 dB/decade, so that the closed-loop steady-state error will be zero. Next, the system was augmented by $P = H^{-1}(0)$ to balance the singular values at DC. The net result is shown in Fig. 10.2-9, which is very suitable.

The entire state vector, including aircraft states and integrator states, is

$$x = [\beta \quad \phi \quad p \quad r \quad \delta_a \quad \delta_r \quad \varepsilon_\phi \quad \varepsilon_\beta]^T. \quad (3)$$

The full state-variable model of the aircraft plus actuators and integrators is of the form

$$\dot{x} = Ax + Bu \quad (4)$$

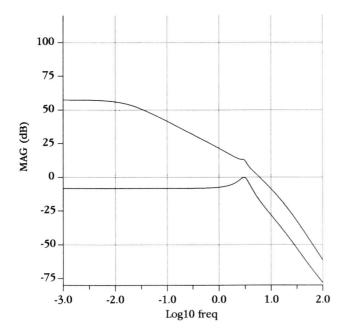

Figure 10.2-7 Singular values of the basic aircraft dynamics

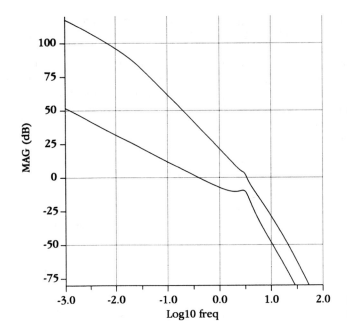

Figure 10.2-8 Singular values of aircraft augmented by integrators

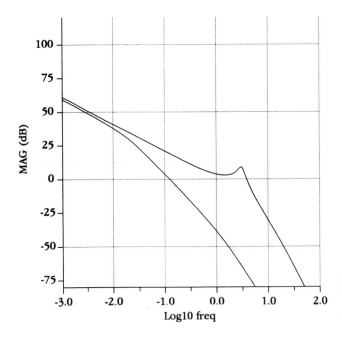

Figure 10.2-9 Singular values of aircraft augmented by integrators and inverse DC gain matrix P

10.2 LQG/Loop-Transfer Recovery Robust Design

with

$$A = \begin{bmatrix} -.3220 & .0640 & .0364 & -.9917 & .0003 & .0008 & 0 & 0 \\ 0 & 0 & 1 & .0037 & 0 & 0 & 0 & 0 \\ -30.6492 & 0 & -3.6784 & .6646 & -.7333 & .1315 & 0 & 0 \\ 8.5395 & 0 & -.0254 & -.4764 & -.0319 & -.0620 & 0 & 0 \\ 0 & 0 & 0 & 0 & -20.2 & 0 & -.01 & -5.47 \\ 0 & 0 & 0 & 0 & 0 & -20.2 & -.168 & 51.71 \\ 0 & 0 & 0 & 0 & 0 & 0 & 0 & 0 \\ 0 & 0 & 0 & 0 & 0 & 0 & 0 & 0 \end{bmatrix} \quad (5)$$

$$B = \begin{bmatrix} 0 & 0 \\ 0 & 0 \\ 0 & 0 \\ 0 & 0 \\ 0 & 0 \\ 0 & 0 \\ 1 & 0 \\ 0 & 1 \end{bmatrix} \quad (6)$$

The output is given by $y = [\phi \ \beta]^T$, or

$$y = \begin{bmatrix} 0 & 57.2958 & 0 & 0 & 0 & 0 & 0 & 0 \\ 57.2958 & 0 & 0 & 0 & 0 & 0 & 0 & 0 \end{bmatrix} x = Cx, \quad (7)$$

where the factor of 57.2958 converts radians to degrees. Then

$$e = r - y. \quad (8)$$

c. Frequency-Domain Robustness Bounds

We now derive the bounds on the loop-gain MIMO Bode magnitude plot that guarantee robustness of the closed-loop system.

Consider first the high-frequency bound. Let us assume that the aircraft model is accurate to within 10% up to a frequency of 2 rad/sec, after which the uncertainty grows without bound at the rate of 20 dB/decade. The uncertainty could be due to actuator modeling inaccuracies, aircraft flexible modes, and so on. This behavior is modeled by

$$m(\omega) = \frac{s + 2}{20}. \quad (9)$$

We assume $m(\omega)$ to be a bound on the multiplicative uncertainty in the aircraft transfer function (section 8.3).

For stability robustness in spite of the modeling errors, we saw in section 8.3 that the loop-gain referred to the output should satisfy

$$\bar{\sigma}(GK(j\omega)) < 1/m(\omega) = \left| \frac{20}{s + 2} \right| \quad (10)$$

when $1/m(\omega) \ll 1$. The function $1/m(\omega)$ is plotted in Fig. 10.2-10.

Turning to the low-frequency bound on the closed-loop loop gain, the closed-loop system should be robust to wind gust disturbances. Using techniques like those in Example 8.4-1, the gust magnitude plot shown in Fig. 10.2-11a may be obtained. According to section

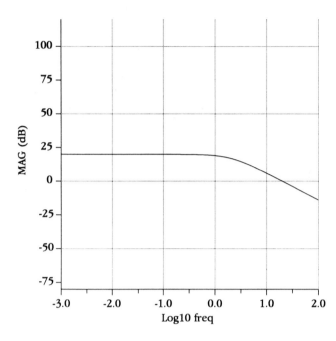

Figure 10.2-10 Multiplicative uncertainty bound $1/m(\omega)$ for the aircraft dynamical model

8.3, for robust performance in spite of wind gusts, the minimum loop-gain singular value $\underline{\sigma}(GK(j\omega))$ should be above this bound.

d. Target Feedback Loop Design

The robustness bounds just derived are expressed in terms of the singular value plots referred to $e(t)$. To recover the loop gain $GK(j\omega)$ at $e(t)$, or equivalently at the output, the Kalman filter should be designed first, so that we should employ LQG/LTR algorithm number two in Table 10.2-1. The Kalman gain L should be selected for good robustness properties as well as *suitable step responses* of the target feedback loop $C\phi(s)L$, where $\phi(s) = (sI - A)^{-1}$.

Using MATLAB, the Kalman filter design equations were solved using

$$Q_f = \text{diag}\{.01, .01, .01, .01, 0, 0, 1, 1\}, \tag{11}$$

$R_f = r_f I$, and various values of r_f. The maximum and minimum singular values of the filter open-loop gain $C\phi(s)L$ for $r_f = 1$ are shown in Fig. 10.2-11a, which also depicts the robustness bounds. The singular values for several values of r_f are shown in Fig. 10.2-11b. Note how the singular value magnitudes increase as r_f decreases, reflecting improved rejection of low-frequency disturbances. The figures show that the robustness bounds are satisfied for $r_f = 1$ and $r_f = 10$, but that the high-frequency bound is violated for $r_f = 0.1$.

The associated step responses of $C\phi(s)L$ with reference commands of $r_\phi = 1$, $r_\beta = 0$ are shown in Fig. 10.2-12. The response for $r_f = 10$ is unsuitable, while the response for $r_f = 0.1$ is too fast and would not be appreciated by the pilot. On the other hand, the response for $r_f = 1$ shows suitable time-of-response and overshoot characteristics, as well as good decoupling between the bank angle $\phi(t)$ and the sideslip $\beta(t)$.

Therefore, the target feedback loop was selected as $C\phi(s)L$ with $r_f = 1$, since this results in a design that has suitable robustness properties and step responses. The corresponding Kalman gain is given by

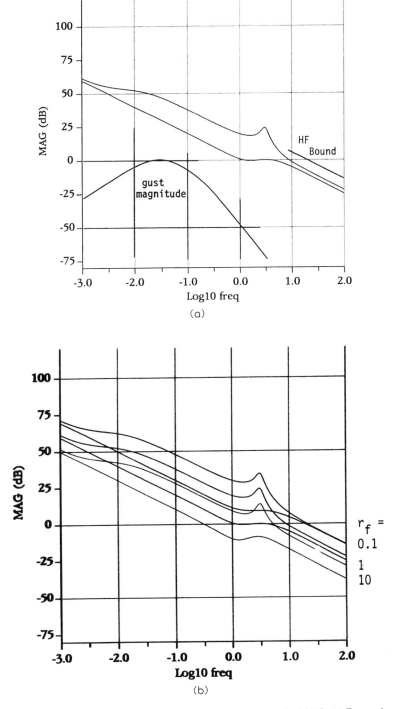

Figure 10.2-11 Singular values of Kalman filter open-loop gain $C\phi(s)L$. (a) For $r_f = 1$, including robustness bounds. (b) For various values of r_f.

$$L = \begin{bmatrix} -0.007 & 0.097 \\ 0.130 & -0.007 \\ 0.199 & -0.198 \\ -0.093 & -0.020 \\ -0.197 & -0.185 \\ 1.858 & 1.757 \\ 0.685 & -0.729 \\ 0.729 & 0.684 \end{bmatrix}. \qquad (12)$$

The Kalman filter poles (e.g., those of $A - LC$) are given by

$$\begin{aligned} s = &-0.002, -0.879, -1.470, \\ &-3.952 \pm j3.589 \\ &-7.205, -20.2, -20.2 \end{aligned} \qquad (13)$$

Although there is a slow pole, the step response is good, so this pole evidently has a small residue.

It is of interest to discuss how the frequency and time responses were plotted. For the frequency response, we used the open-loop system

$$\begin{aligned} \dot{\hat{x}} &= A\hat{x} + Le \\ \hat{y} &= C\hat{x}, \end{aligned} \qquad (14)$$

which has a transfer function of $C\phi(s)L = C(sI - A)^{-1}L$. A program was written which plots the singular values versus frequency for a system given in state-space form. This yielded Fig. 10.2-11.

For the step response, it is necessary to examine the closed-loop system. In this case, the loop is closed by using $e = r - \hat{y}$ in (14), obtaining

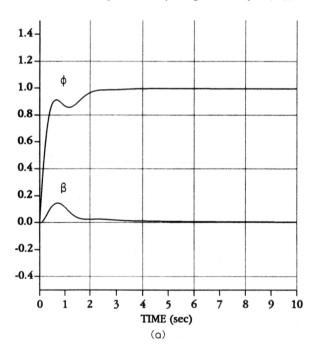

Figure 10.2-12 Step responses of target feedback loop $C\phi(s)L$. (a) $r_f = 10$. (b) $r_f = 1$. (c) $r_f = 0.1$.

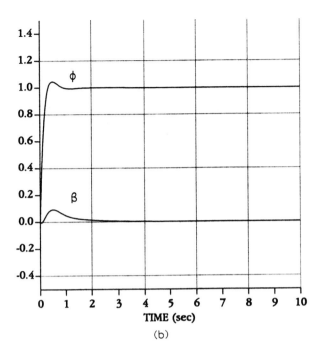

(b)

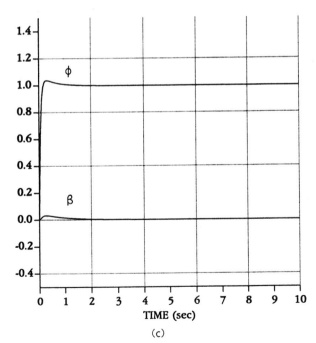

(c)

Figure 10.2-12 (*continued*)

$$\dot{\hat{x}} = (A - LC)\hat{x} + Lr \tag{15}$$
$$\hat{y} = C\hat{x}.$$

Using these dynamics in program TRESP (Appendix A) with $r = [1\ \ 0]^T$ produces the step response plot.

A word on the choice for Q_f is in order. The design parameters Q_f and R_f should be selected so that the target feedback loop $C\phi(s)L$ has good robustness and time-response properties. It is traditional to select $Q_f = BB^T$, which accounts for the last two diagonal entries of (11). However, in this example it was impossible to obtain good step responses using this selection for Q_f. Motivated by the fact that the process noise in the aircraft excites the first four states as well, we experimented with different values for Q_f, plotting in each case the singular values and step responses. After a few iterations, the final choice (11) was made.

e. Loop Transfer Recovery at the Output

The target feedback loop $C\phi(s)L$ using $r_f = 1$ has good properties in both the frequency and time domains. Unfortunately, the closed-loop system with LQG regulator has a loop gain referred to the output of $C\phi(s)BK\phi_r(s)L$, with the regulator resolvent given by

$$\phi_r(s) = [sI - (A - LC - BK)]^{-1}. \tag{16}$$

On the other hand, Algorithm 2 in Table 10.2-1 shows how to select a state-feedback gain K so that the LQG regulator loop gain approaches the ideal loop gain $C\phi(s)L$. Let us now select such a feedback gain matrix.

Using MATLAB, the LQR design problem in Algorithm 2 was solved with $Q_c = C^TC$, $R_c = I$, and various values of $r_c = \rho^2$ to obtain different feedback gains K. Some representative singular values of the LQG loop gain $C\phi(s)BK\phi_r(s)L$ are plotted in Fig. 10.2-13, where L is the target-loop Kalman gain (12). Note how the actual singular values ap-

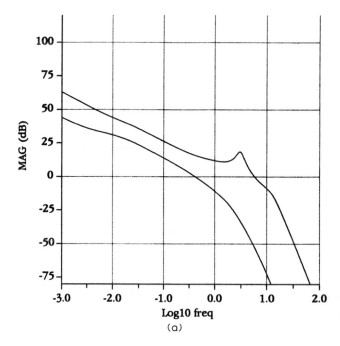

Figure 10.2-13 Singular value plots for the LQG regulator. (a) LQG with $r_c = 10^{-3}$. (b) LQG with $r_c = 10^{-7}$. (c) LQG with $r_c = 10^{-11}$, including robustness bounds.

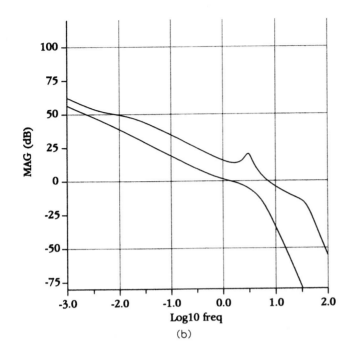

(b)

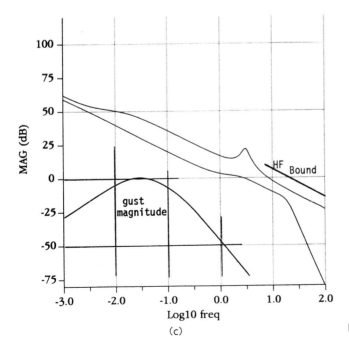

(c)

Figure 10.2-13 (*continued*)

proach the target singular values in Fig. 10.2-11a as r_c decreases. A good match is obtained for $r_c = 10^{-11}$.

Figure 10.2-13c also depicts the robustness bounds, which are satisfied for this choice of $r_c = 10^{-11}$.

The corresponding step responses are given in Fig. 10.2-14. A suitable step response that matches well the target response of Fig. 10.2-12b results when $r_c = 10^{-11}$.

It is of interest to discuss how these plots were obtained. For the LQG singular value plots, the complete dynamics are given by

$$\dot{x} = Ax + Bu$$
$$\dot{\hat{x}} = (A - LC)\hat{x} + Bu + Lw \qquad (17)$$
$$u = -K\hat{x}$$

where $w(t) = -e(t)$. These may be combined into the augmented system

$$\begin{bmatrix} \dot{x} \\ \dot{\hat{x}} \end{bmatrix} = \begin{bmatrix} A & -BK \\ 0 & A-LC-BK \end{bmatrix} \begin{bmatrix} x \\ \hat{x} \end{bmatrix} + \begin{bmatrix} 0 \\ L \end{bmatrix} w \qquad (18)$$

$$y = [C \ 0] \begin{bmatrix} x \\ \hat{x} \end{bmatrix} \qquad (19)$$

which has transfer function $C\phi(s)BK\phi_r(s)L$. The singular values are now easily plotted.

For the step responses, the closed-loop system must be studied. To close the loop, set $w = y - r$ in (18) to obtain the closed-loop dynamics

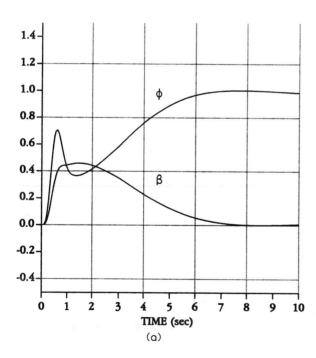

Figure 10.2-14 Closed-loop step responses of the LQG regulator. (a) LQG with $r_c = 10^{-3}$. (b) LQG with $r_c = 10^{-7}$. (c) LQG with $r_c = 10^{-11}$.

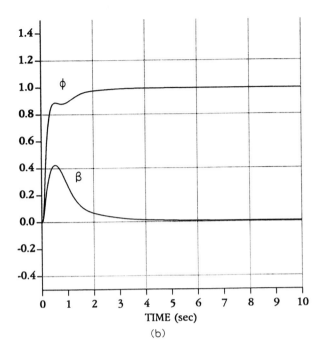

(b)

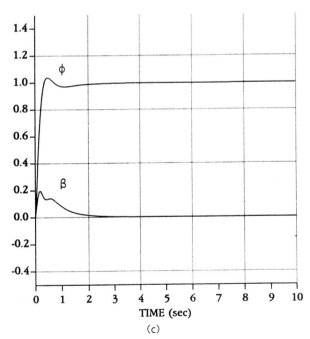

(c)

Figure 10.2-14 (*continued*)

$$\begin{bmatrix} \dot{x} \\ \dot{\hat{x}} \end{bmatrix} = \begin{bmatrix} A & -BK \\ LC & A-LC-BK \end{bmatrix} \begin{bmatrix} x \\ \hat{x} \end{bmatrix} + \begin{bmatrix} 0 \\ -L \end{bmatrix} r \qquad (20)$$

$$y = [C \quad 0] \begin{bmatrix} x \\ \hat{x} \end{bmatrix} \qquad (21)$$

These are used with program TRESP in Appendix A to obtain Fig. 10.2-14.

The final LQG regulator is given by the Kalman gain L in (12) and the feedback gain K corresponding to $r_c = 10^{-11}$.

f. Reduced-Order Regulator

The LQG regulator just designed has order $n = 8$, the same as the plant. This is excessive for an aircraft lateral control system. A reduced-order regulator which produces very good results may easily be determined using the partial-fraction-expansion approach in Example 8.3-2, principal component analysis [Moore 1982], or other techniques. This is easily accomplished using MATLAB. The singular value plots and step response using the reduced-order regulator should be examined to verify robustness and suitable performance.

■

PROBLEMS FOR CHAPTER 10

Problems for Section 10.1

10.1-1 Duality. In Table 10.1-1, replace (A, B, C) by (A^T, C^T, B^T) and so demonstrate that the state-feedback design equations are dual to the filter gain design equations.

10.1-2 Duality. Repeat Problem 10.1-1 for the digital LQG design equations in Table 10.1-2.

10.1-3 Digital LQG with A Posteriori Feedback. Design a digital LQG regulator using the feedback (10.1.29). Derive a separation principle for this case.

10.1-4 LQG Regulator for Disk Drive Head-Positioning Mechanism. A disk drive head-positioning mechanism of the linear actuated sort obeys Newton's laws $F = Ma$. Design a LQG regulator for this system assuming that only position measurements are available. Solve the control and filter algebraic Riccati equations analytically to obtain the feedback gain K and observer gain L. Check the answers using software such as MATLAB [Moler et al. 1987]. Simulate your controller using a digital computer (using, for instance, TRESP in Appendix A). (See Example 3.3-1 and the examples in Chapter 9.)

10.1-5 Digital LQG Regulator. Repeat Problem 10.1-4, now designing a digital LQG regulator using direct discrete design. Use software such as MATLAB to solve the discrete control and filter AREs. Simulate your digital controller using DIGCTL in Appendix A.

10.1-6 LQG Regulator for MIMO System. Design a LQG regulator for the MIMO circuit in Example 4.1-1. Simulate your results.

10.1-7 Digital LQG Regulator for MIMO System. Design a digital LQG regulator for the MIMO circuit in Example 4.1-1. Simulate your results.

Problems for Section 10.2

10.2-1 Show that (10.2.7) implies (10.2.10) (see Laub 1979).

10.2-2 LQG/LTR for Disk Drive Head-Positioning Mechanism. Using LQG/LTR, design a regulator for a disk drive head-positioning system assuming that only position measurements are available. Use software such as MATLAB [Moler et al. 1987]. Obtain singular value plots and step responses for various values of the loop-recovery factor (see Example 10.2-1).

10.2-3 LQG/LTR for MIMO System. Using LQG/LTR, design a regulator for the MIMO circuit in Example 4.1-1. Obtain singular value plots and step responses for various values of the loop-recovery factor (see Example 10.2-1).

10.2-4 LQG/LTR Robust Design. In Example 8.3-2 a reduced second-order model was found for a third-order plant. Perform an LQG/LTR design using the reduced model. Verify that the appropriate high-frequency robustness bounds are satisfied that guarantee stability of the controller when used with the full plant. Examine step responses of the controller on the reduced model as well as the actual plant.

A

Computer Software

This appendix provides some design and simulation software that is used in examples throughout the book. An excellent way to gain an intuitive feel for control systems design and performance is to perform computer simulations. It is conceptually a short step from simulation to actual implementation, since the subroutines that are used on today's digital signal processors are very similar to those used for simulation.

There are some good software packages available for design and simulation of systems, and one should be aware of them and employ them. Samples are MATLAB, MATRIX$_x$, Program CC, and SIMNON. However, it is very instructive to use one's own software during the learning phase, especially since it is sometimes not clear exactly what is going on in some of these packages when dealing with digital control or multirate sampling.

Copies of the programs in the book may be obtained on disk (FORTRAN V for PC only) for $50 by writing

F. L. Lewis
3006 Waterway Court
Arlington, TX 76012

We note we have always found these programs useful, but make no claims on their numerical stability.

A.1 TIME RESPONSE AND CONTROLS SIMULATION

The simulation of discrete-time systems and closed-loop controllers is very simple. It only requires writing a few DO Loops to iterate through the time index.

As for the time-response simulation of continuous systems, a Runge-Kutta integrator works very well. Figure A.1-1 shows a program, TRESP, which uses a fourth-order

A.1 Time Response and Controls Simulation

```
C     FILE DIGCTLF.FOR
C     PROGRAM TO FIND TIME HISTORY
C        USES NONADAPTIVE RUNGE-KUTTA
C     ALLOWS SIMULATION OF SYSTEMS WITH DELAY

C     NEEDS SUBROUTINES:
C        F(TIME,X,XP) FOR CONTINUOUS DYNAMICS
C        DIG(IK,T,X) FOR DISCRETE DYNAMICS
C        SYSINP(IT,Tr,X) FOR CONTINUOUS SYSTEM INPUT (OPTIONAL)

      PROGRAM TRESP
      PARAMETER (NN=20,MM=10)
      REAL Y(0:1024,0:NN+MM),X(NN),SX(NN),XP(NN)
      INTEGER IX(NN+MM),IZ(NN+MM)
      CHARACTER *20 FILNAM,ANS,ANSC
      COMMON/CONTROL/U(MM)
      COMMON/OUTPUT/Z(MM)
      COMMON/DELAY/IT,TS
C
10    WRITE(*,*)'HOW MANY STATES?'
      READ(*,*) NX
      DO 15 I= 1,NX
15    SX(I)= 0.
      WRITE(*,*)'ENTER INITIAL STATES (Def= 0):'
      READ(*,*) (SX(I), I= 1,NX)
C
      WRITE(*,*)'DIGITAL CONTROLLER (C) OR FILTER (F)? (DEF=N)'
      READ(*,'(A)') ANS
      IF(ANS.EQ.'F' .OR. ANS.EQ.'f') ANS= 'F'
      IF(ANS.EQ.'C' .OR. ANS.EQ.'c') ANS= 'C'
      WRITE(*,*)'CONTIN. INPUT SUBROUTINE REQUIRED? (DEF=N)'
      READ(*,'(A)') ANSC
      IF(ANSC.EQ.'Y' .OR. ANSC.EQ.'y') ANSC= 'Y'
30    WRITE(*,*) 'HOW MANY STATES TO BE PLOTTED?'
      READ(*,*) MX
      IF(MX.EQ.0) GO TO 37
      DO 35 I= 1,MX
35    IX(I)= I
      WRITE(*,*)'WHICH ONES? (DEF = IN ORDER)'
      READ(*,*) (IX(I), I= 1,MX)
37    WRITE(*,*)'PLOT HOW MANY OUTPUTS?'
      READ(*,*) MZ
      IF(MZ.EQ.0) GO TO 50
      DO 36 I= 1,MZ
36    IZ(I)= I
      WRITE(*,*)'WHICH ONES? (DEF= IN ORDER)'
      READ(*,*) (IZ(I), I= 1,MZ)
C
50    WRITE(*,*) 'RUN TIME?'
      READ(*,*) TR
      WRITE(*,*)'PRINTING TIME INTERVAL ON SCREEN?'
      READ(*,*) TPR
      IF(ANS.EQ.'F' .OR. ANS.EQ.'C') THEN
         WRITE(*,*)'SAMPLE PERIOD?'
         READ(*,*) TD
      END IF
      WRITE(*,*)'PLOTTING TIME INTERVAL?'
      READ(*,*) TPL
      IF(ANS.NE.'F' .AND. ANS.NE.'C') TD= TPL
      WRITE(*,*)'RUNGE-KUTTA INTEGRATION PERIOD?'
```

Figure A.1-1 Program TRESP for time response of nonlinear continuous systems
(continues)

```
              READ(*,*) TS
              NPR= NINT(TR/TPR)
              NPL= NINT(TPR/TD)
              NT = NINT(TD/TPL)
              NTD= NINT(TPL/TS)
C
              TIME= 0.
              IT= 0
              IP= 0
              IK= 0
              DO 60 I= 1,NX
60            X(I)= SX(I)
              Y(0,0)= TIME
              DO 70 I= 1,MX
70            Y(0,I)= X(IX(I))
              IF(MZ.GT.0) THEN
                  IF(ANS.EQ.'C')   CALL DIG(IK,TD,X)
                  IF(ANSC.EQ.'Y')  CALL SYSINP(IT,TS,X)
                  CALL F(TIME,X,XP)
                  IF(ANS.EQ.'F')   CALL DIG(IK,TD,X)
                  DO 75 I= 1,MZ
75                Y(0,MX+I)= Z(IZ(I))
              END IF
C
              DO 110 I= 1,NPR
              DO 90 J= 1,NPL
              IF(ANS.EQ.'C') CALL DIG(IK,TD,X)
              DO 100 K= 1,NT
              DO 85 KID= 1,NTD
              IF(IT.EQ.0) THEN
                  WRITE(*,*)
                  WRITE(*,80) (IX(IND), IND= 1,MX)
80                FORMAT(35X,'STATES'/'     TIME',10(I12))
                  WRITE(*,'(11(1PE12.3))') (Y(0,IND), IND= 0,MX+MZ)
              END IF
              IF(ANSC.EQ.'Y') CALL SYSINP(IT,TS,X)
              CALL RUNKUT(TIME,TS,X,NX)
              IT= IT+1
85            TIME= FLOAT(IT)*TS
              IP= IP+1
              Y(IP,0)= TIME
              DO 101 L= 1,MX
101           Y(IP,L)= X(IX(L))
              IF(MZ.LE.0) GO TO 100
              DO 105 L= 1,MZ
105           Y(IP,MX+L)= Z(IZ(L))
100           CONTINUE
              IK= IK+1
              IF(ANS.EQ.'F') CALL DIG(IK,TD,X)
90            CONTINUE
              WRITE(*,'(11(1PE12.3))') (Y(IP,L), L= 0,MX+MZ)
110           CONTINUE
C
120           WRITE(*,130)
130           FORMAT(//2X,'ENTER 0 TO FILE ANSWERS'/8X,'1 TO QUIT',
     &               /8X,'2 TO RESTART',/8X,'3 TO PICK NEW STATES',
     &               /8X,'4 TO CHANGE TIME SCALE')
              READ(*,*) I
              GO TO (150,10,30,50) I
C
```

Figure A.1-1 *(continued)*

```
              WRITE(*,*)'OUTPUT FILE NAME?'
              READ(*,'(A)') FILNAM
              OPEN(20,FILE= FILNAM)
              REWIND 20
              WRITE(20,*) MX+MZ
              WRITE(20,*) IP+1
              DO 140 J= 1,MX+MZ
              DO 140 I= 0,IP
140           WRITE(20,'(8(1PE14.6))') Y(I,J)
              REWIND 20
              CLOSE (20)
              GO TO 120
C
150           STOP
              END
C
C
C
C     FOURTH-ORDER RUNGE-KUTTA INTEGRATION SUBROUTINE
C
C     REQUIRES SUBROUTINE F(TIME,X,XP) TO DESCRIBE PLANT DYNAMICS
C
              SUBROUTINE RUNKUT(TIME,TS,X,N)
C
C     TS       SAMPLE PERIOD
C     X        STATE VECTOR
C     N        NUMBER OF STATES
C     XP       DERIVATIVE OF STATE VECTOR
C
              PARAMETER (NDIM=32)
              REAL X(*), XP(NDIM), X1(NDIM), XP1(NDIM)
C
              CALL F(TIME,X,XP)
              DO 10 I= 1,N
10            X1(I)= X(I) + .5*TS*XP(I)
C
              TIME= TIME + .5*TS
              CALL F(TIME,X1,XP1)
              DO 20 I= 1,N
              XP(I)= XP(I) + 2.*XP1(I)
20            X1(I)= X(I) + .5*TS*XP1(I)
C
              CALL F(TIME,X1,XP1)
              DO 30 I= 1,N
              XP(I)= XP(I) + 2.*XP1(I)
30            X1(I)= X(I) + TS*XP1(I)
C
              TIME= TIME + .5*TS
              CALL F(TIME,X1,XP1)
              DO 40 I= 1,N
40            X(I)= X(I) + TS*( XP(I)+XP1(I) )/6.
C
              RETURN
              END
```

Figure A.1-1 (*continued*)

Runge-Kutta subroutine to integrate linear or nonlinear systems in the form

$$\dot{x} = f(x, u, t). \qquad (A.1.1)$$

It requires a subroutine F(time, x, xp) which computes $\dot{x}$ (denoted xp, or "x prime") from the current state $x(t)$ and control input $u(t)$. The control $u(t)$ and any outputs of the form

$$y = h(x, u, t) \qquad (A.1.2)$$

are placed into COMMON storage. This program can, of course, be modified to take the parameters (A, B, C, D) of a linear system from a file. Samples of the use of TRESP are given in examples throughout the book.

It is important to realize the following. To update $x(kT_R)$ to $x((k + 1)T_R)$, the Runge-Kutta integrator calls subroutine F(time, x, xp) four times during each Runge-Kutta integration period T_R. During these four calls, the control input should be *held constant* at $u(kT_R)$. Using subroutine SYSINP to compute $u(kT_R)$ accomplishes this.

In some examples in the text, a slight modification of TRESP, DIGCTL, is used for digital controls purposes. However, the version of TRESP given in the figure incorporates DIGCTL. Note that subroutine DIG(IK, T, x) is called once in every sample period T. Therefore, it is only necessary to place the digital controller into DIG. The time T_R should be selected as an integral divisor of T. Five or ten Runge-Kutta periods within each sample period is usually sufficient. Note that, for digital controls purposes, subroutine DIG is called *before* the Runge-Kutta routine, while for digital filtering, DIG is called *after* the call to Runge-Kutta.

For some systems, the Runge-Kutta integrator in the figure may not work; then, an adaptive step-size Runge-Kutta routine (e.g., Runge-Kutta-Fehlburg) can be used [Press et al. 1986]. (Note: the program given here works for all examples in the book.)

A.2 DISCRETIZATION OF CONTINUOUS-TIME SYSTEMS

The program DISC in Fig. A.1-2 discretizes a continuous-time system

$$\dot{x} = Ax + Bu \qquad (A.2.1)$$

to obtain

$$x_{k+1} = A^s x_k + B^s u_k \qquad (A.2.2)$$

using

$$A^s = e^{AT} \qquad (A.2.3)$$

$$B^s = \int_0^T e^{A\tau} B \, d\tau. \qquad (A.2.4)$$

It uses the series expansions given in section 7.1.

A.2 Discretization of Continuous-Time Systems

```
C     PROGRAM TO DISCRETIZE CT SYSTEM BY SAMPLING AND ZOH

      PROGRAM DISC
      PARAMETER (MM=10,NN=20)
      DIMENSION A(NN,NN),B(NN,MM),C(MM,NN)
      DIMENSION P(NN,NN),Q(NN,NN),R(NN,MM)
      CHARACTER *20 FNAME
C
      WRITE(*,*)'ENTER NAME OF INPUT FILE ("N" FOR NONE)'
      READ(*,'(A)') FNAME
      IF(FNAME.EQ.'N' .OR. FNAME.EQ.'n') GO TO 10
         OPEN(20,FILE=FNAME)
         REWIND 20
         READ(20,*) N,M,L,IDDUM,IGFHDUM
      READ(20,*) ((A(I,J), J=1,N), I=1,N)
      READ(20,*) ((B(I,J), J=1,M), I=1,N)
      READ(20,*) ((C(I,J), J=1,N), I=1,L)
         CLOSE(20)
      GO TO 20

10    WRITE(*,*) ' ENTER # STATE VAR., # INPUTS'
      READ(*,*) N,M
      WRITE(*,*) 'ENTER "A" MATRIX BY ROWS'
      READ(*,*) ((A(I,J), J=1,N), I=1,N)
      WRITE(*,*) 'ENTER "B" MATRIX BY ROWS'
      READ(*,*) ((B(I,J), J=1,M), I=1,N)
      WRITE(*,*) 'ENTER "C" MATRIX BY ROWS'
      READ(*,*) ((C(I,J), J=1,N), I=1,L)
20    WRITE(*,*) 'TIME STEP = ?'
      READ(*,*) T
C
      DO 40 I=1,N
      DO 30 J=1,N
30    A(I,J)=T*A(I,J)
      DO 40 J=1,M
40    B(I,J)=T*B(I,J)
      CALL EXPAT(A,P,Q,N,NCOUNT,*50)
      WRITE(*,*) '# OF EXP. SERIES TERMS USED FOR EXP(A*T) =',NCOUNT
      CALL MATMUL(Q,B,R,M,N)

         WRITE(*,*)'ENTER NAME OF OUTPUT FILE'
         READ(*,'(A)') FNAME
         OPEN(20,FILE=FNAME)
         REWIND 20
         WRITE(20,*) N,M,L,0,0
         DO 45 I= 1,N
45       WRITE(20,*) (P(I,J), J=1,N)
         WRITE(20,*)
         DO 46 I= 1,N
46       WRITE(20,*) (R(I,J), J= 1,M)
         WRITE(20,*)
         DO 47 I= 1,L
47       WRITE(20,*) (C(I,J), J= 1,N)
         REWIND 20
      CLOSE(20)

      STOP
50    WRITE(*,*)'SERIES FOR EXP(AT) HAS NOT CONVERGED'
      GO TO 20
      END
```

Figure A.1-2 Program DISC for discretization of continuous-time systems

(continues)

```
C
C
      SUBROUTINE MATMUL(A,B,C,M,N)
      PARAMETER (MM=10,NN=20)
      DIMENSION A(NN,NN),B(NN,MM),C(NN,MM)
      DO 20 I=1,N
      DO 20 J=1,M
      SUM=0.
      DO 10 K=1,N
   10 SUM=SUM+A(I,K)*B(K,J)
   20 C(I,J)=SUM
      RETURN
      END
C
C
C  SUBROUTINE TO EVALUATE THE MATRIX EXPONENTIAL ***********
C
C  THIS SUBROUTINE EVALUATES THE MATRIX EXPONENTIALS :
C  P=EXP(A*T), Q=(A**-1)*(EXP(A*T)-I)/T
C
      SUBROUTINE EXPAT(A,P,Q,N,NCOUNT,*)
      PARAMETER (MM=10,NN=20)
      DIMENSION A(NN,NN),P(NN,NN),Q(NN,NN),TERM(20,20),TEMP(20)
      TOL=1.E-12
C
C  INITIALISE P, Q, AND TERM MATRICES
C
      DO 20 I=1,N
      DO 10 J=1,N
      P(J,I)=0.
      Q(J,I)=0.
   10 TERM(J,I)=0.
      P(I,I)=1.0
      Q(I,I)=1.0
   20 TERM(I,I)=1.0
      NCOUNT=1
C
C  EVALUATE EXPONENTIAL SERIES
C
   30 DO 60 I=1,N
      DO 50 J=1,N
      SUM=0.
      DO 40 K=1,N
   40 SUM=SUM+TERM(I,K)*A(K,J)
   50 TEMP(J)=SUM
      DO 60 J=1,N
      TERM(I,J)=TEMP(J)
   60 P(I,J)=P(I,J)+TERM(I,J)
      CALL NORMOF(TERM,N,TNORM)
      NCOUNT=NCOUNT+1
      DO 70 I=1,N
      DO 70 J=1,N
      TERM(I,J)=TERM(I,J)/FLOAT(NCOUNT)
   70 Q(I,J)=Q(I,J)+TERM(I,J)
      CALL NORMOF(P,N,PNORM)
      ACC=TNORM/PNORM
      IF(NCOUNT.GE.20) RETURN
      GO TO 30
      END
C
```

Figure A.1-2 *(continued)*

```
      C  SUBROUTINE TO CALCULATE MATRIX ROW NORMS
      C
            SUBROUTINE NORMOF(ARRAY,N,SUM)
            DIMENSION ARRAY(N,N)
            SUM=0.
            DO 10 I=1,N
            S=0.
            DO 20 J=1,N
      20    S=S+ABS(ARRAY(I,J))
      10    IF(S.GT.SUM) SUM=S
            RETURN
            END
```

Figure A.1-2 (*continued*)

A.3 ACKERMANN'S FORMULA FOR POLE PLACEMENT

The program ACKERM in Fig. A.1-3 computes the state-variable feedback K for pole placement using Ackermann's Formula

$$K = e_n^T U_n^{-1} \Delta^D(A), \qquad (A.3.1)$$

where U_n is the reachability matrix, $e_n^T = [0 \ \ldots \ 0 \ 1]$ is the last row of the $n \times n$ identity matrix, and

$$\Delta^D(A) = A^n + \alpha_1 A^{n-1} + \ldots + \alpha_n I. \qquad (A.3.2)$$

$\Delta^D(A)$ is the desired characteristic polynomial evaluated at A (section 2.5). It uses two subroutines from [Press et al. 1986].

A.4 OUTPUT-FEEDBACK DESIGN

Output-feedback design is not an easy problem. Finding the optimal output-feedback gains to minimize a quadratic performance index (PI)

$$J = \frac{1}{2} \int_0^\infty (x^T Q x + u^T R u) \, dt, \qquad (A.4.1)$$

involves solving coupled nonlinear matrix design equations of the form (Chapter 4)

$$0 = A_c^T P + P A_c + C^T K^T R K C + Q \qquad (A.4.2)$$

$$0 = A_c S + S A_c^T + X \qquad (A.4.3)$$

$$0 = RKCSC^T - B^T P S C^T. \qquad (A.4.4)$$

where

$$A_c = A - BKC, \qquad X = x(0)x^T(0).$$

In the design of tracking systems, the equations are even worse.

In section 4.1 we gave the Moerder-Calise algorithm for iterative solution of these equations. This algorithm does not work in the tracker-design case. We have used two

```
C     FILE ACKERM - PROGRAM IMPLEMENTS ACKERMANN'S FORMULA.
C     LINK WITH SUBROUTINES LUDCMP AND LUBKSB [Press et al. 1986]
C
      PROGRAM ACKERM
      PARAMETER(NN=20,MM=5)
      IMPLICIT DOUBLE PRECISION(A-H,O-Z)
      DIMENSION A(NN,NN),B(NN,MM),C(MM,MM),COEFF(NN),POLY(NN,NN)
      DIMENSION BV(NN),CMX(NN,NN),Y(NN,NN),GAIN(NN),ABK(NN,NN)
      DIMENSION IND(NN)
      INTEGER I,J,MP,NP,MPC
      CHARACTER *30 FNAME
C
      MPC= 1
      GO TO 50
 10   WRITE(*,*)'! FILE NOT FOUND, TRY AGAIN'
      GO TO 50
 20   WRITE(*,*)'FILE READ ERROR !'
 50   WRITE(*,*)'Name of system file:'
      READ(*,'(A)') FNAME
      OPEN(UNIT=20,FILE=FNAME,ERR=10)
      REWIND 20
      READ(20,*,ERR=20) NP,MP,LX,ID,M1
      IF (MP.EQ.1) GO TO 2000
 1010 WRITE(*,*)'Which column of B should be used?'
      READ(*,*) MPC
      IF (MPC.GT.MP) GO TO 1010
 2000 CONTINUE
      READ(20,*,ERR=20) ((A(I,J), J=1,NP), I=1,NP)
      READ(20,*,ERR=20) ((B(I,J), J=1,MP), I=1,NP)
      READ(20,*,ERR=20) ((C(I,J), J=1,NP), I=1,LX)
      REWIND 20
      CLOSE(20)
      WRITE(*,'(/,2X,A,/)')
     + 'Enter coeffs. of desired char. pol. (in desc. order):'
      READ(*,*) ONE, (COEFF(I), I= 1,NP)
      DO 200 I= 1, NP
      BV(I) = B(I,MPC)
 200  CONTINUE
      CALL ALPHA(A,COEFF,NN,NP,POLY)
      CALL CTRLMX(A,BV,NN,NP,CMX)

      DO 500 I= 1,NP
        DO 400 J= 1,NP
        Y(I,J)= 0.
 400    CONTINUE
        Y(I,I) = 1.
 500  CONTINUE
      CALL LUDCMP(CMX,NP,NN,IND,D)
      DO 600 J= 1,NP
      CALL LUBKSB(CMX,NP,NN,IND,Y(1,J))
 600  CONTINUE
      DO 700 J= 1,NP
      GAIN(J) = 0.
      DO 700 K= 1,NP
      GAIN(J) = GAIN(J) + Y(NP,K)*POLY(K,J)
 700  CONTINUE

      WRITE(*,*)'Feedback gain is:'
      WRITE(*,'(6(1PE13.4))') (GAIN(I), I= 1,NP)
 1200 WRITE(*,*) 'Name of output file for FB gain (Def= none)'
```

Figure A.1-3 Program ACKERM for pole placement using Ackermann's Formula

```
              READ(*,'(A)') FNAME
              IF(FNAME.EQ.'/') GO TO 100
              OPEN(UNIT=20,FILE=FNAME,ERR=1200)
              WRITE(20,'(6(1PE13.5))') (GAIN(J), J=1,NP)
              CLOSE(20)

C     COMPUTE AND FILE CLOSED-LOOP SYSTEM A-bK, B, C

 100          WRITE(*,*)'File for closed-loop system (Def= none)'
              READ(*,'(A)') FNAME
              IF(FNAME.EQ.'/') GO TO 1000
              OPEN(UNIT=20,FILE=FNAME)

              DO 30 I= 1,NP
              DO 30 J= 1,NP
 30           ABK(I,J)= A(I,J) - BV(I)*GAIN(J)

              WRITE(20,*) NP,MP,LX,0,0
                WRITE(20,'(//)')
                DO 310 I= 1,NP
 310          WRITE(20,'(6(1PE13.5))') (ABK(I,J), J= 1,NP)
                WRITE(20,'(//)')
                DO 311 I= 1,NP
 311          WRITE(20,'(6(1PE13.5))') (B(I,J), J= 1,MP)
                WRITE(20,'(//)')
                DO 312 I= 1,LX
 312          WRITE(20,'(6(1PE13.5))') (C(I,J), J= 1,NP)
              CLOSE(20)
              GO TO 1000
C
 1000 END
C
C
C
C     SUBROUTINE TO MULTIPLY MATRICES C= AB
              SUBROUTINE MUL(A,NA,B,NB,C,NC,N,M,L)
              IMPLICIT DOUBLE PRECISION(A-H,O-Z)
              DIMENSION A(NA,1),B(NB,1),C(NC,1)
C
              DO 100 I= 1,N
              DO 100 J= 1,L
              C(I,J) = 0.
              DO 100 K= 1,M
              C(I,J) = C(I,J) + A(I,K)*B(K,J)
 100          CONTINUE
              RETURN
              END
C
C
C
C     SUBROUTINE TO FIND PA= A*PA
              SUBROUTINE MUL2(A,PA,NMEM,NEFC)
              IMPLICIT DOUBLE PRECISION (A-H,O-Z)
              DIMENSION A(NMEM,1),PA(NMEM,1),TEMP(20)
C
              DO 100 J= 1,NEFC
              DO 200 I= 1,NEFC
              TEMP(I)= 0.
              DO 200 K= 1,NEFC
              TEMP(I)= TEMP(I) + A(I,K)*PA(K,J)
```

Figure A.1-3 (*continued*)

(*continues*)

```
      200 CONTINUE
          DO 100 I= 1,NEFC
          PA(I,J)= TEMP(I)
      100 CONTINUE
          RETURN
          END
C
C
C
C   SUBROUTINE TO EVALUATE MATRIX POLYNOMIAL Alpha(A):
C       Alpha(A) = A^n + Coeff(1)A^(n-1) + ..... + Coeff(n)I
          SUBROUTINE ALPHA(A,COEFF,NMEM,NEFC,POLYA)
          IMPLICIT DOUBLE PRECISION (A-H,O-Z)
          DIMENSION A(NMEM,1),POLYA(NMEM,1),COEFF(NMEM)
C
          DO 200 I= 1,NEFC
          DO 100 J= 1,NEFC
          POLYA(I,J)= A(I,J)
      100 CONTINUE
          POLYA(I,I)= POLYA(I,I) + COEFF(1)
      200 CONTINUE
          DO 300 I= 2,NEFC
          CALL MUL2(A,POLYA,NMEM,NEFC)
          DO 300 J= 1,NEFC
          POLYA(J,J)= POLYA(J,J) + COEFF(I)
      300 CONTINUE
          RETURN
          END
C
C
C
C   SUBROUTINE TO FIND CONTROLLABILITY MATRIX
          SUBROUTINE CTRLMX(A,B,NMEM,NEFC,CMAT)
          IMPLICIT DOUBLE PRECISION (A-H,O-Z)
          DIMENSION A(NMEM,1),B(NMEM),CMAT(NMEM,1)
C
          DO 100 I= 1,NEFC
          CMAT(I,1)= B(I)
      100 CONTINUE
          DO 200 J= 2,NEFC
          JPRED= J-1
          DO 200 I= 1,NEFC
          CMAT(I,J)= 0.
          DO 200 K= 1,NEFC
          CMAT(I,J) = CMAT(I,J) + A(I,K)*CMAT(K,JPRED)
      200 CONTINUE
          RETURN
          END
```

Figure A.1-3 (*continued*)

general approaches to solving such equation sets. In the first, the PI J is computed based on (A.4.2) using

$$J = \frac{1}{2}tr(PX). \quad (A.4.5)$$

A simplex routine was used to minimize J [Press et al. 1986].

In the second approach, a gradient-algorithm (e.g., Davidon-Fletcher-Powell) was used [Press et al. 1986]. There, the gradient $\partial J/\partial K$ is computed using all three design equations (A.4.2)–(A.4.4). Similar remarks hold for discrete-time output-feedback design (section 7.4).

B

Review of Matrix Algebra

In this appendix we review some matrix notions that are used throughout the book. Additional references include Gantmacher [1977], Brogan [1974], and Strang [1980].

B.1 BASIC FACTS

We denote the real numbers by **R** and the complex numbers by **C**. The real n-vectors are $\mathbf{R}^n$, the real $m \times n$ matrices are $\mathbf{R}^{m \times n}$, and similar definitions hold for $\mathbf{C}^n$ and $\mathbf{C}^{m \times n}$. The vectors and matrices used in this appendix are real, not complex, unless otherwise stated.

With $|A|$ symbolizing the determinant of an $n \times n$ matrix and A and B both square, it is true that

$$|A| = |A^T| \qquad (B.1.1)$$

$$|AB| = |A| \cdot |B|, \qquad (B.1.2)$$

with superscript "T" representing transpose. If A is nonsingular, then

$$|A^{-1}| = 1/|A|. \qquad (B.1.3)$$

If $A \in \mathbf{R}^{m \times n}$ and $B \in \mathbf{R}^{n \times m}$ (with m possibly equal to n), then

$$|I_m + AB| = |I_n + BA| \qquad (B.1.4)$$

$$\text{trace}(AB) = \text{trace}(BA) \qquad (B.1.5)$$

where I_n is the $n \times n$ identity matrix, and the trace of any matrix is the sum of its diagonal entries.

If A is nonsingular, we define the shorthand notation

$$A^{-T} \equiv (A^{-1})^T. \qquad (B.1.6)$$

For any matrices A and B

$$(AB)^T = B^T A^T \qquad (B.1.7)$$

and if A and B are nonsingular

$$(AB)^{-1} = B^{-1} A^{-1}. \qquad (B.1.8)$$

We say a real matrix M is *orthogonal* if

$$M^{-1} = M^T. \qquad (B.1.9)$$

Thus, $M^T M = M M^T = I$. We say a complex matrix N is *unitary* if

$$N^{-1} = N^* \qquad (B.1.10)$$

where N^* is the complex conjugate transpose of N. Thus, $NN^* = N^*N = I$.

If $\lambda_i \neq 0$ is an eigenvalue of A with eigenvector v_i, then

$$A v_i = \lambda_i v_i$$

implies that

$$\lambda_i^{-1} v_i = A^{-1} v_i, \qquad (B.1.11)$$

so that $1/\lambda_i$ is an eigenvalue of A^{-1}.

B.2 PARTITIONED MATRICES

If

$$D = \begin{bmatrix} A_{11} & 0 & 0 \\ 0 & A_{22} & 0 \\ 0 & 0 & A_{33} \end{bmatrix} \qquad (B.2.1)$$

where A_{ii} are in general matrices, then we write $D = \text{diag}\{A_{11}, A_{22}, A_{33}\}$ and call D *block diagonal*. If the A_{ii} are square, then

$$|D| = |A_{11}| \cdot |A_{22}| \cdot |A_{33}|, \qquad (B.2.2)$$

and if $|D| \neq 0$ then

$$D^{-1} = \text{diag}\{A_{11}^{-1}, A_{22}^{-1}, A_{33}^{-1}\}. \qquad (B.2.3)$$

If

$$D = \begin{bmatrix} A_{11} & A_{12} & A_{13} \\ 0 & A_{22} & A_{23} \\ 0 & 0 & A_{33} \end{bmatrix} \qquad (B.2.4)$$

where A_{ij} are in general matrices, then D is *upper block triangular* and (B.2.2) still holds.

The inverse of D is also upper block triangular, and its diagonal entries are A_{ii}^{-1}. *Lower block triangular* matrices have the form of the transpose of (B.2.4).

The *well-known matrix inversion lemma* says that

$$(A_{11}^{-1} + A_{12}A_{22}A_{21})^{-1} = A_{11} - A_{11}A_{12}(A_{21}A_{11}A_{12} + A_{22}^{-1})^{-1}A_{21}A_{11}. \quad \text{(B.2.5)}$$

This may be verified by multiplying the right-hand side by $(A_{11}^{-1} + A_{12}A_{22}A_{21})$ to obtain the identity.

B.3 QUADRATIC FORMS AND DEFINITENESS

Given a matrix Q and a vector $x \in \mathbf{R}^n$, we call

$$x^T Q x \quad \text{(B.3.1)}$$

a *bilinear* or *quadratic form*. We assume Q is real.

Every real square matrix Q can be decomposed into a *symmetric part* Q_s (i.e., $Q_s^T = Q_s$) and an *antisymmetric part* Q_a (i.e., $Q_a^T = -Q_a$). Thus

$$Q = Q_s + Q_a \quad \text{(B.3.2)}$$

where

$$Q_s = \frac{Q + Q^T}{2}, \quad Q_A = \frac{Q - Q^T}{2}. \quad \text{(B.3.3)}$$

To verify this decomposition, substitute (B.3.3) into (B.3.2). Therefore, if Q is not symmetric, we may write

$$x^T Q x = x^T (Q_s + Q_a) x = x^T Q_s x + x^T Q_a x.$$

However, both terms are scalars, so they are equal to their transposes. Therefore, $x^T Q_a x = (x^T Q_a x)^T = x^T Q_a^T x = -x^T Q_a x$, which must thus equal zero. For a general real square Q then

$$x^T Q x = x^T Q_s x \quad \text{(B.3.4)}$$

and we may assume without loss of generality that Q in (B.3.1) is symmetric. Let us do so.

We say Q is:

Positive definite ($Q > 0$) if $x^T Q x > 0$ for all nonzero x.
Positive semidefinite ($Q \geq 0$) if $x^T Q x \geq 0$ for all nonzero x.
Negative definite ($Q < 0$) if $x^T Q x < 0$ for all nonzero x.
Negative semidefinite ($Q \leq 0$) if $x^T Q x \leq 0$ for all nonzero x.
Indefinite if $x^T Q x > 0$ for some x, $x^T Q x < 0$ for some x.

If Q is symmetric, we can easily test for its definiteness. Indeed, with λ_i the eigenvalues of Q, we have

$$\begin{aligned} Q > 0 \text{ iff all } \lambda_i > 0, \\ Q \geq 0 \text{ iff all } \lambda_i \geq 0, \\ Q < 0 \text{ iff all } \lambda_i < 0, \\ Q \leq 0 \text{ iff all } \lambda_i \leq 0. \end{aligned} \qquad (B.3.5)$$

Another test is provided as follows. With $Q = [q_{ij}] \; \varepsilon \; \mathbf{R}^{n \times n}$, the *leading minors* of Q are

$$m_1 = q_{11}$$

$$m_2 = \begin{vmatrix} q_{11} & q_{12} \\ q_{21} & q_{22} \end{vmatrix}$$

$$m_3 = \begin{vmatrix} q_{11} & q_{12} & q_{13} \\ q_{21} & q_{22} & q_{23} \\ q_{31} & q_{32} & q_{33} \end{vmatrix} \qquad (B.3.6)$$

$$\vdots$$

$$m_n = |Q|.$$

Then, $Q > 0$ if and only if $m_i > 0$ for all i. To use this test for negative definiteness, note that $Q < 0$ if $-Q > 0$. It is not true that $Q \geq 0$ if $m_i \geq 0$ for all i; for positive semidefiniteness, *all* the minors, not only the leading minors, must be nonnegative.

Any positive semidefinite matrix may be factored into square roots either as

$$Q = \sqrt{Q}\sqrt{Q^T} \qquad (B.3.7)$$

or

$$Q = \sqrt{Q^T}\sqrt{Q}. \qquad (B.3.8)$$

The ("left" and "right") square roots in (B.3.7) and (B.3.8) are not necessarily the same. Additionally, there may be many square roots of each sort. In fact, if M is any orthogonal matrix, then (B.3.8) shows that

$$Q = \sqrt{Q^T} \, M^T \, M \sqrt{Q}$$

so that $M\sqrt{Q}$ is also a square root of Q. If $Q > 0$, then all square roots are nonsingular.

B.4 SINGULAR VALUE DECOMPOSITION

In Chapter 8 we deal with frequency-domain techniques for multivariable systems. There, the following notion is needed.

B.4 Singular Value Decomposition

Given any matrix $A \in \mathbf{C}^{m \times n}$ we define its *singular value decomposition* (SVD) [Strang 1980] as

$$A = U\Sigma V^*, \tag{B.4.1}$$

where $U \in \mathbf{C}^{m \times m}$ and $V \in \mathbf{C}^{n \times n}$ are square unitary matrices and

$$\Sigma = \begin{bmatrix} \sigma_1 & & & & & & \\ & \sigma_2 & & & & & \\ & & \ddots & & & & \\ & & & \sigma_r & & & \\ & & & & 0 & & \\ & & & & & \ddots & \\ & & & & & & 0 \end{bmatrix} \in \mathbf{R}^{m \times n} \tag{B.4.2}$$

with $r = \text{rank}(A)$, where σ_i are nonnegative real numbers. The *singular values of A* are the σ_i, which are ordered so that $\sigma_1 \geq \sigma_2 \geq \ldots \geq \sigma_r$. We define the *maximum singular value* $\bar{\sigma}$ as σ_1, and the *minimum singular value* $\underline{\sigma}$ as σ_r. The SVD may loosely be thought of as the extension to general nonsquare matrices of the Jordan normal form.

Since $AA^* = U\Sigma V^* V \Sigma^T U^* = U\Sigma^2 U^*$, it follows that the singular values of A are simply the (positive) square roots of the nonzero eigenvalues of AA^*. A similar proof shows that the nonzero eigenvalues of AA^* and those of A^*A are the same.

Some useful properties of the SVD follow. A and B are matrices dimensioned so that the following expressions make sense.

Equalities

$$|A| = 0 \text{ iff } \underline{\sigma}(A) = 0$$

$$\underline{\sigma}(A) = 1/\bar{\sigma}(A^{-1})$$

$$\bar{\sigma}(A) = 1/\underline{\sigma}(A^{-1})$$

$$\bar{\sigma}(\alpha A) = |\alpha|\bar{\sigma}(A) \text{ for any scalar } \alpha$$

Inequalities

$$\bar{\sigma}(A + B) \leq \bar{\sigma}(A) + \bar{\sigma}(B), \text{ triangle inequality}$$

$$\bar{\sigma}(AB) \leq \bar{\sigma}(A)\bar{\sigma}(B), \text{ Cauchy-Schwartz inequality}$$

$$\underline{\sigma}(AB) \geq \underline{\sigma}(A)\underline{\sigma}(B)$$

$$\underline{\sigma}(A) - 1 \leq \underline{\sigma}(I + A) \leq \underline{\sigma}(A) + 1$$

$$\underline{\sigma}(A) \leq |\lambda(A)| \leq \bar{\sigma}(A), \text{ with } A \text{ square and } \lambda \text{ any eigenvalue of } A$$

Theorems

$$\bar{\sigma}(B) < \underline{\sigma}(A) \text{ implies } \underline{\sigma}(A + B) > 0, \text{ assuming } \underline{\sigma}(A) > 0$$

$$\bar{\sigma}(A) < 1 \text{ implies } \underline{\sigma}(I + A) \geq 1 - \bar{\sigma}(A).$$

B.5 MATRIX CALCULUS

Throughout the book, in discussing optimal control we shall be interested in minimizing functions of matrices and vectors. For this, the following notions are needed.

Let $x = [x_1 \; x_2 \; \ldots \; x_n]^T \in \mathbf{R}^n$, s be a real scalar, and $f(x) \in \mathbf{R}^m$ be a vector function of x. The differential in x is

$$dx = \begin{bmatrix} dx_1 \\ dx_2 \\ \vdots \\ dx_n \end{bmatrix} \qquad (B.5.1)$$

and the derivative of x with respect to s is

$$\frac{dx}{ds} \equiv \begin{bmatrix} dx_1/ds \\ dx_2/ds \\ \vdots \\ dx_n/ds \end{bmatrix}. \qquad (B.5.2)$$

If s is a function of x, the *gradient* of s with respect to x is

$$s_x \equiv \frac{\partial s}{\partial x} = \begin{bmatrix} \partial s/\partial x_1 \\ \partial s/\partial x_2 \\ \vdots \\ \partial s/\partial x_n \end{bmatrix}. \qquad (B.5.3)$$

(In some references the gradient is defined as a row vector, but in our usage this would add many extra transposes through the equations.) Then the *total differential in s* is

$$ds = \left[\frac{\partial s}{\partial x}\right]^T dx = \sum_{i=1}^{n} \frac{\partial s}{\partial x_i} dx_i. \qquad (B.5.4)$$

If s is a function of two vectors x and y, then

$$ds = \left[\frac{\partial s}{\partial x}\right]^T dx + \left[\frac{\partial s}{\partial y}\right]^T dy.$$

The *jacobian* of f with respect to x is the $m \times n$ matrix

$$f_x \equiv \frac{\partial f}{\partial x} = \left[\frac{\partial f}{\partial x_1} \; \frac{\partial f}{\partial x_2} \; \ldots \; \frac{\partial f}{\partial x_n}\right], \qquad (B.5.5)$$

so that the total differential of f is

$$df = \frac{\partial f}{\partial x} x = \sum_{i=1}^{n} \frac{\partial f}{\partial x_i} dx_i. \qquad (B.5.6)$$

B.5 Matrix Calculus

We shall use the shorthand notation

$$\frac{\partial f^T}{\partial x} \equiv \left[\frac{\partial f}{\partial x}\right]^T. \tag{B.5.7}$$

If s is a scalar, x and y are vectors, and A, B, D, Q are matrices with dimensions so that the following expressions make sense, then we have the following results.

$$\frac{d}{dt}(A^{-1}) = -A^{-1}\dot{A}A^{-1} \tag{B.5.8}$$

Some useful gradients are:

$$\frac{\partial}{\partial x}(y^T x) = \frac{\partial}{\partial x}(x^T y) = y \tag{B.5.9}$$

$$\frac{\partial}{\partial x}(y^T A x) = \frac{\partial}{\partial x}(x^T A^T y) = A^T y \tag{B.5.10}$$

$$\frac{\partial}{\partial x}(y^T f(x)) = \frac{\partial}{\partial x}(f^T(x) y) = f_x^T y \tag{B.5.11}$$

$$\frac{\partial}{\partial x}(x^T A x) = A x + A^T x, \tag{B.5.12}$$

and if Q is symmetric, then

$$\frac{\partial}{\partial x}(x^T Q x) = 2 Q x \tag{B.5.13}$$

$$\frac{\partial}{\partial x}(x - y)^T Q(x - y) = 2 Q(x - y). \tag{B.5.14}$$

The product rule for two vector functions is

$$\frac{\partial}{\partial x}(f^T y) = f_x^T y + y_x^T f. \tag{B.5.15}$$

Some useful jacobians are:

$$\frac{\partial}{\partial x}(A x) = A \tag{B.5.16}$$

and the product rule

$$\frac{\partial}{\partial x}(s f) = \frac{\partial}{\partial x}(f s) = s f_x + f s_x^T. \tag{B.5.17}$$

Some important derivatives involving the trace and the determinant are

$$\frac{\partial}{\partial A}\text{trace}(A) = I \qquad (B.5.18)$$

$$\frac{\partial}{\partial A}\text{trace}(BAD) = B^T D^T \qquad (B.5.19)$$

$$\frac{\partial}{\partial A}\text{trace}(ABA^T) = 2AB, \qquad \text{IF } B = B^T \qquad (B.5.20)$$

$$\frac{\partial}{\partial A}|BAD| = |BAD|A^{-T}. \qquad (B.5.21)$$

C

Review of Probability Theory

In this appendix we review some notions in probability theory that are used in Chapter 9 in connection with the Kalman filter. A good reference is Papoulis [1984].

C.1 MEAN AND VARIANCE

Given a random vector $z \in \mathbf{R}^n$, we denote by $f_z(\zeta)$ the *probability density function* (PDF) of z. The PDF represents the probability that z takes on a value within the differential region $d\zeta$ centered at ζ. Although the value of z may be unknown, it is quite common in many situations to have a good feel for its PDF.

The *expected value* of a function $g(z)$ of a random vector z is defined as

$$E\{g(z)\} = \int_{-\infty}^{\infty} g(\zeta) f_z(\zeta) \, d\zeta. \tag{C.1.1}$$

In particular, the *mean* or *expected value* of z is defined by

$$E\{z\} = \int_{-\infty}^{\infty} \zeta f_z(\zeta) \, d\zeta, \tag{C.1.2}$$

which we shall symbolize by $\bar{z}$ to economize on notation. Note that $\bar{z} \in \mathbf{R}^n$.

Note that the expectation operator is linear, so that, given two random variables x and z and two deterministic scalars a and b

$$\overline{ax + bz} = a\bar{x} + b\bar{z}. \tag{C.1.3}$$

The *covariance* of $z \in \mathbf{R}^n$ is given by

$$P_z = E\{(z - \bar{z})(z - \bar{z})^T\}. \tag{C.1.4}$$

Note that P_z is an $n \times n$ positive definite (constant) matrix. Using the linearity property, it is seen that

$$P_z = E\{zz^T\} - \bar{z}\bar{z}^T. \quad \text{(C.1.5)}$$

We call $E\{zz^T\}$ the *mean-square value* of z.

An important class of random vectors is characterized by the *Gaussian* or *normal* PDF

$$f_z(\zeta) = \frac{1}{\sqrt{(2\pi)^n |P_z|}} e^{-1/2(\zeta - \bar{z})^T P_z^{-1}(\zeta - \bar{z})}, \quad \text{(C.1.6)}$$

where in general $z \in \mathbf{R}^n$. In the scalar case $n = 1$ this reduces to the more familiar

$$f_z(\zeta) = \frac{1}{\sqrt{2\pi P_z}} e^{-(\zeta - \bar{z})^2 / 2P_z}. \quad \text{(C.1.7)}$$

Such random vectors take on values near the mean $\bar{z}$ with greatest probability, and have a decreasing probability of taking on values farther away from $\bar{z}$.

Many naturally occurring random variables are Gaussian. Increased importance is given to Gaussian PDF by *Central-Limit Theorem*, which states that the sum of a large number of random variables has approximately a Gaussian PDF, regardless of the distributions of the individual random variables.

C.2 TWO RANDOM VARIABLES

Given random vectors $z \in \mathbf{R}^n$, $x \in \mathbf{R}^m$ we denote by $f_{z,x}(\zeta, \xi)$ the *joint probability density function* of z and x. The joint PDF represents the probability that z takes on a value within the differential region $d\zeta$ centered at ζ and x takes on a value within the differential region $d\xi$ centered at ξ. An example of a joint PDF of two scalar random variables z_1 and z_2 is provided by (C.1.6) when $z = [z_1 \ z_2]^T$.

The *expected value* of a function $g(z, x)$ of two random vectors z and x is defined as

$$E\{g(z, x))\} = \int_{-\infty}^{\infty} g(\zeta, \xi) f_{z,x}(\zeta, \xi) \, d\zeta d\xi. \quad \text{(C.2.1)}$$

The *cross-covariance* of two random variables $z \in \mathbf{R}^n$ and $x \in \mathbf{R}^m$ is defined as

$$P_{zx} = E\{(z - \bar{z})(x - \bar{x})^T\} \quad \text{(C.2.2)}$$

which is an $n \times m$ constant matrix.

The *conditional PDF* of x given z is given by

$$f_{x/z}(\xi/z = \zeta) = f_{zx}(\zeta, \xi)/f_z(\zeta). \quad \text{(C.2.3)}$$

C.3 Random Processes

The *conditional mean* of x given z is a random variable denoted by $\overline{x/z}$ and defined by the functional dependence

$$\overline{x/z=\zeta} = \int_{-\infty}^{\infty} \xi f_{x/z}(\xi/z=\zeta)\, d\xi. \tag{C.2.4}$$

Two random variables are said to be *independent* if

$$f_{z,x}(\zeta, \xi) = f_z(\zeta) f_x(\xi), \tag{C.2.5}$$

uncorrelated if

$$E\{zx^T\} = \overline{z}\,\overline{x}^T, \tag{C.2.6}$$

and *orthogonal* if

$$E\{zx^T\} = 0. \tag{C.2.7}$$

Independence implies uncorrelatedness. For normal random variables, these two properties are equivalent.

C.3 RANDOM PROCESSES

If the random vector is a time function it is called a *random process*, symbolized as $z(t)$. Then, the PDF may also be time varying and we write $f_z(\zeta, t)$. In this situation, the expected value and covariance matrix are also functions of time, so we write $\overline{z}(t)$ and $P_z(t)$. In discrete time, we write z_k, $\overline{z}_k$, $P_z(k)$ and so on.

Many random processes $z(t)$ of interest to us have a time-invariant PDF. These are *stationary* processes and, even though they are random time functions, they have a constant mean and covariance.

To characterize the relation between two random processes $z(t)$ and $x(t)$ we employ the joint PDF $f_{zx}(\zeta, \xi, t_1, t_2)$, which represents the probability that $(z(t_1), x(t_2))$ is within the differential area $d\zeta \times d\xi$ centered at (ζ, ξ). We shall usually assume that the processes $z(t)$, $x(t)$ are *jointly stationary*, that is, the joint PDF is not a function of both times t_1 and t_2, but depends only on the difference $(t_1 - t_2)$.

In the stationary case, the expected value of the function of two variables $g(z, x)$ is defined as

$$E\{g(z(t_1), x(t_2))\} = \int_{-\infty}^{\infty} g(\zeta, \xi) f_{z,x}(\zeta, \xi, t_1-t_2)\, d\zeta d\xi. \tag{C.3.1}$$

In particular, the *cross-correlation function* is defined by

$$R_{zx}(\tau) = E\{z(t+\tau)x^T(t)\}. \tag{C.3.2}$$

The cross-correlation function of two nonstationary processes is defined as

$$R_{zx}(t, \tau) = E\{z(t)x^T(\tau)\}. \tag{C.3.3}$$

Considering $z(t_1)$ and $z(t_2)$ as two jointly distributed random stationary processes, we may define the *autocorrelation function* of $z(t)$ as

$$R_z(\tau) = E\{z(t + \tau)z^T(t)\} \tag{C.3.4}$$

The autocorrelation function gives us some important information about the random process $z(t)$. For instance

$$\text{trace}[R_z(0)] = \text{trace}[E\{z(t)z^T(t)\}] = E\{\|z(t)\|^2\}$$

is equal to the total energy in the process $z(t)$. (In writing this equation recall that, for any compatible matrices M and N, $\text{trace}(MN) = \text{trace}(NM)$.)

If

$$R_{zx}(\tau) = 0 \tag{C.3.5}$$

we call $z(t)$ and $x(t)$ *orthogonal*. If

$$R_z(\tau) = P\delta(\tau) \tag{C.3.6}$$

where P is a constant matrix and $\delta(t)$ is the dirac delta, then $z(t)$ is orthogonal to $z(t + \tau)$ for any $\tau \neq 0$. What this means is that the value of the process $z(t)$ at one time t is unrelated to its value at another time $\tau \neq t$. Such a process is called *white noise*. An example is the thermal noise in an electric circuit, which is due to the thermal agitation of the electrons in the resistors.

In the discrete-time case, for these quantities we write $R_{zx}(k)$, and so on.

C.4 SPECTRAL DENSITY AND LINEAR SYSTEMS

For a stationary random process $x(t)$, the *spectral density* is defined as

$$\phi_x(s) = \mathbf{L}(R_x(\tau)), \tag{C.4.1}$$

that is, the Laplace transform of the autocorrelation function. In discrete time

$$\phi_x(z) = \mathbf{Z}(R_x(k)), \tag{C.4.2}$$

the Z-transform of the autocorrelation function. The *cross-spectral density* $\phi_{xy}(s)$ of two processes $x(t)$ and $y(t)$ is likewise defined in terms of the cross-correlation function.

Given $u(t) \in \mathbf{R}^m$ and $y(t) \in \mathbf{R}^p$ the input and output respectively of a linear system with impulse response $h(t)$ and transfer function $H(s)$, we have

$$\begin{aligned} R_{yu}(t) &= h(t)*R_u(t) \\ R_y(t) &= h(t)*R_u(t)*h^T(-t) \\ \phi_{yu}(s) &= H(s)\phi_u(s) \\ \phi_y(s) &= H(s)\phi_u(s)H^T(-s), \end{aligned} \tag{C.4.3}$$

with $*$ denoting convolution.

C.4 Spectral Density and Linear Systems

According to (C.3.6), for white noise $P\delta(0)$ is the covariance of $z(t)$, which is unbounded. Since

$$\phi_z(s) = P, \tag{C.4.4}$$

we call P a *spectral density matrix*. It is sometimes loosely referred to as a covariance matrix. In the discrete case, however,

$$R_z(k) = P\delta_k, \tag{C.4.5}$$

with δ_k the Kronecker delta, which is bounded. Therefore, P is a covariance matrix.

D

The Texas Instruments TMS320C25 Digital Signal Processor

The purpose of this appendix is to provide a brief overview of the C25 digital signal processor (DSP) with emphasis on the features that are particularly useful in control applications. A thorough discussion of the C25, including the complete instruction set, may be found in the *Second-Generation TMS320 User's Guide* published by Texas Instruments.

The particular C25 DSP used in the implementation examples is available from

Atlanta Signal Processors, Inc.
770 Spring Street, Suite 208
Atlanta, GA
404-892-7265

D.1 ARCHITECTURAL OVERVIEW

The TMS320C25 is a fixed-point digital signal processor with a 16-bit wordlength. The architecture of the processor is optimized to perform arithmetic operations of the type that are required in linear digital filters. The primary features of the C25 which result in its high-speed arithmetic performance and differentiate it from most general-purpose microprocessors are:

1. a Harvard architecture with separate data and program memory spaces as well as separate data and program buses,
2. internal hardware to implement functions (e.g., multiplication) which are typically performed in microcode,

D.1 Architectural Overview

3. a high degree of parallelism so that, for instance, multiplication, addition, and address manipulation operations can all be performed in the same machine cycle.

The memory of the C25 is organized into a 64K word program memory space and a 64K word data memory space. A logic signal applied to a pin on the C25 selects whether to map a 4K word on-chip program ROM into the lower 4K words of program memory. 544 words of on-chip RAM are available, of which 288 words are always mapped to a fixed location in data memory while the remaining 256 words may be mapped to either program or data memory under software control using the CNFP and CNFD instructions.

The C25 performs two's complement arithmetic using the 32-bit arithmetic logic unit (ALU) and accumulator. The accumulator is 32 bits long and is divided into a high-order word and a low-order word. The accumulator provides one input to the ALU and stores the output from the ALU. The other input to the ALU is either a 16-bit word taken from data memory or an immediate instruction, or the 32-bit result in the multiplier's product register.

The multiplier unit performs a 16 × 16-bit multiplication with a 32-bit result in a single machine cycle. The 32-bit P register always stores the product. The multiplicand comes from either the 16-bit T register or from program memory, while the multiplier is taken from data memory or an immediate instruction.

The arithmetic logic unit, multiplier unit, accumulator, and three shifters are together referred to as the central arithmetic logic unit (CALU). The shifters allow operand scaling to be performed in the same machine cycle as arithmetic or data storage and retrieval operations. The first shifter allows a left shift of 0 to 15 bits on operands which come from data memory before they are passed to the ALU. The second shifter allows a left shift of 0 to 7 bits on operands as they are stored from the high or low accumulator into data memory. The operation of the third shifter is determined by the product shift mode, which may be modified by the SPM instruction. The third shifter allows a left shift of 0, 1, or 4 bits, or a right shift of 6 bits (product modes 0, 1, 2, and 3, respectively), on the 32-bit product in the P register as it is passed to the ALU. All of these shifters must be used with caution, since a left shift can result in an undetectable overflow.

There are two software-selectable modes in addition to the product shift mode which affect the operation of the CALU. The sign-extension mode determines whether 16-bit numbers are treated as two's complement numbers or as unsigned numbers when they are operated upon by the ALU. If sign-extension mode is set, all 16-bit numbers are sign-extended to 32 bits when loaded into the ALU. If sign-extension mode is reset, 16-bit numbers are padded on the left with zeros when loaded into the ALU. The sign-extension mode affects all instructions which include a shift of incoming data, namely ADD, ADDT, ADLK, LAC, LACT, LALK, SBLK, SFR, SUB, and SUBT.

The overflow mode determines whether arithmetic overflows are allowed to wrap around or are saturated at the most positive or most negative value. The overflow mode affects all instructions which add or subtract from the accumulator. However, the ALU does not recognize overflows that occur during a left shift, so the programmer must explicitly check for such overflows as necessary.

The C25 provides a register file that contains eight auxiliary registers (AR0–AR7). The auxiliary registers may be used for indirect addressing of data memory or for temporary data storage. The auxiliary register that is to be used for indirect addressing is pointed to by a 3-bit auxiliary register pointer (ARP) which contains a number from 0 to 7 designating the current auxiliary register. The current auxiliary register is denoted by AR(ARP). The ARP may be directly modified using the LARP instruction, or any instruction that uses indirect addressing may specify a new value to load into the ARP upon completion. The auxiliary register file is provided with its own auxiliary register arithmetic unit (ARAU). The ARAU may automatically index the current auxiliary register by ± 1 or $\pm AR0$ while data memory is being addressed without the intervention of the ALU. This feature allows a table of data to be accessed and the index into the table updated in the same machine cycle. The ARAU may also be used to add or subtract an unsigned 8-bit immediate value to an auxiliary register.

A feature of the C25 that is particularly useful when moving blocks of data or performing repetitive multiply-and-accumulate cycles is the repeat counter. When the 8-bit repeat counter is loaded with a number N, the next single instruction is executed $N + 1$ times. The repeat counter may be loaded using either the RPT or the RPTK instructions. Not only does the use of the repeat counter decrease program memory requirements, it also allows several multicycle instructions, such as MAC and BLKP, to become single-cycle instructions upon repetition. All instructions are repeatable except the repeat instructions themselves, instructions with an immediate operand, branch and call instructions, and the IDLE instruction.

Additional on-chip resources of the C25 include an eight-level hardware stack, a 16-bit timer, three external maskable user interrupts, and a serial port. The timer and the serial port are memory-mapped and also generate maskable interrupts.

D.2 ADDRESSING MODES

The TMS320C25 instruction set provides three memory addressing modes:

1. direct addressing mode
2. indirect addressing mode
3. immediate addressing mode

In direct addressing, the instruction contains the data memory address of the operand. Indirect addressing accesses the operand at the data memory location specified by the current auxiliary register. In immediate addressing, the instruction itself contains the operand.

Direct Addressing Mode

In order to avoid having to use two words for each direct addressing mode instruction, the instruction contains only the 7 least-significant bits of the data memory address. The

D.2 Addressing Modes

upper 9 bits of the address are supplied by the data memory page pointer (DP) register. Thus, the DP register points to one of 512 possible 128-word data memory pages, and the 7-bit address contained in the instruction gives an offset within that page. The DP register can be modified with the LDP and LDPK instructions.

Direct addressing can be used with all instructions except branches and calls, instructions with no operand, and immediate addressing mode instructions. The format of an instruction which uses the direct addressing mode is:

$$\text{ADD dma[,shift]}$$

where "dma" is the 7 least-significant bits of a data memory address and "shift" is an optional number of bits to shift the operand before performing the specified operation. In this example, the 16-bit word at the data memory address specified by the DP register and "dma" is shifted left "shift" bits and then added to the accumulator. The shift operation is only available for a few instructions in the direct addressing mode: a shift of 0 to 15 bits may be specified with the ADD, LAC, and SUB instructions, and a shift of 0 to 7 bits may be specified with the SACH and SACL instructions. Additionally, the ADDT, LACT, and SUBT instructions allow for a shift specified by the T register rather than by an immediate value contained in the instruction itself. No other instructions used in the direct addressing mode may specify a shift.

Indirect Addressing Mode

Since the auxiliary registers are 16 bits wide, any data memory location can be accessed through indirect addressing. The data memory page pointer is not used in the indirect addressing mode. Indirect addressing can be used with all instructions except instructions with no operand and immediate addressing mode instructions.

The format of an instruction that uses the indirect addressing mode is

$$\text{ADD ind[,shift[,next ARP]]}$$

where "ind" specifies the type of indexing, "shift" specifies an optional shift, and "next ARP" optionally specifies a new value to load into the auxiliary register pointer following the completion of the operation indicated by the instruction. In this example, the 16-bit word at the data memory address pointed to by auxiliary register AR(ARP) is shifted left "shift" bits and then added to the accumulator. AR(ARP) is modified according to "ind," and finally the value "next ARP" is loaded into the auxiliary register pointer. The shift operation is only available for the same instructions as in the direct addressing mode.

The type of indexing specified by "ind" must be one of the following:

*	no indexing,
*+	increment AR(ARP)
*−	decrement AR(ARP)
*0+	add AR0 to AR(ARP)
*0−	subtract AR0 from AR(ARP)

There is also a type of bit-reversed indexing which is useful when performing a fast

Fourier transform (FFT). Remember that indexing is performed *after* the data memory is accessed but *before* the auxiliary register pointer is updated.

An important point to note is that the auxiliary register pointer is only used to specify the current auxiliary register for the purpose of indirect addressing. Any instruction that directly accesses an auxiliary register (such as LAR—load auxiliary register) accepts the auxiliary register number as its first argument. Thus, the instruction operates on the auxiliary register specified within the instruction itself rather than the auxiliary register specified by the auxiliary register pointer.

Immediate Addressing Mode

The TMS320C25 has both short (single word) and long (two word) immediate addressing instructions. The short instructions have an 8-bit operand, except for the MPYK (multiply short immediate) instruction, which has a 13-bit operand. The long instructions all have a 16-bit operand, and all long instructions except LRLK (load auxiliary register long immediate) allow an optional left shift of 0 to 15 bits. Notice that the mnemonic for each of the immediate instructions ends in the letter "K."

D.3 INSTRUCTION SET

The TMS320C25 assembly language instruction set consists of 121 instructions. Only the instructions that we shall use most frequently are listed here. Consult the *Second-Generation TMS320 User's Guide* for a complete description of the instruction set.

ACCUMULATOR AND MEMORY REFERENCE INSTRUCTIONS

ABS	absolute value of accumulator
ADD	add to accumulator with shift
ADDH	add to high accumulator
ADDK	add to accumulator short immediate
ADDS	add to low accumulator with sign extension suppressed
ADDT	add to accumulator with shift specified by T register
ADLK	add to accumulator long immediate with shift
LAC	load accumulator with shift
LACK	load accumulator short immediate
LACT	load accumulator with shift specified by T register
LALK	load accumulator long immediate with shift
NEG	negate accumulator
NORM	normalize contents of accumulator
SACH	store high accumulator with shift
SACL	store low accumulator with shift

D.3 Instruction Set

SBLK	subtract from accumulator long immediate with shift
SFL	shift accumulator left
SFR	shift accumulator right
SUB	subtract from accumulator with shift
SUBH	subtract from high accumulator
SUBK	subtract from accumulator short immediate
SUBS	subtract from low accumulator with sign extension suppressed
SUBT	subtract from accumulator with shift specified by T register
ZAC	zero accumulator (same as LACK 0)
ZALH	zero low accumulator and load high accumulator
ZALR	zero low accumulator and load high accumulator with rounding
ZALS	zero accumulator and load low accumulator with sign extension suppressed

AUXILIARY REGISTERS AND DATA PAGE POINTER INSTRUCTIONS

ADRK	add to auxiliary register short immediate
LAR	load auxiliary register
LARK	load auxiliary register short immediate
LARP	load auxiliary register pointer
LDP	load data memory page pointer
LDPK	load data memory page pointer immediate
LRLK	load auxiliary register long immediate
MAR	modify auxiliary register
SAR	store auxiliary register
SBRK	subtract from auxiliary register short immediate

MULTIPLY INSTRUCTIONS

APAC	add P register to accumulator
LT	load T register
LTA	load T register and accumulate previous product
LTD	load T register, accumulate previous product, and move data
LTP	load T register and store P register in accumulator
LTS	load T register and subtract previous product
MAC	multiply and accumulate
MACD	multiply and accumulate with data move
MPY	multiply with T register
MPYA	multiply with T register and accumulate previous product
MPYS	multiply with T register and subtract previous product
MPYU	multiply with T register unsigned

PAC   load accumulator with *P* register
SPAC   subtract *P* register from accumulator
SPM   set *P* register output shift mode

BRANCH AND CALL INSTRUCTIONS

B   branch unconditionally
BANZ   branch on auxiliary register not zero
BGEZ   branch if accumulator ≥ 0
BGZ   branch if accumulator > 0
BLEZ   branch if accumulator ≤ 0
BLZ   branch if accumulator < 0
BNZ   branch if accumulator $\neq 0$
BZ   branch if accumulator $= 0$
CALL   call subroutine
RET   return from subroutine

DATA MOVEMENT INSTRUCTIONS

BLKD   block move from data memory to data memory
BLKP   block move from program memory to data memory
DMOV   data move in data memory
TBLR   table read from program memory to data memory
TBLW   table write from data memory to program memory

CONTROL INSTRUCTIONS

CNFD   configure on-chip RAM block as data memory
CNFP   configure on-chip RAM block as program memory
ROVM   reset overflow mode
RPT   repeat instruction as specified by data memory value
RPTK   repeat instruction as specified by immediate value
RSXM   reset sign-extension mode
SOVM   set overflow mode
SSXM   set sign-extension mode

D.4 USING THE TMS320C25 IN DIGITAL CONTROL

The architecture and instruction set of the C25 are optimized to perform the type of calculations that are common in digital signal processing applications. Fortunately, the calculations required in linear digital control algorithms are quite similar to digital signal processing algorithms. This means that we can efficiently utilize many of the specialized features of the C25. Specifically, we shall discuss the multiply-and-accumulate operation, coefficient and variable scaling, and overflow management.

In both digital signal processing and digital control, it is convenient to interpret two's complement numbers as fractions rather than integers. The reason for this is that the product of two fractions in the range $[-1, 1]$ is another fraction in the range $[-1, 1]$. We shall use the notation that a two's complement fraction with n bits to the right of the binary point is a Qn number. Since the C25 has a 16-bit word length, most coefficients and states are stored in memory as Q15 numbers. (Recall that the most significant bit is a sign bit.)

When it is desired to work with numbers that do not fall in the range $[-1, 1]$, we can still interpret all two's complement numbers as fractions where each fraction is multiplied by a scale factor. For instance, if a state variable is known to be confined to the range $[-16, 16]$, then we can represent the state as a Q15 number with an implied scale factor of 2^4. Notice that the scale factor is not actually encoded in the number representation—if it was we would be on our way to performing floating-point arithmetic. Rather, the scale factor is simply understood by the programmer, whose responsibility it is to correctly interpret the results of calculations that are affected by it.

D.5 MULTIPLY-AND-ACCUMULATE

The most common operation in digital signal processing algorithms is the multiply-and-accumulate. For instance, to compute the output y_k of the auto-regressive moving average (ARMA) filter

$$y_k = -\sum_{i=1}^{n} a_i y_{k-i} + \sum_{i=0}^{m} b_i u_{k-i},$$

it is necessary to accumulate, that is add up, a series of products. The multiply-and-accumulate operation also occurs frequently in control algorithms.

As noted previously, the product of two 2's complement fractions is also a fraction. However, the number of bits in the product is one less than the sum of the number of bits in the multiplicand and the multiplier. For example, the product of two Q15 numbers is a Q30 number. Since the product is stored in a double-length (i.e., 32-bit) register on the C25, bit 30 of the product is simply a duplicate copy of the sign bit (bit 31). The purpose of product mode 1 is to automatically shift out the extra copy of the sign bit when the product is passed from the P register to the accumulator, thus converting the product from a Q30 to a Q31 number. Product mode 2 (left shift of 4 bits) is used in conjunction with the MPYK instruction, where a Q12 immediate value is multiplied by a Q15 data memory value to yield a Q27 product with 4 duplicate copies of the sign bit. Products are normally accumulated as Q31 numbers until a complete sum-of-products expression is evaluated, at which time the result can be truncated to a Q15 number simply by storing the high accumulator with the SACH instruction. If it is desired to round the result, this can be accomplished by first adding 1 to the most significant of the bits which will be discarded.

All DSP chips are fundamentally optimized to perform the multiply-and-accumulate operation at high speed, and the C25 is no exception. The C25 has two basic multiplication instructions: MPY and MAC. The MPY instruction multiplies the contents of the T register by the contents of the addressed data memory location. The MAC instruction multiplies a program memory value by a data memory value and accumulates the previous product.

The MPY instruction and the LT (load T register) instruction have several variations which can be used to reduce code size and execution time when performing repetitive multiply-and-accumulate operations. This is best illustrated by an example. Suppose we wish to evaluate the sum-of-products

$$x = a_1 b_1 + a_2 b_2 + a_3 b_3,$$

where a_i are Q15 coefficients, b_i are Q15 variables, and the result x should be truncated to a Q15 number. Assuming that labels have been defined for all the coefficients and variables and that the data page register has been initialized appropriately, the following two code fragments below which evaluate this expression are functionally equivalent.

SUM-OF-PRODUCTS CODE FRAGMENT 1:

```
SPM    1         ; shift out extra sign bit
LT     a1        ; T = a1
MPY    b1        ; P = (a1)(b1) [Q30]
PAC              ; Acc = (a1)(b1) [Q31]
LT     a2        ; T = a2
MPY    b2        ; P = (a2)(b2) [Q30]
APAC             ; Acc = (a1)(b1)+(a2)(b2) [Q31]
LT     a3        ; T = a3
MPY    b3        ; P = (a3)(b3) [Q30]
APAC             ; Acc = (a1)(b1)+(a2)(b2)+(a3)(b3) [Q31]
SACH   x         ; store truncated result [Q15]
```

SUM-OF-PRODUCTS CODE FRAGMENT 2:

```
SPM    1         ; shift out extra sign bit
LT     a1        ; T = a1
MPY    b1        ; P = (a1)(b1) [Q30]
LTP    a2        ; Acc = (a1)(b1) [Q31]
                 ; T = a2
MPY    b2        ; P = (a2)(b2) [Q30]
LTA    a3        ; Acc = (a1)(b1)+(a2)(b2) [Q31]
                 ; T = a3
MPY    b3        ; P = (a3)(b3) [Q30]
APAC             ; Acc = (a1)(b1)+(a2)(b2)+(a3)(b3) [Q31]
SACH   x         ; store truncated result [Q15]
```

D.5 Multiply-And-Accumulate

Notice that the second code fragment makes use of the LTP and LTA instructions to combine the loading of the multiplicand into the T register with the transfer of the previous product from the P register into the accumulator. The LTP and LTA instructions (as well as LTS, MPYA, and MPYS) should be used whenever possible, since each instance essentially reduces program memory requirements by one word and reduces execution time by one machine cycle.

It is important to note that the MPY instruction always performs two's complement signed multiplication regardless of the sign-extension mode. The MPYU instruction provides a method of performing unsigned multiplication. The contents of the T register and the data memory value addressed by the MPYU instruction are both treated as unsigned 16-bit numbers. Unsigned multiplication is required when calculating double-precision products, such as when two Q31 numbers are multiplied to yield a Q62 product. Product mode 0 (no shifting) is provided for use with the MPYU instruction.

The MPY instruction and its variations require that the multiplicand be loaded into the T register before the MPY instruction is executed. Since the instructions that load a value into the T register address data memory only, it is often necessary to move coefficient tables from program memory ROM into data memory RAM. The block data movement commands, BLKP and TBLR, are useful in this regard when used in conjunction with the RPTK instruction.

Unlike the MPY instruction, the MAC instruction does not require that coefficient tables be moved from program memory to data memory. The MAC instruction accepts two arguments: an immediate program memory address and a direct or indirect data memory address. The operation of the MAC instruction is to add the previous product in the P register to the accumulator and multiply the number stored at the program memory address by the number at the data memory address. The program memory address is automatically incremented when the MAC instruction is repeated, and the data memory address can be modified via indirect addressing. For instance, our previous sum-of-products example could be implemented using the MAC instruction as shown below, assuming that the coefficients are stored in a contiguous block of program memory and the variables are stored in a contiguous block of data memory.

SUM-OF-PRODUCTS CODE FRAGMENT 3:

```
SPM  1; shift out extra sign bit
LARP AR1; make auxiliary register AR1 current
LRLK AR1,b; AR1 points to variable table
ZAC  ; Acc = 0
MPYK 0; P = 0
RPTK 2; execute the following instruction 3 times
MAC  a,*+; Acc = (a1)(b1)+(a2)(b2) [Q31]
     ; P = (a3)(b3)   [Q30]
APAC ; Acc = (a1)(b1)+(a2)(b2)+(a3)(b3) [Q31]
SACH x; store truncated result [Q15]
```

The MAC instruction takes four machine cycles for the first execution but becomes a single cycle instruction when repeated. Compare this to the LTA/MPY instruction pair, which takes two machine cycles to perform the multiply-and-accumulate operation. Comparison of the MAC multiply-and-accumulate implementation with the LTA/MPY implementation reveals that the MAC implementation is more efficient in terms of both program memory and execution time for long sum-of-products expressions. Specifically, the MAC implementation requires less program memory if more than four multiply-and-accumulates are performed and executes in fewer machine cycles if more than nine multiply-and-accumulates are performed. However, the LTA/MPY implementation may have an advantage even in long sum-of-products expressions if many of the coefficients are zero, since zero coefficients can be omitted in the LTA/MPY implementation but must be included in the MAC implementation. Also, the MAC implementation cannot be used if it is necessary to rescale some of the partial sums in the sum-of-products expression.

D.6 DELAYING AND STORING SIGNALS

Returning to the ARMA filter

$$y_k = -\sum_{i=1}^{n} a_i y_{k-i} + \sum_{i=0}^{m} b_i u_{k-i},$$

notice that implementation of the filter requires that the n most recent values of the output and the $m + 1$ most recent values of the input be available. This requirement to store a certain number of previous values of a variable arises in many other types of digital filters as well. The traditional computer science approach to this problem would be to use a ring buffer, in which the newest value overwrites the oldest value which is no longer needed. The use of a ring buffer entails a certain amount of overhead to maintain the pointer which indicates the location of the oldest value.

The TMS320C25 provides a special instruction to facilitate the signal delaying operation without resorting to a ring buffer. The DMOV instruction copies the contents of a directly or indirectly addressed data memory location to the next higher location. DMOV executes in a single instruction cycle and does not require the use of the accumulator or any auxiliary register. Unfortunately, *DMOV works only within the on-chip RAM*.

Two additional instructions are provided which incorporate the data move operation: LTD and MACD. LTD is functionally equivalent to the pair of instructions LTA and DMOV, except it executes in a single machine cycle. MACD is functionally equivalent to the pair of instructions MAC and DMOV, except MACD executes in a single machine cycle when repeated. These two instructions allow for the signal delaying operation to be performed in conjunction with the multiply-and-accumulate operation with absolutely no memory or speed penalty as compared to performing the multiply-and-accumulate operation alone. Note that, due to the fact that DMOV copies data to the *next higher* data memory location, the signal that is to be delayed must be stored with the oldest value at the highest data memory address, and the oldest value must be operated on first.

D.7 COEFFICIENT AND VARIABLE SCALING

The dynamic range of a particular number representation scheme is defined as the ratio of the largest number that can be represented to the smallest number that can be represented. The dynamic range of a fixed-point two's complement number is strictly a function of the number of bits in the mantissa. For instance, a Q15 number has a dynamic range of 2^{15}. Contrast this with the comparatively unlimited dynamic range of 2^{254} for a standard 32-bit floating-point number. The limited dynamic range of fixed-point numbers is the reason that fixed-point filters must be scaled. Scaling is simply the process of placing the binary point in all filter variables such that the range of each variable is accommodated by the range of the number representation for that variable, while simultaneously providing adequate resolution.

Notice that there is a trade-off between range and resolution. For instance, a Q15 number has a range from -2^0 to almost 2^0 and a resolution of 2^{-15}. If we apply a scaling factor of 2^3 then the range becomes -2^3 to almost 2^3, but the resolution is now just 2^{-12}. The dynamic range remains constant at 2^{15}. If we need more dynamic range for a fixed-point two's complement variable, we must use more bits to represent it.

Input and Output Scaling

During the controller design process, inputs and outputs of the controller are usually considered to be in engineering units such as inches or m/s². In practice, the inputs are normally measured by transducers so that the actual value measured is, for instance, a voltage. The outputs of the controller (inputs to the plant) are also often not in the units for which they were designed. For instance, suppose a control algorithm for a positioning system was designed for an output voltage in the range -10 Volts to 10 Volts, which would be applied to the input of the motor power amplifier. The actual implementation of the controller might require that the output of the control algorithm be a Q15 number which will be sent to a digital-to-analog converter to produce the desired voltage. The moral is that many control systems have implicit gains associated with transducers and actuators which are ignored during the design process. These implicit gain factors must be taken into account when the control algorithm is implemented. There are at least two ways of handling implicit gains.

1. Make the implicit gains explicit and redesign the control algorithm.
2. Multiply the inputs and outputs of the controller by the reciprocal of the implicit gain factors in order to convert from the normalized units to the engineering units for which the control algorithm was designed.

Notice that the inputs and outputs are internally represented in normalized units in the former method and in engineering units in the latter. The internal number representations of the inputs and outputs must be scaled to accommodate their ranges in whichever units are chosen.

State Vector Scaling

Just as with the inputs and outputs, a choice of units for each state must be made before the internal number representation for the state can be scaled. The digital signal processing literature contains a great deal of information on automatic scaling techniques which are widely used in practice to scale digital filters. These techniques may sometimes be useful in scaling the states of a digital controller, but often they are too conservative and result in inadequate resolution. Since the states of a digital controller usually correspond to physical quantities, it is often possible to scale the states based on knowledge of the physical system which is being controlled. When physical insight is insufficient, floating-point simulations of the controller can be useful in determining the necessary range of each state.

Coefficient Scaling

Once all of the states have been scaled, the coefficients of a digital controller are scaled automatically. However, some rescaling may have to be performed if a coefficient falls outside of the desired range. An example will illustrate this point. Suppose state x is a Q15 number with a scale factor of 2^3, state y is a Q15 number with a scale factor of 2^7, and the state equation for x is

$$x_{k+1} = ay_k.$$

Clearly, coefficient a should be scaled as a Q15 number with a scale factor of 2^{-4} in order for the binary point in the product ay_k to be properly aligned following the multiplication, without the need for any additional shifting. The following code fragment implements this state equation.

COEFFICIENT SCALING CODE FRAGMENT 1:

```
SPM    1        ; shift out extra sign bit
LT     a        ; T = a [Q15 * 2^-4]
MPY    y        ; P = (a)(y) [Q30 * 2^3]
PAC             ; Acc = (a)(y) [Q31 * 2^3]
SACH   x        ; store x [Q15 * 2^3]
```

A problem arises in our previous example if the value of coefficient a does not fall within the range -2^{-4} to 2^{-4}, since it will not fit in the available number range. This problem occurs frequently when implementing digital filters. If the coefficient greatly exceeds the available number range, that is usually an indication that rescaling of some of the states is in order. In our example we could either scale up state x (i.e., increase its range) or scale down state y. If the coefficient exceeds the available number range by less than a factor of two, which is frequently the case, we can solve the problem by dividing the coefficient by two and accumulating its product twice. The following code fragment illustrates the application of this useful technique on our previous example.

D.8 Overflow Management

COEFFICIENT SCALING CODE FRAGMENT 2:

```
SPM    1        ; shift out extra sign bit
LT     a        ; T = a/2 [Q15 * 2⁻¹]
MPY    y        ; P = (a/2)(y) [Q30 * 2³]
PAC             ; Acc = (a/2)(y) [Q31 * 2³]
APAC            ; Acc = (a)(y) [Q31 * 2³]
SACH   x        ; store x [Q15 * 2³]
```

Occasionally it happens that a coefficient exceeds the available number range and it is not feasible to rescale the states or use the double APAC technique. Under these circumstances, the last resort is to scale up the number representation for the coefficient just enough to accommodate its value. This will result in the need to scale down products which involve the coefficient by shifting the products left by the number of bits the coefficient was scaled up. Since the overflow mode of the C25 will not detect overflows that result from a shifting operation, it is necessary to explicitly check for overflows and saturate the product as appropriate. The shifting-with-saturation operation is quite time consuming compared to other arithmetic operations and should be avoided except when absolutely necessary. In a typical control algorithm, it is often necessary to scale down the output of the controller since its allowed range is constrained by the capabilities of the actuator. The LQG regulator for the inverted pendulum in Chapter 10 includes a shift-with-saturation subroutine which takes advantage of the NORM instruction.

D.8 OVERFLOW MANAGEMENT

The shifting-with-saturation subroutine we have just described is one method of overflow management. The C25 has several specialized features which can minimize the need to resort to explicit overflow detection and saturation. The most useful of these features is the overflow mode, which is enabled with the SOVM instruction and disabled with the ROVM instruction. When overflow mode is enabled, all additions or subtractions from the accumulator which result in an overflow will automatically be forced to saturate at the most positive or most negative value as appropriate. If overflow mode is disabled, overflows will be allowed to wrap around.

A frequently cited feature of two's complement arithmetic is that overflows in intermediate partial sums do not matter as long as the final result is within range. Unfortunately, this fact is of little practical use since it is rarely the case to be able to guarantee that the result of a sum-of-products expression is in range under all circumstances. Considering the potentially disastrous effects of allowing an overflow to wrap around, the overflow mode should normally be enabled when the C25 is used as a controller.

The disadvantage of enabling the overflow mode is, of course, that the saturation of intermediate partial sums in a long sum-of-products expression can cause the calculated result to be wildly inaccurate even if the actual result is in range. The standard method of dealing with this problem is to scale up the product to provide additional bits of "headroom." The headroom allows intermediate partial sums which would otherwise overflow to be evaluated correctly without saturation, while still maintaining the protection of the overflow mode for partial sums which overflow even the additional bits of headroom. The final result of the sum-of-products will have to be scaled down to its desired range by the shift-with-saturation operation.

The product shifter of the C25 provides for a right shift of 6 bits (product mode 3) to facilitate the down-scaling of intermediate partial sums in order to provide headroom. If 6 bits of headroom are enough for a particular application, the product mode shifter can provide the headroom without any additional computational overhead. Note that the right shift of the product mode shifter always sign-extends regardless of the sign-extension mode; hence it can be used only with signed products.

Finally, a cautionary note is in order concerning the asymmetry of the two's complement number range. Since the number -1 is included in the range of a two's complement fraction but the number 1 is not, the multiplication of -1 by -1 results in an overflow. This is the only situation in which an overflow can result from fractional multiplication. If product mode 1 is in use, this overflow will go by undetected and will not be saturated regardless of the overflow mode. Hence, the number -1 should be considered to be outside the fractional two's complement number range. Any Q15 coefficient which has the value -1 should be handled by the same methods as any other coefficient which exceeds the available number range, such as by the double APAC technique.

REFERENCES

ABDEL-MONEIM, T. M., and N. N. SORIAL, "On the Design of Optimal Regulators Using Time-Multiplied Performance Index with Prescribed Closed-Loop Eigenvectors," *IEEE Trans. Automat. Control*, vol. AC-27, no. 5, pp. 1128–1129, Oct. 1982.

AL-SUNNI, F. M., B. L. STEVENS, and F. L. LEWIS, "Negative State Weighting in the Linear Quadratic Regulator for Aircraft Control," *J. of Guidance, Control, and Dynamics*, to appear, 1992.

ANDERSON, B. D. O., and J. B. MOORE, "Linear System Optimization with Prescribed Degree of Stability," *Proc. IEE*, vol. 116, no. 12, pp. 2083–2087, 1969.

ANDERSON, B. D. O., and Y. LIU, "Controller Reduction: Concepts and Approaches," *IEEE Trans. Automat. Control*, vol. AC-34, no. 8, pp. 802–812, Aug. 1989.

ARMSTRONG, E. S., *ORACLS, A Design System for Linear Multivariable Control*, New York: Dekker, 1980.

ÅSTRÖM, K. J., and B. WITTENMARK, *Computer Controlled Systems*, Englewood Cliffs, N.J.: Prentice Hall, 1984.

ATHANS, M., "A Tutorial on the LQG/LTR Method," *Proc. American Control Conf.*, pp. 1289–1296, June 1986.

ATHANS, M., and P. FALB, *Optimal Control*, New York: McGraw-Hill, 1966.

ATHANS, M., P. KAPSOURIS, E. KAPPOS, and H. A. SPANG III, "Linear-Quadratic Gaussian with Loop-Transfer Recovery Methodology for the F-100 Engine," *J. Guidance*, vol. 9, no. 1, pp. 45–52, Jan.–Feb. 1986.

BELL, R. F., E. W. JOHNSON, R. K. WHITAKER, and R. V. WILCOX, "Head Positioning in a Large Disk Drive," *Hewlett-Packard J.*, pp. 14–20, Jan. 1984.

BIERMAN, G. J., *Factorization Methods for Discrete Sequential Estimation*, New York: Academic, 1977.

BIRDWELL, J. D., "Evolution of a Design Methodology for LQG/LTR," *IEEE Control Systems Magazine*, pp. 73–77, April 1989.

BLAKELOCK, J. H., *Automatic Control of Aircraft and Missiles*, New York: Wiley, 1965.

BROGAN, W. L., *Modern Control Theory*, New York: Quantum, 1974.

BROUSSARD, J., and N. HALYO, "Active Flutter Control Using Discrete Optimal Constrained Dynamic Compensators," *Proc. Amer. Control Conf.*, pp. 1026–1034, June 1983.

BRYSON, A. E., JR., and Y.-C. HO, *Applied Optimal Control*, New York: Hemisphere, 1975.

CASTI, J., *Dynamical Systems and Their Applications: Linear Theory*, New York: Academic, 1977.

CHANG, S. S. L., *Synthesis of Optimum Control Systems*, New York: McGraw-Hill, 1961.

CHEN, C.-T., *Linear System Theory and Design*, New York: Holt, Rinehart, and Winston, 1984.

CHOI, S. S., and H. R. SIRISENA, "Computation of Optimal Output Feedback Gains for Linear Multivariable Systems," *IEEE Trans. Automat. Control*, pp. 257–258, June 1974.

CHYUNG, D. H., "Optimal Tracking Controller," *Proc. American Control Conf.*, pp. 1211–1212, June 1987.

CLARKE, D. W., "Introduction to Self-Tuning Controllers," *Self-Tuning and Adaptive Control*, eds. C. J. Harris and S. A. Billings, pp. 36–71, London: Perigrinus, 1981.

DAVISON, E. J., and I. J. FERGUSON, "The Design of Controllers for the Multivariable Robust Servomechanism Problem Using Parameter Optimization Methods," *IEEE Trans. Automat. Control*, vol. AC-26, no. 1, pp. 93–110, Feb. 1981.

D'AZZO, J. J., and C. H. HOUPIS, *Linear Control System Analysis and Design*, third ed., New York: McGraw-Hill, 1988.

DE SILVA, C. W., *Control Actuators and Sensors*, Englewood Cliffs, N.J.: Prentice Hall, 1989.

DIDUCH, C. P., and R. DORAISWAMI, "Robustness of Optimally Designed Sampled Data Control Systems," *Proc. American Control Conf.*, pp. 1247–1252, June 1987.

DONGARRA, J. J., C. B. MOLER, J. R. BUNCH, and G. W. STEWART, *LINPACK User's Guide*, Philadelphia: SIAM Press, 1979.

DORSEY, J. F., *Systems Analysis Software*, School of Elect. Eng., Ga. Tech., Atlanta, GA, 30332, 1987

DOYLE, J. C., "Guaranteed Margins for LQG Regulators," *IEEE Trans. Automat. Control*, pp. 756–757, Aug. 1978.

DOYLE, J. C., and G. STEIN, "Robustness with Observers," *IEEE Trans. Automat. Control*, vol. AC-24, no. 4, pp. 607–611, Aug. 1979.

DOYLE, J. C., and G. STEIN, "Multivariable Feedback Design: Concepts for a Classical/Modern Synthesis," *IEEE Trans. Automat. Control*, vol. AC-26, no. 1, pp. 4–16, Feb. 1981.

DOYLE, J. C., K. GLOVER, P. P. KHARGONEKAR, and B. FRANCIS, "State-Space Solutions to Standard H_2 and H_∞ Control Problems," *IEEE Trans. Automat. Control*, vol. AC-34, no. 8, pp. 831–847, Aug. 1989.

FORTIN, P., and G. PARKINS, "Comments on 'The Design of Linear Regulators Optimal for a Time-Multiplied Performance Index'," *IEEE Trans. Automat. Control*, p. 176, Feb. 1972.

FRANCIS, B., J. W. HELTON, and G. ZAMES, "H_∞ Optimal Feedback Controllers for Linear Multivariable Systems," *IEEE Trans. Automat. Control*, vol. AC-29, no. 10, pp. 888–900, Oct. 1984.

FRANKLIN, G. F., and J. D. POWELL, *Digital Control*, Reading, Mass.: Addison-Wesley, 1980.

FRANKLIN, G. F., J. D. POWELL, and A. EMAMI-NAEINI, *Feedback Control of Dynamic Systems*, Reading, Mass.: Addison-Wesley, 1986.

FRANKLIN, G. F., J. D. POWELL, and M. L. WORKMAN, *Digital Control of Dynamic Systems*, second ed., Reading, Mass.: Addison-Wesley, 1990.

FRIEDLAND, B., *Control System Design*, New York: McGraw-Hill, 1986.

Fu Y., and G. A. DUMONT, "Choice of Sampling Period to Ensure Minimum-Phase Behavior," *IEEE Trans. Automat. Control*, vol. AC-34, no. 5, pp. 560–563, May 1989.

FUKUTA, S., and H. TAMURA, "The Evaluation of Time-Weighted Quadratic Performance Indices for Discrete and Sampled-Data Linear Systems," *Int. J. Control*, vol. 39, no. 1, pp. 135–142, 1984.

GANTMACHER, F. R., *The Theory of Matrices*, New York: Chelsea, 1977.

GELB, A., ed., *Applied Optimal Estimation*, Cambridge, Mass.: MIT Press, 1974.

GRIMBLE, M. J., and M. A. JOHNSON, *Optimal Control and Stochastic Estimation: Theory and Applications*, Vol. 1, New York: Wiley, 1988.

HALYO, N., and J. R. BROUSSARD, "A Convergent Algorithm for the Stochastic Infinite-Time Discrete Optimal Output Feedback Problem," *Proc. American Control Conf.*, paper WA-1E, 1981.

HANSELMANN, H., "Implementation of Digital Controllers—a survey," *Automatica*, vol. 23, no. 1, pp. 7–32, 1987.

HARVEY, C. A., and G. STEIN, "Quadratic Weights for Asymptotic Regulator Properties," *IEEE Trans. Automat. Control*, vol. AC-23, no. 3, pp. 378–387, June 1978.

HOSTETTER, G. H., *Digital Control System Design*, New York: Holt, Rinehart, and Winston, 1988.

IMSL, *Library Contents Document*, 8th ed., International Mathematical and Statistical Libraries, Inc., 7500 Bellaire Blvd., Houston, TX 77036.

KAILATH, T., *Linear Systems*, Englewood Cliffs, N.J.: Prentice Hall, 1980.

KALMAN, R. E., "Contributions to the Theory of Optimal Control," *Bol. Soc. Mat. Mexicana*, vol. 5, pp. 102–119, 1960a.

KALMAN, R. E., "A New Approach to Linear Filtering and Prediction Problems," *Trans. ASME J. Basic Eng.*, vol. 82, pp. 34–35, 1960b.

KALMAN, R. E., "When Is a Linear Control System Optimal?," *J. Basic Eng., Trans. ASME*, Series D, vol. 86, pp. 51–60, 1964.

KALMAN, R. E., and R. S. BUCY, "New Results in Linear Filtering and Prediction Theory," *Trans. ASME J. Basic Eng.*, vol. 83, pp. 95–108, 1961.

KIRK, D. E., *Optimal Control Theory*, Englewood Cliffs, N.J.: Prentice Hall, 1970.

KLEINMAN, D. L., "An Easy Way to Stabilize a Linear Constant System," *IEEE Trans. Automat. Control*, pp. 692–693, Dec. 1970.

KREINDLER, E., and D. ROTHSCHILD, "Model-Following in Linear-Quadratic Optimization," *AIAA Journal*, vol. 14, no. 7, pp. 835–842, July 1976.

KWAKERNAAK, H., and R. SIVAN, *Linear Optimal Control Systems*, New York: Wiley, 1972.

KWON, W. H., and A. E. PEARSON, "A Modified Quadratic Cost Problem and Feedback Stabilization of a Linear System," *IEEE Trans. Automat. Control*, vol. AC-22, no. 5, pp. 838–842, Oct. 1977.

KWON, B.-H., and M.-J. YOUN, "Optimal Regulators Using Time-Weighted Quadratic Performance Index with Prescribed Closed-Loop Eigenvalues," *IEEE Trans. Automat. Control*, vol. AC-31, no. 5, pp. 449–451, May 1986.

KWON, B.-H., and M.-J. YOUN, "Eigenvalue-Generalized Eigenvector Assignment by Output Feedback," *IEEE Trans. Automat. Control*, vol. AC-32, no. 5, pp. 417–421, May 1987.

LAUB, A. J., "An Inequality and Some Computations Related to the Robust Stability of Linear Dynamic Systems," *IEEE Trans. Automat. Control*, vol. AC-24, no. 2, pp. 318–320, April 1979.

LAUB, A. J., "Efficient Multivariable Frequency Response Computations," *IEEE Trans. Automat. Control*, vol. AC-26, no. 2, pp. 407–408, April 1981.

LEVINE, W. S., and M. ATHANS, "On the Determination of the Optimal Constant Output Feedback Gains for Linear Multivariable Systems," *IEEE Trans. Automat. Control*, vol. AC-15, no. 1, pp. 44–48, Feb. 1970.

LEWIS, F. L., *Optimal Control*, New York: Wiley, 1986a.

LEWIS, F. L., *Optimal Estimation*, New York: Wiley, 1986b.

LY., U.-L., A. E. BRYSON, and R. H. CANNON, "Design of Low-Order Compensators Using Parameter Optimization," *Automatica*, vol. 21, no. 3, pp. 315–318, 1985.

MACFARLANE, A. G. J., "The Calculation of Functionals of the Time and Frequency Response of a Linear Constant Coefficient Dynamical System," *Quart. J. Mech. and Appl. Math.*, vol. 16, Pt. 2, pp. 259–271, 1963.

MACFARLANE, A. G. J., "Return-Difference and Return-Ratio Matrices and Their Use in the Analysis and Design of Multivariable Feedback Control Systems," *Proc. IEE*, vol. 117, no. 10, pp. 2037–2049, Oct. 1970.

MACFARLANE, A. G. J., and B. KOUVARITAKIS, "A Design Technique for Linear Multivariable Feedback Systems," *Int. J. Control*, vol. 25, pp. 837–874, 1977.

MACIEJOWSKI, J. M., "Asymptotic Recovery for Discrete-Time Systems," *IEEE Trans. Automat. Control*, vol. AC-30, no. 6, pp. 602–605, June 1985.

MARION, J. B., *Classical Dynamics*, New York: Academic, 1965.

MARTENSSON, K., "On the Matrix Riccati Equatiron," *Informat. Sci.*, vol. 3, pp. 17–50, Jan. 1971.

MATRIX$_x$, Integrated Systems, Inc., 2500 Mission College Blvd., Santa Clara, CA 95054, 1989.

MAYBECK, P. S., *Stochastic Models, Estimation, and Control*, New York: Academic, 1979.

MCCLAMROCH, N. H., *State Models of Dynamic Systems*, New York: Springer-Verlag, 1980.

MEDANIC, J., H. S. THARP, and W. R. PERKINS, "Pole Placement by Performance Criterion Modification," *IEEE Trans. Automat. Control*, vol. 33, no. 5. pp. 469–472, May 1988.

MIDDLETON, R. H., and G. C. GOODWIN, *Digital Control and Estimation*, Englewood Cliffs, N.J.: Prentice Hall, 1990.

Mil. Spec. 1797, *Flying Qualities of Piloted Vehicles*, 1987.

MITA, T., "Optimal Digital Feedback Control Systems Counting Computation Time of Control Laws," *IEEE Trans. Automat. Control*, vol. AC-30, no. 6, pp. 542–548, 1985.

MOERDER, D. D., and A. J. CALISE, "Convergence of a Numerical Algorithm for Calculating Optimal Output Feedback Gains," *IEEE Trans. Automat. Control*, vol. AC-30, no. 9, pp. 900–903, Sept. 1985.

MOLER, C., J. LITTLE, and S. BANGERT, *PC-Matlab*, The Mathworks, Inc., 20 North Main St., Suite 250, Sherborn, MA 01770, 1987.

MOORE, B. C., "On the Flexibility Offered by State Feedback in Multivariable Systems Beyond Closed-Loop Eigenvalue Assignment," *IEEE Trans. Automat. Control*, pp. 689–692, Oct. 1976.

MOORE, B. C., "Principal Component Analysis in Linear Systems: Controllability, Observability, and Model Reduction," *IEEE Trans. Automat. Control*, vol. AC-26, no. 1, pp. 17–32, 1982.

MORARI, M., and E. ZAFIRIOU, *Robust Process Control*, Englewood Cliffs, N.J.: Prentice Hall, 1989.

O'BRIEN, M. J., and J. R. BROUSSARD, "Feedforward Control to Track the Output of a Forced Model," *Proc. IEEE Conf. Decision and Control*, Dec. 1978.

OHTA, H., M. NAKINUMA, and P. N. NIKIFORUK, "Use of Negative Weights in Linear Quadratic Regulator Synthesis," preprint, 1990.

OPPENHEIM, A. V., and R. W. SCHAFER, *Digital Signal Processing*, Englewood Cliffs, N.J.: Prentice Hall, 1975.

PAPOULIS, A., *Probability, Random Variables, and Stochastic Processes*, second ed., New York: McGraw-Hill, 1984.

PARKS, T. W., and C. S. BURRUS, *Digital Filter Design*, New York: Wiley, 1987.

PHILLIPS, C. L., and H. T. NAGLE, JR., *Digital Control System Analysis and Design*, Englewood Cliffs, N.J.: Prentice Hall, 1984.

PONTRYAGIN, L. S., V. G. BOLTYANSKII, R. V. GAMKRELIDZE, and E. F. MISHCHENKO, *The Mathematical Theory of Optimal Processes*, New York: Wiley-Interscience, 1962.

POSTLETHWAITE, I., J. M. EDMUNDS, and A. G. J. MACFARLANE, "Principal Gains and Principal Phases in the Analysis of Linear Multivariable Systems," *IEEE Trans. Automat. Control*, vol. AC-26, no. 1, pp. 32–46, Feb. 1981.

PRESS, W. H., B. P. FLANNERY, S. A. TEUKOLSKY, and W. T. VETTERLING, *Numerical Recipes*, New York: Cambridge, 1986.

RAO, C. R., and S. K. MITRA, *Generalized Inverse of Matrices and Its Applications*, New York: Wiley, 1971.

ROSENBROCK, H. H., *Computer-Aided Control System Design*, New York: Academic, 1974.

SAFONOV, M. G., and M. ATHANS, "Gain and Phase Margin for Multiloop LQG Regulators," *IEEE Trans. Automat. Control*, vol. AC-22, no. 2, pp. 173–178, April 1977.

SAFONOV, M. G., A. J. LAUB, and G. L. HARTMANN, "Feedback Properties of Multivariable Systems: The Role and Use of the Return Difference Matrix," *IEEE Trans. Automat. Control*, vol. AC-26, no. 1, pp. 47–65, Feb. 1981.

SCHULTZ, D. G., and J. L. MELSA, *State Functions and Linear Control Systems*, New York: McGraw-Hill, 1967.

SHAKED, U., "Singular and Cheap Optimal Control: The Minimum and Nonminimum Phase Cases," Report TWISK 181, Nat. Res. Inst. for Math. Sci., Pretoria, Rep. S. Africa, 1980.

SHIN, V., and C. CHEN, "On the Weighting Factors of the Quadratic Criterion in Optimal Control," *Int. J. Control*, vol. 19, pp. 947–955, May 1974.

SLIVINSKY, C., and J. BORNINSKI, "Control System Compensation and Implementation with the TMS32010," in *Digital Signal Processing Applications*, K.-S. Lin ed., Englewood Cliffs, N.J.: Prentice Hall, 1987.

SÖDERSTRÖM, T., "On Some Algorithms for Design of Optimal Constrained Regulators," *IEEE Trans. Automat. Control*, vol. AC-23, no. 6, pp. 1100–1101, Dec. 1978.

STEIN, G., "Generalized Quadratic Weights for Asymptotic Regulator Properties," *IEEE Trans. Automat. Control*, vol. AC-24, no. 4, pp. 559–566, Aug. 1979.

STEIN, G., and M. ATHANS, "The LQR/LTR Procedure for Multivariable Feedback Control Design," *IEEE Trans. Automat. Control*, vol. AC-32, no. 2, pp. 105–114, Feb. 1987.

STEVENS, B. L., P. VESTY, B. S. HECK, and F. L. LEWIS, "Loop Shaping with Output Feedback," *Proc. American Control Conf.*, pp. 146–149, June 1987.

STEVENS, B. L., and F. L. LEWIS, *Aircraft Modelling, Dynamics, and Control*, New York: Wiley, to appear, 1991.

STEVENS, B. L., F. L. LEWIS, and F. AL-SUNNI, "Aircraft Flight Controls Design Using Output Feedback," *J. of Guidance, Control, and Dynamics*, 1991.

STICH, M. C., "Digital Servo Algorithm for Disk Actuator Control," *Proc. Conf. Appl. Motion Control*, pp. 35–41, 1987.

STRANG, G., *Linear Algebra and Its Applications*, second ed., New York: Academic, 1980.

SU, K. L., *Time-Domain Synthesis of Linear Networks*, Englewood Cliffs, N.J.: Prentice Hall, 1971.

SUBBAYYAN, R., and M. C. VAITHILINGAM, "On the Suboptimal Design of Regulators Using Time-Multiplied Performance Index," *IEEE Trans. Automat. Control*, vol. AC-22, no. 5, pp. 864–866, Oct. 1977.

SYRMOS, V. L., *Feedback Design Techniques in Linear System Theory: Geometric and Algebraic Approaches*, Ph.D. Thesis, School of Electrical Eng., Ga. Tech., Atlanta, GA 30332, June 1991.

THOMPSON, P. M., *Program CC User's Guide*, 1985, Systems Technol., Inc., 13766 So. Hawthorne Blvd., Hawthorne, CA 90250.

VERRIEST, E. I., and F. L. LEWIS, "On the Linear Quadratic Minimum-Time Problem," *IEEE Trans. Automat. Control*, pp. 859–863, July 1991.

YOUNG, P. C., and J. C. WILLEMS, "An Approach to the Linear Multivariable Servomechanism Problem," *Int. J. Control*, vol. 15, no. 5, pp. 961–979, 1972.

YOUSUFF, A., and R. E. SKELTON, "A Note on Balanced Controller Reduction," *IEEE Trans. Automat. Control*, vol. AC-29, no. 3, pp. 254–257, March 1984.

ZIEGLER, J. G., and N. B. NICHOLS, "Optimum Settings for Automatic Controllers," *Trans. ASME*, vol. 64, pp. 759–768, 1942.

Index

A

Ackermann's formula, 94, 462, 581
Actuator saturation, 299
Additive uncertainty, 438
Aircraft control example, 447, 560
Aliasing, 279
Antiwindup compensation, 300
ARMA filter, 605
Autocorrelation, 596
    initial, in LQR with output feedback, 195

B

Balancing the singular values, 434
Ball balancer, 110
Bandwidth, 430
Bang-bang control, 178
Bilinear transformation, 255
Bode plots, 11
    discrete, 389
    gain-phase relation, 57
    multivariable, 424

C

Calculus, matrix, 590
Canonical forms, 67
    Jordan form, 72, 327
    observable, 85
    reachable, 82
Cayley-Hamilton Theorem, 61
Central-limit Theorem, 594
Characteristic polynomial, 34, 51, 71
Circuit, electric, 29, 101, 201, 214, 407
Command-generator tracker, 233, 242
Compensator with desired structure, 204, 378
Compound interest, 112
Computation delay, 287
Computer simulation, 43, 149, 574
    of digital controller, 252
Conservation of energy, 124
Constrained:
    feedback design, 208
    input design, 175
Controllability. *See* Reachability
Controller implementation, 298
    realization structures, 79, 326
Consensitivity, 434
Costate, 123
Cost function. *See* Performance index
Covariance, 593
    cross-, 483, 594
    error, 488
    propagation of, 483
Cross-correlation, 595

D

DC gain, 232, 435
DC motor:
    armature-controlled, 128, 152, 219, 319, 392
    simplified model, 140, 150, 163, 216, 303
    with compliant coupling, 30, 47
Deadbeat:
    control, 381
    observer, 462, 477
Decoupling, 101
Delay:
    computation, 287

Delay (*continued*)
  digital control of systems with, 270
  sampling of system with, 363, 369
  system, 57
Delta form, 333
Detectability, 105
Deviation system, 207, 230, 391
Deyst filter, 515
Digital:
  control, 16, 251, 355
  Kalman filter, 481, 493
  LQG regulator, 532
  observer, 460
  signal processor (DSP), 336, 598
Digital filter:
  ARMA, 605
  FIR, 337
  IIR, 339
  Kalman, 493
Direct discrete-time design, 355
Discretization of:
  continuous controller, 254
  continuous system, 356, 578
  performance index, 369
  stochastic system, 493
  transfer function, 365
Disk drive, 237, 465, 471, 476, 479, 497
Disturbance, 421
  rejection, 239
Dither, 310
Duality, 463
Dynamic compensator, 20, 527

E

Eigenvalues, 73
Eigenvectors, 73
  assignment, 98
  LQR design, 167
  PBH test, 77
Error:
  covariance, 488
  estimation, 461, 484
  model mismatch, 241
  steady-state, 207, 433
  tracking, 207, 398

Estimate:
  linear mean-square, 485
  of state vector, 459, 485
Euler:
  approximation, 513
  equations, 123
Expected value, 593
  conditional, 595
  propagation of, 483

F

Feedback:
  control, 3
  output, 20, 106
  state, 19, 92
Filter. *See* Digital filter
Final:
  condition, 123
  -value theorem, 379
First-order hold, 283, 361
Fixed-point arithmetic, 313
Fractional representation for numbers, 313
Frequency:
  domain, 10
  Nyquist, 256, 280
  prewarping, 259
  sampling, 255, 280

G

Gain margin, 550
Gaussian pdf, 594
Gilbert's realization method, 89

H

Hamiltonian, 121, 177, 194, 210
  matrix, 144
  system, 123, 167
Harmonic oscillator, 40, 46, 361, 517
Hold device, 282
Hydraulic transmission, 84

I

Impulse response, 36
Independent random variables, 595

Intersample behavior, 253, 387
Inverted pendulum, 95, 109, 244
  digital control of, 261, 287, 291, 310, 344, 401, 533

J

Jordan normal form, 72, 327
Joseph stabilized form of Riccati equation, 148

K

Kalman filter:
  continuous, 512
  digital, 493
  discrete, 481
  implementation 491, 508
  steady-state, 492, 505, 516
  suboptimal, 492
Kalman gain, 145, 489

L

L_2 operator gain, 430
Lagrange multiplier, 121, 177, 194, 210
Limit cycle, 308
Linear quadratic gaussian (LQG) dynamic regulator, 527
  digital, 532
  implementation, 537
  loop transfer recovery (LTR), 548
  transfer function, 531, 554
Linear quadratic regulator (LQR):
  discrete, 406, 411
  loop transfer recovery (LTR), 548
  with output feedback, 192, 406
  with state feedback, 144, 411
Loop:
  gain, 93, 421
  transfer recovery (LTR), 548, 553
Lyapunov equation, 138, 194, 400, 483

M

Matched pole-zero (MPZ) discretization, 258
Matrix inversion lemma, 587
Mean. *See* Expected value

Mean-square value, 594
Measurement, 485
  update, 485
Minimality, 87, 375, 387
Minimum-phase system, 53
Minimum-time control, 124, 172, 295
Modal decomposition, 80
Model:
  internal, 236
  reduction, 440
Model following, explicit, 240
Modes. *See* Natural modes
Multiplicative uncertainty, 438, 552
Multiply-and-accumulate, 605

N

Natural modes, 37, 52
Newton's system, 38
  controls design, 142, 159, 165, 170, 179, 381
  discretization of, 360, 364
  observer, 465, 471, 476, 479, 497
Noise:
  covariance, 482, 494
  measurement, 319, 421, 434, 464, 482, 513
  nonwhite, 496
  process, 464, 482, 513
  white, 596
Non-minimum-phase:
  LTR design, 558
  zeros, 53, 373
Normal:
  pdf, 594
  system, 179
Numerical solution of output-feedback design equations, 198, 581
Nyquist:
  criterion, 11
  frequency, 256, 280

O

Observability, 63, 72
  gramian, 65
  in PI weighting matrices, 158, 197, 209, 213, 401, 413

Observability (*continued*)
    loss of on sampling, 375, 387
    matrix, 64
Observer, 460
    deadbeat, 462
    reduced-order, 474
    robustness, 464
Open-loop control, 136
Operator gain, L_2, 430
Optimal control, general solution, 121
Optimal return difference relation, LQR, 549
Orthogonal random processes, 595, 596
Output:
    injection, 106, 460
    measured, 204
    performance, 204
Output feedback, 20, 106
    eigenvector assignment, 108
    LQR, 192, 406, 581
    robust, 447
    stabilizable, 107
    tracker, 204, 396
Overflow, 313
    management, 611
    oscillations, 314
    protection, 314

P

Padé approximant, 290
Parameter variation, 419, 446, 558
Partial fraction expansion, 36, 77, 327, 366, 440
Pendulum, 27
Performance index, LQR, 14, 19, 120
    constrained feedback, 208
    discrete, 399
    discretization of, 369
    explicit model-following, 241
    time weighting, 208
    tracker, 208
Performance specifications, frequency-domain, 429
Phase margin, 551
Phase plane, 182
PID controller, 11
    antiwindup protection, 301
    digital, 266, 342
Pole, 34
    of sampled system, 358
Pole-placement design, 94, 386. *See also* Ackermann's formula
Pontryagin's minimum principle, 177
Positive definite matrix, 587
Probability:
    density function (PDF), 593
    theory, review, 593

Q

Quantization, 306, 328

R

Random process, 595
Reachability, 59, 71
    gramian, 60, 62, 138
    matrix, 59
Realization:
    structures for controller, 79, 326
    theory, 66
Reduced-order:
    model, 440
    observer, 474
    regulator, 440
Regulator:
    dynamic, 20, 527
    reduced-order, 440
Relative degree, 256, 366
Resolvent matrix, 34, 50
Return difference, 93, 424, 549
Riccati equation, 20, 145
    algebraic, 157, 516
    discrete, 412
    observer/filter, 463, 489, 514
Rise time, 281
Robust design, 419
    with output feedback, 447
Robustness:
    of LQR, 548
    of observer/filter, 464
    performance, 420, 432

to plant parameter variations, 419, 446
stability, 419, 440
to unmodeled dynamics, 419, 437
using LQG/LTR design, 548
Root locus design, 12, 170
discrete, 380
Roundoff, 306
Runge-Kutta integration, 253, 577

S

Sampling, 279
frequency, 255, 280
period, 252, 377
the system, 356, 493
theorem, 280
the transfer function, 365
Saturation:
actuator, 299
overflow, 314
Scaling:
filter coefficient, 318, 610
signal, 316, 609
Second-order modules for controller realization, 328
Sensitivity, 424, 550
Separation principle, 529
Servo design, 206, 229
Shaping filter, 496
Shooting point method, 128
Signum function, 177
Singular condition in bang-bang control, 178
Singular vallue, 431
balancing, 434
decomposition, 425, 588
LQR constraint on, 550
Smith predictor, 271
Spectral:
density, 494, 596
factorization, 496
Stability, 9, 37, 52, 359
in observer/filter design, 464, 471
using LQR design, 20, 158, 196, 413
Stabilizability, 105
using output feedback, 107, 197
State equations, 18

continuous-time, 26
discrete-time, 50
with feedback,
nonlinear, 43
solution, 33, 50
State feedback, 19
eigenvector assignment, 98
LQR, 19
State transition matrix, 36, 52
Static loop sensitivity, 58
Stationary random process, 595
Steady-state:
control, 157
error, 207, 379, 433
Steering system example, 124
Step invariance discretization, 365
Stochastic:
process, 595
system, 482
Suboptimal control, 161
Sweep method, 145
Switching:
curve, 182
function, 178
time, 181
System:
closed-loop, 3
deviation, 207, 230, 391
stochastic, 482
theory, 10

T

Tangent control law, linear, 125
Target feedback loop, LTR, 557
Time constant, 42
Time update in Kalman filter, 485
Time weighting in LQ performance index, 208
Tracker, LQ:
digital, 396
with model following, 242
with output feedback, 204
by regulator redesign, 229, 390
Transfer function, 34, 51, 71
closed-loop, 93, 423

Transfer function (*continued*)
　discretization of, 365
Transformation, state-space, 67
Truncation, 306
Tustin's approximation, 255
Two-body problem, 69
Two-point boundary-value problem, 123, 127
Two's-complement number system, 314
Type, feedback control system, 209

U

Unbiased estimate, 488
Uncorrelated random variables, 595
Unit solution method, 133
Unmodeled dynamics, 437

V

Van de Pol oscillator, 44
Variance, 593
Velocity form filter implementation, 335

W

W-plane design, 389
Windup, integrator, 299
Wraparound, 314

Z

Z-transform table, 367
Zero-input response, 34, 51
Zero-order hold (ZOH), 252, 282, 358, 365
Zeros, 34
　in LQR design, 206, 229
　non-minimum-phase, 54, 373, 558
Ziegler-Nichols tuning, 269